MW01054680

Photosynthesis, Respiration, and Climate Change

Upper left: A view of the whole tree chambers (WTCs) at the Hawkesbury Institute for the Environment at Western Sydney University. The WTCs are used to study CO_2, warming and drought impacts on trees (including photosynthesis and respiration) in the field. (Credit: Dani Way) *Upper right*: A view of the OzFACE Free Air CO_2 Enrichment site at the Hawkesbury Institute for the Environment at Western Sydney University. The view is taken from a canopy crane, and shows two other rings, with each ring being either an ambient CO_2 control plot, or a plot where elevated CO_2 is blown into an intact Eucalyptus forest. This type of experiment allows us to study leaf-level responses to rising CO_2, but also ecosystem-level responses. (Credit: Dani Way) *Lower left*: A photo of a Nothofagus forest in Patagonia, Chile. Natural ecosystems like this one play a key role in determining the global carbon cycle, but rising temperatures and CO_2 concentrations will alter the carbon fluxes of vegetation in ways that are not currently well understood. (Credit: Dani Way) *Lower right*: A collection of Crassulacean Acid metabolism (CAM) species. Plants that rely on carbon-concentrating mechanisms (i.e., CAM and C4 species) play an outsized role in global productivity and have greater environmental stress tolerance than C3 species, and thus provide important insights into how we might generate more climate-resilient plants. (Credit: Dani Way)

Advances in Photosynthesis and Respiration
Including Bioenergy and Related Processes

VOLUME 48

The book series Advances in Photosynthesis and Respiration – Including Bioenergy and Related Processes provides a comprehensive and state-of-the-art account of research in photosynthesis, respiration, bioenergy production and related processes. Virtually all life on our planet Earth ultimately depends on photosynthetic energy capture and conversion to energy-rich organic molecules. These are used for food, fuel, and fiber. Photosynthesis is the source of almost all Bioenergy on Earth. The fuel and energy uses of photosynthesized products and processes have become an important area of study and competition between food and fuel has led to resurgence in photosynthesis research. This series of books spans topics from physics to agronomy and medicine; from femtosecond processes through season-long production to evolutionary changes over the course of the history of the Earth; from the photophysics of light absorption, excitation energy transfer in the antenna to the reaction centers, where the highly-efficient primary conversion of light energy to charge separation occurs, through intermediate electron transfer reactions, to the physiology of whole organisms and ecosystems; and from X-ray crystallography of proteins to the morphology of organelles and intact organisms. In addition to photosynthesis in natural systems, genetic engineering of photosynthesis and artificial photosynthesis is included in this series. The goal of the series is to offer beginning researchers, advanced undergraduate students, graduate students, and even research specialists, a comprehensive, up-to-date picture of the remarkable advances across the full scope of research on photosynthesis and related energy processes. This series is designed to improve understanding of photosynthesis and bioenergy processes at many levels both to improve basic understanding of these important processes and to enhance our ability to use photosynthesis for the improvement of the human condition.

For more information, please contact the Series Editors Thomas D. Sharkey, Michigan State University, East Lansing, MI, U.S.A. E-mail: tsharkey@msu.edu; phone 1-517-353-3257 or Julian J. Eaton-Rye, Department of Biochemistry, University of Otago, New Zealand, E-mail: julian.eaton-rye@otago.ac.nz. A complete list of references listed per volume can be found following this link: http://www.life.uiuc.edu/govindjee/Reference-Index.htm

More information about this series at http://www.springer.com/series/5599

Photosynthesis, Respiration, and Climate Change

Edited by

Katie M. Becklin
Department of Biology, Syracuse University, Syracuse, NY, USA

Joy K. Ward
College of Arts and Sciences, Department of Biology, Case Western Reserve University, Cleveland, OH, USA

and

Danielle A. Way
Department of Biology, University of Western Ontario, London, ON, Canada
Nicholas School of the Environment, Duke University, Durham, NC, USA
Terrestrial Ecosystem Science & Technology Group, Environmental & Climate Sciences Department, Brookhaven National Laboratory, Upton, NY, USA

Editors
Katie M. Becklin
Department of Biology
Syracuse University
Syracuse, NY, USA

Danielle A. Way
Department of Biology
University of Western Ontario
London, ON, Canada

Joy K. Ward
College of Arts and Sciences,
Department of Biology
Case Western Reserve University
Cleveland, OH, USA

ISSN 1572-0233 ISSN 2215-0102 (electronic)
Advances in Photosynthesis and Respiration
ISBN 978-3-030-64925-8 ISBN 978-3-030-64926-5 (eBook)
https://doi.org/10.1007/978-3-030-64926-5

This Springer imprint is published by the registered company Springer Nature Switzerland AG
The registered company address is: Gewerbestrasse 11, 6330 Cham, Switzerland

From the Series Editors

Advances in Photosynthesis and Respiration Including Bioenergy and Related Processes

Volume 48: Photosynthesis, Respiration, and Climate Change

The Earth initially had very high levels of carbon dioxide. Over time, biology and geology conspired to lower carbon dioxide levels so that now the Earth has very low amounts of gaseous carbon dioxide. Biological and geological processes have altered the Earth in other fundamental ways, in some cases, catastrophically. The best example is the evolution of the capacity to make molecular oxygen during photosynthesis. This set the stage for an explosion of biology, but also the demise of most of the species that were on Earth at that time. We are now in the midst of an extremely rapid (almost instantaneous on geological time scales) alteration of the Earth's climate because of man-made processes. This may set the stage for an explosion of new life forms, but could be the end of the extreme dominance by humans of Earth systems and the delicate ecological balance that allows a great diversity of life forms.

Climate change deniers like to point out that carbon dioxide can be considered a fertilizer; many greenhouse (glasshouse) operators enrich the air in their greenhouses with carbon dioxide in order to improve crop growth. Fertilization is not always a good thing: fertilization of aquatic systems by phosphorus leads to algal growth that then leads to eutrophication and degradation of aquatic systems for human needs. It also can alter the balance between algae and cyanobacteria, which can sometimes be toxic and has famously shut off municipal water supplies on occasion. Thus the question, will carbon dioxide fertilization have consequences we would prefer to avoid? How will plants respond to the changes in carbon dioxide? More importantly, how will plants respond to the accompanying changes in temperature and water availability? This has been the topic of study for many researchers, and the series editors are pleased that three outstanding scientists have put together this edited volume about how plants will respond to the global changes that are occurring and are likely to continue, even if human activities are adjusted to reduce the very rapid global changes we are now experiencing (and causing).

Large scale human responses to improve the atmosphere have been possible in the past. In 1952 London, England was gripped by a major air pollution event that may have killed over 10,000 people. This led to a switch away from coal burning for domestic heat and London air quality improved. In the 1970s acid rain was causing significant forest decline over large regions of the Earth. Aggressive reductions in sulfur dioxide and NO_x emissions are credited with an 80% reduction in acid rain. Ozone destruction by chlorinated fluorocarbon molecules was reversed by the Montreal protocol of 1990 that limited the production of the molecules most damaging to the ozone layer. However, we are still in the middle of the efforts to address the global changes brought about by atmospheric gases that absorb radiation, leading to increased energy in the atmosphere, which is typically called global warming. Recent attempts by the US government to prohibit the use of the term global warming are reminiscent of orders in the 1980s to prohibit US government employees from using the term "acid rain." We need more information on the climate changes that are occurring at an alarming rate. With luck this could lead to action to limit further increases in carbon dioxide. In any case, we need to be able to predict how global changes in climate and atmospheric conditions will affect plants because we depend on plants not only for the air we breathe, but also for ecosystem services and food, fuel, and fiber.

In volume 46 of the Advances in Photosynthesis and Respiration (AIPR) series, Katie Becklin, Joy Ward, and Dani Way have assembled a team of experts on photosynthetic and respiration responses to global change. We are grateful that they have produced a volume for this series that describe how photosynthetic processes, and plant processes in general, are affected by both the direct effects and indirect effects of global change.

Authors of volume 46

This volume represents an international collaboration among plant scientists. We are especially proud of the near even distribution of men and women authors who represent the following six countries: Australia (6); Canada (9), England (5), Germany (2), Spain (3), and the United States (12). This volume includes 37 authors, who are experts in modeling, empirical, and applied research on photosynthesis and respiration responses to climate change. Three of these authors are also editors of this volume. Alphabetically (by last names), they are: Michael J. Aspinwall, Hermann Bauwe, Katie M. Becklin (editor), Kristyn Bennett, Brittany B. Blair, Timothy E. Burnette, Marc Carriquí, Jacob M. Carter, John C. Cushman, André G. Duarte, Mirindi E. Dusenge, Michele Faralli, Alisdair R. Fernie, James M. Fischer, Jaume Flexas, Oula Ghannoum, Thomas E. Juenger, Tracy Lawson, Karen Lemon, Benedict M. Long, Sarah McDonald, Miquel Nadal, Nicholas A. Niechayev, Paula N. Pereira, Julianne Radford, Alexis Rodgers, Paul D. Rymer, Rowan F. Sage, Robert E. Sharwood, John D. Stamford, Matt Stata, Jim Stevens, David T. Tissue, Shellie Wall, Joy K. Ward (editor), Alexander Watson-Lazowski, Danielle A. Way (editor). We are grateful for their efforts in making this insightful volume.

Our Books

We list below information on the 45 volumes that have been published thus far (see http://www.springer.com/series/5599 for the series web site). Electronic access to individual chapters depends on subscription (ask your librarian) but Springer provides free downloadable front matter as well as indexes for nearly all volumes. The available web sites of the books in the Series are listed below.

- **Volume 45 (2020) Photosynthesis in Algae,** Edited by Anthony Larkum, Arthur Grosman, and John Raven
- **Volume 44 (2018) The Leaf: A Platform for Performing Photosynthesis**, Edited by William W. Adams III from USA and Ichiro Terashima from Japan. Eighteen chapters, 575 pp, Hardcover ISBN 978-3-319-93592-8, eBook ISBN 978-3-319-93594-2 [https://www.springer.com/gp/book/9783319935928]
- **Volume 43 (2018) Plant Respiration: Metabolic Fluxes and Carbon Balance**,

Edited by Guillaume Tcherkez from Australia and Jaleh Ghashghaie. Fourteen chapters, 302 pp, Hardcover ISBN 978-3-319-68701-8, eBook ISBN 978-3-319-68703-2 [https://www.springer.com/gp/book/9783319687018]

- **Volume 42 (2016) Canopy Photosynthesis: From Basics to Applications**, Edited by Kouki Hikosaka from Japan, Ülo Niinemets from Estonia, and Neils P.R. Anten from the Netherlands. Fifteen chapters, 423 pp, Hardcover ISBN 978-94-017-7290-7, eBook ISBN 978-94-017-7291-4 [http://www.springer.com/book/9789401772907]
- **Volume 41 (2016) Cytochrome Complexes: Evolution, Structures, Energy Transduction, and Signaling**, Edited by William A. Cramer and Tovio Kallas from the USA. Thirty-five chapters, 734 pp, Hardcover ISBN 978-94-017-7479-6, eBook ISBN 978-94-017-7481-9 [http://www.springer.com/book/9789401774796]
- **Volume 40 (2014) Non-Photochemical Quenching and Energy Dissipation in Plants, Algae and Cyanobacteria**, edited by Barbara Demmig-Adams, Győző Garab, William W. Adams III, and Govindjee from USA and Hungary. Twenty-eight chapters, 649 pp, Hardcover ISBN 978-94-017-9031-4, eBook ISBN 978-94-017-9032-1 [http://www.springer.com/life+sciences/plant+sciences/book/978-94-017-9031-4]
- **Volume 39 (2014) The Structural Basis of Biological Energy Generation**, edited by Martin F. Hohmann-Marriott from Norway. Twenty-four chapters, 483 pp, Hardcover ISBN 978-94-017-8741-3, eBook ISBN 978-94-017-8742-0 [http://www.springer.com/life+sciences/book/978-94-017-8741-3]
- **Volume 38 (2014) Microbial BioEnergy: Hydrogen Production**, edited by Davide Zannoni and Roberto De Phillipis, from Italy. Eighteen chapters, 366 pp, Hardcover ISBN 978-94-017-8553-2, eBook ISBN 978- 94-017-8554-9 [http://www.springer.com/life+sciences/plant+sciences/book/978-94-017-8553-2]
- **Volume 37 (2014) Photosynthesis in Bryophytes and Early Land Plants**, edited by David T. Hanson and Steven K. Rice, from USA. Eighteen chapters, approx. 342 pp, Hardcover ISBN 978-94-007-6987-8, eBook ISBN 978-94-007-6988-5 [http://www.springer.com/life+sciences/plant+sciences/book/978-94-007-6987-8]
- **Volume 36 (2013) Plastid Development in Leaves during Growth and Senescence**, edited by Basanti Biswal, Karin Krupinska and Udaya Biswal, from India and Germany. Twenty-eight chapters, 837 pp, Hardcover ISBN 978-94-007-5723-33, eBook ISBN 978-94-007-5724-0 [http://www.springer.com/life+sciences/plant+sciences/book/978-94-007-5723-3]
- **Volume 35 (2012) Genomics of Chloroplasts and Mitochondria**, edited by Ralph Bock and Volker Knoop, from Germany. Nineteen chapters, 475 pp, Hardcover ISBN 978-94-007-2919-3 eBook ISBN 978-94-007-2920-9 [http://www.springer.com/life+sciences/plant+sciences/book/978-94-007-2919-3]
- **Volume 34 (2012) Photosynthesis - Plastid Biology, Energy Conversion and Carbon Assimilation**, edited by Julian Eaton-Rye, Baishnab C. Tripathy, and Thomas D. Sharkey, from New Zealand, India, and USA. Thirty-three chapters, 854 pp, Hardcover, ISBN 978-94-007-1578-3, eBook ISBN 978-94-007-1579-0 [http://www.springer.com/life+sciences/plant+sciences/book/978-94-007-1578-3]
- **Volume 33 (2012): Functional Genomics and Evolution of Photosynthetic Systems**, edited by Robert L. Burnap and Willem F.J. Vermaas, from USA. Fifteen chapters, 428 pp, Hardcover ISBN 978-94-007-1532-5, Softcover ISBN 978-94-007-3832-4, eBook ISBN 978-94-007-1533-2 [http://www.springer.com/life+sciences/book/978-94-007-1532-5]
- **Volume 32 (2011): C_4 Photosynthesis and Related CO_2 Concentrating Mechanisms,** edited by Agepati S. Raghavendra and Rowan Sage, from India and Canada. Nineteen chapters, 425 pp, Hardcover ISBN 978-90-481-9406-3, Softcover ISBN 978-94-007-3381-7, eBook ISBN 978-90-481-9407-0 [http://

www.springer.com/life+sciences/plant+sciences/book/978-90-481-9406-3]

- **Volume 31 (2010): The Chloroplast: Basics and Applications,** edited by Constantin Rebeiz (USA), Christoph Benning (USA), Hans J. Bohnert (USA), Henry Daniell (USA), J. Kenneth Hoober (USA), Hartmut K. Lichtenthaler (Germany), Archie R. Portis (USA), and Baishnab C. Tripathy (India). Twenty-five chapters, 451 pp, Hardcover ISBN 978-90-481-8530-6, Softcover ISBN 978-94-007-3287-2, eBook ISBN 978-90-481-8531-3 [http://www.springer.com/life+sciences/plant+sciences/book/978-90-481-8530-6]
- **Volume 30 (2009): Lipids in Photosynthesis: Essential and Regulatory Functions,** edited by Hajime Wada and Norio Murata, both from Japan. Twenty chapters, 506 pp, Hardcover ISBN 978-90-481-2862-4, Softcover ISBN 978-94-007-3073-1 eBook ISBN 978-90-481-2863-1 [http://www.springer.com/life+sciences/plant+sciences/book/978-90-481-2862-4]
- **Volume 29 (2009): Photosynthesis in Silico: Understanding Complexity from Molecules,** edited by Agu Laisk, Ladislav Nedbal, and Govindjee, from Estonia, The Czech Republic, and USA. Twenty chapters, 525 pp, Hardcover ISBN 978-1-4020-9236-7, Softcover ISBN 978-94-007-1533-2, eBook ISBN 978-1-4020-9237-4 [http://www.springer.com/life+sciences/plant+sciences/book/978-1-4020-9236-7]
- **Volume 28 (2009): The Purple Phototrophic Bacteria,** edited by C. Neil Hunter, Fevzi Daldal, Marion C. Thurnauer and J. Thomas Beatty, from UK, USA and Canada. Forty-eight chapters, 1053 pp, Hardcover ISBN 978-1-4020-8814-8, eBook ISBN 978-1-4020-8815-5 [http://www.springer.com/life+sciences/plant+sciences/book/978-1-4020-8814-8]
- **Volume 27 (2008): Sulfur Metabolism in Phototrophic Organisms** , edited by Christiane Dahl, Rüdiger Hell, David Knaff and Thomas Leustek, from Germany and USA. Twenty-four chapters, 551pp, Hardcover ISBN 978-4020-6862-1, Softcover ISBN 978-90-481-7742-4, eBook ISBN 978-1-4020-6863-8 [http://www.springer.com/life+sciences/plant+sciences/book/978-1-4020-6862-1]
- **Volume 26 (2008): Biophysical Techniques Photosynthesis** , Volume II, edited by Thijs Aartsma and Jörg Matysik, both from The Netherlands. Twenty-four chapters, 548 pp, Hardcover, ISBN 978-1-4020-8249-8, Softcover ISBN 978-90-481-7820-9, eBook ISBN 978-1-4020-8250-4 [http://www.springer.com/life+sciences/plant+sciences/book/978-1-4020-8249-8]
- **Volume 25 (2006): Chlorophylls and Bacteriochlorophylls: Biochemistry, Biophysics, Functions and Applications,** edited by Bernhard Grimm, Robert J. Porra, Wolfhart Rüdiger, and Hugo Scheer, from Germany and Australia. Thirty-seven chapters, 603 pp, Hardcover, ISBN 978-1-40204515-8, Softcover ISBN 978-90-481-7140-8, eBook ISBN 978-1-4020-4516-5 [http://www.springer.com/life+sciences/plant+sciences/book/978-1-4020-4515-8]
- **Volume 24 (2006): Photosystem I: The Light-Driven Plastocyanin: Ferredoxin Oxidoreductase,** edited by John H. Golbeck, from USA. Forty chapters, 716 pp, Hardcover ISBN 978-1-40204255-3, Softcover ISBN 978-90-481-7088-3, eBook ISBN 978-1-4020-4256-0 [http://www.springer.com/life+sciences/plant+sciences/book/978-1-4020-4255-3]
- **Volume 23 (2006): The Structure and Function of Plastids**, edited by Robert R. Wise and J. Kenneth Hoober, from USA. Twenty-seven chapters, 575 pp, Softcover, ISBN: 978-1-4020-6570–6; Hardcover ISBN 978-1-4020-4060-3, Softcover ISBN 978-1-4020-6570-5, eBook

ISBN 978-1-4020-4061-0 [http://www.springer.com/life+sciences/plant+sciences/book/978-1-4020-4060-3]

- **Volume 22 (2005): Photosystem II: The Light-Driven Water:Plastoquinone Oxidoreductase**, edited by Thomas J. Wydrzynski and Kimiyuki Satoh, from Australia and Japan. Thirty-four chapters, 786 pp, Hardcover ISBN 978-1-4020-4249-2, eBook ISBN 978-1-4020-4254-6 [http://www.springer.com/life+sciences/plant+sciences/book/978-1-4020-4249-2]
- **Volume 21 (2005): Photoprotection, Photoinhibition, Gene Regulation, and Environment,** edited by Barbara Demmig-Adams, William W. Adams III and Autar K. Mattoo, from USA. Twenty-one chapters, 380 pp, Hardcover ISBN 978-14020-3564-7, Softcover ISBN 978-1-4020-9281-7, eBook ISBN 978-1-4020-3579-1 [http://www.springer.com/life+sciences/plant+sciences/book/978-1-4020-3564-7]
- **Volume 20 (2006): Discoveries in Photosynthesis**, edited by Govindjee, J. Thomas Beatty, Howard Gest and John F. Allen, from USA, Canada and UK. One hundred and eleven chapters, 1304 pp, Hardcover ISBN 978-1-4020-3323-0, eBook ISBN 978-1-4020-3324-7 [http://www.springer.com/life+sciences/plant+sciences/book/978-1-4020-3323-0]
- **Volume 19 (2004): Chlorophyll *a* Fluorescence: A Signature of Photosynthesis**, edited by George C. Papageorgiou and Govindjee, from Greece and USA. Thirty-one chapters, 820 pp, Hardcover, ISBN 978-1-4020-3217-2, Softcover ISBN 978-90-481-3882-1, eBook ISBN 978-1-4020-3218-9 [http://www.springer.com/life+sciences/biochemistry+%26+biophysics/book/978-1-4020-3217-2]
- **Volume 18 (2005): Plant Respiration: From Cell to Ecosystem,** edited by Hans Lambers and Miquel Ribas-Carbo, from Australia and Spain. Thirteen chapters, 250 pp, Hardcover ISBN978-14020-3588-3, Softcover ISBN 978-90-481-6903-0, eBook ISBN 978-1-4020-3589-0 [http://www.springer.com/life+sciences/plant+sciences/book/978-1-4020-3588-3]
- **Volume 17 (2004): Plant Mitochondria: From Genome to Function**, edited by David Day, A. Harvey Millar and James Whelan, from Australia. Fourteen chapters, 325 pp, Hardcover, ISBN: 978-1-4020-2399-6, Softcover ISBN 978-90-481-6651-0, eBook ISBN 978-1-4020-2400-9 [http://www.springer.com/life+sciences/cell+biology/book/978-1-4020-2399-6]
- **Volume 16 (2004): Respiration in Archaea and Bacteria: Diversity of Prokaryotic Respiratory Systems,** edited by Davide Zannoni, from Italy. Thirteen chapters, 310 pp, Hardcover ISBN 978-14020-2002-5, Softcover ISBN 978-90-481-6571-1, eBook ISBN 978-1-4020-3163-2 [http://www.springer.com/life+sciences/plant+sciences/book/978-1-4020-2002-5]
- **Volume 15 (2004): Respiration in Archaea and Bacteria: Diversity of Prokaryotic Electron Transport Carriers**, edited by Davide Zannoni, from Italy. Thirteen chapters, 350 pp, Hardcover ISBN 978-1-4020-2001-8, Softcover ISBN 978-90-481-6570-4 (no eBook at this time) [http://www.springer.com/life+sciences/biochemistry+%26+biophysics/book/978-1-4020-2001-8]
- **Volume 14 (2004): Photosynthesis in Algae**, edited by Anthony W. Larkum, Susan Douglas and John A. Raven, from Australia, Canada and UK. Nineteen chapters, 500 pp, Hardcover ISBN 978-0-7923-6333-0, Softcover ISBN 978-94-010-3772-3, eBook ISBN 978-94-007-1038-2 [http://www.springer.com/life+sciences/plant+sciences/book/978-0-7923-6333-0]
- **Volume 13 (2003): Light-Harvesting Antennas in Photosynthesis**, edited by Beverley R. Green and William W. Parson, from Canada and USA. Seventeen chapters, 544 pp, Hardcover ISBN 978-07923-6335-4, Softcover ISBN 978-90-481-5468-5, eBook

ISBN 978-94-017-2087-8 [http://www.springer.com/life+sciences/plant+sciences/book/978-0-7923-6335-4]

- **Volume 12 (2003): Photosynthetic Nitrogen Assimilation and Associated Carbon and Respiratory Metabolism,** edited by Christine H. Foyer and Graham Noctor, from UK and France. Sixteen chapters, 304 pp, Hardcover ISBN 978-07923-6336-1, Softcover ISBN 978-90-481-5469-2, eBook ISBN 978-0-306-48138-3 [http://www.springer.com/life+sciences/plant+sciences/book/978-0-7923-6336-1]
- **Volume 11 (2001): Regulation of Photosynthesis,** edited by Eva-Mari Aro and Bertil Andersson, from Finland and Sweden. Thirty-two chapters, 640 pp, Hardcover ISBN 978-0-7923-6332-3, Softcover ISBN 978-94-017-4146-0, eBook ISBN 978-0-306-48148-2 [http://www.springer.com/life+sciences/plant+sciences/book/978-0-7923-6332-3]
- **Volume 10 (2001): Photosynthesis: Photobiochemistry and Photobiophysics,** edited by Bacon Ke, from USA. Thirty-six chapters, 792 pp, Hardcover ISBN 978-0-7923-6334-7, Softcover ISBN 978-0-7923-6791-8, eBook ISBN 978-0-306-48136-9 [http://www.springer.com/life+sciences/plant+sciences/book/978-0-7923-6334-7]
- **Volume 9 (2000): Photosynthesis: Physiology and Metabolism,** edited by Richard C. Leegood, Thomas D. Sharkey and Susanne von Caemmerer, from UK, USA and Australia. Twenty-four chapters, 644 pp, Hardcover ISBN 978-07923-6143-5, Softcover ISBN 978-90-481-5386-2, eBook ISBN 978-0-306-48137-6 [http://www.springer.com/life+sciences/plant+sciences/book/978-0-7923-6143-5]
- **Volume 8 (1999): The Photochemistry of Carotenoids,** edited by Harry A. Frank, Andrew J. Young, George Britton and Richard J. Cogdell, from USA and UK. Twenty chapters, 420 pp, Hardcover ISBN 978-0-7923-5942-5, Softcover ISBN 978-90-481-5310-7, eBook ISBN 978-0-306-48209-0 [http://www.springer.com/life+sciences/plant+sciences/book/978-0-7923-5942-5]
- **Volume 7 (1998): The Molecular Biology of Chloroplasts and Mitochondria in *Chlamydomonas*,** edited by Jean David Rochaix, Michel Goldschmidt-Clermont and Sabeeha Merchant, from Switzerland and USA. Thirty-six chapters, 760 pp, Hardcover ISBN 978-0-7923-5174-0, Softcover ISBN 978-94-017-4187-3, eBook ISBN 978-0-306-48204-5 [http://www.springer.com/life+sciences/plant+sciences/book/978-0-7923-5174-0]
- **Volume 6 (1998): Lipids in Photosynthesis: Structure, Function and Genetics**, edited by Paul-André Siegenthaler and Norio Murata, from Switzerland and Japan. Fifteen chapters, 332 pp. Hardcover ISBN 978-0-7923-5173-3, Softcover ISBN 978-90-481-5068-7, eBook ISBN 978-0-306-48087-4 [http://www.springer.com/life+sciences/plant+sciences/book/978-0-7923-5173-3]
- **Volume 5 (1997): Photosynthesis and the Environment,** edited by Neil R. Baker, from UK. Twenty chapters, 508 pp, Hardcover ISBN 978-07923-4316-5, Softcover ISBN 978-90-481-4768-7, eBook ISBN 978-0-306-48135-2 [http://www.springer.com/life+sciences/plant+sciences/book/978-0-7923-4316-5]
- **Volume 4 (1996): Oxygenic Photosynthesis: The Light Reactions**, edited by Donald R. Ort and Charles F. Yocum, from USA. Thirty-four chapters, 696 pp, Hardcover ISBN 978-0-7923-3683-9, Softcover ISBN 978-0-7923- 3684–6, eBook ISBN 978-0-306-48127-7 [http://www.springer.com/life+sciences/plant+sciences/book/978-0-7923-3683-9]
- **Volume 3 (1996): Biophysical Techniques in Photosynthesis**, edited by Jan Amesz and Arnold J. Hoff, from The Netherlands. Twenty-four chapters, 426 pp, Hardcover ISBN 978-0-7923-3642-6, Softcover ISBN

978-90-481-4596-6, eBook ISBN 978-0-306-47960-1 [http://www.springer.com/life+sciences/plant+sciences/book/978-0-7923-3642-6]

- **Volume 2 (1995): Anoxygenic Photosynthetic Bacteria**, edited by Robert E. Blankenship, Michael T. Madigan and Carl E. Bauer, from USA. Sixty-two chapters, 1331 pp, Hardcover ISBN 978-0-7923-3682-8, Softcover ISBN 978-0-7923-3682-2, eBook ISBN 978-0-306-47954-0 [http://www.springer.com/life+sciences/plant+sciences/book/978-0-7923-3681-5]
- **Volume 1 (1994): The Molecular Biology of Cyanobacteria**, edited by Donald R. Bryant, from USA. Twenty-eight chapters, 916 pp, Hardcover, ISBN 978-0-7923-3222-0, Softcover ISBN 978-0-7923-3273-2, eBook ISBN 978-94-011-0227-8 [http://www.springer.com/life+sciences/plant+sciences/book/978-0-7923-3222-0]

Further information on these books and ordering instructions is available at http://www.springer.com/series/5599. Contents of volumes 1–31 can also be found at <http://www.life.uiuc.edu/govindjee/photosynSeries/ttocs.html>. (For volumes 33-35, pdf files of the entire Front Matter are available.)

Special 25% discounts are available to members of the International Society of Photosynthesis Research, ISPR http://www.photosynthesisresearch.org/. See http://www.springer.com/ispr.

Advances in Photosynthesis and Respiration Including Bioenergy and Related Processes

The readers of the current series are encouraged to watch for the publication of the forthcoming books (not necessarily arranged in the order of future appearance)

- Cyanobacteria (Editor: Donald Bryant)
- Photosynthesis in Algae: Biofuels and Value-Added Products (Editors: Anthony Larkum and Arthur Grossman)

In addition to the above books, the following topics are under consideration:

Algae, Cyanobacteria: Biofuel and Bioenergy
Artificial Photosynthesis
ATP Synthase: Structure and Function
Bacterial Respiration II
Evolution of Photosynthesis
Green Bacteria and Heliobacteria
Interactions between Photosynthesis and other Metabolic Processes
Limits of Photosynthesis: Where do we go from here?
Photosynthesis, Biomass and Bioenergy
Photosynthesis under Abiotic and Biotic Stress

If you have any interest in editing/co-editing any of the above listed books, or being an author, please send an e-mail to Tom Sharkey (tsharkey@msu.edu) and/ or to Julian Eaton-Rye (julian.eaton-rye@otago.ac.nz). Suggestions for additional topics are also welcome. Instructions for writing chapters in books in our series are available by sending e-mail requests to one or both of us.

We take this opportunity to thank and congratulate Katie Becklin, Joy Ward, and Dani Way for their outstanding editorial work. Our editorial teams have been either all male or including one female with other male editors until now. We are glad that this volume breaks that mold and is edited by a team of all women, each an outstanding scientist in the field of photosynthesis research. We thank all the 37 authors of this book (see the list given earlier); without their authoritative chapters, there would be no such volume. We give special thanks to Mr. Prasad Gurunadham of SPi Global, India, for directing the typesetting of this book; his/her expertise has been crucial in bringing this book to completion. We owe Zuzana Bernhart and Mariska van der Stigchel much thanks for their work at Springer/Nature that made this book possible.

October 19, 2020

Thomas D. Sharkey
Department of Biochemistry
and Molecular Biology
Michigan State University
East Lansing, MI, USA
email: tsharkey@msu.edu

Julian J. Eaton-Rye
Department of Biochemistry
University of Otago
Dunedin, New Zealand
email: julian.eaton-rye@otago.ac.nz

Series Editors

A 2017 informal photograph of Govindjee (right) and his wife Rajni (left) in Champaign-Urbana, Illinois; Photograph by Dilip Chhajed.

Govindjee, who uses one name only, was born on October 24, 1932, in Allahabad, India. Since 1999, he has been professor emeritus of biochemistry, biophysics, and plant biology at the University of Illinois at Urbana-Champaign (UIUC), Urbana, IL, USA. He obtained his B.Sc. (chemistry, botany and zoology) and M.Sc. (botany, plant physiology) in 1952 and 1954, from the University of Allahabad. He learned his Plant Physiology from Shri Ranjan, who was a student of Felix Frost Blackmann (of Cambridge, UK). Then, Govindjee studied *Photosynthesis* at the UIUC, under two giants in the field, Robert Emerson (a student of Otto Warburg) and Eugene Rabinowitch (who had worked with James Franck), obtaining his Ph.D., in biophysics, in 1960.

Govindjee is best known for his research on excitation energy transfer, light emission (prompt and delayed fluorescence, and thermoluminescence), primary photochemistry, and electron transfer in *Photosystem II* (PS II, water-plastoquinone oxidoreductase). His research, with many others, includes the discovery of a short-wavelength form of chlorophyll (Chl) *a* functioning in PS II; of the two-light effect in Chl *a* fluorescence; and, with his wife Rajni Govindjee, of the two-light effect (Emerson Enhancement) in $NADP^+$ reduction in chloroplasts. His major achievements, together with several others, include an understanding of the basic relationship between Chl *a* fluorescence and photosynthetic reactions; a unique role of bicarbonate/carbonate on the electron acceptor side of PS II, particularly in the protonation events involving the Q_B binding region; the theory of thermoluminescence in plants; the first picosecond measurements on the primary photochemistry of PS II; and the use of fluorescence lifetime imaging microscopy (FLIM) of Chl *a* fluorescence in understanding photoprotection by plants against excess light. His current focus is on the *History of Photosynthesis Research*, and in *Photosynthesis Education*. He has served on the faculty of the UIUC for ~40 years.

Govindjee's honors include: Fellow of the American Association of Advancement of Science (AAAS); distinguished lecturer of the School of Life Sciences, UIUC; fellow and lifetime member of the National Academy of Sciences (India); president of the American Society for Photobiology (1980–1981); Fulbright scholar (1956), Fulbright senior lecturer (1997), and Fulbright specialist (2012); honorary president of the 2004 International Photosynthesis Congress (Montréal, Canada); the first recipient of the Lifetime Achievement Award of the Rebeiz Foundation for Basic Biology, 2006; recipient of the Communication Award of the International Society of Photosynthesis Research, 2007; and of the Liberal Arts and Sciences Lifetime Achievement Award of the UIUC, (2008). Further, Govindjee has been honored many times: (1) In 2007, through 2 special volumes of Photosynthesis Research, celebrating his 75th birthday and for his 50-year dedicated research in Photosynthesis (guest editor: Julian J. Eaton-Rye); (2) In 2008, through a special International Symposium on "Photosynthesis in a Global Perspective", held in November 2008, at the University of Indore, India; this was followed by a book "*Photosynthesis: Basics and Applications*" (edited by S. Itoh, P. Mohanty and K.N. Guruprad); (3) In 2012, through "*Photosynthesis – Plastid Biology, Energy Conversion and Carbon Assimilation*", edited by Julian J. Eaton-Rye, Baishnab C. Tripathy, and Thomas D. Sharkey; (4) In 2013, through special issues of Photosynthesis Research (volumes 117 and 118), edited by Suleyman Allakhverdiev, Gerald Edwards and Jian-Ren Shen celebrating his 80th (or rather 81st) birthday; (5) In 2014, through celebration of his 81st birthday in Třeboň, the Czech Republic (O. Prasil [2014] Photosynth Res 122: 113–119); (6) In 2016, through the Award of the prestigious Prof. B.M. Johri Memorial Award of the Society of Plant Research, India. In 2018, *Photosynthetica* published a special issue to celebrate his 85th birthday (Editor: Julian J. Eaton-Rye).

Govindjee's unique teaching of the Z-scheme of photosynthesis, where students act as different intermediates, has been published in two papers (1) P.K. Mohapatra and N.R. Singh [2015] Photosynth Res 123:105–114); (2) S. Jaiswal, M. Bansal, S. Roy, A, Bharati and B, Padhi [2017] Photosynth Res 131: 351—359. Govindjee is a coauthor of a classic and highly popular book "*Photosynthesis*" (with E.I. Rabinowitch, 1969), and of a historical book "*Maximum Quantum Yield of Photosynthesis: Otto Warburg and the Midwest Gang*" (with K. Nickelsen, 2011). He is editor (or coeditor) of many books including: *Bioenegetics of Photosynthesis* (1975); *Photosynthesis*, 2 volumes (1982); *Light Emission by Plants and Bacteria* (1986); *Chlorophyll a Fluorescence: A Signature of Photosynthesis* (2004); *Discoveries in Photosynthesis* (2005); and *Non-Photochemical Quenching and Energy Dissipation in Plants, Algae and Cyanobacteria* (2015).

Since 2007, each year a **Govindjee and Rajni Govindjee Award** is given to graduate students, by the Department of Plant Biology (odd years) and by the Department of Biochemistry (even years), at the UIUC, to recognize excellence in biological sciences. For further information on Govindjee, see his website at http://www.life.illinois.edu/govindjee.

Thomas D. (Tom) Sharkey, obtained his Bachelor's degree in Biology in 1974 from Lyman Briggs College, a residential science college at Michigan State University, East Lansing, Michigan, USA. After 2 years as a research technician, Tom entered a Ph.D. program in the Department of Energy Plant Research Laboratory at Michigan State University under the mentorship of Klaus Raschke and finished in 1979. Post-doctoral research was carried out with Graham Farquhar at the Australian National University, in Canberra, where he coauthored a landmark review on photosynthesis and stomatal conductance. For 5 years he worked at the Desert Research Institute, Reno, Nevada. After Reno, Tom spent 20 years as Professor of Botany at the University of Wisconsin in Madison. In 2008, Tom became Professor and Chair of the Department of Biochemistry and Molecular Biology at Michigan State University. In 2017 Tom stepped down as department chair and moved to the MSU-DOE Plant Research Laboratory completing a 38-year sojourn back to his beginnings. Tom's research interests center on the exchange of gases between plants and the atmosphere and carbon metabolism of photosynthesis. The biochemistry and biophysics underlying carbon dioxide uptake and isoprene emission from plants form the two major research topics in his laboratory. Among his contributions are measurement of the carbon dioxide concentration inside leaves, an exhaustive study of short-term feedback effects in carbon metabolism, and a significant contribution to elucidation of the pathway by which leaf starch breaks down at night. In the isoprene research field, his laboratory has cloned many of the genes that underlie isoprene synthesis and he has published many important papers on the biochemical regulation of isoprene synthesis. Tom's work has been cited over 35,000 times according to Google Scholar in 2021. He has been named an Outstanding Faculty member by Michigan State University and in 2015, he was named a University Distinguished Professor. He is a fellow of the American Society of Plant Biologists and of the American Association for the Advancement of Science. Tom has co-edited three books, the first on trace gas emissions from plants in 1991 (with Elizabeth Holland and Hal Mooney), volume 9 of this series (with Richard Leegood and Susanne von Caemmerer) on the *Physiology of Carbon Metabolism of Photosynthesis* in 2000, and volume 34 (with Julian J. Eaton-Rye and Baishnab C. Tripathy) entitled *Photosynthesis: Plastid Biology, Energy Conversion and Carbon Assimilation*. Tom has been co-editor of this series since volume 31.

Julian J. Eaton-Rye, is a professor in the Department of Biochemistry at the University of Otago, New Zealand. He received his undergraduate degree in botany from the University of Manchester in the U.K. in 1981 and his Ph.D. from the University of Illinois in 1987, where he worked with Govindjee on the role of bicarbonate in the regulation of electron transfer through Photosystem II. Before joining the Biochemistry Department at Otago University in 1994, he was a postdoctoral researcher focusing on various aspects of Photosystem II protein biochemistry with Professor Norio Murata at the National Institute of Basic Biology in Okazaki, Japan, with Professor Wim Vermaas at Arizona State University, and with Dr. Geoffrey Hind at Brookhaven National Laboratory. His current research interests include structure-function relationships of Photosystem II proteins both in biogenesis and electron transport as well as the role of additional protein factors in the assembly of Photosystem II. Julian has been a Consulting Editor for the *Advances in Photosynthesis and Respiration* series since 2005, and edited volume 34 (with Baishnab C. Tripathy and Thomas D. Sharkey) entitled *Photosynthesis: Plastid Biology, Energy Conversion and Carbon Assimilation*. He is also an associate editor for the *New Zealand Journal of Botany* and an Associate Editor for the Plant Cell Biology section of *Frontiers in Plant Science*. He edited a Frontiers Research Topic on the *Assembly of the Photosystem II Membrane-Protein Complex of Oxygenic Photosynthesis* (with Roman Sobotka) in 2016 and this is available as an eBook [ISBN 978-2-88945-233-0]. Julian has served as the president of the New Zealand Society of Plant Biologists (2006–2008) and as president of the New Zealand Institute of Chemistry (2012). He has been a member of the International Scientific Committee of the Triennial International Symposium on Phototrophic Prokaryotes (2009–2018) and is currently the Secretary of the International Society of Photosynthesis Research.

Contents

Part III Population- and Community-Level Responses of Photosynthesis and Respiration to Climate Change

Part IV Responses of Plants with Carbon-Concentrating Mechanisms to Climate Change

Preface: Photosynthesis and Respiration in a Changing World

Currently, terrestrial photosynthesis absorbs ~120 GT of carbon every year, while autotrophic respiration returns half of this carbon back to our atmosphere, fueling the global carbon cycle and helping determine our atmospheric CO_2 concentration. But while these physiological pathways affect our climate, the climate also affects photosynthesis and respiration, setting up the potential for feedbacks in the Earth System. Understanding how climate change, including rising atmospheric CO_2 concentrations, climate warming and drought, will alter carbon exchange in plants is therefore critical for predicting plant responses to future conditions, and for meeting the rapidly increasing demand for food and biofuels by what may soon be >9 billion people. This volume seeks to provide readers with an integrative perspective on photosynthetic and respiration responses to global change with an emphasis on the impacts of increasing atmospheric CO_2 concentrations, rising temperatures and increasing water stress. Over 12 chapters, a range of experts from around the globe synthesize our understanding of how leaf carbon fluxes respond to climate change, and present new data on the fundamental mechanisms that facilitate acclimation and adaptation to changing conditions.

In Chap. 1, we set the scene by discussing the importance of understanding how leaf carbon fluxes will respond to climate change, including leaf-level responses that are critical for Earth System models, but also the importance of considering diverse plant functional types and ecological diversity. The next three chapters focus on the physiological mechanisms that constrain carbon assimilation and the potential for C3 photosynthesis and leaves to acclimate to climate change. For CO_2 to be fixed in photosynthesis, it must first diffuse into the leaf through stomata, which control vegetation-atmosphere exchanges of water and CO_2, and the effects of climate change on stomatal behavior are covered in Chap. 2. Once CO_2 reaches the intercellular airspaces of the leaf, it must then diffuse into the chloroplast where it can be fixed by Ribulose-1,5-bisphosphate carboxylase/oxygenase (Rubisco). The suite of physical and biochemical processes enabling the movement of CO_2 along this pathway are known as mesophyll conductance, and the effects of environmental change on mesophyll conductance are discussed in Chap. 3. Once CO_2 reaches the photosynthetic machinery, the rate of CO_2 fixation depends on the concentration of photosynthetic enzymes and pigments, as well as the relative allocation of resources to different photosynthetic components. Plants acclimate both the production of photosynthetic proteins and the relative abundance of different photosynthetic components in response to changes in growth temperature and CO_2 concentration. Chapter 4 therefore explores the strength of this acclimation and the degree to which changes in leaf N concentration (a common proxy for photosynthetic enzyme concentrations) are related to changes in photosynthetic performance under future climate conditions. And while we often concentrate on stomatal and biochemical responses to climate change when considering photosynthesis and leaf performance, leaf trichomes play a critical role in regulating leaf carbon and water fluxes through their effects on the leaf boundary layer and its energy balance, as well as being an important defense against herbivory. In Chap. 5, we therefore review what is known about how elevated CO_2 concentrations alter leaf trichomes, and discuss the implications of these alterations for leaf function in a future climate.

Evolutionary studies provide insight into genetic changes in photosynthetic traits, the emergence of novel plant strategies, and the potential for rapid evolutionary responses to future climate conditions. Chapter 6 examines how variation in the responses of individuals to rising CO_2, warming and drought can lead to evolutionary changes in response to climate change. Chapter 7 then uses a case study to demonstrate how populations and communities can be altered by extreme climate events when there is considerable intraspecific variation in plant responses to environmental stress.

The next set of chapters considers the potential responses of plants to cope with climate change when they use a carbon-concentrating mechanism, such as C4 photosynthesis or Crassulacean Acid metabolism (CAM). The ability to concentrate CO_2 around Rubisco gives these species an advantage in the low CO_2 conditions that have prevailed on Earth for millenia, as well as in dry, hot environments. This begs the question of whether these plants, including some of our most important crop species, will be negatively affected by future high CO_2 concentrations or whether the warm, dry conditions that will accompany rising CO_2 will make C4 and CAM species even more important for providing food and energy to a growing human population. Chapter 8 considers the future of terrestrial species that use carbon-concentrating mechanisms as the climate changes, while Chap. 9 hones in on how warming, rising CO_2 concentrations and increasing drought stress will specifically affect key C4 crops. CAM species are particularly known for their high water use efficiency, a result of closing their stomata during the day and instead initially fixing CO_2 during cool nights, and the ability of these plants to cope with climate change is discussed in Chap. 10.

Finally, in the last two chapters, we take a broad timescale approach, looking forward to novel approaches for engineering photosynthesis to improve future plant productivity (Chap. 11), and looking into the past to understand how the close coupling between metabolic processes, especially photorespiration, benefited plants in past low CO_2 environments (Chap. 12).

The question of how vegetation will respond to rising atmospheric CO_2 concentrations and warming temperatures underpins our ability to predict the trajectory of future climate change, to feed the billions of people on Earth and to maintain critical ecosystem services for both ourselves and the multitude of species on our planet. Our goals with this book were to help identify what we know about how photosynthesis and respiration, two of the most important carbon fluxes on Earth, will be affected by climate change, and to identify key challenges for understanding how these physiological processes will change in coming decades. The authors of this volume have risen to the challenge, and we sincerely thank them for their dedication in making this book possible.

Katie M. Becklin
Syracuse, NY, USA

Joy K. Ward
Cleveland, OH, USA

Danielle A. Way
London, ON, Canada

The Editors

Katie M. Becklin obtained her Ph.D. from the University of Missouri in 2010. Working with Candance Galen, she explored the dynamics of plant-fungal symbioses across environmental gradients. Katie then joined Joy Ward's lab at the University of Kansas for her postdoctoral research, focusing on plant physiological responses across a broad CO_2, ranging from low levels experienced during the last glacial maximum to predicted future concentrations. During this time Katie was awarded an NIH IRACDA postdoctoral fellowship to further test the role of mycorrhizal fungi in mediating plant responses to CO_2. In 2017, Katie joined the Biology Department at Syracuse University as an Assistant Professor. She has published multiple papers in prominent plant physiology and ecology journals, and is a Review Editor for *Global Change Biology* and *Frontiers in Forests and Global Change*. In 2019, Katie chaired the Gordon Research Seminar on CO_2 Assimilation in Plants, supporting the work of early career plant biologists.

Joy K. Ward received her BS degree in biology from Penn State University and earned her masters and doctorate degrees at Duke University under Professor Boyd Strain. She also held a post-doctoral position with Distinguished Professor James Ehleringer at the University of Utah. She was a long-term professor at the University of Kansas for 17 years prior to becoming dean of the College of Arts and Sciences at Case Western Reserve. In 2021, she became a Fellow of the American Association for the Advancement of Science (AAAS). She was the recipient of a Presidential Early Career Award for Scientists and Engineers (PECASE) from the White House in 2009 and a CAREER award from the National Science Foundation. The Kavli Foundation and National Academy of Sciences also named her a Kavli Fellow. She has served as U.S. chair and planning member for the National Academy's *Frontiers of Science* in Saudi Arabia, Japan, Oman, Kuwait and has been a scientific delegate to Uzbekistan through the State Department. Joy is internationally recognized for her studies on how plants respond to changing atmospheric CO_2 over geologic and contemporary time. By incorporating the fossil record, she has provided novel insights into how plants have responded to long-term environmental changes since the last glacial period, as well as understanding how plants will respond to environments of the future. Ward has mentored many successful students at the undergraduate, graduate, and post-doctoral levels, and has been a mentor for several programs to enhance diversity in STEM fields.

Danielle (Dani) Way obtained her BSc from the University of Toronto in 2002 and then began her PhD at the same institution in Rowan Sage's lab, studying the effects of elevated growth temperatures on black spruce, graduating in 2008. Dani did her post-doctoral work at Duke University with Rob Jackson, focusing on how elevated CO_2 altered photosynthesis and plant performance, including work on forests and prairie systems at the Duke Free Air CO_2 Enrichment (FACE) site and the Lysimeter CO_2 Gradient (LYCOG) study in Texas. She then became a Research Associate at the Nicholas School of the Environment at Duke, working closely with Ram Oren and Gaby Katul. In 2012, she joined the Department of Biology at the University of Western Ontario, where she was granted tenure in 2018 and currently has her lab. Dani has also been an Adjunct Assistant Professor at Duke University since 2012, and joined the Terrestrial Ecosystem Science & Technology research group at Brookhaven National Laboratory as a Joint Appointee in 2019. Dani has published nearly 80 papers focusing on how warming and rising atmospheric CO_2 concentrations alter photosynthesis, respiration, growth and survival in plants, with a focus on high latitude tree species. Dani was elected to the College of the Royal Society of Canada in 2018 and was awarded the C.D. Nelson Award by the Canadian Society of Plant Biologists in 2019 for outstanding research in plant biology within 10 years of an independent position. She is the Reviews Editor for *Global Change Biology*, a Topic Editor for *Plant, Cell & Environment*, and worked from 2012-2020 as an Editor and Associate Editor-in-Chief for *Tree Physiology*.

Contributors

Michael J. Aspinwall
School of Forestry and Wildlife Sciences, Auburn University, Auburn, AL, USA

Hermann Bauwe
Plant Physiology Department, University of Rostock, Rostock, Germany

Katie M. Becklin
Department of Biology, Syracuse University, Syracuse, NY, USA

Kristyn Bennett
Department of Biology, University of Western Ontario, London, ON, Canada

Brittany B. Blair
Department of Biochemistry and Molecular Biology, University of Nevada, Reno, NV, USA

Timothy E. Burnette
Organismal Biology, Ecology, & Evolution, Division of Biological Sciences, University of Montana, Missoula, MT, USA

Department of Ecology and Evolutionary Biology, University of Kansas, Lawrence, KS, USA

Marc Carriquí
Research Group in Plant Biology under Mediterranean Conditions, Universitat de les Illes Balears - Instituto de Agroecología y Economía del Agua (INAGEA), Palma, Illes Balears, Spain

School of Natural Sciences, University of Tasmania, Hobart, TAS, Australia

Jacob M. Carter
Center for Science & Democracy, Union of Concerned Scientists, Washington, DC, USA

Department of Ecology and Evolutionary Biology, University of Kansas, Lawrence, KS, USA

John C. Cushman
Department of Biochemistry and Molecular Biology, University of Nevada, Reno, NV, USA

André G. Duarte
Department of Biology, University of Western Ontario, London, ON, Canada

Mirindi E. Dusenge
Department of Biology, University of Western Ontario, London, ON, Canada

Michele Faralli
School of Life Sciences, University of Essex, Colchester, UK

Alisdair R. Fernie
Max Planck Institute of Molecular Plant Physiology, Potsdam, Golm, Germany

James M. Fischer
Carl R. Woese Institute for Genomic Biology, University of Illinois at Urbana Champaign, Urbana, IL, USA

Department of Ecology and Evolutionary Biology, University of Kansas, Lawrence, KS, USA

Jaume Flexas
Research Group in Plant Biology under Mediterranean Conditions, Universitat de les Illes Balears – Instituto de Agroecología y Economía del Agua (INAGEA), Palma, Illes Balears, Spain

Oula Ghannoum
ARC Centre of Excellence for Translational Photosynthesis, Hawkesbury Institute for the

Environment, Western Sydney University, Penrith, NSW, Australia

Thomas E. Juenger
Department of Integrative Biology, University of Texas at Austin, Austin, TX, USA

Tracy Lawson
School of Life Sciences, University of Essex, Colchester, UK

Karen Lemon
Department of Biology, University of Western Ontario, London, ON, Canada

Benedict M. Long
ARC Centre of Excellence for Translational Photosynthesis, Australian National University, Canberra, ACT, Australia

Sarah McDonald
Department of Biology, University of Western Ontario, London, ON, Canada

Miquel Nadal
Research Group in Plant Biology under Mediterranean Conditions, Universitat de les Illes Balears – Instituto de Agroecología y Economía del Agua (INAGEA), Palma, Illes Balears, Spain

Nicholas A. Niechayev
Department of Biochemistry and Molecular Biology, University of Nevada, Reno, NV, USA

Paula N. Pereira
Department of Biochemistry and Molecular Biology, University of Nevada, Reno, NV, USA

Julianne Radford
Department of Biology, University of Western Ontario, London, ON, Canada

Alexis Rodgers
Department of Biology, University of North Florida, Jacksonville, FL, USA

Paul D. Rymer
Hawkesbury Institute for the Environment, Western Sydney University, Penrith, NSW, Australia

Rowan F. Sage
Department of Ecology and Evolutionary Biology, University of Toronto, Toronto, ON, Canada

Robert E. Sharwood
ARC Centre of Excellence for Translational Photosynthesis, Australian National University, Canberra, ACT, Australia
Hawkesbury Institute for the Environment, Western Sydney University, Richmond, NSW, Canberra, ACT, Australia

John D. Stamford
School of Life Sciences, University of Essex, Colchester, UK

Matt Stata
Department of Ecology and Evolutionary Biology, University of Toronto, Toronto, ON, Canada

Jim Stevens
School of Life Sciences, University of Essex, Colchester, UK

David T. Tissue
Hawkesbury Institute for the Environment, Western Sydney University, Penrith, NSW, Australia

Shellie Wall
School of Life Sciences, University of Essex, Colchester, UK

Joy K. Ward
College of Arts and Sciences, Department of Biology, Case Western Reserve University, Cleveland, OH, USA
Department of Ecology and Evolutionary Biology, University of Kansas, Lawrence, KS, USA

Alexander Watson-Lazowski
ARC Centre of Excellence for Translational Photosynthesis, Hawkesbury Institute for the Environment, Western Sydney University, Penrith, NSW, Australia

Danielle A. Way
Department of Biology, University of Western Ontario, London, ON, Canada

Nicholas School of the Environment, Duke University, Durham, NC, USA

Terrestrial Ecosystem Science & Technology Group, Environmental & Climate Sciences Department, Brookhaven National Laboratory, Upton, NY, USA

Part I

Introduction

Chapter 1

Leaf Carbon Flux Responses to Climate Change: Challenges and Opportunities

Danielle A. Way*
Department of Biology, University of Western Ontario, London, ON, Canada

Nicholas School of the Environment, Duke University, Durham, NC, USA

Terrestrial Ecosystem Science & Technology Group, Environmental & Climate Sciences Department, Brookhaven National Laboratory, Upton, NY, USA

Katie M. Becklin
Department of Biology, Syracuse University, Syracuse, NY, USA

and

Joy K. Ward
College of Arts and Sciences, Department of Biology, Case Western Reserve University, Cleveland, OH, USA

Summary

To better predict how terrestrial vegetation will respond to climate change and to accelerate our breeding programs to produce climate-resilient plants, we need to understand plant responses to rising atmospheric CO_2 concentrations, temperature and water stress. Achieving

*Author for correspondence, e-mail: dway4@uwo.ca

K. M. Becklin et al. (eds.), *Photosynthesis, Respiration, and Climate Change*, Advances in Photosynthesis and Respiration 48, https://doi.org/10.1007/978-3-030-64926-5_1

this goal requires integrating our knowledge of the physiology and acclimation potential for all photosynthetic pathways, considering the role of evolutionary processes under climate stress, and embracing opportunities and technological advances to create more climate-resilient plants. Here, we discuss key constraints that may limit plant performance in future climates, and discuss the importance of climate change experiments for furthering our understanding of plant-atmosphere interactions, with a focus on photosynthesis and respiration. Lastly, we provide an overview of current research challenges with regard to leaf carbon fluxes and climate change, and highlight how the chapters in this book help us to overcome those challenges.

I. Introduction

Since the Industrial Revolution, anthropogenic activities, mainly fossil fuel burning and land use change, have caused atmospheric CO_2 concentrations to increase by nearly 50% (Joos and Spahni 2008; Ciais et al. 2013). As a powerful greenhouse gas, these increases in CO_2 concentration have already raised global mean annual temperatures by ~1 °C since 1900 (NOAA 2018). The Intergovernmental Panel on Climate Change predicts that if we remain on our current socioeconomic trajectory, CO_2 levels could reach over 1000 ppm by the year 2100, which would lead to a planetary warming of ~4 °C (Ciais et al. 2013). These changes will in turn affect many other aspects of our climate system, including altering precipitation patterns with more frequent and severe droughts in many places, increasing extreme weather events such as heat waves, and shifting major disturbance regimes like fire occurrence and insect outbreaks (Kurz et al. 1995; Seidl et al. 2017). Additionally, other global change factors, such as eutrophication and biodiversity loss, are simultaneously modifying Earth's ecosystems, and in many cases can create positive feedbacks with climate change (Moss et al. 2011; Cardinale et al. 2017). While these changes to our Earth system are immense, CO_2 concentrations would already be considerably higher (and temperatures much warmer) if the natural world, and vegetation in particular, was not mitigating the effects of anthropogenic CO_2 emissions. Only about half of the CO_2 emitted via human activities accumulates in the atmosphere every year, with the oceans absorbing 25% of our emissions and another 30% of anthropogenic CO_2 emissions being taken up by terrestrial plants (Friedlingstein et al. 2019).

The ability of land plants to continue subsidizing our fossil fuel addiction rests largely on the extent to which the physiological processes underpinning plant CO_2 fluxes are affected by climate change. Plants help regulate atmospheric CO_2 concentrations (and thereby play a role in controlling global and regional temperatures) by fixing CO_2 in photosynthesis and releasing CO_2 via respiration and photorespiration (Friedlingstein et al. 2019). Yet these physiological processes themselves are sensitive to both CO_2 concentrations and temperature, setting up complex feedbacks between plants and the atmosphere (Dusenge et al. 2019). As CO_2 is the substrate for carbon fixation in the Calvin-Benson cycle, one might expect that elevated CO_2 concentrations will stimulate photosynthesis, thus generating a larger carbon sink in a high CO_2 world (Drake et al. 1997). And because photosynthesis has a peaked response to leaf temperature and respiration increases exponentially as temperatures warm, one might also predict that warming will increase photosynthesis in cool regions, suppress CO_2 uptake in hot environments (Huang et al. 2019), and stimulate respiration globally (King et al. 2006). But the reality is more complicated: more frequent extreme events will cause mortality and stress, thus reducing plant performance (Niu et al. 2014), species will migrate and communities will reshuffle as the climate changes (Thullier 2004), plants acclimate to new environmental conditions (King et al. 2006; Dusenge et al. 2019), different photosynthetic pathways respond differently to rising CO_2 and temperatures (Sage and Kubien 2007), and biotic interactions will alter plant responses to climate (Classen et al. 2015). For all of these reasons (and many more), these simplistic predictions are unlikely to hold across the diverse range of natural and managed ecosystems found across the globe.

II. Constraints on Plant Responses to Climate Change

One of the primary reasons why vegetation may not respond in a simple, predictable way to climate change is the increase in extreme events brought about by climate change. Although climate change is often described with a single value for the expected change in global mean annual temperatures by the end of this century, the changes in temperature experienced by plants will be much more complex. Warming does not merely increase temperatures everywhere equally at all times, but rather has much more pronounced effects on temperatures at high latitudes, during the winter, and at night (Collins et al. 2013). Thus, a global increase in mean annual temperatures of 4 °C might translate into a winter nighttime warming of >12 °C for the Arctic and only a 1 °C warming for daytime temperatures at the equator (Collins et al. 2013). While even 1 °C warming may be detrimental to tropical regions (Doughty and Goulden 2008), the degree of warming predicted for high latitude regions will have enormous implications for their functioning and species composition. Additionally, warmer average temperatures are correlated with more frequent and severe heat waves (Russo et al. 2014; Carter et al. 2021). For species that can tolerate a warmer growing climate, these extreme events may be enough to push plants over the edge, killing them or causing serious damage that reduces their growth, fecundity and carbon sink capacity (Teskey et al. 2015).

Not only will higher temperatures directly impact vegetation, but a warmer climate will also be, in many regions, a drier climate (Dai 2012). Ecosystems and the species that comprise them are tuned to historical cycles of water availability (Huxman et al. 2004; Engelbrecht et al. 2007). Changes in precipitation, be it alterations in the amount of total water received, the timing of water availability, or the manner in which the water falls (i.e., intense storms versus gentle rain) will therefore also affect the growth and survival of vegetation (Knapp et al. 2002; Engelbrecht et al. 2007). Warming will likely lead to more frequent and intense droughts in many parts of the world, which will also constrain the ability of plants to respond positively to higher growth temperatures (Adams et al. 2009; Allen et al. 2010).

III. A Brief History of Global Change Biology Research in Photosynthesis and Respiration

We have known for decades that human activities were increasing atmospheric CO_2 concentrations: C. David Keeling began monitoring atmospheric CO_2 concentrations at Mauna Loa in 1958 (Keeling 1998) and showed that atmospheric CO_2 concentrations rose and fell in a seasonal pattern that was correlated with plant activity. He also showed that atmospheric CO_2 levels were slowly rising, in concert with rising fossil fuel emissions (Keeling 1960). We have known for even longer, indeed for over a century, that CO_2 is a greenhouse gas that can warm the planet (Arrhenius 1896). Keeling's work and its implications for climate warming soon motivated interest in determining how plants will respond to these global changes. Particularly, there was considerable interest in how photosynthesis and respiration will respond to climate change, given the role of these two processes in controlling atmospheric CO_2 concentrations.

Early experiments characterizing how plant growth, photosynthesis and respiration respond to changes in temperature and CO_2 concentration were mostly limited to laboratory and growth cabinet studies on small potted plants or to natural environments where plants from different thermal conditions could be compared (e.g., Rook 1969; McCree 1974; Pearcy 1977; Mooney et al. 1978; Berry and Björkman 1980; Carlson and Bazzaz 1980; Patterson and Flint 1980). The need for more realistic ecological conditions to confirm and expand on these early studies led to the development of closed and open-top chambers (e.g., Drake et al. 1989) and passive warming chambers (e.g., Chapin and Shaver 1985) where CO_2 concentrations and temperatures could be altered in a field setting. These approaches allowed researchers to investigate plant responses to climate change while growing in natural communities in native soils. However, both types of chambers altered the precipitation, humidity and wind conditions inside the plots and were only feasible on a relative small scale (Hendrey and Kimball 1994). The invention of Free Air CO_2 Enrichment (FACE) technology (described in Hendrey 1992) allowed for even more realistic growth conditions, whereby crop plots and entire forest stands could be exposed to elevated CO_2 concentrations without using any walls that would alter other environmental conditions. For warming, a similar open-air approach was adopted that relies on a set of infrared heaters to generate an even and reliable increase in plant and soil temperature (Kimball 2005). More recently, large-scale experiments investigating the joint effects of elevated CO_2 and temperature on plant physiology have been designed (Hanson et al. 2017).

The data from these experiments have honed our understanding of how key plant physiological processes are altered in plants that grow under future climate conditions. But to understand how shifts in plant performance under elevated CO_2 and temperature will feed back onto the Earth system and help to mitigate or accelerate climate change, we need to use these data to parameterize models of vegetation-atmosphere interactions (Medlyn et al. 2015; Norby et al. 2016; Bonan and Doney 2018). There has therefore been considerable interest in the last decade in leveraging the results from large-scale global change experiments, such as FACE sites, to test the performance of, and assumptions built into, terrestrial ecosystem models and Earth System models (ESMs) (De Kauwe et al. 2013, 2014; Walker et al. 2014; Zaehle et al. 2014; Medlyn et al. 2015). These efforts have brought modelers and plant physiologists together, helping experimentalists understand the knowledge gaps in models where more data are needed and allowing modelers to discuss the current state of knowledge for relevant physiological processes and incorporate this into their models (e.g., Rogers et al. 2017). Additionally, there has been a shift towards incorporating modeling perspectives into global change studies from the start, rather than only at the end of an experiment (Norby et al. 2016). By using ecosystem-level models to generate hypotheses for studies, field researchers can refocus the measurements they should make, while the data gathered can be used to test model performance under future climate conditions (Leuzinger and Thomas 2011). Our ability to capture biosphere-atmosphere interactions has been identified as one of the most pressing challenges for ESM development (Bonan and Doney 2018). By working with modelers, global change biologists and plant scientists can now directly help improve and refine the ESMs and other models we rely on to predict future climate change (Leuzinger and Thomas 2011; Medlyn et al. 2015).

IV. Current Research Challenges

As we move into a warmer, higher CO_2 world, there are a number of key challenges that we must overcome in order to better predict the future of our planet and ensure the ability of plants to provide food, fuel and fiber for a growing human population, habitat for the world's biodiversity, and critical ecosystem services (such as carbon sinks and climate change mitigation potential) for a changing world.

To better predict the trajectory of climate change, we must improve the ability of ESMs to accurately capture vegetation-atmosphere interactions (Leuzinger and Thomas 2011). One way to do this is to reduce uncertainty in the parameters used to model key physiological processes that affect atmospheric CO_2 concentrations, such as photosynthesis, stomatal conductance and respiration (Huntington et al. 2017; Rogers et al. 2017). These physiological parameters are often sensitive to both short- and long-term changes in both CO_2 concentration and temperature (Dusenge et al. 2019; Stevens et al. 2021; Nadal et al. 2021; Duarte et al. 2021). Incorporating realistic acute and acclimation responses of plant processes to each of these climate change drivers remains a challenge, although much progress has been made (Smith and Dukes 2013; Smith et al. 2016; Mercado et al. 2018). However, accurately representing how photosynthetic and respiratory processes will be altered under conditions of *combined* elevated CO_2 and temperature likely requires moving beyond thinking about these global change drivers separately. Additionally, where ESMs use a semi-empirical approach to estimate some parameter values (such as stomatal conductance), moving towards a more mechanistic approach or using an optimization framework may improve model performance (e.g., Franks et al. 2018).

Additionally, we must embrace the diversity of plant form and function found in nature when studying plant response to climate change. This requires working to understand how differences in life history strategies, functional traits, and growth forms impact the performance of vegetation in a warmer, higher CO_2 world. With regard to plant carbon fluxes, photosynthetic pathway is one of the biggest differences among plants. Although the majority of plant species use the C3 pathway, species that use a carbon-concentrating mechanism (CCMs) have an outsized impact on plant productivity and have numerous traits that make them well suited for future climates, but also potentially more vulnerable to them. C4 plants comprise only 3% of the world's flowering plant species, but are responsible for ~23% of global terrestrial productivity (Ehleringer et al. 1997) and include some of the world's most important crops. The high heat tolerance and water use efficiency of C4 plants means that they could play an even more important role in feeding the world this century than they do now (Watson-Laxowski and Ghannoum 2021; Sage and Stata 2021). Similarly, the ability of species that use Crassulacean Acid Metabolism (CAM) to limit water loss by fixing CO_2 at night may give them an edge in a hotter, drier world where droughts become more severe (Pereira et al. 2021). The question remains, however, whether plants that use energy to concentrate CO_2 internally to suppress photorespiration will still have a significant advantage in a high CO_2 world where C3 plants get a similar physiological benefit without any biochemical or anatomical cost.

Physiological differences between plant species or functional groups aren't the only types of diversity that must be considered when trying to determine how climate change will alter vegetation. Plants exist in complex webs of relationships with other organisms, both beneficial and detrimental. Over 90% of terrestrial plants form mutualisms with mycorrhizal fungi, providing photosynthate to the fungi in exchange for water and nutrients (Brundrett and Tedersoo 2018). The dynamics of resource exchange and growth benefits to the plant often vary among plant species and with environmental conditions (Hoeksema et al. 2010). Yet, the fate of this mutualism under both elevated CO_2 and warmer temperatures (with accompanying changes in water supply) has received relatively little attention until recently (e.g., Terrer et al. 2016). Other soil microbes are also likely to play important, but currently poorly understood, roles in plant responses to climate change. Pathogens and diseases may become more prevalent under higher temperatures, and drier conditions may make infections more deleterious (Garrett et al. 2006). In contrast, the low stomatal conductance induced by high CO_2, along with changes in leaf chemistry, could reduce plant vulnerability to some types of infection (McElrone et al. 2005). Herbivory pressures may also be greater in a warmer, high CO_2 world where insect life cycles are accelerated and the higher C:N ratio of leaf tissues encourages herbivores to consume more tissue to meet their nitrogen needs (Bale et al. 2002; Massad and Dyer 2010). In one of the few studies to evaluate the impact of herbivory on a multi-factorial global change experiment, Post and Pederson (2008) found that shifts in species abundance seen under warming treatments were reversed when the plots were grazed, emphasizing the need to embrace complex food webs in our studies. Additionally, the mutualisms that many flowering plants share with pollinators will be affected by climate changes via mismatches in the shifts in phenology, migration and population dynamics between the sets of species, which may reduce the fertility of these plants (Hegland et al. 2009; Miller-Struttmann et al. 2015). Lastly, plant-plant interactions, including competition and facilitation, will help dictate the community composition and functioning of future ecosystems (Olsen et al. 2016). Accounting for biological complexity in a manner that encompasses this diversity while still allowing for the simplifications necessary for modeling vegetation across the entire planet therefore remains a challenge, and research into how these species interactions will affect vegetation is a pressing issue for the future.

V. Opportunities for Developing Climate-Resilient Plants

While climate change presents us with arguably the greatest challenge humans have faced, our ability to adapt to, mitigate and reverse the effects of climate change on natural and managed systems has never been greater. To meet the challenge of producing climate-resilient plants, we need to take an integrated approach, combining a detailed understanding of physiological and anatomical responses to climate change with an appreciation for the diversity that exists between individual plants, between plant functional types and between plants and the broader communities in which they exist. That integration is the goal of this book. We start by bringing together expert reviews on the effects of climate change on stomatal conductance (Stevens et al. 2021), mesophyll conductance (Nadal et al. 2021), photosynthesis (Duarte et al. 2021), and trichomes (Fischer and Ward 2021). Only by understanding the connections between these leaf-level processes and structures and thinking holistically about how elevated CO_2 and temperatures will alter plant carbon uptake and energy balance from the atmosphere to the chloroplast will we be able to breed and engineer plants that can thrive in a future warmer, high CO_2 world.

We then move from the leaf level to discuss whole plant and species level responses to climate change, by examining the intraspecific variation in plant responses to climate change. This includes both a review of intraspecific variation in reproduction, growth and physiology with rising CO_2, temperatures and water stress (Aspinwall et al. 2021) and a case study of variation within a single species in tolerating an extreme heat event (Carter et al. 2021). A better understanding of the genetic variation that exists within species will be critical for predicting how well species will fare as the climate changes, and will also provide us with a deeper understanding of the material available in natural populations for selecting climate-resilient plants.

The next set of chapters focus on plants with CCMs. Many of our most productive species have CCMs, including maize and sugarcane, and both C4 and CAM photosynthesis increase the abiotic stress tolerance of plants, making them important food sources in hot, dry regions. These chapters examine how rising CO_2 will impact plants that rely on these mechanisms (Sage and Stata 2021), and then more specifically discuss C4 crops (Watson-Laxowski and Ghannoum 2021) and the future of CAM species (Pereira et al. 2021).

Lastly, we look at possible tools and opportunities for producing more productive and stress tolerant plants for future climates. While this topic is also covered in the various chapters of the book (e.g., the final sections of Pereira et al. 2021), our last two chapters focus on how we can adapt to future climate change and mitigate its effects on our natural and managed systems. First, we examine avenues for improving photosynthetic CO_2 assimilation in future climates by targeting aspects of the Calvin-Benson cycle (Sharwood and Long 2021). Secondly, we look at the role of photorespiration in leaf metabolism and consider how altering the photorespiratory pathway might lead to more climate-resilient plants (Bauwe and Fernie 2021).

VI. Conclusions

Photosynthesis and respiration are key fluxes in the global carbon cycle, and understanding how these processes will respond to climate change is critical for predicting how rapidly climate change will occur. While climate change represents an enormous challenge for both natural ecosystems and food production, global change experiments have provided a wealth of data that have shaped our understanding of plant responses to rising CO_2, warming and drought stress. Future research should focus on incorporating this knowledge base into ESMs, embracing biological complexity, and building on opportunities to develop climate-resilient plants through both natural breeding and synthetic biology approaches.

Acknowledgments

DAW acknowledges support from the Natural Sciences and Engineering Research Council of Canada in the form of a Discovery Grant. DW was also supported in part by the United States Department of Energy contract No. DE-SC0012704 to Brookhaven National Laboratory.

References

Adams HD, Guardiola-Claramonte M, Barron-Gafford GA, Camilo Villegas J, Breshears DD, Zou CB, Troch PA, Huxman TE (2009) Temperature sensitivity of drought-induced tree mortality portends increased regional die-off under global-change-type drought. Proc Natl Acad Sci U S A 106:7063–7066

Allen CD, Macalady AK, Chenchouni H, Bachelet D, McDowell N, Vennetier M, Kitzberger T, Rigling A, ..., Cobb N (2010) A global overview of drought and heat-induced tree mortality reveals emerging climate change risks for forests. For Ecol Manage 259:660–684.

Arrhenius S (1896) On the influence of carbonic acid in the air upon the temperature of the ground. Philos Magaz J Sci Series 5 41:237–276

Aspinwall MJ, Juenger TE, Rymer PD, Tissue DT (2021) Intraspecific variation in plant responses to atmospheric CO_2, temperature, and water availability. In: Becklin KM, Ward JK, Way DA (eds) Photosynthesis, Respiration, and Climate Change. Advances in Photosynthesis and Respiration, Including Bioenergy and Related Processes, vol 46. Springer, Dordrecht

Bale JS, Masters GJ, Hodkinson ID, Awmack C, Bexemer TM, Brown VK, Butterfield J, Buse A, ...,

Whittaker JB (2002) Herbivory in global climate change research: direct effects of rising temperature on insect herbivores. Glob Chang Biol 8:1–16.

Bauwe H, Fernie AR (2021) With a little help from my friends: The central role of photorespiration and related metabolic processes in the acclimation and adaptation of plants to oxygen and low-CO_2 stress. In: Becklin KM, Ward JK, Way DA (eds) Photosynthesis, Respiration, and Climate Change. Advances in Photosynthesis and Respiration, Including Bioenergy and Related Processes, vol 46. Springer, Dordrecht

Berry J, Björkman O (1980) Photosynthetic response and adaptation to temperature in higher plants. Annu Rev Ecol Syst 31:491–543

Bonan GB, Doney SC (2018) Climate, ecosystems, and planetary futures: The challenge to predict life in Earth system models. Science 359:eaam8328

Brundrett MC, Tedersoo L (2018) Evolutionary history of mycorrhizal symbioses and global host plant diversity. New Phytol 220:1108–1115

Cardinale BJ, Duffy JE, Gonzalez A, Hooper DU, Perrings C, Venail P, Narwani A, Mace GM, …, Naeem S (2017) Biodiversity loss and its impact on humanity. Nature 486:59–67.

Carlson RW, Bazzazz FA (1980) The effects of elevated CO_2 concentrations on growth, photosynthesis, transpiration, and water use efficiency of plants. Environmental and climatic impact of coal utilization. In: Singh JJ, Deepak A (eds) Proceedings of a Symposium, Wiilliamsburg, Virginia, USA, April 17–19, 1979, pp 609–623

Carter JM, Burnette TE, Ward JK (2021) Tree physiology and intraspecific responses to extreme events: Insights from the most extreme heat year in U.S. history. In: Becklin KM, Ward JK, Way DA (eds) Photosynthesis, Respiration, and Climate Change. Advances in Photosynthesis and Respiration, Including Bioenergy and Related Processes, vol 46. Springer, Dordrecht

Chapin FS III, Shaver GR (1985) Individualistic growth response of tundra plant species to environmental manipulations in the field. Ecology 66:564–576

Ciais P, Sabine C, Bala G, Bopp L, Brovkin V, Canadell J, Chhabra A, De Fries R, …, Jones C (2013) Carbon and other biogeochemical cycles. In: Stocker TF, Qin D, Plattner G-K, Tignor M, Allen SK, Boschung J, Nauels A, Xia Y, …, Midgley PM (eds). Climate Change 2013: The Physical Science Basis. Contribution of Working Group I to the Fifth Assessment Report of the Intergovernamental Panel on Climate Change, Cambridge University Press, Cambridge

Classen AT, Sundqvist MK, Henning JA, Newman GS, Moore JAM, Cregger MA, Moorhead LC, Patterson CM (2015) Direct and indirect effects of climate change on soil microbial and soil microbial-plant interactions: What lies ahead? Ecosphere 6:1–21

Collins M, Knutti R, Arblaster J, Dufresne J-L, Fichefet T, Friedlingstein P, Gao X, Gutowski WJ, … Wehner M (2013) Long-term climate change: Projections, commitments and irreversibility. In: Stocker TF, Qin D, Plattner G-K, Tignor M, Allen SK, Boschung J, Nauels A, Xia Y, …, Midgley PM (eds). Climate Change 2013: The Physical Science Basis. Contribution of Working Group I to the Fifth Assessment Report of the Intergovernmental Panel on Climate Change, Cambridge University Press, Cambridge

Dai A (2012) Increasing drought under global warming in observations and models. Nat Climate Change 3:52–58

De Kauwe MG, Medlyn BE, Zaehle S, Walker AP, Dietze MC, Hickler T, Jain AK, Luo Y, …, Norby RJ (2013) Forest water use and water use efficiency at elevated CO2: a model–data intercomparison at two contrasting temperate forest FACE sites. Glob Chang Biol 19:1759–1779

De Kauwe MG, Medlyn BE, Zaehle S, Walker AP, Dietze MC, Wang YP, Luo Y, Jain AK, …, Norby RJ (2014) Where does the carbon go? A model–data intercomparison of vegetation carbon allocation and turnover processes at two temperate forest free-air CO_2 enrichment sites. New Phytol 203:883–899

Doughty CE, Goulden ML (2008) Are tropical forests near a high temperature threshold? J Geophys Res Biogeo 113:G00B07

Drake BG, Leadley PW, Arp WJ, Nassiry D, Curtis PS (1989) An open top chamber for field studies of elevated atmospheric CO_2 concentration on saltmarsh vegetation. Funct Ecol 3:363–371

Drake BG, Gonzalez-Meler MA, Long SP (1997) More efficient plants: A consequence of rising atmospheric CO_2? Annu Rev Plant Physiol Plant Mol Biol 48:609–639

Duarte AG, Dusenge ME, McDonald S, Bennett K, Lemon K, Radford J, Way DA (2021) Photosynthetic acclimation to temperature and CO_2: the role of leaf nitrogen. In: Becklin KM, Ward JK, Way DA (eds) Photosynthesis, Respiration, and Climate Change. Advances in Photosynthesis and Respiration, Including Bioenergy and Related Processes, vol 46. Springer, Dordrecht

Dusenge ME, Duarte AG, Way DA (2019) Plant carbon metabolism and climate change: elevated CO_2 and temperature impacts on photosynthesis, photorespiration and respiration. New Phytol 221:32–49

Ehleringer JR, Cerling TE, Helliker BR (1997) C_4 photosynthesis, atmospheric CO_2 and climate. Oecologia 112:285–299

Engelbrecht BMJ, Comita LS, Condit R, Kursar TA, Tyree MT, Turner BL, Hubbell SP (2007) Drought sensitivity shapes species distribution patterns in tropical forests. Nature 447:80–82

Fischer JM, Ward JK (2021) Trichome responses to elevated atmospheric CO_2 of the future. In: Becklin KM, Ward JK, Way DA (eds) Photosynthesis, Respiration, and Climate Change. Advances in Photosynthesis and Respiration, Including Bioenergy and Related Processes, vol 46. Springer, Dordrecht

Franks PJ, Bonan GB, Berry JA, Lombardozzi DL, Holbrook NM, Herold N, Oleson KW (2018) Comparing optimal and empirical stomatal conductance models for application in Earth system models. Glob Chang Biol 24:5708–5723

Friedlingstein P, Jones MW, O'Sullivan M, Andrew RM, Hauck J, Peters GP, Peters W, Pongratz J, ..., Zaehle S (2019) Global carbon budget 2019. Earth Sys Sci Data 11:1783–1838.

Garrett KA, Dendy SP, Frank EE, Rouse MN, Travers SE (2006) Climate change effects on plant disease: genomes to ecosystems. Annu Rev Phytopathol 44:489–509

Hanson PJ, Riggs JS, Nettles IV WR, Phillips JR, Krassovski MB, Hook LA, Gu L, Richardson AD, ..., Barbier CN (2017) Attaining whole-ecosystem warming using air and deep-soil heating methods with an elevated CO_2 atmosphere. Biogeosciences 14:861–883.

Hegland SJ, Nielsen A, Lázaro A, Bjerknes A-L, Totland Ø (2009) How does climate warming affect plant-pollinator interactions? Ecol Lett 12:184–195

Hendrey GR (ed) (1992) FACE: Free-Air CO_2 Enrichment for Plant Research in the Field. CRC Press, Boca Raton, 308 pp

Hendrey GR, Kimball BA (1994) The FACE program. Agric For Meteorol 70:3–14

Hoeksema JD, Chaudhary VB, Gehring CA, Johnson NC, Karst J, Koide RT, Pringle A, Zabinski C, ..., Umbanhowar, J (2010). A meta-analysis of context-dependency in plant response to inoculation with mycorrhizal fungi. Ecol Lett 13:394–407.

Huang M, Piao S, Ciais P, Peñuelas J, Wang X, Keenan TF, Peng S, Berry JA, ..., Janssens IA (2019) Air temperature optima of vegetation productivity across global biomes. Nat Ecol Evol 5:772–779.

Huntington C, Atkin OK, Martinez-de la Torre A, Mercado LM, Heskel MA, Harper AB, Bloomfield KJ, O'Sullivan OS, ..., Malhi Y (2017) Implications of improved representations of plant respiration in a changing climate. Nat Commun 8:1602.

Huxman TE, Snyder KA, Tissue D, Leffler AJ, Ogle K, Pockman WT, Sandquist DR, Potts DL, Schwinning S (2004) Precipitation pulses and carbon fluxes in semiarid and arid ecosystems. Oecologia 141:254–268

Joos F, Spahni R (2008) Rates of change in natural and anthropogenic radiative forcing over the past 20,000 years. Proc Natl Acad Sci U S A 105:1425–1430

Keeling CD (1960) The concentration and isotopic abundances of carbon dioxide in the atmosphere. Tellus 12:200–203

Keeling CD (1998) Rewards and penalties of monitoring the Earth. Ann Rev Energy Environ 23:25–82

Kimball BA (2005) Theory and performance of an infrared heater for ecosystem warming. Glob Chang Biol 11:2041–2056

King AW, Gunderson CA, Post WM, Weston DJ, Wullschleger SD (2006) Plant respiration in a warmer world. Science 312:536–537

Knapp AK, Fay PA, Blair JM, Collins SL, Smith MD, Carlisle JD, Harper CW, Danner BT, ..., McCarron JK (2002) Rainfall variability, carbon cycling, and plant species diversity in a mesic grassland. Science 298: 2202–2205.

Kurz WA, Apps MJ, Stocks BJ, Volney WJA. (1995) Global climate change: disturbance regimes and biospheric feedbacks of temperate and boreal forests. In: Biotic Feedbacks in the Global Climatic System: Will the Warming Feed the Warming? GM Woodwell, FT Mackenzie, Groupe d'experts intergouvernemental sur l'évolution du climat, eds. Science working group. Oxford University Press, 1995. pp 119–133.

Leuzinger S, Thomas RQ (2011) How do we improve Earth system models? Integrating Earth system models, ecosystem models, experiments and long-term data. New Phytol 191:15–18

Brian Moss, Sarian Kosten, Mariana Meerhoff, Richard W. Battarbee, Erik Jeppesen, Néstor Mazzeo, Karl Havens, Gissell Lacerot, Zhengwen Liu, Luc De Meester, Hans Paerl & Marten Scheffer (2011) Allied attack: climate change and eutrophication, Inland Waters, 1:2, 101-105

Massad TJ, Dyer LA (2010) A meta-analysis of the effects of global environmental change on plant-herbivore interactions. Arthropod-Plant Interactions 4:181–188

McCree KJ (1974) Equations for the rate of dark respiration of white clover and grain sorghum, as functions of dry weight, photosynthetic rate, and temperature. Crop Sci 14:509–514

McElrone AJ, Reid CD, Hoye KA, Hart E, Jackson RB (2005) Elevated CO_2 reduces disease incidence and severity of a red maple fungal pathogen via changes in host physiology and leaf chemistry. Glob Chang Biol 11:1828–1836

Medlyn BE, Zaehle S, De Kauwe MG, Walker AP, Dietze MC, Hanson PJ, Hickler T, Jain A, …, Norby RJ (2015) Using ecosystem experiments to improve vegetation models. Nat Climate Change 5:528–534
Mercado LM, Medlyn BE, Huntingford C, Oliver RJ, Clark DB, Sitch S, Zelazowski P, Kattge J, …, Cox PM (2018) Large sensitivity in land carbon storage due to geographical and temporal variation in the thermal response of photosynthetic capacity. New Phytol 218:1462–1477.
Miller-Struttmann NE, Geib JC, Franklin JD, Kevan PG, Holdo RM, Ebert-May D, Lynn AM, Kettenbach JA, …, Galen C (2015). Functional mismatch in a bumble bee pollination mutualism under climate change. Science 349:1541–1544.
Mooney HA, Björkman O, Collatz GJ (1978) Photosynthetic acclimation to temperature in the desert shurb, *Larrea divaricata*. 1. Carbon dioxide exchange characteristics of intact leaves. Plant Physiol 61:406–410
Nadal M, Carriquí M, Flexas J (2021) Mesophyll conductance to CO_2 diffusion in a climate change scenario: effects of elevated CO_2, temperature, and water stress. In: Becklin KM, Ward JK, Way DA (eds) Photosynthesis, Respiration, and Climate Change. Advances in Photosynthesis and Respiration, Including Bioenergy and Related Processes, vol 46. Springer, Dordrecht
Niu S, Luo Y, Li D, Cao S, Xia J, Li J, Smith M (2014) Plant growth and mortality under climatic extremes: An overview. Environ Exp Bot 98:13–19
NOAA National Centers for Environmental Information, State of the Climate: Global Climate Report for Annual 2018. https://www.ncdc.noaa.gov/sotc/global/201813. Accessed 16 June 2020
Norby RJ, DeKauwe MG, Domingues TF, Duursma RA, Ellsworth DS, Goll DS, Lapola DM, Luus KA, …, Zaehle S (2016) Model–data synthesis for the next generation of forest free-air CO_2 enrichment (FACE) experiments. New Phytol 209:17–28.
Olsen SL, Töpper JP, Skarpaas O, Vandvik V, Klanderud K (2016) From facilitation to competition: temperature-driven shift in dominant plant interactions affects population dynamics in seminatural grasslands. Glob Chang Biol 22:1915–1926
Patterson DT, Flint EP (1980) Potential effects of global atmospheric CO_2 enrichment on the growth and competitiveness of C3 and C4 weed crop plants. Weed Sci 28:71–75
Pearcy RW (1977) Acclimation of photosynthetic and respiratory carbon dioxide exchange to growth temperature in *Atriplex lentiformis* (Torr.) Wats. Plant Physiol 59:795–799
Pereira PN, Niechayev NA, Blair BB, Cushman JC (2021) Climate change responses and adaptations in Crassulacean Acid metabolism (CAM) plants. In: Becklin KM, Ward JK, Way DA (eds) Photosynthesis, Respiration, and Climate Change. Advances in Photosynthesis and Respiration, Including Bioenergy and Related Processes, vol 46. Springer, Dordrecht
Post E, Pederson C (2008) Opposing plant community responses to warming with and without herbivores. Proc Natl Acad Sci U S A 105:12353–12358
Rogers AR, Medlyn BE, Dukes JS, Bonan G, von Caemmerer S, Dietze MC, Kattge J, Leakey ADB et al (2017) A roadmap for improving the representation of photosynthesis in Earth system models. New Phytol 213:22–42
Rook DA (1969) The influence of growing temperature on photosynthesis and respiration of *Pinus radiata* seedlings. New Zealand J Bot 7:43–55
Russo S, Dosio A, Graversen RG, Sillmann J, Carrao H, Dunbar MB, Singleton A, Montagna P, …, Vogt JV (2014) Magnitude of extreme heat waves in present climate and their projection in a warming world. J Geophys Res Atmos 119:12500–12512.
Sage RF, Kubien DS (2007) The temperature response of C_3 and C_4 photosynthesis. Plant Cell Environ 30:1086–1106
Sage RF, Stata M (2021) Terrestrial CO_2 concentrating mechanisms in a high CO_2 world. In: Becklin KM, Ward JK, Way DA (eds) Photosynthesis, Respiration, and Climate Change. Advances in Photosynthesis and Respiration, Including Bioenergy and Related Processes, vol 46. Springer, Dordrecht
Seidl RS, Thom D, Kautz M, Martin-Benito D, Peltoniemi M, Vacchiano G, Wild J, Ascoli D, …, Reyer CPO (2017) Forest disturbances under climate change. Nat Climate Change 7:395–402.
Sharwood RE, Long BM (2021) Engineering photosynthetic CO_2 assimilation to develop new crop varieties to cope with future climate. In: Becklin KM, Ward JK, Way DA (eds) Photosynthesis, Respiration, and Climate Change. Advances in Photosynthesis and Respiration, Including Bioenergy and Related Processes, vol 46. Springer, Dordrecht
Smith NG, Dukes JS (2013) Plant respiration and photosynthesis in global-scale models: incorporating acclimation to temperature and CO_2. Glob Chang Biol 19:45–63
Smith NG, Malyshev SL, Shevliakova E, Kattge J, Dukes JS (2016) Foliar temperature acclimation reduces simulated carbon sensitivity to climate. Nat Climate Change 6:407–411
Stevens J, Faralli M, Wall S, Stamford JD, Lawson T (2021) Stomatal responses to climate change. In: Becklin KM, Ward JK, Way DA (eds) Photosynthesis, Respiration, and Climate Change. Advances in Photosynthesis and Respiration,

Including Bioenergy and Related Processes, vol 46. Springer, Dordrecht

Terrer C, Vicca S, Hungate BA, Prentice IC (2016) Mycorrhizal association as a primary control of the CO_2 fertilization effect. Science 353:72–74

Teskey R, Wertin T, Bauweraerts I, Ameye M, McGuire MA, Steppe K (2015) Responses of tree species to heat waves and extreme heat events. Plant Cell Environ 38:1699–1712

Thullier W (2004) Patterns and uncertainties of species' range shifts under climate change. Glob Chang Biol 10:2020–2027

Walker AP, Hanson PJ, De Kauwe MG, Medlyn BE, Zaehle S, Asao S, Dietze M, Hickler T, …, Norby RJ (2014) Comprehensive ecosystem model–data synthesis using multiple data sets at two temperate forest free-air CO_2 enrichment experiments: model performance at ambient CO_2 concentration. J Geophys Res Biogeosci 119:2013JG002553

Watson-Laxowski A, Ghannoum O (2021) The outlook for C_4 crops in future climate scenarios. In: Becklin KM, Ward JK, Way DA (eds) Photosynthesis, Respiration, and Climate Change. Advances in Photosynthesis and Respiration, Including Bioenergy and Related Processes, vol 46. Springer, Dordrecht

Zaehle S, Medlyn BE, De Kauwe MG, Walker AP, Dietze MC, Hickler T, Luo Y, Wang YP, …, Norby RJ (2014) Evaluation of eleven terrestrial carbon–nitrogen cycle models against observations from two temperate Free-Air CO_2 enrichment studies. New Phytol 202:803–822

Part II

Leaf-level Responses to Climate Change

Chapter 2

Stomatal Responses to Climate Change

Jim Stevens · Michele Faralli · Shellie Wall
John D. Stamford · Tracy Lawson*
School of Life Sciences, University of Essex, Colchester, UK

*Author for correspondence, e-mail: tlawson@essex.ac.uk

K. M. Becklin et al. (eds.), *Photosynthesis, Respiration, and Climate Change*, Advances in Photosynthesis and Respiration 48, https://doi.org/10.1007/978-3-030-64926-5_2

Summary

There remains an ongoing need to make crops more productive in the face of further increases in atmospheric CO_2 concentrations as predicted under climate change, along with higher global surface temperatures and more prolonged, severe and frequent periods of drought. With over 90 % of water transpired by plants diffusing through stomata, studying these small, morphologically varied valves in leaf surfaces remains critical to our understanding the consequences of climate change on stomatal responses, and by extension crop productivity. In the short term, stomata adjust aperture in response to changes in environmental variables affecting carbon assimilation and water loss. In the longer term, adjustments to stomatal density and size may occur, in conjunction with a range of other responses from the plant.

While increasing CO_2 concentration under climate change had been shown to raise yields and reduce water use by partial stomatal closure, the extent of the fertilisation effect has not always been as strong as expected. Meanwhile, higher temperatures and decreasing water availability are likely to have negative yield consequences, with divergent expectations for stomatal aperture and consequently plant water use. However, changes in environmental factors will not occur in isolation and therefore stomatal and plant responses to a combination of these changes will be hierarchical and involve multiple and possibly unique signalling pathways. Predicting stomatal responses to several simultaneous abiotic stresses such as those outlined above adds a layer of complexity, notably where the stresses produce antagonistic responses in the plant. Targeting steady-state stomatal behaviour has been a successful breeding tactic to date, and continues to generate new insights under interacting stresses. Meanwhile, the emerging field of dynamic stomatal responses to environmental stresses offers new phenotypic targets, and the possibility for enhancing water use efficiency by targeting novel signalling and molecular pathways in stomatal responses.

I. General Introduction

The current global population is over 7.2 billion, and is projected to increase to 9.6 billion by 2050 (McGuire 2013), which means that crop yields need to double in the next 30 years to meet increased demand for plant based products (Ray et al. 2013). In addition to population growth, changing diets and bio-fuel use are further drivers of the need to increase biomass and yield (Amin et al. 2006). The growing population is driving greater urbanisation, (Jones and Kandel 1992) meaning that the increased demands for food and fuel cannot be met with more land use for crop production and will have to be delivered by greater crop productivity (Ray et al. 2013). World food production is heavily reliant on a small number of crops. Rice, wheat and maize are staple foods for 80 % of humanity and rice alone is consumed by almost 50 % of the global population (Maclean et al. 2013). Against the backdrop of the requirement for increasing productivity with less available land, the changing climate presents additional challenges. The current IPCC 5th assessment (IPCC 2014) predicts higher concentrations of greenhouse gases leading to higher global mean surface temperatures. In addition, heat waves, drought, and heavy and sporadic pre-

Abbreviations: A – Assimilation rate of CO_2 (μmol m^{-2} s^{-1}); ABA – Abscisic acid; C_i – Leaf internal CO_2 concentration; C_3 – Plants exhibiting C_3 photosynthetic pathway; C_4 – Plants exhibiting C_4 photosynthetic pathway; [CO_2] – CO_2 concentration; e[CO_2] – Elevated CO_2 concentration; GC – Guard cell(s); GMC – Guard mother cell(s); g_s – Stomatal conductance to water vapour (mmol or mol m^{-2} s^{-1}); ROS – Reactive oxygen species; VPD – Vapour pressure deficit; WUE_i – Intrinsic water use efficiency (A g_s^{-1})

cipitation events are expected to occur with greater frequency or duration (IPCC 2014). Atmospheric CO_2 concentration [CO_2] is predicted to rise to 500–1000 ppm by the end of the century (Meehl et al. 2007), which could potentially increase yields. Global mean surface temperature is likely to exceed 1.5°C by the end of the century under most climate change projections (IPCC 2014) and rising temperatures will be detrimental to productivity of C_3 crops which face higher rates of photorespiration and a decrease in photosynthesis (Teskey et al. 2015), with further impacts from greater evapotranspiration, leading to reduced soil moisture content, particularly where night temperatures are high (Hatfield and Prueger 2015).

Higher temperatures are also closely associated with water availability through the impact on transpiration. Transpiration accounts for over 90 % of land-based water losses, and therefore, a major goal of crop science is to manage water use more-effectively (Morison et al. 2008; FAO 2015; Li et al. 2009; Lawson et al. 2003). Stomata, microscopic valves on the leaf surface (Hetherington and Woodward 2003) enable gaseous diffusion and are influenced by a number of environmental factors including water availability, leaf temperature and [CO_2] (Lawson and Blatt 2014). Higher temperatures raise evaporative demand, thus decreasing water availability, driving the need for crops that can tolerate multiple stress conditions (Atkinson and Urwin 2012). Stomatal pores are at the nexus of the global carbon (Sugano et al. 2010) and hydrological cycles (Betts et al. 2007) and as such are critical to managing the impact of changing temperature, water availability and [CO_2] on crop species.

II. Introduction to Stomata

Stomata are found in large numbers over the predominantly-waterproof and CO_2-impermeable leaf cuticle. Stomata comprise two specialized guard cells (GC) flanking a central pore which facilitate diffusional gaseous flux between the interior of the leaf and the atmosphere (Hetherington and Woodward 2003; Lawson and Weyers 1999). When stomatal pores open to maintain CO_2 supply to mesophyll cells for photosynthetic carbon assimilation (*A*), water is lost through transpiration as a consequence (Morison et al. 2008). Transpirational water loss plays a key role in nutrient uptake from the plant roots as well as evaporative cooling of the leaf tissue (Hetherington and Woodward 2003; McAusland et al. 2016; Murray et al. 2016; Peterson et al. 2010; Raven 1977; Raven 2002). Dynamic stomatal movement acting in response to environmental cues and internal signals in an attempt to optimise the trade-off between *A* and maintaining plant water status (Farquhar and Sharkey 1982; Mansfield et al. 1990; Lawson and Von Caemmerer 2010; Buckley and Mott 2013; Buckley et al. 2017) often referred to as stomatal behavior (Lawson and Blatt 2014; Zhu et al. 2018). Stomatal behavior is frequently measured as stomatal conductance (g_s), the capacity for the gaseous exchange of water vapor from the leaf to atmosphere, in mole of flux per unit area per second (mol m^{-2} s^{-1}) (Vialet-Chabrand et al. 2017c) and is determined by both stomatal density and behaviour (Lawson and Vialet-Chabrand 2019). Wong et al. (1979) described the relationship of *A* and g_s as proportional, although the rate of water leaving the leaf through stomata is an order of magnitude greater than CO_2 entering the leaf for *A* (Drake et al. 2013; Lawson and Vialet-Chabrand 2019). While photosynthesis is occurring, there is a constant demand for CO_2 influx into the leaf as chloroplast [CO_2] declines. Carbon dioxide entering the leaf encounters a series of resistance points *en route* to the chloroplast, including boundary layer, stomatal and mesophyll resistance. Of these, stomatal resistance can create the largest impediment to CO_2 influx with g_s being the inverse of stomatal resistance.

A. *Factors Influencing Stomatal Responses*

Abiotic variables such as light intensity and spectral quality, [CO_2] and temperature are considered to have the greatest direct and indirect impacts on stomatal behaviour (Blatt 2000), although there is disparity in stomatal sensitivity and responsiveness among different species (Lawson et al. 2012; Lawson et al. 2003; Lawson and Vialet-Chabrand 2019). Typically, stomata in C3 and C4 species open in response to increasing or high light intensity, low internal [CO_2], high temperatures and low VPD. Conversely, stomatal closure is driven by low or decreasing light, high internal [CO_2] and high VPD as well as hormones such as ABA (Outlaw 2003; Berry et al. 2010; Elliott-Kingston et al. 2016; Franks and Farquhar 2001; Inoue and Kinoshita 2017; Mott and Peak 2013; Poole et al. 2000; Shimizu et al. 2015; Vialet-Chabrand et al. 2017b; Vialet-Chabrand et al. 2017c; Wang et al. 2008; Woodward 1987).

The close positive relationship between A and g_s has been well documented in the laboratory (Wong et al. 1979) and a positive correlation between steady-state g_s and yield reported by Fischer et al. (1998) in the field. In contrast, in the short term (minutes) stomata can be 'sluggish' in response to environmental factors and internal stimuli leading to non-synchronous behaviour between A and g_s, which under dynamic conditions can lead to either a limit in A or an unnecessary loss of water (Farquhar and Sharkey 1982; Hetherington and Woodward 2003; Franks and Farquhar 2007; Brodribb et al. 2010; Brodribb et al. 2009; Lawson and Von Caemmerer 2010; Lawson et al. 2012; McAusland et al. 2016; Lawson and Vialet-Chabrand 2019) leading to sub-optimal water use efficiencies (WUE= A/g_s; (Lawson and Vialet-Chabrand 2019)). For instance, Vialet-Chabrand et al. (2017b) reported 18.8 % lower than expected A under fluctuating light during the course of a day, which was attributed to stomatal limitation.

B. *Stomatal Anatomy: Intra- & Inter-Specific Variation*

Guard cells (GC) together with the stomatal pore, and if relevant, subsidiary cells make up the stomatal complex (Fig. 2.1), with stomatal pores ranging in size (10 to 80 μm in length) and density (between 5 and 1,000 mm^{-2}), depending on the species and the environmental growth conditions (Knapp 1993; Willmer and Fricker 1996; Beerling et al. 1997; Hetherington and Woodward 2003). Guard cells are responsible for determining stomatal aperture although the signalling pathways are complex and many remain contentious and are not fully understood (Lawson et al. 2014). By adjusting vacuolar volume through the movement of osmotica including malate, sucrose and K^+, and water, total GC volume is changed (Sussmilch et al. 2019). The change in GC volume and therefore turgor pressure alters stomatal aperture as microfibrils preventing a change in GC width cause curvature of the cell when the pressure potential becomes positive (Willmer and Fricker 1996; Blatt 2000; Ziegler 1987).

Although modern stomata are directly comparable in morphology and function with those in the fossil record from 410 million years ago the density of stomata per leaf area and size has changed considerably through time, as a result of changing climate, most markedly [CO_2] (McElwain and Chaloner 1995; Beerling and Woodward 1997; Edwards et al. 1998; Hetherington and Woodward 2003). Early examples from the Silurian and basal Devonian rock indicate that stomatal complex was either circular or oval in shape (Edwards et al. 1998). Early stomata were anomocytic, i.e., they had no distinct subsidiary cells (specialised epidermal cells that surround GCs that assist the process of opening and closing stomata), whereas modern species of terrestrial plants can have many subsidiary

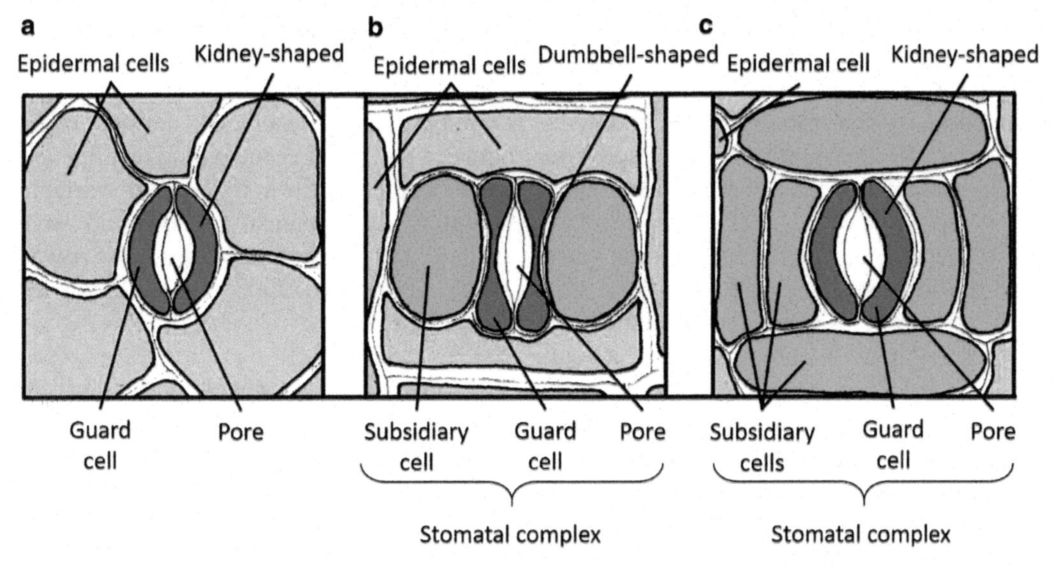

Fig. 2.1. Morphological characteristics of stomatal complexes. (**a**) Elliptical (kidney-shaped – KS) stomatal complex typical of soybean and *A. thaliana* and (**b**) the graminaceous (dumbbell-shaped, DBS) stomatal complex, paired with two main subsidiary cells, typical of rice and wheat. (**c**) Elliptical (KS) stomatal complex with multiple subsidiary cells and kidney-shaped guard cells typical of *Commelina communis*. (Adapted and extended from Taiz et al. 2008)

cells of varying shapes and sizes (Fig. 2.1) which surround a pair of morphologically and mechanistically diverse guard cells (Edwards et al. 1998; Franks and Farquhar 2007; Ziegler 1987). Subsidiary cells differ from epidermal cells due to their ability to shuttle ions (particularly K^+) and water rapidly between themselves and guard cells, modulated by the activity of aquaporins, facilitating a mechanical advantage in guard cells and increased responsiveness of stomatal movement through rapid changes in turgor pressure (Hachez et al. 2017; Franks and Farquhar 2007; Raissig et al. 2017; Schafer et al. 2018). The number of subsidiary cells that surround a pair of guard cells differs between species although it is conserved within species, and ranges from zero (anomocytic - for example sporophytes of some extant hornworts and mosses) to six (hexacytic - for example *Commelina communis*) (Weyers and Paterson 1987; Rudall et al. 2013; Ziegler 1987). The significance of subsidiary cells in the efficiency of stomatal functioning in grasses is supported by a recent study by Raissig et al. (2017) who manipulated the levels of a transcription factor (BdMUTE) necessary for subsidiary cell formation in *Brachypodium distachyon*. Plants lacking subsidiary cells (known as subsidiary cell identity defective (*sid*)) had reduced stomatal kinetics, lower g_s and impaired growth (Hughes et al. 2017; Raissig et al. 2017; Hepworth et al. 2018). The evolutionary conservation of the stomatal complex suggests that the pairing of GC and subsidiary cells is integral for the efficiency of stomatal aperture control, highlighting the importance of further studying how the heterogeneity of stomatal complex morphologies affects plant physiology (Raissig et al. 2017; Bertolino et al. 2019; Hepworth et al. 2018).

Two main types of GC distinguished by shape are found in terrestrial plants; graminaceous or dumbbell-shaped (often paired with two main subsidiary cells) (Fig. 2.1b); and elliptical or kidney-shaped (Hetherington and Woodward 2003) (Fig. 2.1a). Dumbbell-shaped GC are typical of grasses and other monocots such as palms, while kidney-shaped GC are found in all dicots, in several monocots as well as mosses, ferns, and gymnosperms (Hetherington and Woodward 2003). Dumbbell-shaped GC evolved later than kidney-shaped GC, between 70 and 50 million years ago, roughly 350 million years after the first perforation that evolved into their kidney-shaped counterparts (Hetherington and Woodward 2003; Kellogg 2013). It has been proposed (Hetherington and Woodward 2003; Franks and Farquhar 2007) that species such as wheat (*Triticum aestivum*) which possess the characteristic dumbbell-shaped GC have a faster movement of water across semi-permeable cell-membranes, facilitating superior dynamic performance. The mechanism is thought to be due to the high surface area to volume ratio of the dumbbell-shaped GC's and the close relationship with the adjacent subsidiary cells (Hetherington and Woodward 2003), with less water needed in order for dumbbell-shaped GCs to increase turgor relative to kidney-shaped GC. As a result, more-rapid opening and closing is facilitated, reducing stomatal response time and supporting higher rates of photosynthetic gas exchange and higher WUE in favourable environments (Johnsson et al. 1976; Willmer and Fricker 1996; Hetherington and Woodward 2003; Roelfsema and Hedrich 2005; Taiz et al. 2018). Two recent reports by McAusland et al. (2016) and Lawson &Vialet-Chabrand (2019) support this by demonstrating slower stomatal responses reduce photosynthesis by ca. 10% as well as leading to greater unnecessary water loss during stomatal closure in plants which have kidney-shaped GC's compared with those that exhibit dumbbell-shaped GCs.

C. Stomatal Density and Size

Stomatal anatomical characteristics including the number of stomata per unit area (i.e., density), stomatal size and pore aperture together determine g_s (Lawson and Blatt 2014) and therefore changes in any one of these variables has a direct influence on g_s. Developmentally, stomatal density differs between and within species and is influenced by a number of environmental variables including light intensity, [CO_2] and water availability (Gay and Hurd 1975; Woodward 1987; Gray et al. 2000; Hetherington and Woodward 2003; Doheny-Adams et al. 2012).

Stomata are morphologically diverse (Willmer and Fricker 1996; Ziegler 1987) and species with high densities often have smaller stomata and *vice-versa* (Franks and Beerling 2009; Hetherington and Woodward 2003; Lawson and Blatt 2014). Hetherington and Woodward (2003) suggested that smaller stomata have a more-rapid opening and closing time (speed of response) partly due to less water and solute movement between GC and subsidiary cells (osmotic shuttling) (Franks and Farquhar 2007) and shorter diffusional pathways (Franks and Beerling 2009). Smaller stomata therefore provide the capacity for rapid increases in g_s, allowing faster diffusion of CO_2 into the leaf for photosynthesis during favourable conditions (Aasamaa et al. 2001). However, species with different stomatal features may have distinct mechanisms influencing the speed of response independently of size (Franks and Farquhar 2007). Although smaller stomata are not always faster to respond to changing environmental conditions, this relationship typically holds within closely-related species although it is less strong across taxa (Drake et al. 2013; Elliott-Kingston et al. 2016; Lawson and Vialet-Chabrand 2019; McAusland et al. 2016).

Manipulating gene expression of key components in the stomatal developmental pathway has proven to be a powerful tool in modifying stomatal density, size and stomatal patterning (Franks et al. 2015). For example, the epidermal patterning factors (EPF) are a family of 11 related small, secreted peptides found to regulate stomatal density in *Arabidopsis thaliana* (Franks et al. 2015). Reducing the expression of EPF1 and EPF2 results in higher stomatal densities while constitutive overexpression produced a similar phenotype to wild type but with reduced numbers of stomata (Hara et al. 2007, 2009; Franks et al. 2015; Hunt and Gray 2009). Differences in the spacing of cells or the number of stomatal clusters were also demonstrated. A lack of EPF1, which is expressed in GCs of young stomata and their precursors led to high clustering, whereas plants with EPF2 expressed at slightly earlier stages of stomatal development showed almost no clustering (Hara et al. 2007, 2009; Hunt and Gray 2009; Franks et al. 2015). These studies also indicated that a strong correlation between stomatal density and size was maintained within these plants and the changes in these parameters have the potential to influence A, g_s, and WUE (Doheny-Adams et al. 2012). Both Doheny-Adams et al. (2012) and Mohammed et al. (2019) found that plants with reduced density and larger stomata also had reduced g_s yet greater plant biomass. The increase in biomass could be attributed to improved water use and lower metabolic costs associated with GC development (Lawson and Blatt 2014).

D. *Stomatal Development and Patterning*

Stomatal initiation is a complex series of patterning events laid down in the leaf primordia, split into two fundamentally different modes of growth and development. Eudicotyledon leaf cells divide at multiple points, with clones of new cells forming throughout leaf growth, adding new cells to the leaf, and leading to subsequent expansion (Croxdale 2000). Committed cells divide asymmetrically, each with an innate ability to both propagate and influence stomatal development (Dow and Bergmann 2014; Han and Torii 2016; Pillitteri et al. 2011; Vaten and Bergmann 2012). The larger cell undergoes asymmetric (amplifying) divisions before differentiating into a guard mother cell (GMC). The GMC divides symmetrically to create two GC. In dumbbell-shaped species, guard mother cells (GMC – guard cell precursors) recruit neighbouring epidermal cells to form subsidiary cells (Dow and Bergmann 2014; Peterson et al. 2010).

The leaves of monocotyledons on the other hand, have polarized growth from a single source of cells at the leaf base, creating a leaf blade with the oldest cells at the tip. The epidermis consists of regular longitudinal files of cells along the blade length (Croxdale 2000). New blade cells and stomatal precursors originate in polarized fashion at the leaf base and are present in a continuum of stages from the immature leaf base to the mature leaf tip (Croxdale 2000; Dow and Bergmann 2014). Asymmetric cell divisions produce GMC without a meristemoid stage. Files of protodermal cells flank GMC which polarize towards the GMC and divide asymmetrically producing subsidiary cells. Arrest of the developing stomata is a common phenomenon in monocotyledons and is known to yield more-regularly ordered stomata adhering to the one-cell-spacing rule (Serna et al. 2002; Bergmann and Sack 2007). Dicot stomata typically adhere to a one-cell-spacing rule where adjacent stomata are separated by at least one intervening epidermal cell; the rule is thought to be important for efficient gas exchange (Nadeau and Sack 2002; Sachs 1991).

E. *Steady State and Kinetic Stomatal Responses*

Steady state measurements of g_s have remained the core technique for understanding stomatal physiology, (see Ainsworth and Roger 2007). Fischer et al. (1998) for example, made point measurements in the early afternoon over the course of a season, and showed a strong correlation between yield and g_s, underlining the value of steady-state g_s as a measure of breeding success.

Successive rounds of breeding had produced wheat cultivars with increased g_s which reduced diffusional constraints, reducing leaf temperature and increasing A (Fischer et al. 1998). Not only is steady-state g_s important in determining yield, but also the kinetics and magnitude of change. To explore stomatal responses in more detail, the rate at which stomata open and close under changing environmental conditions has recently been investigated as a novel target for manipulation (Lawson and Blatt 2014; Lawson and Vialet-Chabrand 2019; Raven 2014), with light-induced changes in g_s the main focus of this work, but with water availability as well as temperature and VPD also being variables whose manipulation might be of interest. Stomata open and close much more slowly than the rate at which environmental inputs vary and an order of magnitude slower than photosynthetic responses (Lawson and Blatt 2014; McAusland et al. 2016; Qu et al. 2016), and these response rates can be parameterised and modelled (Vialet-Chabrand et al. 2013, 2016, 2017a, c). Environmental conditions are expected to become more-variable under climate change (IPCC 2014) driving the need to develop new breeding targets under climate change constraints. Latterly, there have been attempts to understand the impact not just of simple step changes to changing environmental parameters, but of naturalistic fluctuations in them (Matthews et al. 2018; Vialet-Chabrand et al. 2017b), although this work remains in its infancy.

III. Stomatal Responses to Changing Atmospheric CO_2 Concentration

A. [CO_2], g_s and Yield

A large body of evidence has highlighted the effect of [CO_2] on C_3 crop yield, mainly driven by positive effects on photosynthesis (Leakey 2009; Gray et al. 2016; Hatfield et al. 2011). Several reasons for this enhancement of photosynthesis by elevated [CO_2] (e[CO_2]) have been posited. First, elevated [CO_2] increases C_i and therefore the carboxylation efficiency of Rubisco by reducing photorespiration, adding further to the enhancement of photosynthesis (Leakey 2009). Second, the rate of electron transport increases, as ATP and NADPH product removal improves reaction kinetics, increasing the efficiency of PS II and PS I and enhancing electron transfer (Zhang et al. 2008). Overall, stimulation of the photosynthetic machinery leads to increases in biomass or yield of between 10 % and 30 % depending on the species, environmental conditions and / or experimental [CO_2] (Leakey 2009). Additionally, several lines of experimental evidence have demonstrated a mean reduction of g_s under e[CO_2] by 22 % as reported by Ainsworth and Rogers (Ainsworth and Rogers 2007) depending on species and photosynthetic pathway (Fig 2.2a), with the smallest change half that of the largest. More recently, molecular mechanisms have been explored which go some way to explaining the anatomical and physiological changes triggered by increasing atmospheric [CO_2] and involved in the reduction of g_s which go beyond short term changes in aperture and developmental adjustment of stomatal density in plants (Gamage et al. 2018), and which include impacts on nitrogen metabolism, hormonal regulation and the cell cycle (Gamage et al. 2018; Ainsworth and Rogers 2007).

B. Physiological and Anatomical Changes in Stomatal Responses to [CO_2]

In the short term, stomatal responses to e[CO_2] are initially seen as changes in stomatal aperture (Bertolino et al. 2019); the physiological mechanisms involved in partial stomatal closure under e[CO_2] (Fig. 2.2) are relatively well understood and have recently been reviewed (Gamage et al. 2018; Engineer et al. 2016). Reduction of g_s under

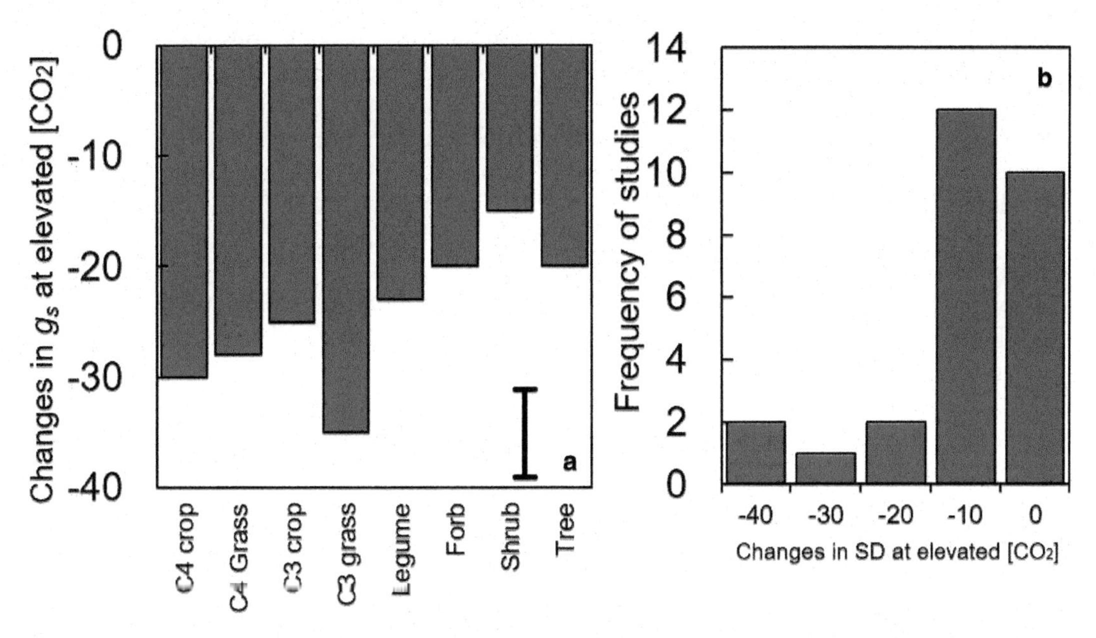

Fig. 2.2. The response of stomatal conductance (g_s) to e[CO_2] in free-air CO_2 enrichment experiments (**a**) and the histogram of observations from free-air CO_2 enrichment experiments of the change in stomatal density at elevated [CO_2] (**b**). (Data were redrawn from Ainsworth and Rogers 2007)

e[CO_2] is mainly due to increased activity of K^+ channels, the stimulation of Cl^- release from guard cells and increases in Ca^{2+} concentration causing stomatal closure (Brearley et al. 1997). Genes directly affecting signalling pathways under e[CO_2], include the SLAC1 (Slow Anion Channel Associated 1) gene which has been extensively associated with stomatal closure (Laanemets et al. 2013; Vahisalu et al. 2008). Several recent studies have also suggested a central role by carbonic anhydrase as a key regulatory factor in stomatal dynamics, with the bicarbonate ion activating SLAC1 anion channels (Xue et al. 2011; Hu et al. 2010). The hormone ABA is also involved, mainly through triggering the activation of the OST1 (*O*pen *ST*omata 1) gene, a positive downstream regulator of ABA signalling that ultimately modulates ion channel activity in the guard cell (Chater et al. 2015; Merilo et al. 2015) although there is evidence of further ABA-independent pathways (Yamamoto et al. 2016). Other hormones known to be involved in partial stomatal closure under e[CO_2] include jasmonic acid (Geng et al. 2016).

Over developmental timeframes, plants sense e[CO_2] which induces a well-established anatomical effect on g_s via a reduction in stomatal density (Woodward 1987). A reduction in density has been reported for a wide range of species grown at e[CO_2] (Casson and Gray 2008; Casson and Hetherington 2010; Woodward 1987; Woodward and Kelly 1995). Different degrees of reduction in stomatal densities have been also shown (Ainsworth and Rogers 2007) with some work showing reductions up to 40 % (Fig. 2.2b), however species-level responses vary; for example woody species typically demonstrate little change in stomatal density with e[CO_2] (Ainsworth and Rogers 2007; Xu et al. 2016). Therefore, the mechanisms of e[CO_2] on stomatal density are species-specific and genotype-dependent.

Gray et al. (2000) initially explored the genetic mechanism of stomatal density reduction, and demonstrated that the HIC

(*HI*gh *C*arbon dioxide) gene downregulated stomatal development and therefore density under e[CO_2]. Elsewhere the SDD1 (*S*tomatal *D*ensity & *D*istribution *1*) gene, which encodes for a key protein involved in guard cell development, is also downregulated under e[CO_2] (Kim et al. 2006), reducing stomatal density.

Elevated [CO_2] is known to increase photosynthetic efficiency by increasing C_i and accelerating electron transport (Leakey 2009; Zhang et al. 2008). On the other hand, there appear to be both symplastic and apoplastic feedback routes by which higher photosynthetic activity under e[CO_2] would initiate a negative feedback loop by concentrating sugars at the guard cell driven by the transpiration stream, limiting stomatal opening, g_s and decreasing stomatal dynamics (Kang et al. 2007; Kelly et al. 2013). Translocation of excess sugars via phloem and the apoplast to stomata may be responsible for the feedback phenomenon via increased apoplastic osmolarity (Kang et al. 2007) or via GC sensing of internal sucrose concentrations, and the expression of ABA-related genes (Kelly et al. 2013; Bauer et al. 2013; Waadt et al. 2015). Sucrose may not be the only carbohydrate signal, and a number of other signals originating in the mesophyll may be involved in the process, including malate, pH changes, redox state signalling or vapor-phase ions (Lawson et al. 2014). Downstream of e[CO_2] are stomatal closure and reduction in density, leading to lower transpiration, higher temperatures and improved WUE, which will also be discussed in following sections.

IV. Stomatal Responses to Water Stress

One of the consequences of climate change is the increasing risk of drought and/or heavy precipitation events, with the additional possibility of greater contrast between wet and dry regions (IPCC 2014). Mild to moderate drought stress lacks a consistent definition, but broadly considered, it is the reduction in soil water content to a level at which recovery of plant function is possible post-drought. Harb et al. (2010) define the value as 30 % of field capacity in Arabidopsis, but this may vary in other species.

A. *Phenotypic Variability in Responses to Low Water Availability*

Plant responses to water stress fall into a two categories: conservative or non-conservative (Caine et al. 2019; Chapin 1980; Valladares et al. 2000). Conservative plants may mature rapidly before late-season drought risks become extreme, thereby *escaping* mortality risks, but there is a cost in terms of potential yield. Alternatively, the plant may *avoid* drought through mechanisms that reduce water loss by closing stomata as well as maintaining turgor (Farooq et al. 2009), with yield consequences. The non-conservative phenotype *tolerates* drought by withstanding lower water potential through detoxification of reactive oxygen species, the production of LEA-proteins (which appear to protect membranes) and producing osmolytes / osmoprotectants such as proline (Claeys and Inze 2013; Harb et al. 2010). In the context of climate change, the conservative vs non-conservative behaviours are associated with differences in stomatal responses, both within genotypes and across species (Munns et al. 2010). Non-conservatism in the field has been closely related to slower opening and closing of stomata and higher biomass under drought in rice, while conservative cultivars that exhibited faster g_s closing responses were better able to manage water deficit (Qu et al. 2016). The conservative/non-conservative paradigm is analogous to isohydry/anisohydry. In the former case, stomata closed rapidly in response to water stress, maintaining higher water potential (Skelton et al. 2015) while anisohydric plants attempt to maintain carbon assimilation by retaining more-open stomata (Skelton et al.

2015). Crops such as wheat and barley are non-conservative under early stress conditions (Munns et al. 2010) (representing early-to-mid season drought) while rice appears highly non-conservative under elevated [CO_2], using more water than it has under ambient [CO_2] (Kumar et al. 2017). Under expected future conditions of greater water stress, the non-conservative phenotype may not be appropriate, and there may be pressure to develop ideotypes that have more-conservative water-use strategies, particularly if yield costs can be minimised (Bertolino et al. 2019).

B. Physiological and Genetic Consequences of Low Soil Water Availability

Stomata close progressively as water stress increases, restricting CO_2 diffusion into the leaf leading to a decline in photosynthetic rate (Farquhar and Sharkey 1982). Stomatal limitation on A may be small under non-stressed conditions, but stomatal closure becomes the major limitation under moderate drought (Farooq et al. 2009). Under mild drought, soybean, winter barley, winter wheat and spring triticale respond with a slow decline in transpiration of between 40 % and 53 % with a concurrent decrease in A of 40–54 % (Lipiec et al. 2013). In some circumstances, the reduction in biomass under mild drought is not linked to lower A in crops as diverse as barley, rice and maize (Munns et al. 2010; Rollins et al. 2013), but is an adaptive response to stress implying a conservative phenotype (Rollins et al. 2013). To complicate matters, crops such as wheat and barley are generally non-conservative under water deficit, but may switch to conservative behaviour as stress increases in severity (Munns et al. 2010). Furthermore, phenotypes are highly variable across cultivars and within cultivars over time. Stomata in some modern spring wheat cultivars are more sensitive to drought than historical varieties, showing greater stomatal closure than older varieties under drought, with associated reductions in C_i and PS II activity (Guan et al. 2015). Elsewhere, soil moisture can be a stronger driver of stomatal responses than leaf water status in younger leaves, but this situation is reversed as those leaves age (Chen et al. 2013).

There have been a number of successful approaches to the challenge of breeding enhanced drought resistance into crops in order to combat the effects of climate change. Hundreds of genes induced by drought have been identified (Chaves et al. 2003), some of which may relate to stomatal responses. Yet the relationship between drought and stomatal behaviour is complex. GM approaches have had some favourable outcomes, with perennial grass *L. chinensis* incorporating the wheat LEA gene, (*L*ate *E*mbryogenesis *A*bundant, a family of genes whose products are linked to ABA responses (Hundertmark and Hincha 2008)) and in oilseed rape, by downregulating ABA sensitivity; in both cases, increased drought tolerance was achieved (Wan et al. 2009; Wang et al. 2009). One problem for GM approaches has been the multifactorial responses to drought: resistant GM plants may have thicker, smaller leaves and lower g_s overall, leading to yield performance comparable to that of selection-bred cultivars (Lawlor 2013).

C. Improving Dynamic Water Use Efficiency

Productivity gains in crops have already been made by targeting steady-state stomatal responses and overall g_s (Fischer et al. 1998), although steady state conditions rarely persist in nature (Lawson and Blatt 2014). In this context, short-term stomatal kinetics offer another option by which breeders can seek adaptive responses to climate change. Stomatal kinetics refers to the rate of change in stomatal aperture or g_s in response to a change in an environmental variable (e.g., light intensity) (McAusland et al. 2016). Under environmental perturbation there is an order-of-magnitude difference between the rate at which the photosynthetic machinery can be activated relative to speed of stomatal opening, leading to potential yield losses or equally, excess water use (Lawson and Blatt 2014). Kinetic responses to changing light levels are readily modelled by estimating the time taken to reach 63% of maximal g_s (*tau,* Fig. 2.3). There is wide disparity even between C3 grass crops (Fig. 2.3) in terms of *tau* for the opening and closing of stomata in response to step changes in light intensity. Meanwhile, there is a pronounced difference between the slow speed of opening and faster closing speed for stomata in the case of the legumes, *Vicia faba* and *Pisum sativum* (Fig. 2.3), suggesting a range of strategies for managing dynamic water use needs.

D. *Breeding for Improved Performance Under Drought*

Plants show wide variability in the speed of stomatal responses to changing light levels under drought vs well-watered conditions (Fig. 2.4; Lawson and Blatt 2014). Tobacco is notable for its high sensitivity to drought, with large differences between the time constant for opening versus closing depending on the presence or absence of drought. However, barley shows no real change in the speed of g_s response under drought or control conditions under similar conditions (Fig. 2.4). Rice demonstrated a preference for slow closing over slow opening under control conditions, suggesting a prioritising of *A* over g_s (Fig. 2.3). Under drought, time constants decreased, but the decrease was greater for opening than closing, indicating water was being saved rather than *A* maintained (Qu et al. 2016). Breeding for g_s kinetic traits should result in increased fitness with respect to drought, but could be at the cost of lower biomass (Lawson and Blatt 2014; Qu et al. 2016; Faralli et al. 2019a). Therefore, farmers facing water stress will have to trade-off productivity and yield stability when selecting varieties in the face of drought that is likely to increase in severity, duration or frequency (Macholdt and Honermeier 2016; IPCC 2014).

E. *Physiological and Genetic Consequences of Excess Water Availability*

Excess water availability (waterlogging) is an important issue in many areas (Box 1986) and could become a greater problem with climate change, with predictions of greater

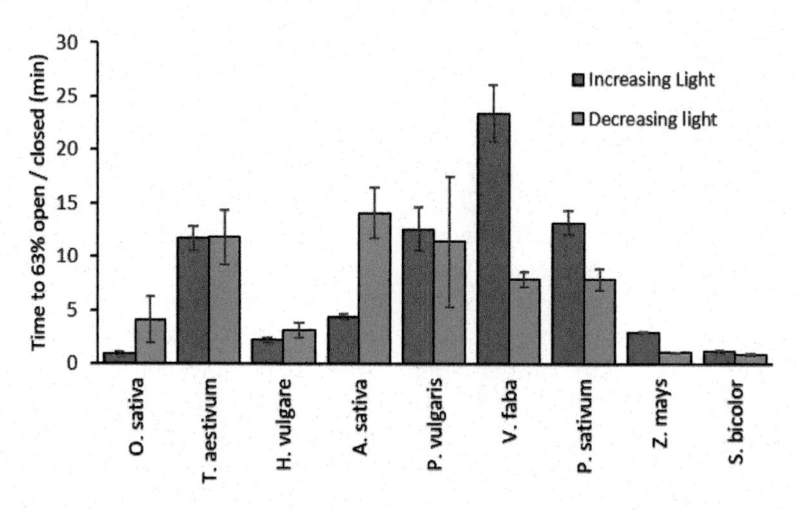

Fig. 2.3. Time constants of kinetic responses of stomata to step changes in light levels for a range of crop species. (Data redrawn from McAusland et al. 2016. Means shown +/− se, n = 3–5)

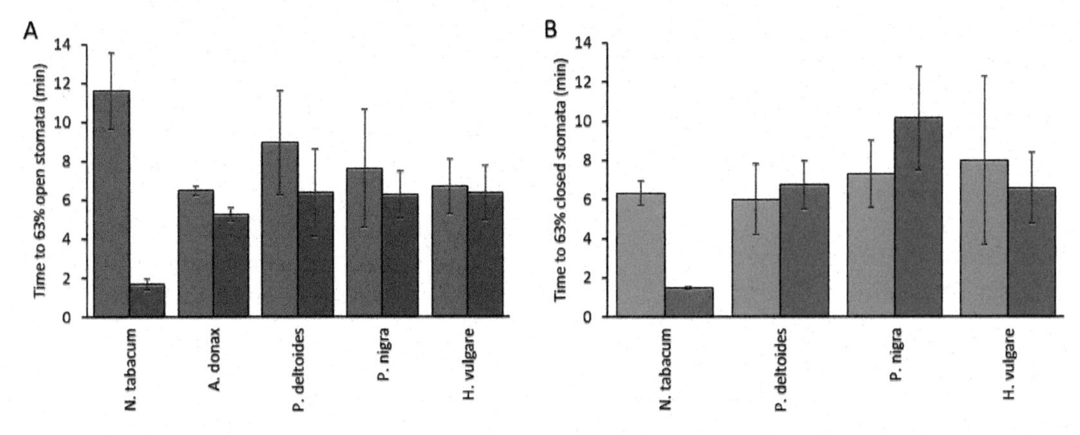

Fig. 2.4. Modelled time constants (tau) to reach 63% of max/min of the increase (**a**) in stomatal conductance for a step increase in PAR for a range of spp. under control (light red) and drought (dark red) conditions. (**b**) Time constants for the decrease in stomatal conductance for a step decrease in PAR for a range of spp. under control (light blue) and drought (dark blue) conditions. (Data for tobacco from Gerardin et al. 2018, for Arundo from Haworth et al. 2018, for Populus spp. from Durand et al. 2019, for Hordeum from Stevens J. (unpublished data). Means shown +/– error bars)

risks from more frequent and more extreme weather events, including precipitation (Porter and Semenov 2005; IPCC 2014). Stomata are known to close in response to waterlogging, with reductions in g_s by 30–40 % with 24 h of stress application reported by Bradford and Hsiao (1982) and in legumes, reductions in g_s and A during flooding were correlated (Pociecha et al. 2008). There is clear phenotypic variation across species. Huang et al. (1994) reported a range of g_s and A responses to waterlogging in wheat, and noted that reduction in yield potential could be offset by increasing nutrient application.

Signalling from root to shoot induces stomatal closure via a mechanism that is not entirely clear (Box 1986; Bradford and Hsiao 1982; Chaves et al. 2003; Chaves et al. 2010). Although stress response hormones such as ABA are known to be involved in the process of stomatal closure, alternative pathways exist, including responses mediated by ethylene, cytokinin concentration and xylem pH, while direct hydraulic signals are also thought to be important (Najeeb et al. 2015). Teasing apart gene expression data remains complex, with constitutively expressed gene abundance overlapping with stress response genes under water stress (Chaves et al. 2003).

The timing of waterlogging can affect the severity of the stress, for example in barley g_s was reduced by early (but not late) waterlogging and yield was only reduced by 15 %. Late waterlogging reduced barley yield by as much as 62 %, mainly through non-stomatal constraints (Ploschuk et al. 2018). A similar pattern was also observed for wheat with between 14 % yield reduction under early and 29 % under late waterlogging while in peas, waterlogging in general was poorly resisted, with yield declines of up to 92 % (Ploschuk et al. 2018). The risk of heavy precipitation events under climate change is a cause for concern for legumes in particular given their susceptibility to waterlogging (Ploschuk et al. 2018), but also for crops such as wheat, which is at risk of significant yield losses under conditions of excess water availability (Herzog et al. 2016). While root-level responses, particularly the formation of aerenchyma, are major drivers of tolerance of waterlogging, stomatal conductance shows wide variability in sensitive (e.g., pea) vs tolerant (e.g., rice, wheat, barley) species (Ploschuk et al. 2017; Herzog et al. 2016; Mohammed et al. 2019). The range of stomatal responses to waterlogging suggest an opportunity for breeders to identify traits for varieties tolerant of the expected increase in duration, severity and frequency of heavy precipitation events (IPCC 2014).

V. Stomatal Responses to Temperature Stress

A. *The Direct Impact of Increasing Temperature*

Higher future global temperatures are predicted under all carbon emissions scenarios considered in the most recent assessment of the effects of global climate change, with temperature rises of 1.5°C or more likely by the end of this century (IPCC 2014). For every 1°C increase in global temperature, yields of wheat would on average decline by 6%, rice by 3.2%, maize by 7.4% and soybean by 3.1% (Zhao et al. 2017). Thus increasing global temperatures poses a significant risk to global food security.

High temperature can directly affect stomatal behavior and development. Stomatal responses to increasing temperature varies depending on species (Sage and Sharkey 1987) and linked to changes in C_i which is an important determinant of stomatal responses (Mott 1988). Internal [CO_2] varies according to temperature (Von Caemmerer and Evans 2015), and the diffusion pathway from the intercellular airspace to the chloroplast is also sensitive to changes in temperature which influences both the gas and liquid phases of resistances (Evans et al. 2009). Whether g_s increases, decreases or is unaffected by temperature remains unclear overall (Sage and Sharkey 1987; Urban et al. 2017b) although the situation is complicated by vapour pressure deficit (VPD, the difference between the water moisture content of air and its saturation point) (Sage and Sharkey 1987; Merilo et al. 2018). When faced with increasing temperatures, some C3 plants encounter an increase in photorespiration due to a reduction in Rubisco specificity for CO_2 (Peterhansel et al. 2010), an increase in affinity for O_2, along with reduced gaseous fluxes from stomatal closure (Sage and Sharkey 1987; Mott 1988). Yet other species have instead shown increases in stomatal conductance in response to increased temperature (Sage and Sharkey 1987), which may be attributed to temperature-dependent effects on guard cell metabolism (Lu et al. 2000).

B. *Indirect Impact of Increasing Temperature: Vapor Pressure Deficit*

While high temperatures *per se* can affect stomata behaviour, responses are often considered indirect, with second-order effects such as transpiration rate, VPD, plant water status and assimilation rate all having an impact on stomatal responses (Urban et al. 2017b). VPD, which is an important consideration in the relationship between leaf temperature and stomata, is a key factor influencing transpiration. An increase in mean temperature, such as that predicted under climate change (IPCC 2014), is closely associated with increased VPD, and the regulation of transpiration through g_s is a key response to variation in VPD (Lawson and Blatt 2014). Stomatal responses to VPD are generally well-characterised, with increases in VPD eliciting stomatal closure to conserve water, while decreases in VPD lead to stomata re-opening (Merilo et al. 2018). Stomatal closure through increases in VPD occurs through reduction of leaf turgor and guard cell turgor (Mott and Peak 2013; Oren et al. 1999). This is also the reason for midday depression in g_s, as high temperatures produce a demand in transpiration that is too great for the plant, and thus a loss of turgor results in the closure of stomata (Balota et al. 2008).

Stomatal responses to VPD are regulated by a variety of mechanisms; for instance, there are active and passive hydraulic responses by stomata to a change in VPD

(McAdam and Brodribb 2014). Temperature and humidity conditions external to the leaf drive the primary factor regulating stomatal conductance - passive stomatal responses to VPD - by affecting leaf and guard cell turgor, and by extension, transpiration (McAdam and Brodribb 2015). An active response, mediated by ABA, is also present in many species in addition to the passive response. (McAdam and Brodribb 2015). Guard cells have been shown to synthesise ABA autonomously in sufficient concentrations to induce stomatal closure (Bauer et al. 2013). Knock out of ABA synthesis except in guard cells maintained stomatal response to increased VPD, suggesting that the stomatal responses to VPD are controlled by guard cell ABA synthesis (Bauer et al. 2013), even where foliar [ABA] is low. However, it has also been shown that ABA promotes stomatal closure through a decrease in water permeability in vascular tissues (Pantin et al. 2013). Recently, it was shown that phloem and guard cell ABA production were functionally redundant, and that ABA is involved in multiple possible pathways in the control of stomatal opening (Merilo et al. 2018). However, the pathway via the protein kinase OST1 remains an important source of stomatal control under high VPD (Merilo et al. 2018; Xue et al. 2011). Increased temperatures due to climate change will result in a higher VPD between leaves and atmosphere, and thus this will decrease stomatal conductance through stomatal closure. This will have an effect on growth and yields, as stomatal conductance is a major limiting factor for carbon assimilation (Farquhar and Sharkey 1982).

Increased stomatal conductance has been reported under elevated temperature. For example, higher temperature increased g_s by 163%, and transpiration by 83% in maize (*Zea mays*) (Zheng et al. 2013) and has also been reported in C3 plants (Crawford et al. 2012; Urban et al. 2017a). The effect of increasing temperature on stomatal anatomy depends on the species and the magnitude of change in temperature. Increased stomatal density at higher temperatures has been reported in soybean (*Glycine max*), oak (*Quercus robur*), tobacco (*Nicotiana tabacum*), shrubs and grapevines (*Vitis vinifera*) with no effect reported for maize, and a decrease was reported in Arabidopsis (Beerling and Chaloner 1993; Jumrani et al. 2017; Hu et al. 2014; Hill et al. 2014; Rogiers et al. 2011; Zheng et al. 2013; Crawford et al. 2012; Vile et al. 2012). These findings suggest the employment of different strategies for managing leaf cooling under higher temperatures to counteract the negative impact of heat on photosynthesis and yield (Crafts-Brandner and Salvucci 2002; Sage and Kubien 2007).

C. *Night-Time Temperature Increases Also Affect Crop Productivity*

Global temperature increases due to climate change will also drive greater minimum night-time temperature in addition to higher day temperatures, with night time temperature reducing yield in rice and wheat (Easterling et al. 1997; Shi et al. 2016; Narayanan et al. 2015; Prasad et al. 2008; Peng et al. 2004). Although the effect of night temperature on g_s and the consequences for photosynthesis are not fully understood (Peng et al. 2004; Prasad et al. 2008), some reports suggest rice yield declines by 10% for every 1°C increase in minimum temperatures (Peng et al. 2004). While the precise mechanism affecting yield is not clear, it has been shown that higher night temperatures can affect stomatal opening and photosynthesis during the day perhaps owing to greater water deficit in the leaves (Prasad et al. 2008; Pasternak and Wilson 1972; Drew and Bazzaz 1982), highlighting the importance of increases in both the minimum and maximum temperature on crop yield (Welch et al. 2010).

The impact of increased temperature, mediated by behavioural and developmental responses of stomata on crop performance remains a concern under all climate change scenarios, with wheat and maize already showing yield losses (IPCC 2014; Zhao et al. 2017). There is variability in the consequences of temperature change by region, crop and depending on climate model used, although overall impact assessments remain consistent (Zhao et al. 2017; Iizumi et al. 2017). Furthermore, higher temperatures will drive higher ET rates, which may lead to reduced water availability and greater susceptibility to yield loss in addition to those impacts described above (Condon et al. 2002; Mueller et al. 2012; Van Ittersum et al. 2013).

D. The Impact of Low Temperatures on Physiology

Stomatal response to low temperatures or chilling depend on whether the species is cold tolerant or cold sensitive. Plants can be considered cold tolerant when stomatal closure occurs before the onset of any water deficit due to decreases in hydraulic conductance and root activity, while cold sensitive plants have less capacity to increase Ca^{2+} uptake by guard cells and therefore stomatal closure is delayed (Hussain et al. 2018; Wilkinson et al. 2001).

Cold sensitive plants, such as *Phaseolus vulgaris* and maize, can have a lowered stomatal conductance after exposure to 24 h of chilling conditions (i.e., less than 8°C day and 4°C night temperatures), or when grown under cool (i.e., 18°C/12°C) conditions (Wolfe 1991). However in these cold-sensitive plants, which also include cucumber, tomato bean and soybean (Allen and Ort 2001) stomata can remain open for a period after chilling, which may lead to severe water stress as hydraulic activity in the roots decreases (Eamus et al. 1983). Root temperature, ambient humidity and recovery of hydraulic conductance post-chill appears to be important to the subsequent stomatal response (Allen and Ort 2001; Bloom et al. 2004), putting these plants at risk of less responsive stomata, lower g_s and by extension, reduced yield.

Cold-tolerant species such as pea and spinach, show 'normal' stomatal opening rates after a period of chilling, and some have been reported to increase overall g_s (Wolfe 1991). Chilling of Spinach roots to 5°C initially resulted in lower g_s, along with lower root hydraulic conductance, which recovered after a few days of higher temperature to a g_s higher than before the initial chilling, but lower than that observed at higher (i.e., 20°C) root temperatures (Fennell and Markhart 1998). Overall, cold-tolerant species exhibit smaller variation in leaf water potential, demonstrating resistance to low-temperature induced water stress (Bloom et al. 2004; Wolfe 1991), unlike cold-sensitive plants which exhibit decreases in leaf water potential and lower hydraulic conductivity due to their stomatal response (Hussain et al. 2018).

Low temperature affects ABA synthesis (Pardossi et al. 1992) and reduces guard cell sensitivity to ABA (Honour et al. 1995; Wilkinson et al. 2001) regardless of whether they are cold-tolerant or cold-sensitive, thereby slowing responses. However, in cold-tolerant species, stomata have been reported to close after a few hours with an increase in root ABA synthesis (Melkonian et al. 2004). Low temperature can affect maintenance of guard cell osmotic potential with stomatal closure the result of increased apoplastic calcium uptake by guard cells (Ilan et al. 1995). Apoplastic calcium influx into the guard cell cytosol has been shown to be responsible for stomatal closure in cold tolerant species, but not in cold sensitive species, through increased sensitivity to Ca^{2+} (Wilkinson et al. 2001). Calcium acts as an intracellular secondary messenger, regulating ion transporter activity in plasma and vacuolar membranes in the guard cell, which determines guard cell turgor (Assmann and Shimazaki 1999; Wilkinson et al. 2001). Cold hardening, by which process plants become more tolerant of low temperatures through prior exposure, may be due to the cold-induced uptake of calcium into guard cells, which primes the plant for exposure to ABA and thus rapid stomatal closure (Wilkinson et al. 2001). For instance, cold-hardened *Phaseolus vulgaris*, when exposed to low temperatures, reduced stomatal aperture and maintained a positive leaf turgor (Eamus et al. 1983).

The threat from lower temperatures persists even under general conditions of global warming, as increased variability in temperatures is expected (IPCC 2014). The impact on crop productivity remains uncertain and will be dependent on both the local temperature history and the sensitivity of the crop species being grown.

VI. Interactions Between Factors Related to Climate Change

Under climate change, plants face variability in [CO_2], temperature and water availability. However, the stress caused to plants by these factors does not occur in isolation, and the interaction between them is less predictable. In some circumstances, stresses driving stomatal responses may be additive, in others, antagonistic. Stresses may not combine arithmetically and the impact may be greater than sum of the individual stresses alone (Atkinson and Urwin 2012). Meanwhile, the genetic regulation of responses varies according to both the nature and extent of the combination of stresses (Vile et al. 2012).

A. CO_2 Concentration and Drought

The positive impact of e[CO_2] may not only relate to increased photosynthesis, but also to a reduction in g_s and transpiration leading to soil water conservation and water stress mitigation. Many studies have shown that the increased [CO_2] can substantially mitigate the effect of drought, potentially offsetting the reduction in crop productivity relating to climate change. A review by Leakey et al. (2009), clearly highlighted that a reduction in ET (followed by reduction in g_s) preserves soil moisture and led to yield maintenance under e[CO_2] in both C3 and C4 crops. Yield maintenance combined with water stress tolerance induced by e[CO_2] was shown in sorghum, wheat, soybean and maize, thus suggesting an optimistic outlook for future crop production under climate change (Kimball et al. 2002). However, recent work focusing on CO_2 x stress interactions has reduced confidence in this prediction. In particular, multi-year and multi-location Free-Air CO_2 Enrichment (FACE) experiments have led to a partial rethinking of earlier beliefs about the positive interaction of e[CO_2] with water stress (Gray et al. 2016; Osborne 2016). For example, Gray et al. (2016), reported that several recent FACE experiments found that the CO_2-stress mitigation hypothesis (mainly driven by the reduction in g_s) is not fully supported. For instance, aerodynamically smooth canopies (such as those seen for soybean and other agriculturally-important species) are likely to see high ET, despite reductions in g_s, arising from a dryer boundary layer driving ET, and an increase in leaf temperature driving higher VPD and thus ET (Gray et al. 2016; Field et al. 1995).

The same study by Gray et al. (2016) also showed that e[CO_2] did not stimulate deeper rooting and led to a significant increase in stomatal sensitivity to ABA. The consequence of this was that decreasing g_s response to water stress overrode the stimulative effect of increased sub-stomatal CO_2 concentration. Other long-term FACE and controlled-environment studies supported the lack of [CO_2]-stress mitigation findings in other species (e.g., grasses, canola) with significant negative interactions between leaf area index, g_s, leaf temperature and root-to-shoot signalling for instance (Faralli et al. 2017; McGranahan and Poling 2018; Osborne 2016).

B. CO_2 Concentration and Elevated Temperature

The impact of the interaction of e[CO_2] with elevated temperature is controversial with contrasting experimental evidence (Table 2.1). The expected reduction in g_s under e[CO_2] suggests that crops will inevitably reduce transpirational leaf cooling thus leading to significant increases in leaf temperature (with potential negative downstream effects on photosynthesis). Elevated [CO_2] and temperature can reduce yield by 10–12 % in wheat, while for rice the interaction between [CO_2] *and* heat led to modest losses of −1.6 % (Hatfield et al. 2011; Cai et al. 2016). Yet in FACE experiments under well-watered conditions, elevated

Table 2.1. Example field experiments measuring yield or physiological consequences of an interaction of [CO_2] with water or temperature stress

Crop	Stress	Physiology or Yield Component	% Change in yield vs control	References
Zea mays (C_4)	Elevated [CO_2] and precipitation	Yield	14% increase from 390ppm to 450ppm, 11% increase from 450ppm to 550pppm. On average, increasing precipitation increases yield by 14.57%	Meng et al. (2014) PLoS ONE 9
Zea mays (C_4)	Elevated [CO_2] and precipitation	Dry Mass	No significant increase under elevated CO_2. Dry soil results in elevated CO_2 having from 20%-54% higher dry mass.	Samarakoon et al. (1996) Australian Journal of Plant Physiology 23
Zea mays (C_4)	Elevated [CO_2] and temperature	Grain Yield	Temperature reduces yield by 19% under ambient [CO_2], 38% under e[CO_2]	Abebe et al. (2016) Agriculture, Ecosystems and Environment 218
Zea mays (C_4)	Elevated [CO_2], well-watered	Assimilation	No significant effect of elevated [CO_2] on photosynthesis under well-watered conditions	Leakey et al. (2006) Plant Physiology 140
Triticum aestivum	Elevated [CO_2] and semi-arid precipitation	Yield	13% yield increase	Fitzgerald et al. (2016) Global Change Biology 22
Glycine max (Soybean)	Elevated [CO_2], Temperature, Soil Moisture and precipitation	Assimilation	5% increase in assimilation under elevated [CO_2] and temperature, no change under low precipitation/soil moisture	Rosenthal et al. (2014) Plant Science
C_3 (Mainly Wheat + Soybean)	Elevated [CO_2], Precipitation and Irrigation	Yield Response Ratio	26% increase in Yield Response Ratio with 200mm to 700mm precipitation for FACE experiments, 26% increase in Yield Response Ratio with 1000mm to 200mm precipitation for OTC experiments	Bishop et al. (2014) Food and Energy Security 3
Oryza sativa	Elevated [CO_2] + Temperature	Leaf Area, Plant Height	Decreases in leaf area up to 12%, reduced plant height at tillering stage	Liu et al. (2017) PLoS ONE 12
Oryza sativa	Elevated [CO_2] and drought stress	Photosynthesis	Higher photosynthesis for elevated [CO_2] plants under drought, than those at ambient [CO_2]	Widodo et al. (2003) Environmental and Experimental Biology 49

[CO_2] had a positive effect on wheat yield (Fitzgerald et al. 2016), as soil water counteracted the negative effect of e[CO_2] and heat on leaf temperature by supporting high g_s. Therefore, under heat stress the positive effect of [CO_2] on cereals is not only dependent on the timing of stress (i.e., post or pre-anthesis) (Fitzgerald et al. 2016) but also the maintenance of adequate g_s and subsequently optimal leaf temperature (IPCC 2014). Figure 2.5 shows the hypothetical pattern of g_s responses (which will vary by species) of conservative and non-conservative crops under varying combinations of environmental stresses over the course of a season, with the overall integral of g_s at the right hand side of each panel. The latter we link to final yield following Fischer et al. (1998). Under developing water stress after establishment (Fig. 2.5a) g_s responses for conservative genotypes are predicted to involve early stomatal closure (Negin and Moshelion 2017) compared to non-conservative genotypes (Faralli et al. 2019c), leading to greater integrated g_s over the season for the non-conservative phenotype, and by extension, a yield advantage. Elevated [CO_2] could exacerbate stress-related reduction in yield as described above (Fig. 2.5b). Yields can fall if no e[CO_2]-induced soil water conservation is achieved and mainly due to the increase in biomass (giving a higher transpirational surface area) and elevated leaf temperature due to reduced g_s (Faralli et al. 2017; Gray et al. 2016; McGranahan and Poling 2018; Osborne 2016). Under the drought & elevated e[CO_2] scenario, lower average g_s is achieved, threatening final yields, while the non-conservative phenotype retains a modest advantage in total g_s. The effects of combined water stress and high temperature under e[CO_2] remain to be elucidated by long term FACE experiments although the outcome may be to increase g_s and reduce yield (Fig. 2.5c) relative to Fig. 2.5b. The consequences of the combined stresses on soil water availability are likely to be exacerbated by the behaviour of the non-conservative plant, which delays g_s responses with serious consequences as the effects of drought are felt. Permanent damage occurs, and the plant is never able to recover to its prior g_s position. Although damage also occurs to the conservative plant, the extent is relatively lower, and post-drought the plant is able to recover to somewhere nearer its previous output. There are yield consequences for both phenotypes, but now the conservative plant outperforms as it is able to deliver higher g_s throughout the season.

In general, the overall inconsistencies in interactions between elevated [CO_2] and environmental stresses are probably derived from variability in treatment applications (e.g., [CO_2] and the timing, degree and length of applied stress). It appears likely that responses are species-specific and that further large-scale multi-year experiments are required to fill this knowledge gap.

C. *Drought and Heat Stress*

There has been extensive work on interactions between water deficit and high temperature (Table 2.1). The impact varies according to the species, growth conditions, conservatism or non-conservatism of the plant, and degree, extent and timing of the stress. One of the difficulties plants face, in particular with reference to combined heat and drought stress is that the initial (and ongoing) plant response – adjustment of stomatal aperture – may involve countervailing and mutually antagonistic hormone pathways (Atkinson and Urwin 2012). In general, combined stresses lead to lower g_s as minimising water loss is prioritised over heat damage. Transpiration was reduced by 82 % under combined heat and drought stress in spring wheat, and *A* was reduced by 69 % (Lipiec et al. 2013). Combined water and heat stresses reduced *A* regardless of the impact on g_s, with drought, and combined drought & heat affecting yield more than heat alone as total light intercep-

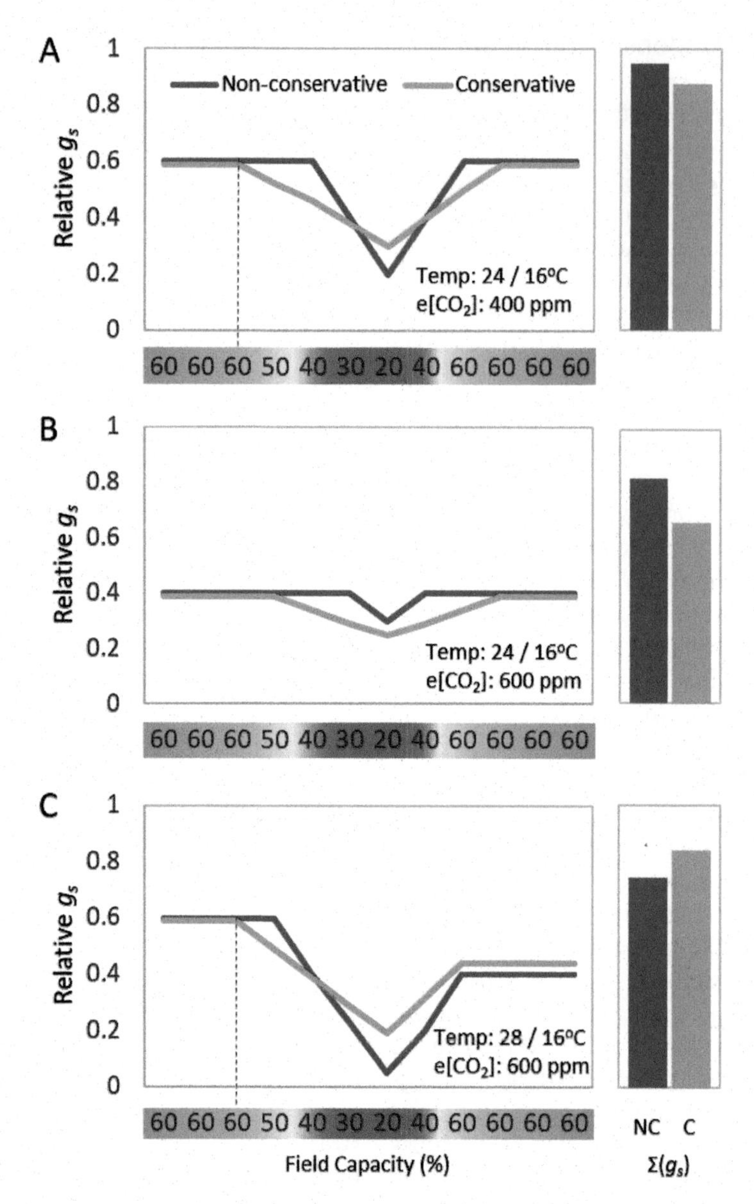

Fig. 2.5. Hypothetical consequences of environmental stress on plants with conservative (green) and non-conservative (magenta) phenotypes during the course of a growing season. In all three instances (**a**–**c**) a mid-seasonal drought is applied after establishment progressively reaches 20% of field capacity. For each combination of stresses, the left side describes the relative *gs* response to the stress over time, while the bar chart on the right integrates *gs* over the season and is proportional to final yield. The dotted grey line shows where the conservative plant initiates stress responses. (**a**) Under temperature and e[CO_2] conditions that approximate a Northern European growing season, the conservative phenotype reduces *gs* as soil moisture falls while the non-conservative phenotype delays action. As a result, the non-conservative plant is eventually forced to reduce *gs* below that of the conservative plant, but at the end of the drought both phenotypes are able to recover *gs* to the prior level. The non-conservative plant is able to maintain a greater integral *gs* and by extension, higher yield. (**b**) Under a situation of e[CO_2], *gs* is reduced overall, and the onset of drought responses are delayed. The extent of the reduction in *gs* during the drought is affected by other factors (higher total ET from greater biomass, higher leaf temperatures from lower *gs* etc.) leading to negative yield consequences overall. (**c**) When both temperature and e[CO_2] are elevated, *gs* is initially higher for both phenotypes, perhaps in the range seen in (**a**) above. Once again, the conservative plant responds early to the changing soil moisture availability, and is able to manage its water needs. The non-conservative plant is rapidly forced to decrease *gs* to offset declining availability and increased demand. Both phenotypes may sustain damage (reflected in lower post-drought *gs*) but the conservative phenotype is likely to recover better from the stress than the non-conservative. Under these condition, it is probable that the conservative plant will be able to achieve greater integrated *gs*, and therefore higher yields under possible and imminent climate change scenarios

tion is reduced over the shortened life cycle, although g_s is also expected to decline (Lipiec et al. 2013). A similar outcome has been reported in tobacco, where stomata opened under heat stress but closed under drought and the combined stresses (Rizhsky et al. 2002). It appears that drought is the 'greater risk' to the plant, and hierarchically dominates responses. Work in barley has underlined the difficulty in comparing across species, showing a reduction in biomass and yield but not photosynthesis under drought, while heat or combined drought and heat stresses led to lower photosynthesis (Rollins et al. 2013). Morphological changes such as reduced spike number were more evident under drought, while under heat treatment, physiological effects dominated (Rollins et al. 2013), notably in terms of grain yield relative to water use. Underlying these mechanisms are pathways and the expression of cascades of genes; for instance, *Arabidopsis* subjected to combined heat and drought stress had 454 transcripts only elicited under the combined stress compared to the individual stresses alone, and understanding the interaction of multiple abiotic stresses on physiology remains a complex endeavour (Rizhsky et al. 2004; Bechtold et al. 2018). Gene expression had remarkably little overlap under combined heat and drought stress, underlining the validity of the different physiological pathways in a comparison of two durum wheat cultivars (Aprile et al. 2013; Zandalinas et al. 2018). Combined heat and drought stress elicits differential responses of pathways involved in photosynthesis and antioxidant mechanisms, hormone signalling and transcription factor abundance, meaning the link to stomatal responses remains complex and dependent on the balance of stresses (Zandalinas et al. 2018).

There has been relatively little published on responses to a combination of excess water availability and heat stress. In general, flooding and low temperature responses appear aligned (Klay et al. 2018), and can lead to stomatal closure and photosystem damage in wheat (Li et al. 2014), although at lower levels of waterlogging, A was relatively improved (Li et al. 2014). Responses to flooding and high temperatures were antagonistic in some genes but aligned in others (Klay et al. 2018).

Multi-location and multi-year field trials will remain of critical significance (Ainsworth and Long 2005; Gray et al. 2016; Osborne 2016) in the unpacking of stomatal phenotypes and in the understanding of GxE interactions (Claeys and Inze 2013). A desire to understand ever more complex combinations of factors – such as $[CO_2]$ x Water stress x Heat stress – will gain in importance as climate change will bring about a wide range of stressors that will vary by region (IPCC 2014). Current understanding accepts that stomatal responses are complex and possibly hierarchical (Lawson and Blatt 2014), and much more work needs to be done to understand which stomatal response will dominate and under what set of circumstances.

VII. Conclusion

Stomata are the gatekeepers of gas exchange and the primary determinants of ET and CO_2 assimilation rates (Hetherington and Woodward 2003). They have complex developmental histories, and are under genetic control that is heavily influenced by environmental conditions experienced during development (Doheny-Adams et al. 2012). From day-to-day, stomata respond to changing variables such as temperature and water availability, as well as atmospheric $[CO_2]$ (Blatt 2000). Thus managing the size, number and responsiveness of stomata offers breeders the potential to manage the interaction of g_s and A; by extension, there will be an impact on yield (Lawson and Blatt 2014).

For these reasons, climate change offers potential opportunities as well as threats to plant productivity, and these threats will vary by region across the world. Climate change is expected to drive higher average temperatures. Heat waves, droughts and heavy rain are likely to occur more often and for greater duration. The exact extent of the combination of these factors for any given location remains highly uncertain (IPCC 2014).

Rising [CO_2] should lead to higher yields with less water use as stomata can reduce aperture while maintaining internal [CO_2] (Gamage et al. 2018), and over longer periods, reduce stomatal density (Woodward 1987). Higher temperatures are likely to result in stomatal closure as a result of higher internal [CO_2] from photorespiration (Mott 1988). Lower water availability is likely to force stomata to reduce aperture on average (Farooq et al. 2009), although varieties and cultivars that adopt a non-conservative approach may maintain yield better than conservative genotypes (Bertolino et al. 2019). Meanwhile the link between stomatal density and drought tolerance is contested (Jones 1977); developmentally, altered stomatal size and density under water stress may be offset by other changes such as to leaf size and thickness, and total leaf area (Lawlor 2013).

Elevated [CO_2] was thought to offset the negative implications of drought (Leakey 2009). Recent work puts this conclusion in doubt, particularly in terms of unintended consequences such as shallower roots or stomatal sensitivity to ABA (Gray et al. 2016). The interaction between higher temperatures and e[CO_2] might also be expected to lead to higher leaf temperatures, as g_s declines, with a knock-on consequence for photosynthesis and hence yield (Cai et al. 2016). However, in well-watered plots, high temperatures can be overcome and yield maintained under elevated [CO_2] (Fitzgerald et al. 2016). The combination of drought and heat stress involves potentially antagonistic pathways of stomatal responses (Atkinson and Urwin 2012), with water stress dominating heat stress, and with unique transcription pathways evoked under the combined stresses (Rollins et al. 2013; Zandalinas et al. 2018). However, A is not always affected, even if g_s and yield are (Rollins et al. 2013). In a three-way interaction, the situation is more-complicated again, and there is little clarity over whole-plant and field-level responses as mediated by stomata. We might expect these responses to be hierarchical, and in the context of genetic control, to be epistatic. This is an area that certainly needs more work if we are to understand better the link between expression and response to multiple interacting factors.

Meanwhile, expression patterns and physiological and anatomical responses to single stresses in stomata are relatively well-understood. Measuring kinetic rather than steady-state responses clearly needs more work, with limited understanding to date of the rate of stomatal response to fluctuating light, let alone diurnal patterns of fluctuating temperature (Vialet-Chabrand et al. 2017c; Faralli et al. 2019b) or water availability. Furthermore, one of the predictions of climate change is an increased variability in environmental stressors such as water availability or temperature (IPCC 2014). Once again, the acclimation of stomata in response to these fluctuations is poorly understood, and is an active area of research. Overall, there is a great heterogeneity of research, and we call for some agreement between groups in defining protocols and experimental conditions which probe some of these complicated interactions in a consistent and reasoned manner.

There has been widespread realisation in recent years that plant physiology, notably in stomatal responses, is critical to driving ongoing increases in yield in crop species under climate change. Much work remains to be done to understand the genetic and metabolic pathways underpinning responses to such complex interactions. We have discovered that stomatal responses to changing environmental variables appeared straightforward, but like Churchill's comments on pre-war Russia, remain 'a riddle wrapped in a mystery inside an enigma'. We look forward to unravelling this conundrum.

VIII. Outlook

FACE, greenhouse and growth chamber experimental work are now providing detailed insights into our understanding of stomatal responses to climate change, and forcing some revision of earlier expectations. The emerging study of stomatal kinetics under varying environmental conditions in addition to steady-state measurements gives additional opportunities to discover phenotypes of interest. The interaction of multiple abiotic stresses such as e[CO_2], heat and drought simultaneously remain difficult to model and poorly understood, with a lack of clarity over experimental methods. An agreement among researchers over definitions of stress levels and applications (e.g., mild drought, high temperature) that allow consistency across studies and greater confidence in the robustness of results would be extremely beneficial.

Acknowledgements

JS acknowledges The Perry Foundation and The School of Life Sciences, the University of Essex for supporting his PhD. MF was supported by Biotechnology and Biological Sciences Research Council (BBSRC) grants to TL (BB/NO16898/1) and JC (BB/N016931/1), with IPA co-funding from BASF. JDS was funded by a BBSRC Industrial Case studentship award (1721987) awards to Essex (TL) with industrial partner Dr Iain Cameron from Environment System Ltd. SW was supported through a BBSRC Industrial Studentship (1775930) award to BASF (JVR), Essex (TL), and NAIB (JC).

References

Aasamaa K, Sober A, Rahi M (2001) Leaf anatomical characteristics associated with shoot hydraulic conductance, stomatal conductance and stomatal sensitivity to changes of leaf water status in temperate deciduous trees. Aus J Plant Physiol 28:765–774

Ainsworth EA, Long SP (2005) What have we learned from 15 years of free-air CO_2 enrichment (FACE)? A meta-analytic review of the responses of photosynthesis, canopy. New Phytol 165:351–371

Ainsworth EA, Rogers A (2007) The response of photosynthesis and stomatal conductance to rising CO_2: mechanisms and environmental interactions. Plant Cell Environ 30:258–270

Allen DJ, Ort DR (2001) Impacts of chilling temperatures on photosynthesis in warm-climate plants. Trends Plant Sci 6:36–42

Amin S, Bongaarts J, Mcnicoll G, Todaro MP (2006) 2006 state of the future. Population Develop Rev 32:787–787

Aprile A, Havlickova L, Panna R, Mare C, Borrelli GM, Marone D, Perrotta C, Rampino P, …, Cattivelli L (2013) Different stress responsive strategies to drought and heat in two durum wheat cultivars with contrasting water use efficiency. BMC Genomics 14:1–18

Assmann SM, Shimazaki K (1999) The multisensory guard cell. Stomatal responses to blue light and abscisic acid. Plant Physiol 119:809–815

Atkinson NJ, Urwin PE (2012) The interaction of plant biotic and abiotic stresses: from genes to the field. J Exp Bot 63:3523–3543

Balota M, William AP, Evett SR, Peters TR (2008) Morphological and physiological traits associated with canopy temperature depression in three closely related wheat lines. Crop Sci 48:1897–1910

Bauer H, Ache P, Lautner S, Fromm J, Hartung W, Al-Rasheid KAS, Sonnewald S, Sonnewald U, …, Hedrich R (2013) The Stomatal Response to Reduced Relative Humidity Requires Guard Cell-Autonomous ABA Synthesis. Curr Biol 23:53–57

Bechtold U, Ferguson JN, Mullineaux PM (2018) To defend or to grow: lessons from Arabidopsis C24. J Exp Bot 69:2809–2821

Beerling DJ, Chaloner WG (1993) The impact of atmospheric CO_2 and temperature change on stomatal density - observations from Quercus robur lammas leaves. Ann Bot 71:231–235

Beerling DJ, Woodward FI (1997) Changes in land plant function over the Phanerozoic: Reconstructions based on the fossil record. Botanic J Linnean Soc 124:137–153

Beerling DJ, Woodward FI, Lomas M, Jenkins AJ (1997) Testing the responses of a dynamic global vegetation model to environmental change: a comparison of observations and predictions. Glob Ecol Biogeogr Lett 6:439–450

Bergmann DC, Sack FD (2007) Stomatal development. Annu Rev Plant Biol 58:163–181

Berry JA, Beerling DJ, Franks PJ (2010) Stomata: key players in the earth system, past and present. Curr Opin Plant Biol 13:232–239

Bertolino LT, Caine RS, Gray JE (2019) Impact of Stomatal Density and Morphology on Water-Use Efficiency in a Changing World. Front Plant Sci 10

Betts RA, Boucher O, Collins M, Cox PM, Falloon PD, Gedney N, Hemming DL, Huntingford C, …, Webb MJ 2007) Projected increase in continental runoff due to plant responses to increasing carbon dioxide. Nature 448:1037–10U5

Blatt MR (2000) Cellular signaling and volume control in stomatal movements in plants. Annu Rev Cell Dev Biol 16:221–241

Bloom AJ, Zwieniecki MA, Passioura JB, Randall LB, Holbrook NM, ST Clair DA (2004) Water relations under root chilling in a sensitive and tolerant tomato species. Plant Cell Environ 27:971–979

Box JE (1986) Winter-wheat grain-yield responses to soil oxygen diffusion rates. Crop Sci 26:355–361

Bradford KJ, Hsiao TC (1982) Stomatal behavior and water relations of waterlogged tomato plants. Plant Physiol 70:1508–1513

Brearley J, Venis MA, Blatt MR (1997) The effect of elevated CO_2 concentrations on K^+ and anion channels of Vicia faba L. guard cells. Planta 203:145–154

Brodribb TJ, McAdam SAM, Jordan GJ, Feild TS (2009) Evolution of stomatal responsiveness to CO_2 and optimization of water-use efficiency among land plants. New Phytol 183:839–847

Brodribb TJ, Feild TS, Sack L (2010) Viewing leaf structure and evolution from a hydraulic perspective. Funct Plant Biol 37:488–498

Buckley TN, Mott KA (2013) Modelling stomatal conductance in response to environmental factors. Plant Cell Environ 36:1691–1699

Buckley TN, Sack L, Farquhar GD (2017) Optimal plant water economy. Plant Cell Environ 40:881–896

Cai C, Yin XY, He SQ, Jiang WY, Si CF, Struik PC, Luo WH, Li G, …, Pan GX (2016) Responses of wheat and rice to factorial combinations of ambient and elevated CO_2 and temperature in FACE experiments. Glob Chang Biol 22:856–874

Caine RS, Yin XJ, Sloan J, Harrison EL, Mohammed U, Fulton T, Biswal AK, Dionora J, …, Gray JE (2019) Rice with reduced stomatal density conserves water and has improved drought tolerance under future climate conditions. New Phytol 221:371–384

Casson S, Gray JE (2008) Influence of environmental factors on stomatal development. New Phytol 178:9–23

Casson SA, Hetherington AM (2010) Environmental regulation of stomatal development. Curr Opin Plant Biol 13:90–95

Chapin FS (1980) The mineral nutrition of wild plants. Annu Rev Ecol Syst 11:233–260

Chater C, Peng K, Movahedi M, Dunn JA, Walker HJ, Liang YK, Mclachlan DH, Casson S, …, Hetherington AM (2015) Elevated CO_2-induced responses in stomata require ABA and ABA signaling. Curr Biol 25:2709–2716

Chaves MM, Maroco JP, Pereira JS (2003) Understanding plant responses to drought – from genes to the whole plant. Funct Plant Biol 30:239–264

Chaves MM, Zarrouk O, Francisco R, Costa JM, Santos T, Regalado AP, Rodrigues ML, Lopes CM (2010) Grapevine under deficit irrigation: hints from physiological and molecular data. Ann Bot 105:661–676

Chen L, Dodd IC, Davies WJ, Wilkinson S (2013) Ethylene limits abscisic acid- or soil drying-induced stomatal closure in aged wheat leaves. Plant Cell Environ 36:1850–1859

Claeys H, Inze D (2013) The agony of choice: how plants balance growth and survival under water-limiting conditions. Plant Physiol 162:1768–1779

Condon AG, Richards RA, Rebetzke GJ, Farquhar GD (2002) Improving intrinsic water-use efficiency and crop yield. Crop Sci 42:122–131

Crafts-Brandner SJ, Salvucci ME (2002) Sensitivity of photosynthesis in a C4 plant, maize, to heat stress. Plant Physiol 129:1773–1780

Crawford AJ, Mclachlan DH, Hetherington AM, Franklin KA (2012) High temperature exposure increases plant cooling capacity. Curr Biol 22:R396–R397

Croxdale JL (2000) Stomatal patterning in angiosperms. Am J Bot 87:1069–1080

Doheny-Adams T, Hunt L, Franks PJ, Beerling DJ, Gray JE (2012) Genetic manipulation of stomatal density influences stomatal size, plant growth and tolerance to restricted water supply across a growth carbon dioxide gradient. Philos Trans R Soc B Biol Sci 367:547–555

Dow GJ, Bergmann DC (2014) Patterning and processes: how stomatal development defines physiological potential. Curr Opin Plant Biol 21:67–74

Drake PL, Froend RH, Franks PJ (2013) Smaller, faster stomata: scaling of stomatal size, rate of response, and stomatal conductance. J Exp Bot 64:495–505

Drew AP, Bazzaz FA (1982) Effects of night temperature on daytime stomatal conductance in early and late successional plants. Oecologia 54:76–79

Eamus D, Fenton R, Wilson JM (1983) Stomatal behavior and water relations of chilled Phaseolus vulgaris L and Pisum sativum L. J Exp Bot 34:434–441

Easterling DR, Horton B, Jones PD, Peterson TC, Karl TR, Parker DE, Salinger MJ, Razuvayev V, …, Folland CK (1997) Maximum and minimum temperature trends for the globe. Science 277:364–367

Edwards D, Kerp H, Hass H (1998) Stomata in early land plants: an anatomical and ecophysiological approach. J Exp Bot 49:255–278

Elliott-Kingston C, Haworth M, Yearsley JM, Batke SP, Lawson T, McElwain JC (2016) Does size matter? Atmospheric CO_2 may be a stronger driver of stomatal closing rate than stomatal size in taxa that diversified under low CO_2. Front Plant Sci 7:1253

Engineer CB, Hashimoto-Sugimoto M, Negi J, Israelsson-Nordstrom M, Azoulay-Shemer T, Rappel W-J, Iba K, Schroeder JI (2016) CO_2 Sensing and CO_2 peculation of Stomatal Conductance: Advances and Open Questions. Trends Plant Sci 21:16–30

Evans JR, Kaldenhoff R, Genty B, Terashima I (2009) Resistances along the CO_2 diffusion pathway inside leaves. J Exp Bot 60:2235–2248

FAO (2015) Towards a Water and Food Secure Future: Critical Perspectives for Policy-Makers. Natural Resources and Environment Department.

Faralli M, Grove IG, Hare MC, Kettlewell PS, Fiorani F (2017) Rising CO_2 from historical concentrations enhances the physiological performance of Brassica napus seedlings under optimal water supply but not under reduced water availability. Plant Cell Environ 40:317–325

Faralli M, Cockram J, Ober E, Wall S, Galle A, van Rie J, Raines C, Lawson T (2019a) Genotypic, developmental and environmental effects on the rapidity of g_s in wheat: impacts on carbon gain and water-use efficiency. Front Plant Sci 10:492

Faralli M, Matthews J, Lawson T (2019b) Exploiting natural variation and genetic manipulation of stomatal conductance for crop improvement. Curr Opin Plant Biol 49:1–7

Faralli M, Williams K, Han J, Corke F, Doonan J, Kettlewell P (2019c) Water-saving traits can protect wheat grain number under progressive soil drying at the meiotic stage: a phenotyping approach. J Plant Growth Regul:1–12

Farooq M, Wahid A, Kobayashi N, Fujita D, Basra SMA (2009) Plant drought stress: effects, mechanisms and management. Agron Sustain Dev 29:185–212

Farquhar GD, Sharkey TD (1982) Stomatal conductance and photosynthesis. Annu Rev Plant Physiol Plant Mol Biol 33:317–345

Fennell A, Markhart AH (1998) Rapid acclimation of root hydraulic conductivity to low temperature. J Exp Bot 49:879–884

Field CB, Jackson RB, Mooney HA (1995) Stomatal responses to increased CO_2 - Implications from the plant to the global scale. Plant Cell Environ 18:1214–1225

Fischer RA, Rees D, Sayre KD, Lu ZM, Condon AG, Saavedra AL (1998) Wheat yield progress associated with higher stomatal conductance and photosynthetic rate, and cooler canopies. Crop Sci 38:1467–1475

Fitzgerald GJ, Tausz M, O'Leary G, Mollah MR, Tausz-Posch S, Seneweera S, Mock I, Low M, …, Norton RM (2016) Elevated atmospheric CO_2 can dramatically increase wheat yields in semi-arid environments and buffer against heat waves. Glob Chang Biol 22:2269–2284

Franks PJ, Beerling DJ (2009) Maximum leaf conductance driven by CO_2 effects on stomatal size and density over geologic time. Proc Natl Acad Sci U S A 106:10343–10347

Franks PJ, Farquhar GD (2001) The effect of exogenous abscisic acid on stomatal development, stomatal mechanics, and leaf gas exchange in Tradescantia virginiana. Plant Physiol 125:935–942

Franks PJ, Farquhar GD (2007) The mechanical diversity of stomata and its significance in gas-exchange control. Plant Physiol 143:78–87

Franks PJ, Doheny-Adams TW, Britton-Harper ZJ, Gray JE (2015) Increasing water-use efficiency directly through genetic manipulation of stomatal density. New Phytol 207:188–195

Gamage D, Thompson M, Sutherland M, Hirotsu N, Makino A, Seneweera S (2018) New insights into the cellular mechanisms of plant growth at elevated atmospheric carbon dioxide concentration. Plant Cell Environ 41:1233–1246

Gay AP, Hurd RG (1975) Influence of light on stomatal density on tomato. New Phytol 75:37–46

Geng SS, Misra BB, De Armas E, Huhman DV, Alborn HT, Sumner LW, Chen SX (2016) Jasmonate-mediated stomatal closure under elevated CO_2 revealed by time-resolved metabolomics. Plant J 88:947–962

Gwray JE, Holroyd GH, van der Lee FM, Bahrami AR, Sijmons PC, Woodward FI, Schuch W, Heterington AM (2000) The HIC signalling pathway links CO_2 perception to stomatal development. Nature 408:713–716

Gray SB, Dermody O, Klein SP, Locke AM, McGrath JM, Paul RE, Rosenthal DM, Ruiz-Vera UM, …, Leakey ADB (2016) Intensifying drought eliminates the expected benefits of elevated carbon dioxide for soybean. Nat Plants 2

Guan XK, Song L, Wang TC, Turner NC, Li FM (2015) Effect of drought on the gas exchange, chlorophyll fluorescence and yield of six different-era spring wheat cultivars. J Agronomy Crop Sci 201:253–266
Hachez C, Milhiet T, Heinen RB, Chaumont F (2017) Roles of Aquaporins in Stomata. Plant Acquaporins. Springer, Netherlands
Han SK, Torii KU (2016) Lineage-specific stem cells, signals and asymmetries during stomatal development. Development 143:1259–1270
Hara K, Kajita R, Torii KU, Bergmann DC, Kakimoto T (2007) The secretory peptide gene EPF1 enforces the stomatal one-cell-spacing rule. Genes Dev 21:1720–1725
Hara K, Yokoo T, Kajita R, Onishi T, Yahata S, Peterson KM, Torii KU, Kakimoto T (2009) Epidermal Cell Density is Autoregulated via a Secretory Peptide, EPIDERMAL PATTERNING FACTOR 2 in Arabidopsis Leaves. Plant Cell Physiol 50:1019–1031
Harb A, Krishnan A, Ambavaram MMR, Pereira A (2010) Molecular and physiological analysis of drought stress in arabidopsis reveals early responses leading to acclimation in plant growth. Plant Physiol 154:1254–1271
Hatfield JL, Prueger JH (2015) Temperature extremes: effect on plant growth and development. Weather Climate Extrem 10:4–10
Hatfield JL, Boote KJ, Kimball BA, Ziska LH, Izaurralde RC, Ort D, Thomson AM, Wolfe D (2011) Climate impacts on agriculture: implications for crop production. Agron J 103:351–370
Hepworth C, Caine RS, Harrison EL, Sloant J, Gray JE (2018) Stomatal development: focusing on the grasses. Curr Opin Plant Biol 41:1–7
Herzog M, Striker GG, Colmer TD, Pedersen O (2016) Mechanisms of waterlogging tolerance in wheat - a review of root and shoot physiology. Plant Cell Environ 39:1068–1086
Hetherington AM, Woodward FI (2003) The role of stomata in sensing and driving environmental change. Nature 424:901–908
Hill KE, Guerin GR, Hill RS, Watling JR (2014) Temperature influences stomatal density and maximum potential water loss through stomata of Dodonaea viscosa subsp angustissima along a latitude gradient in southern Australia. Aus J Botany 62:657–665
Honour SJ, Webb AAR, Mansfield TA (1995) The responses of stomata to abscisic acid and temperature are interrelated. Philos Trans R Soc B Biol Sci 259:301–306
Hu HH, Boisson-Dernier A, Israelsson-Nordstrom M, Bohmer M, Xue SW, Ries A, Godoski J, Kuhn JM, Schroeder JI (2010) Carbonic anhydrases are upstream regulators of CO_2-controlled stomatal movements in guard cells. Nat Cell Biol 12:87–U234
Hu J, Yang QY, Huang W, Zhang SB, Hu H (2014) Effects of temperature on leaf hydraulic architecture of tobacco plants. Planta 240:489–496
Huang BR, Johnson JW, Nesmith S, Bridges DC (1994) Growth, physiological and anatomical responses of 2 wheat genotypes to waterlogging and nutrient supply. J Exp Bot 45:193–202
Hughes J, Hepworth C, Dutton C, Dunn JA, Hunt L, Stephens J, Waugh R, Cameron DD, Gray JE (2017) Reducing stomatal density in barley improves drought tolerance without impacting on yield. Plant Physiol 174:776–787
Hundertmark M, Hincha DK (2008) LEA (Late Embryogenesis Abundant) proteins and their encoding genes in Arabidopsis thaliana. BMC Genomics 9
Hunt L, Gray JE (2009) The signaling peptide EPF2 controls asymmetric cell divisions during stomatal development. Curr Biol 19:864–869
Hussain HA, Hussain S, Khaliq A, Ashraf U, Anjum SA, Men SN, Wang LC (2018) Chilling and drought stresses in crop plants: implications, cross talk, and potential management opportunities. Front Plant Sci 9
Iizumi T, Furuya J, Shen ZH, Kim W, Okada M, Fujimori S, Hasegawa T, Nishimori M (2017) Responses of crop yield growth to global temperature and socioeconomic changes. Sci Rep 7
Ilan N, Moran N, Schwartz A (1995) The role of potassium channels in the temperature control of stomatal aperture. Plant Physiol 108:1161–1170
Inoue S, Kinoshita T (2017) Blue light regulation of stomatal opening and the plasma membrane H+-ATPase. Plant Physiol 174:531–538
IPCC (2014) Climate Change 2014: Synthesis Report. Contribution of Working Groups I, II and III to the Fifth Assessment Report of the Intergovernmental Panel on Climate Change. IPCC, Geneva
Johnsson M, Issaias S, Brogardh T, Johnsson A (1976) Rapid blue-light-induced transpiration response restricted to plants with grass-like stomata. Physiol Plant 36:229–232
Jones HG (1977) Transpiration in barley lines with differing stomatal frequencies. J Exp Bot 28:162–168
Jones BG, Kandel WA (1992) Population-growth, urbanization, and disaster risk and vulnerability in metropolitan areas – a conceptual framework. Environ Management Urban Vulnerability 168:51–76
Jumrani K, Bhatia VS, Pandey GP (2017) Impact of elevated temperatures on specific leaf weight, stomatal density, photosynthesis and chloro-

phyll fluorescence in soybean. Photosynth Res 131:333–350
Kang Y, Outlaw WH, Andersen PC, Fiore GB (2007) Guard-cell apoplastic sucrose concentration – a link between leaf photosynthesis and stomatal aperture size in the apoplastic phloem loader Vicia faba L. Plant Cell Environ 30:551–558
Kellogg EA (2013) C-4 photosynthesis. Curr Biol 23:R594–R599
Kelly G, Moshelion M, David-Schwartz R, Halperin O, Wallach R, Attia Z, Belausov E, Granot D (2013) Hexokinase mediates stomatal closure. Plant J 75:977–988
Kim SH, Sicher RC, Bae H, Gitz DC, Baker JT, Timlin DJ, Reddy VR (2006) Canopy photosynthesis, evapotranspiration, leaf nitrogen, and transcription profiles of maize in response to CO_2 enrichment. Glob Chang Biol 12:588–600
Kimball BA, Kobayashi K, Bindi M (2002) Responses of Agricultural Crops to Free-Air CO_2 Enrichment. Elsevier, Amsterdam
Klay I, Gouia S, Lu M, Mila I, Khoudi H, Bernadac A, Bouzayen M, Pirrello J (2018) Ethylene Response Factors (ERF) are differentially regulated by different abiotic stress types in tomato plants. Plant Sci 274:137–145
Knapp AK (1993) Gas-exchange dynamics in C-3 and C-4 grasses – consequences of differences in stomatal conductance. Ecology 74:113–123
Kumar U, Quick WP, Barrios M, Cruz PCS, Dingkuhn M (2017) Atmospheric CO_2 concentration effects on rice water use and biomass production. Plos One 12
Laanemets K, Brandt B, Li JL, Merilo E, Wang YF, Keshwani MM, Taylor SS, Kollist H, Schroeder JI (2013) Calcium-dependent and -independent stomatal signaling network and compensatory feedback control of stomatal opening via Ca2+ sensitivity priming. Plant Physiol 163:504–513
Lawlor DW (2013) Genetic engineering to improve plant performance under drought: physiological evaluation of achievements, limitations, and possibilities. J Exp Bot 64:83–108
Lawson T, Blatt MR (2014) Stomatal size, speed, and responsiveness impact on photosynthesis and water use efficiency. Plant Physiol 164:1556–1570
Lawson T, Vialet-Chabrand S (2019) Speedy stomata, photosynthesis and plant water use efficiency. New Phytol 221:93–98
Lawson T, Von Caemmerer S (2010) Photosynthesis and stomatal behaviour. In: Baroli I (ed) Progress in Botany. Springer, Netherlands
Lawson T, Weyers J (1999) Spatial and temporal variation in gas exchange over the lower surface of Phaseolus vulgaris L. primary leaves. J Exp Bot 50:1381–1391
Lawson T, Oxborough K, Morison JIL, Baker NR (2003) The responses of guard and mesophyll cell photosynthesis to CO_2, O_2, light, and water stress in a range of species are similar. J Exp Bot 54:1743–1752
Lawson T, Kramer DM, Raines CA (2012) Improving yield by exploiting mechanisms underlying natural variation of photosynthesis. Curr Opin Biotechnol 23:215–220
Lawson T, Simkin AJ, Kelly G, Granot D (2014) Mesophyll photosynthesis and guard cell metabolism impacts on stomatal behaviour. New Phytol 203:1064–1081
Leakey ADB (2009) Rising atmospheric carbon dioxide concentration and the future of C-4 crops for food and fuel. Philos Trans R Soc B Biol Sci 276:2333–2343
Li Y, Ye W, Wang M, Yan X (2009) Climate change and drought: a risk assessment of crop-yield impacts. Climate Res 39:31–46
Li XN, Cai J, Liu FL, Dai TB, Cao WX, Jiang D (2014) Physiological, proteomic and transcriptional responses of wheat to combination of drought or waterlogging with late spring low temperature. Funct Plant Biol 41:690–703
Lipiec J, Doussan C, Nosalewicz A, Kondracka K (2013) Effect of drought and heat stresses on plant growth and yield: a review. Int Agrophysic 27:463–477
Lu ZM, Quinones MA, Zeiger E (2000) Temperature dependence of guard cell respiration and stomatal conductance co-segregate in an F-2 population of Pima cotton. Aus J Plant Physiol 27:457–462
Macholdt J, Honermeier B (2016) Variety choice in crop production for climate change adaptation: farmer evidence from Germany. Outlook Agriculture 45:117–123
Maclean J, Hardy B, Hettel G (2013) Rice Almanac: Source Book for One of the Most Important Economic Activities on Earth. IRRI, Los Baños
Mansfield TA, Hetherington AM, Atkinson CJ (1990) Some current aspects of stomatal physiology. Annu Rev Plant Physiol Plant Mol Biol 41:55–75
Matthews JSA, Vialet-Chabrand S, Lawson T (2018) Acclimation to fluctuating light impacts the rapidity of response and diurnal rhythm of stomatal conductance. Plant Physiol 176:1939–1951

McAdam SAM, Brodribb TJ (2014) Separating active and passive influences on stomatal control of transpiration. Plant Physiol 164:1578–1586

McAdam SAM, Brodribb TJ (2015) The evolution of mechanisms driving the stomatal response to vapor pressure deficit. Plant Physiol 167:833–843

McAusland L, Vialet-Chabrand S, Davey P, Baker NR, Brendel O, Lawson T (2016) Effects of kinetics of light-induced stomatal responses on photosynthesis and water-use efficiency. New Phytol 211:1209–1220

McElwain JC, Chaloner WG (1995) Stomatal density and index of fossil plants track atmospheric carbon dioxide in the paleozoic. Ann Bot 76:389–395

McGranahan DA, Poling BN (2018) Trait-based responses of seven annual crops to elevated CO_2 and water limitation. Renew Agriculture Food System 33:259–266

McGuire S (2013) WHO, World Food Programme, and International Fund for Agricultural Development. 2012. The State of Food Insecurity in the World 2012. Economic growth is necessary but not sufficient to accelerate reduction of hunger and malnutrition. Rome, FAO. Adv Nutrition 4:126–127

Meehl GA, Covey C, Delworth T, Latif M, Mcavaney B, Mitchell JFB, Stouffer RJ, Taylor KE (2007) The WCRP CMIP3 multimodel dataset – a new era in climate change research. Bull Am Meteorol Soc 88:1383–1394

Melkonian J, Yu LX, Setter TL (2004) Chilling responses of maize (Zea mays L.) seedlings: root hydraulic conductance, abscisic acid, and stomatal conductance. J Exp Bot 55:1751–1760

Merilo E, Jalakas P, Kollist H, Brosche M (2015) The role of ABA recycling and transporter proteins in rapid stomatal responses to reduced air humidity, elevated CO_2, and exogenous ABA. Mol Plant 8:657–659

Merilo E, Yarmolinsky D, Jalakas P, Parik H, Tulva I, Rasulov B, Kilk K, Kollist H (2018) Stomatal VPD response: there is more to the story than ABA. Plant Physiol 176:851–864

Mohammed U, Caine RS, Atkinson JA, Harrison EL, Wells D, Chater CC, Gray JE, Swarup R, Murchie EH (2019) Rice plants overexpressing OsEPF1 show reduced stomatal density and increased root cortical aerenchyma formation. Sci Rep 9

Morison JIL, Baker NR, Mullineaux PM, Davies WJ (2008) Improving water use in crop production. Philos Trans R Soc B Biol Sci 363:639–658

Mott KA (1988) Do stomata respond to CO_2 concentrations other than intercellular. Plant Physiol 86:200–203

Mott KA, Peak D (2013) Testing a vapour-phase model of stomatal responses to humidity. Plant Cell Environ 36:936–944

Mueller ND, Gerber JS, Johnston M, Ray DK, Ramankutty N, Foley JA (2012) Closing yield gaps through nutrient and water management. Nature 490:254–257

Munns R, James RA, Sirault XRR, Furbank RT, Jones HG (2010) New phenotyping methods for screening wheat and barley for beneficial responses to water deficit. J Exp Bot 61:3499–3507

Murray RR, Emblow MSM, Hetherington AM, Foster GD (2016) Plant virus infections control stomatal development. Sci Rep 6

Nadeau JA, Sack FD (2002) Stomatal development in Arabidopsis. The Arabidopsis Book 1:e0066–e0066

Najeeb U, Bange MP, Tan DKY, Atwell BJ (2015) Consequences of waterlogging in cotton and opportunities for mitigation of yield losses. Aob Plants 7

Narayanan S, Prasad PVV, Fritz AK, Boyle DL, Gill BS (2015) Impact of high night-time and high daytime temperature stress on winter wheat. J Agronomy Crop Sci 201:206–218

Negin B, Moshelion M (2017) The advantages of functional phenotyping in pre-field screening for drought-tolerant crops. Funct Plant Biol 44:107–118

Oren R, Sperry JS, Katul GG, Pataki DE, Ewers BE, Phillips N, Schafer KVR (1999) Survey and synthesis of intra- and interspecific variation in stomatal sensitivity to vapour pressure deficit. Plant Cell Environ 22:1515–1526

Osborne CP (2016) Crop yields CO_2 fertilization dries up. Nature Plants 2

Outlaw WH (2003) Integration of cellular and physiological functions of guard cells. Crit Rev Plant Sci 22:503–529

Pantin F, Monnet F, Jannaud D, Costa JM, Renaud J, Muller B, Simonneau T, Genty B (2013) The dual effect of abscisic acid on stomata. New Phytol 197:65–72

Pardossi A, Vernieri P, Tognoni F (1992) Involvement of abscisic acid in regulating water status in Phaseolus vulgaris L during chilling. Plant Physiol 100:1243–1250

Pasternak D, Wilson GL (1972) Aftereffects of night temperatures on stomatal behaviour and photosynthesis of sorghum. New Phytol 71:683

Peng SB, Huang JL, Sheehy JE, Laza RC, Visperas RM, Zhong XH, Centeno GS, Khush GS, Cassman KG (2004) Rice yields decline with higher night temperature from global warming. Proc Natl Acad Sci U S A 101:9971–9975

Peterhansel C, Horst I, Niessen M, Blume C, Kebeish R, Kürkcüoglu S, Kreuzaler F (2010) Photorespiration. The Arabidopsis Book. Am SocPlant Biol, Rockville.

Peterson KM, Rychel AL, Torii KU (2010) Out of the mouths of plants: the molecular basis of the evolution and diversity of stomatal development. Plant Cell 22:296–306

Pillitteri LJ, Peterson KM, Horst RJ, Torii KU (2011) Molecular profiling of stomatal meristemoids reveals new component of asymmetric cell division and commonalities among stem cell populations in Arabidopsis. Plant Cell 23:3260–3275

Ploschuk RA, Grimoldi AA, Ploschuk EL, Striker GG (2017) Growth during recovery evidences the waterlogging tolerance of forage grasses. Crop Pasture Sci 68:574–582

Ploschuk RA, Miralles DJ, Colmer TD, Ploschuk EL, Striker GG (2018) Waterlogging of winter crops at early and late stages: impacts on leaf physiology, growth and yield. Front Plant Sci 9

Pociecha E, Koscielniak J, Filek W (2008) Effects of root flooding and stage of development on the growth and photosynthesis of field bean (Vicia faba L. minor). Acta Physiologiae Plantarum 30:529–535

Poole I, Lawson T, Weyers JDB, Raven JA (2000) Effect of elevated CO_2 on the stomatal distribution and leaf physiology of Alnus glutinosa. New Phytol 145:511–521

Porter JR, Semenov MA (2005) Crop responses to climatic variation. Philos Trans R Soc B Biol Sci 360:2021–2035

Prasad PVV, Pisipati SR, Ristic Z, Bukovnik U, Fritz AK (2008) Impact of nighttime temperature on physiology and growth of spring wheat. Crop Sci 48:2372–2380

Qu MN, Hamdani S, Li WZ, Wang SM, Tang JY, Chen Z, Song QF, Li M, …, Zhu XG (2016) Rapid stomatal response to fluctuating light: an under-explored mechanism to improve drought tolerance in rice. Funct Plant Biol 43:727–738

Raissig MT, Matos JL, Gil MXA, Kornfeld A, Bettadapur A, Abrash E, Allison HR, Badgley G, …, Bergmann DC (2017) Mobile MUTE specifies subsidiary cells to build physiologically improved grass stomata. Science 355:1215–1218

Raven JA (1977) The Evolution of Vascular Land Plants in Relation to Supracellular Transport Processes. Academic Press, London

Raven JA (2002) Selection pressures on stomatal evolution. New Phytol 153:371–386

Raven JA (2014) Speedy small stomata? J Exp Bot 65:1415–1424

Ray DK, Mueller ND, West PC, Foley JA (2013) Yield trends are insufficient to double global crop production by 2050. Plos One 8

Rizhsky L, Liang HJ, Mittler R (2002) The combined effect of drought stress and heat shock on gene expression in tobacco. Plant Physiol 130:1143–1151

Rizhsky L, Liang HJ, Shuman J, Shulaev V, Davletova S, Mittler R (2004) When defense pathways collide. The response of Arabidopsis to a combination of drought and heat stress. Plant Physiol 134:1683–1696

Roelfsema MRG, Hedrich R (2005) In the light of stomatal opening: new insights into 'the Watergate'. New Phytol 167:665–691

Rogiers SY, Hardie WJ, Smith JP (2011) Stomatal density of grapevine leaves (Vitis vinifera L.) responds to soil temperature and atmospheric carbon dioxide. Aus J Grape Wine Res 17:147–152

Rollins JA, Habte E, Templer SE, Colby T, Schmidt J, Von Korff M (2013) Leaf proteome alterations in the context of physiological and morphological responses to drought and heat stress in barley (Hordeum vulgare L.). J Exp Bot 64:3201–3212

Rudall PJ, Hilton J, Bateman RM (2013) Several developmental and morphogenetic factors govern the evolution of stomatal patterning in land plants. New Phytol 200:598–614

Sachs T (1991) Pattern Formation in Plant Tissues. Cambridge University Press, Cambridge

Sage RF, Kubien DS (2007) The temperature response of C-3 and C-4 photosynthesis. Plant Cell Environ 30:1086–1106

Sage RF, Sharkey TD (1987) The effect of temperature on the occurrence of O_2 and CO_2 insensitive photosynthesis in field-grown plants. Plant Physiol 84:658–664

Schafer N, Maierhofer T, Herrmann J, Jorgensen ME, Lind C, Von Meyer K, Lautner S, Fromm J, …, Hedrich R (2018) A Tandem Amino Acid Residue Motif in Guard Cell SLAC1 Anion Channel of Grasses Allows for the Control of Stomatal Aperture by Nitrate. Curr Biol 28:1370–U145

Serna L, Torres-Contreras J, Fenoll C (2002) Clonal analysis of stomatal development and patterning in Arabidopsis leaves. Dev Biol 241:24–33

Shi W, Yin X, Struik PC, Xie F, Schmidt RC, Jagadish KSV (2016) Grain yield and quality responses of tropical hybrid rice to high night-time temperature. Field Crop Res 190:18–25

Shimizu H, Katayama K, Koto T, Torii K, Araki T, Endo M (2015) Decentralized circadian clocks process thermal and photoperiodic cues in specific tissues. Nature Plants 1:15163

Skelton RP, West AG, Dawson TE (2015) Predicting plant vulnerability to drought in biodiverse regions using functional traits. Proc Natl Acad Sci U S A 112:5744–5749

Sugano SS, Shimada T, Imai Y, Okawa K, Tamai A, Mori M, Hara-Nishimura I (2010) Stomagen positively regulates stomatal density in Arabidopsis. Nature 463:241–U130

Sussmilch FC, Roelfsema MRG, Hedrich R (2019) On the origins of osmotically driven stomatal movements. New Phytol 222:84–90

Taiz L, Zeiger E, Møller IM, Murphy AS (2018) Plant Physiology and Development. Oxford University Press, Oxford

Teskey R, Wertin T, Bauweraerts I, Ameye M, McGuire MA, Steppe K (2015) Responses of tree species to heat waves and extreme heat events. Plant Cell Environ 38:1699–1712

Urban J, Ingwers M, McGuire MA, Teskey RO (2017a) Stomatal conductance increases with rising temperature. Plant Signal Behav 12:e1356534

Urban J, Ingwers MW, Mcguire MA, Teskey RO (2017b) Increase in leaf temperature opens stomata and decouples net photosynthesis from stomatal conductance in Pinus taeda and Populus deltoides x nigra. J Exp Bot 68:1757–1767

Vahisalu T, Kollist H, Wang YF, Nishimura N, Chan WY, Valerio G, Lamminmaki A, Brosche M, …, Kangasjarvi J (2008) SLAC1 is required for plant guard cell S-type anion channel function in stomatal signalling. Nature 452:487–U15

Valladares F, Martinez-Ferri E, Balaguer L, Perez-Corona E, Manrique E (2000) Low leaf-level response to light and nutrients in Mediterranean evergreen oaks: a conservative resource-use strategy? New Phytol 148:79–91

Van Ittersum MK, Cassman KG, Grassini P, Wolf J, Tittonell P, Hochman Z (2013) Yield gap analysis with local to global relevance – a review. Field Crop Res 143:4–17

Vaten A, Bergmann DC (2012) Mechanisms of stomatal development: an evolutionary view. Evodevo 3

Vialet-Chabrand S, Dreyer E, Brendel O (2013) Performance of a new dynamic model for predicting diurnal time courses of stomatal conductance at the leaf level. Plant Cell Environ 36:1529–1546

Vialet-Chabrand S, Matthews JSA, Brendel O, Blatt MR, Wang Y, Hills A, Griffiths H, Rogers S, Lawson T (2016) Modelling water use efficiency in a dynamic environment: an example using Arabidopsis thaliana. Plant Sci 251:65–74

Vialet-Chabrand S, Hills A, Wang YZ, Griffiths H, Lew VL, Lawson T, Blatt MR, Rogers S (2017a) Global sensitivity analysis of Onguard models identifies key hubs for transport interaction in stomatal dynamics. Plant Physiol 174:680–688

Vialet-Chabrand S, Matthews JSA, Simkin AJ, Raines CA, Lawson T (2017b) Importance of fluctuations in light on plant photosynthetic acclimation. Plant Physiol 173:2163–2179

Vialet-Chabrand SRM, Matthews JSA, McAusland L, Blatt MR, Griffiths H, Lawson T (2017c) Temporal dynamics of stomatal behavior: modeling and implications for photosynthesis and water use. Plant Physiol 174:603–613

Vile D, Pervent M, Belluau M, Vasseur F, Bresson J, Muller B, Granier C, Simonneau T (2012) Arabidopsis growth under prolonged high temperature and water deficit: independent or interactive effects? Plant Cell Environ 35:702–718

Von Caemmerer S, Evans JR (2015) Temperature responses of mesophyll conductance differ greatly between species. Plant Cell Environ 38:629–637

Waadt R, Manalansan B, Rauniyar N, Munemasa S, Booker MA, Brandt B, Waadt C, Nusinow DA, …, Schroeder JI (2015) Identification of Open Stomatal-Interacting Proteins Reveals Interactions with Sucrose Non-fermenting1-Related Protein Kinases2 and with Type 2A Protein Phosphatases That Function in Abscisic Acid Responses. Plant Physiol 169:760

Wan JX, Griffiths R, Ying JF, Mccourt P, Huang YF (2009) Development of drought-tolerant canola (Brassica napus L.) through genetic modulation of ABA-mediated stomatal responses. Crop Sci 49:1539–1554

Wang Y, Noguchi K, Terashima I (2008) Distinct light responses of the adaxial and abaxial stomata in intact leaves of Helianthus annuus L. Plant Cell Environ 31:1307–1316

Wang LJ, Li XF, Chen SY, Liu GS (2009) Enhanced drought tolerance in transgenic Leymus chinensis plants with constitutively expressed wheat TaLEA (3). Biotechnol Lett 31:313–319

Welch JR, Vincent JR, Auffhammer M, Moya PF, Dobermann A, Dawe D (2010) Rice yields in tropical/subtropical Asia exhibit large but opposing sensitivities to minimum and maximum temperatures. Proc Natl Acad Sci U S A 107:14562–14567

Weyers JDB, Paterson NW (1987) Responses of Commelina communis stomata in vitro. J Exp Bot 38:631–641

Wilkinson S, Clephan AL, Davies WJ (2001) Rapid low temperature-induced stomatal closure occurs in cold-tolerant Commelina communis leaves but not in cold-sensitive tobacco leaves, via a mechanism

that involves apoplastic calcium but not abscisic acid. Plant Physiol 126:1566–1578
Willmer C, Fricker M (1996) Stomata. Springer, Netherlands
Wolfe DW (1991) Low-temperature effects on early vegetative growth, leaf gas-exchange and water potential of chilling-sensitive and chilling-tolerant crop species. Ann Bot 67:205–212
Wong SC, Cowan IR, Farquhar GD (1979) Stomatal conductance correlates with photosynthetic capacity. Nature 282:424–426
Woodward FI (1987) Stomatal Numbers are sensitive to increases in CO_2 from preindustrial levels. Nature 327:617–618
Woodward FI, Kelly CK (1995) The influences of CO_2 concentration on stomatal density. New Phytol 131:311–327
Xu Z, Jiang Y, Jia B, Zhou G (2016) Elevated-CO_2 response of stomata and its dependence on environmental factors. Front Plant Sci 7:657
Xue SW, Hu HH, Ries A, Merilo E, Kollist H, Schroeder JI (2011) Central functions of bicarbonate in S-type anion channel activation and OST1 protein kinase in CO_2 signal transduction in guard cell. EMBO J 30:1645–1658
Yamamoto Y, Negi J, Wang C, Isogai Y, Schroeder JI, Iba K (2016) The transmembrane region of guard cell SLAC1 channels perceives CO_2 signals via an ABA-independent pathway in Arabidopsis. Plant Cell 28:557–567
Zandalinas SI, Mittler R, Balfagon D, Arbona V, Gomez-Cadenas A (2018) Plant adaptations to the combination of drought and high temperatures. Physiol Plant 162:2–12
Zhang DY, Chen GY, Gong ZY, Chen J, Yong ZH, Zhuo JG, Xu DQ (2008) Ribulose-1,5-bisphosphate regeneration limitation in rice leaf photosynthetic acclimation to elevated CO_2. Plant Sci 175:348–355
Zhao C, Liu B, Piao SL, Wang XH, Lobell DB, Huang Y, Huang MT, Yao YT, …, Asseng S (2017) Temperature increase reduces global yields of major crops in four independent estimates. Proc Natl Acad Sci USA 114:9326–9331
Zheng YP, Xu M, Hou RX, Shen RC, Qiu S, Ouyang Z (2013) Effects of experimental warming on stomatal traits in leaves of maize (Zea may L.). Ecol Evol 3:3095–3111
Zhu XC, Cao QJ, Sun LY, Yang XQ, Yang WY, Zhang H (2018) Stomatal Conductance and Morphology of Arbuscular Mycorrhizal Wheat Plants Response to_Elevated CO_2 and NaCl Stress. Front Plant Sci 9
Ziegler H (1987) The evolution of stomata. Stomatal Function 29

Chapter 3

Mesophyll Conductance to CO_2 Diffusion in a Climate Change Scenario: Effects of Elevated CO_2, Temperature and Water Stress

Miquel Nadal and Jaume Flexas*
Research Group in Plant Biology under Mediterranean Conditions, Universitat de les Illes Balears – Instituto de Agroecología y Economía del Agua (INAGEA), Palma, Illes Balears, Spain

and

Marc Carriquí
Research Group in Plant Biology under Mediterranean Conditions, Universitat de les Illes Balears – Instituto de Agroecología y Economía del Agua (INAGEA), Palma, Illes Balears, Spain

School of Natural Sciences, University of Tasmania, Hobart, TAS, Australia

*Author for correspondence, e-mail: jaume.flexas@uib.es

K. M. Becklin et al. (eds.), *Photosynthesis, Respiration, and Climate Change*, Advances in Photosynthesis and Respiration 48, https://doi.org/10.1007/978-3-030-64926-5_3

Summary

Mesophyll conductance to CO_2 (g_m) is a key component of leaf photosynthesis: it describes the inverse of the resistance of the CO_2 diffusion from the substomatal cavity to the chloroplast, and thus it is highly determined by anatomical and biochemical features along that path. Over the years, g_m has been acknowledged to be a key factor in limiting photosynthesis across plant groups, explaining part of the interspecific variation of photosynthetic capacity. However, g_m has been scarcely considered in photosynthesis studies under a climate change scenario, especially regarding its response to [CO_2] and temperature acclimation. Nonetheless, increasing evidence highlights the importance of g_m in determining photosynthesis response to those factors, and of considering g_m for further modelling of plant performance under future climate conditions. This chapter focuses on both short- and long-term responses of g_m – and its effect on net CO_2 assimilation – to [CO_2] and temperature, among other climate change-related factors (water stress, ozone and nutrient availability), with special emphasis on the mechanistic basis that drive the g_m response. Despite few studies reporting data on them, cell wall thickness, chloroplast distribution and aquaporins appear to be the key underlying factors of g_m response to environmental changes. Nonetheless, g_m response seems to be a species-specific feature, highlighting current uncertainties in the prediction of photosynthetic responses under future climate scenarios and the need to account for the mechanics and variability of g_m in future research.

I. Introduction to Mesophyll Conductance

At the current CO_2 atmospheric concentration (ca. 400 ppm), photosynthesis in C_3 plants is a non-saturated process, partially limited by CO_2 availability in the chloroplast. Whereas stomatal conductance (g_s) was thought to be the only factor constraining CO_2 diffusion in the original Farquhar, von Caemmerer and Berry (FvCB) model of C_3 net CO_2 assimilation (A_n; Farquhar et al.

Abbreviations: A_n – Net CO_2 assimilation rate; C_a – CO_2 concentration in the atmosphere; C_c – CO_2 concentration at the sites of carboxylation; C_i – CO_2 concentration in the inter-cellular air spaces; FACE – Free-Air CO_2 Enrichment; g_{liq} – Conductance to CO_2 diffusion through the liquid phase (cell wall, cytoplasm and chloroplast stroma); g_m – Mesophyll conductance to CO_2 diffusion; g_{mem} – Conductance to CO_2 diffusion across biological membranes (plasmalemma and chloroplast envelopes); g_s – Stomatal conductance to CO_2 diffusion; J – Electron transport rate; J_{max} – Maximum electron transport rate; LMA – Leaf mass per area; NRH – Non-rectangular hyperbolic; OTC – Open Top Chamber; S_c/S – Ratio of exposed chloroplasts to mesophyll surface areas; S_m/S – Ratio of exposed mesophyll cells to leaf intercellular space; T_{cw} – Cell wall thickness; TDLAS – Tunable Diode Laser Absorption Spectroscopy V_{cmax} – Maximum Rubisco carboxylation rate; WTC – Whole-tree chamber

1980), nowadays CO_2 diffusion conductance through the leaf from the substomatal cavity to the sites of carboxylation within the chloroplast (i.e., mesophyll conductance to CO_2, g_m), is acknowledged as an additional constraint to CO_2 diffusion (Flexas et al. 2008; Warren 2008a). g_m can be divided into several components, each accounting for the different steps CO_2 molecules go through until reaching the enzyme Rubisco (ribulose-1,5-bisphosphate carboxylase/oxygenase). It is worth noting that this definition is slightly different when considering g_m in C_4 plants, where it accounts for CO_2 transport from the intercellular air spaces to the mesophyll cell cytosol (Evans and von Caemmerer 1996). In C_3 plants, g_m encompasses both gaseous and aqueous phases, the former represented by the intercellular air spaces of the leaf mesophyll (Evans et al. 1994; Earles et al. 2018), and the latter accounting for the diffusion of CO_2 across cell walls, lipid membranes, and the cytosolic and chloroplastic stroma (Evans et al. 2009). The aqueous phase usually accounts for a large fraction of g_m, as cell wall characteristics – mainly cell wall thickness (T_{cw}) and chloroplast surface area exposed to intercellular airspaces per unit of leaf area (S_c/S) – have been widely reported to be the main anatomical drivers of g_m (Terashima et al. 2011; Tomàs et al. 2013; Onoda et al. 2017; Flexas et al. 2018a). These features partially explain photosynthstic variability across plant groups, where low g_m species usually present thick cell walls and low chloroplast exposure (Carriquí et al. 2015; Tosens et al. 2016; Veromann-Jürgenson et al. 2017; Carriquí et al. 2019b). However, g_m also varies in response to short-term stimuli such as water stress, temperature, CO_2 and light irradiance (see Flexas et al. 2008, 2018a); these short-term variations are partially attributed to the biochemical components of g_m. Some aquaporins (or 'cooporins') facilitate CO_2 diffusion through lipid membranes (Hanba et al. 2004; Flexas et al. 2006), whereas carbonic anhydrases catalyze the conversion of CO_2 to bicarbonate, enhancing CO_2 gradients along the CO_2 diffusion pathway, and consequently facilitating CO_2 diffusion (Flexas et al. 2018a). Changes in the activity and/or expression of aquaporins and carbonic anhydrases may explain g_m decreases in response to abiotic stress (Pérez-Martin et al. 2014; Han et al. 2016).

Mesophyll conductance is a rather difficult parameter to estimate, as current methods cannot directly measure the CO_2 concentration inside the chloroplast. Thus, g_m studies rely on the combination of several methods and assumptions for determining the nature and dynamics of g_m (see Pons et al. 2009 and Flexas et al. 2018a for a complete depiction of the methods). Briefly, the currently accepted methods are the variable and constant J, based on coupled measurements of gas exchange and electron transport rate (J) using chlorophyll fluorescence (Harley et al. 1992); curve-fitting, constraining fitted to observed values from A_n-C_i curves (Ethier and Livingston 2004; Sharkey et al. 2007); isotope discrimination, which accounts for the $^{13}CO_2/^{12}CO_2$ discrimination across photosynthetic tissues (Evans et al. 1986); anatomical methods, to determine g_m from the ultrastructural features of the leaf mesophyll (Niinemets and Reichstein 2003; Tomàs et al. 2013); and reaction-diffusion models, which are partially based on mass conservation laws and consider additional factors such as the 3D structure of the mesophyll or the re-assimilation of (photo)respired CO_2 (Ho et al. 2016; Berghuijs et al. 2018). All current methods present their own assumptions; consequently, precise g_m measurements require the use of two or more methods to account for potential artifacts, especially when measuring under non-optimal conditions (Galmés et al. 2006; Tholen et al. 2012; Gu and Sun 2014).

The nature, estimation methods, mechanistic basis and variability of g_m have been recently reviewed (see Flexas et al. 2018a). In this chapter, we will focus on the response of g_m to the main climate change components (CO_2, temperature, water availability) and its importance in both photosynthesis response to those factors and for accurately modelling plant response to future climate change scenarios.

II. Climate Change: CO_2

A. Short Term Response

1. Controversies Regarding g_m Changes Under Variable $[CO_2]$

Photosynthesis is very sensitive to short-term variations in environmental stimuli, particularly changes in $[CO_2]$. It is widely accepted that the regulation of stomatal aperture, allowing or restricting the entry of CO_2 into the substomatal cavities, is the major cause of the CO_2 sensitivity of A_n. In most angiosperms, g_s decreases when the leaf is exposed to short-term $[CO_2]$ increases (Brodribb et al. 2009). In other vascular plant groups, the presence of mechanisms for detecting and responding to increases in $[CO_2]$ is less certain (Brodribb and McAdam 2013). On the other hand, the short-term response of g_m to a rise in CO_2, and the subsequent effects on A_n in C_3 plants are not so clear. Both the first reports of the short-term response of g_m to $[CO_2]$ fluctuations (Harley et al. 1992; Loreto et al. 1992) and later studies (Flexas et al. 2007; Hassiotou et al. 2009; Yin et al. 2009; Douthe et al. 2011, 2012; Xiong et al. 2015; Carriquí et al. 2019a; Mizokami et al. 2019a), reported a several-fold change in g_m along the range of substomatal $[CO_2]$ concentrations (C_i), with g_m decreasing as C_i increases, but with a species-dependent pattern. However, the veracity of these g_m responses are currently strongly questioned since: (1) some studies reported g_m to be stable in response to changes in $[CO_2]$ (von Caemmerer and Evans 1991; Tazoe et al. 2009); (2) several researchers have suggested that the observed variation might be a methodological artefact (Tholen et al. 2012; Gu and Sun 2014); and (3) there is no consensus on the mechanistic basis that explains the data (Xiong et al. 2015; Carriquí et al. 2019a). Most structural features that constrain g_m (e.g., leaf or cell wall thickness) are considered to be constant in the short-term (Evans et al. 2009; Terashima et al. 2011), and other anatomical drivers (such as chloroplast arrangement, one of the determinants of S_c/S) have not been found to change despite the observed short-term variations in g_m (Carriquí et al. 2019a). Overall, these studies suggest that the reported variation of g_m to $[CO_2]$ is at least partially artefactual.

2. Mesophyll Conductance in C_4 Plants: Response to $[CO_2]$

As mentioned earlier, g_m in C_4 plants describes a shorter pathway than in C_3 plants: it includes the CO_2 path from the intercellular air spaces to the cytosol of mesophyll cells through mesophyll cell wall and plasmalemma (Evans and von Caemmerer 1996; Barbour et al. 2016). Very few reports of g_m in C_4 species are available due to the complexity of C_4 photosynthesis modelling (Bellasio et al. 2016) and its low absolute biochemical fractionation, which leads to inadequate carbon isotope discrimination resolution for g_m estimation (von Cammerer 2013). However, in recent years the incorporation of oxygen isotope measurements of CO_2 and H_2O has allowed for reliable estimations of g_m in C_4 species (Barbour et al. 2016; Ogée et al. 2018). This new method is based on the isotopic exchange between CO_2 and H_2O during the reaction catalyzed by carbonic anhydrases, and thus allows estimation of CO_2 partial pressure in the cytosol of mesophyll cells (Barbour et al. 2016). New studies on C_4 species have shown that C_4 plants present g_m values at the higher end of those reported for C_3 plants (Barbour et al. 2016; Ubierna et al. 2018). The short-term response of g_m to rapid changes in $[CO_2]$ in the C_4 species *Zea mays* and *Setaria viridis* follows a similar response to that reported in C_3 plants, with a significant increase (up to four-fold) in response to low C_i conditions (Osborn et al. 2017; Kolbe and Cousins 2018; Ubierna et al. 2018) and a slight decrease under high C_i (Ubierna et al. 2018). Nonetheless, it is thought that g_m is a significant factor – together with carbonic anhydrase activity – in limiting net CO_2 assimilation under low C_i (Osborn et al. 2017; Ubierna et al. 2018). Using three different modelling approaches for estimating g_m, Kolbe and Cousins (2018) point towards an important role of carbonic anhydrases and phosphoenolpyruvate carboxylase activity in determining g_m, as well as some discrepancies between estimations that may reflect the "incomplete chemical and isotopic equilibrium between CO_2 and bicarbonate" under low C_i. Similar approaches are needed for further understanding of the potential dynamic nature of g_m in C_4 plants.

B. Long Term Impact of [CO_2]

1. Photosynthesis Under Increased [CO_2]

Carbon dioxide constitutes the primary substrate for photosynthesis. Consequently, the response of photosynthesis to elevated $[CO_2]$ is crucial to understand the response of plants and overall ecosystem primary production to future climate change scenarios. A great number of studies have evaluated the potential 'fertilization' effect on the main photosynthesis components (A_n, g_s and biochemistry – Rubisco and chlorophyll activity and/or content). C_3 plants, when exposed to high levels of CO_2 (500–700 ppm) during a growing period (>1 month) show a significant increase in A_n when measured at growth $[CO_2]$, ranging from 5% to 50% depending on plant type (Saxe et al. 1998; Long et al. 2004; Ainsworth and Rogers 2007). However, when measured at either ambient (350–400 ppm – same as 'control' treatments) or saturated (>1500 ppm) $[CO_2]$, plants grown under elevated $[CO_2]$ usually show a down-regulated assimilation capacity compared to ambient CO_2-grown plants (Smith et al. 2012; Crous et al. 2013; Meng et al. 2013; Lewis et al. 2015). Indeed, most studies report a decrease in the main mechanisms underlying photosynthesis: g_s (by 15–35%), maximum velocity of Rubisco carboxylation (V_{cmax}; by around 15%) and Rubisco content (by 5–35%) and activity (Medlyn et al. 2001; Long et al. 2004; Aranjuelo et al. 2011, 2013; Meng et al. 2013; Sharwood et al. 2017).

2. Mesophyll Conductance Response to Increased [CO_2]

g_m Studies: Main Response to [CO_2]

While stomatal and biochemical responses to elevated $[CO_2]$ have been largely studied, g_m is not usually included in studies aiming to understand photosynthetic acclimation to elevated CO_2. So far, a total of only 20 studies of 18 species have reported the response of g_m to high $[CO_2]$ – most of them being herbaceous annual crops (Table 3.1). The first study reporting g_m in plants grown at high $[CO_2]$ (from both chamber and FACE environments), performed by Singsaas et al. (2003) using chlorophyll fluorescence techniques, did not observe any significant modification in most of the studied species except for *Cucumis sativus*, which presented a decreased g_m. However, Singsaas et al. (2003) emphasized the necessity of incorporating g_m in high CO_2 studies to provide

Table 3.1. Mesophyll conductance response to elevated $[CO_2]$. List of studied species

Species	Group[a]/growth form	Design[b]	$[CO_2]$ (ppm)	Study duration	Combined treatments	g_m estimation method	g_m response to ↑ $[CO_2]$[c]	Mechanisms	References
Cucumis sativus	A/annual herb crop	Chamber	360, 700	5–7 weeks	–	Constant J	↓		Singsaas et al. (2003)
Spinacia oleracea	A/annual herb crop	Chamber	360, 700	5–7 weeks	–	Constant J	=		Singsaas et al. (2003)
Phaseolus vulgaris	A/annual herb crop	Chamber	360, 700	5–7 weeks	–	Constant J	=		Singsaas et al. (2003)
Populus tremuloides	A/deciduous tree	Face	360, 560	3 years	–	Constant J	=		Singsaas et al. (2003)
Liquidambar styraciflua	A/deciduous tree	Face	360, 560	3 years	Sun/shade leaves	Constant J	↑ (sun) = (shade)		Singsaas et al. (2003)
Glycine max	A/annual herb crop	Face	370, 550	3–4 months	Time (DOY)	Variable J	=		Bernacchi et al. (2005)
Acacia nigrescens	A/deciduous tree	Chamber	200, 390, 600	120 days	–	Curve fitting	= (high), ↑ (low)		Possell and Hewitt (2009)
Platanus orientalis	A/deciduous tree	Chamber	380, 800	1 month	New/pre-existing leaves, heat stress	Constant J	↓ (both new and pre-existing) = (heat)		Velikova et al. (2009)
Vitis vinifera	A/deciduous vine	Greenhouse	375, 700	2, 4 weeks	+4 °C, water stress	Variable J	↓ (14 d) = (28 d) = (WS)		Salazar-Parra et al. (2012)
Oryza sativa	A/annual herb crop	Face	400, 600	Not specified[d]	Leaf age	Curve fitting	↓ (advanced stages)	↑ T_{cw}	Zhu et al. (2012)
Triticum aestivum	A/annual herb crop	Face	400, 600	Not specified[d]	Leaf age	Curve fitting	=	= T_{cw}	Zhu et al. (2012)
Eucalyptus globulus	A/evergreen tree	WTC	400, 640	10–11 months	–	TDLAS[e]	=		Crous et al. (2013)
Gossypium hirsutum	A/annual herb crop	Chamber	400, 800	35–40 days	Nutrients (P)	Constant J	=		Singh et al. (2013)
Nicotiana tabacum	A/annual herb crop	Chamber	400, 800	3 weeks	Thioredoxin *f* OX lines	Variable J	↓ (WT) = (OX)		Aranjuelo et al. (2015)
Betula platyphylla	A/deciduous tree	Chamber	360, 720	5 weeks	Nutrients (N)	Variable J, curve fitting	= (high N) ↓ (low N)	↑ S_c/S	Kitao et al. (2015)

Wollemia nobilis	G/evergreen tree	Greenhouse	290, 400, 650	17 months	+ 4 °C	Online isotope discrimination	↓ (high) = (low) = (T)		Lewis et al. (2015)
Vitis vinifera	A/deciduous vine	Greenhouse	400, 700	6 months	+ 4 °C, UV-B radiation, stages	Variable J	↓ (veraison) ↓ (UV-B)		Martínez-Lüscher et al. (2015)
Coffea arabica	A/evergreen tree crop	Face	390, 550	18–24 months	Cultivar, season	Variable J, curve fitting	= (cv) = (season)		DaMatta et al. (2016)
Glycine max	A/annual herb crop	Chamber	400, 800	48 days	Nutrients (P)	Various[f]	↓ (no P interaction)		Singh and Reddy (2016)
Glycine max	A/annual herb crop	Chamber	400, 800	46–50 days	Temperature (22, 28, 36 °C)	Variable J	= (no temp interaction)		Xue et al. (2016)
Nicotiana tabacum	A/annual herb crop	Face	400, 600	45 days	Cultivar, N	Variable J	↓ (only 1 cv; no N interaction)		Ruiz-Vera et al. (2017)
Oryza sativa	A/annual herb crop	Face	400, 500–600	1–3 months	Cultivar, +1–2 °C	NRH-A[g]	cv-dependent; ↓ at elongation		Cai et al. (2018)
Arabidopsis thaliana	A/annual herb	Chamber	390, 780	5 weeks	Mutant lines	Online isotope discrimination	=	= S_c/S	Mizokami et al. (2019a)
Salicornia ramosissima	A/annual herb	Chamber	400, 700	1 month	Salinity (NaCl)	Curve fitting	↓, = under NaCl		Pérez-Romero et al. (2018)
Glycine max	A/annual herb crop	Chamber	400, 800	48–56 days	Nutrients (K)	Curve fitting	↓, ↓ (K, no interaction)		Singh and Reddy (2018)

[a]Group: A (Angiosperms), G (Gymnosperms)
[b]Field studies: *FACE* Free-Air CO_2 Enrichment, *OTC* Open Top Chambers, *WTC* Whole-Tree Chambers
[c]Relative response compared to control conditions (ambient [CO_2])
[d]Three leaf stages were measured: 'the newly matured, fully expanded flag leaf, at mid-anthesis and at the end of grain filling'
[e]*TDLAS* Tunable Diode Laser Absorption Spectroscopy
[f]Methods used for g_m estimation: curve fitting, 'Bunce' (Bunce 2009), constant J and variable J – both 'Valentini' and 'Yin' approaches (Martins et al. 2013)
[g]NRH-A method: fitting a non-rectangular hyperbolic (NRH) equation for the RuBP-limited part of the C_i-based FvCB model (Yin and Struik 2009)

proper estimations of V_{cmax} and J_{max} (maximum electron transport rate), and thus improving the agreement between gas exchange-measured and biochemically assayed estimates of these parameters. Other early studies also found no differences in g_m in *Glycine max* grown under elevated [CO_2] (Bernacchi et al. 2005). Similarly, Possell and Hewitt (2009) found that photosynthestic down-regulation in the tropical tree *Acacia nigrescens* was not due to g_m modifications, and attributed changes in A_n to lower levels of N and Rubisco activity. More recent studies – using an extensive array of g_m-estimation techniques – have systematically found either a trend towards lower g_m under high [CO_2] (Velikova et al. 2009; Aranjuelo et al. 2015; Lewis et al. 2015; Singh and Reddy 2016, 2018; Pérez-Romero et al. 2018) or no significant change in g_m (Crous et al. 2013; DaMatta et al. 2016; G. Xu et al. 2016). Interestingly, the response could be species- or even cultivar-dependent. For example, g_m decreased in rice whereas it was unaffected in wheat in a FACE study at 600 ppm during a full growing season (Zhu et al. 2012) and cultivars of tobacco and rice displayed different g_m response in two recent FACE studies, with some cultivars showing a decrease in g_m whereas in others no effect was observed (Ruiz-Vera et al. 2017; Cai et al. 2018). This down-regulation of CO_2 diffusion observed in some species is in the same range compared to short-term [CO_2] changes, suggesting the absence of g_m acclimation to long-term exposure to elevated [CO_2] (Mizokami et al. 2019a).

Despite the disparity of g_m responses observed between different species and/or genotypes studied, the overall long-term [CO_2] response shows a trend towards lower g_m (up to −75%) at 700–800 ppm atmospheric [CO_2] in growth chamber (Fig. 3.1a). But this g_m down-regulation is not clear under a 500–600 ppm [CO_2] growth range, where most FACE studies take place. g_m

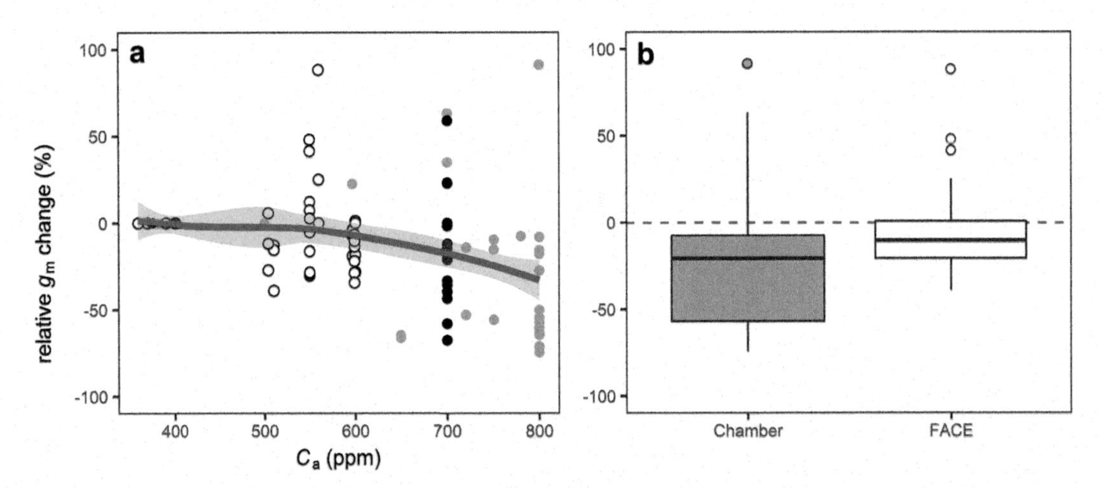

Fig. 3.1. Long-term response of mesophyll conductance to CO_2 diffusion (g_m) to ambient [CO_2] (C_a) displayed over a continuous range of C_a (**a**) and as a boxplot depending on the CO_2 enrichment system (**b**); growth chamber *vs* Free Air CO_2 Enrichment, FACE). g_m response is expressed as a percentage relative to the value for each species/condition/study at 400 ppm. Red line and grey area correspond to a local regression fit using the LOESS method and its 95% confidence interval, respectively ('ggplot2'; Wickham 2016). White and grey points correspond to FACE and chamber studies, respectively, while black points correspond to data from Salazar-Parra et al. (2012) and Martínez-Lüscher et al. (2015), where a 4 °C higher temperature treatment was also applied. All other data are from the following references (see Table 3.1 for details): Singsaas et al. (2003), Bernacchi et al. (2005), Possell and Hewitt (2009), Velikova et al. (2009), Zhu et al. (2012), Crous et al. (2013), Singh et al. (2013), Aranjuelo et al. (2015), Kitao et al. (2015), Lewis et al. (2015), DaMatta et al. (2016), Singh and Reddy (2016, 2018), G. Xu et al. (2016), Ruiz-Vera et al. (2017), Cai et al. (2018) and Mizokami et al. (2019a)

responses to increased [CO_2] in FACE studies range from zero to almost −50% (Fig. 3.1b). Only in sun leaves of the deciduous tree *Liquidambar styraciflua* was reported a significant positive response of g_m to elevated [CO_2] (Singsaas et al. 2003). The downward trend of g_m under elevated [CO_2] is of a similar degree to that observed for g_s (Medlyn et al. 2001; Ainsworth and Rogers 2007); moreover, g_m seems to increase under low [CO_2] (200–290 ppm; Possell and Hewitt 2009; Lewis et al. 2015), which has also been observed for g_s in angiosperms (Overdieck 1989; Maherali et al. 2002). This similar response of both conductances to long-term [CO_2] changes may suggest a common mechanistic basis – possibly at the hydraulic-cell wall level (Flexas et al. 2013b, 2018b; Xiong et al. 2017) – as observed for their short-term responses (Flexas et al. 2013a; Sorrentino et al. 2016; Nadal and Flexas 2018). However, some studies have observed a decoupled response of g_m and g_s under elevated [CO_2] (Velikova et al. 2009; Kitao et al. 2015; Lewis et al. 2015; Mizokami et al. 2019a), highlighting the particularities of these two diffusion processes and their differential relationship with certain structural and/or functional metabolites (Gago et al. 2016). Photosynthetic limitations analysis allows to disentangle which of the main components (diffusion or biochemistry) is determining A_n (Grassi and Magnani 2005), enabling the identification of potential targets for A_n optimization (Nadal and Flexas 2019). Under elevated [CO_2], the trend towards lower g_m (and g_s) could imply a strong diffusive limitation to A_n, although elevated [CO_2] tends to reduce overall diffusive limitations in comparison to ambient [CO_2] conditions (Singh et al. 2013). In the few studies reporting photosynthestic limitations, CO_2 diffusion was the main factor constraining A_n in coffee cultivars (DaMatta et al. 2016), although the percentage of stomatal vs. mesophyll limitation at elevated [CO_2] depended on the cultivar, with mesophyll limitation accounting for up to 73% of total limitations in one cultivar. However, other studies accounting for A_n limitations have yielded inconclusive results so far: in soybean, different reports show either a greater diffusive constraint under elevated [CO_2] (Singh and Reddy 2018) or no differences among limitations (G. Xu et al. 2016).

Mechanistic Basis of g_m Modification

Leaf structure responds to elevated [CO_2]: leaf mass per area (LMA) shows up to a 20% increase in C_3 species grown under elevated CO_2 conditions (Roumet et al. 1996; Long et al. 2004; Poorter et al. 2009). LMA sets an upper limit for g_m (Flexas et al. 2008; Niinemets et al. 2009), presumably due to its relationship with cell wall content and overall leaf density (Poorter et al. 2009; Onoda et al. 2017). However, the increase of LMA under elevated [CO_2] is a product of increased leaf thickness and/or non-structural carbohydrate concentrations (Roumet et al. 1996; Poorter et al. 2009; Smith et al. 2012), which are not likely to strongly influence g_m. Therefore, the general decrease of g_m under elevated [CO_2] could be related to subtler anatomical adjustments that would increase the CO_2 diffusion resistance; in this regard, T_{cw} and chloroplast traits are the most likely parameters affecting g_m.

Some studies have reported that chloroplast exposure along the mesophyll cells and chloroplast size increase under elevated [CO_2] (Robertson and Leech 1995; Uprety et al. 2001). Nonetheless, these observations may not strictly translate into greater S_c/S (Kitao et al. 2015; Mizokami et al. 2019a), as chloroplast size increases are more related to greater chloroplast thickness, mainly due to starch accumulation and/or increased stromal surface (Oksanen et al. 2005; Teng et al. 2006). On the other hand, some studies have reported that high [CO_2] increases chloroplast number 18–70% compared to ambient [CO_2]-grown plants (Bockers et al. 1997; Z. Wang et al. 2004; Teng et al. 2006), thus possibly contributing to greater surface exposure available for CO_2 diffusion (C.Y. Xu et al. 2012). Chloroplast effects on g_m may be indirectly affected by leaf N content (Onoda et al. 2017): in a recent FACE study, Cai et al. (2018) found a strong relationship between g_m and leaf N in rice cultivars under elevated $[CO_2]$. There, the authors suggest that long-term high $[CO_2]$ exposure resulted in a strong N limitation, which could decrease S_c/S and thus g_m. Nonetheless, most studies suggest that chloroplast exposure is either unaffected or enhanced under a high $[CO_2]$ growth environment (mainly through increased number of chloroplasts), although this would suggest we should expected increased g_m under elevated $[CO_2]$, which is not supported by experimental data (Kitao et al. 2015). Therefore, other components affecting g_m may exert a greater limitation to CO_2 diffusion at elevated $[CO_2]$.

Regarding the consequences of increased $[CO_2]$ on T_{cw}, there is little information available. Oksanen et al. (2005) reported a 'subtle' increase in average mesophyll T_{cw} in *Betula pendula* grown under high $[CO_2]$. The observed increase in T_{cw} was coupled with increases in chloroplast number, and it is possible that these two modifications cancel the effect of each other on g_m, thus maintaining a similar conductance. Teng et al. (2006) showed a very similar result in *Arabidopsis thaliana*: at 700 ppm, T_{cw} increased by 24%, concomitant with a higher number of chloroplasts (18%). These preliminary studies did not report g_m values, although they suggested a potential negative effect of increased T_{cw} on g_m. Indeed, Mizokami et al. (2019b) showed that the increase in T_{cw} (and not starch-related changes in chloroplast ultrastructure) was the main factor responsible for g_m decreases under high $[CO_2]$ in *Arabidopsis thaliana*. Zhu et al. (2012) also combined measurements of leaf ultrastructure with g_m estimations in both rice and wheat. In this study, while g_m decreased by 19% and T_{cw} increased by 23% under elevated $[CO_2]$ in rice, neither g_m nor T_{cw} were altered in wheat. Although scarce, evidence suggests a potential key role of cell wall thickness in determining the lower g_m at elevated $[CO_2]$. The differential response of cell wall thickness could reflect divergent species-specific strategies in the use of the cell wall as a sink of non-structural carbohydrates (Sugiura et al. 2017).

Biochemical components, such as aquaporins or carbonic anhydrases, could also play a role in the response of g_m to $[CO_2]$ increases. However, few studies report the activity or expression of carbonic anhydrases, and the available data are not conclusive. Carbonic anhydrase may show higher expression levels (Aranjuelo et al. 2011), lower activity (Porter and Grodzinski 1984; Majeau and Coleman 1996) or present no significant response (Sicher et al. 1994) to an increase in growth $[CO_2]$. To the best of our knowledge, no studies on aquaporins have been conducted so far in response to increased growth $[CO_2]$.

3. *Plant Performance Under Increased $[CO_2]$ Could Be Affected by g_m*

Long-term projections of primary production in a climate change scenario are mostly assessed using the FvCB model, where net CO_2 assimilation can be predicted for any CO_2 concentration (Farquhar et al. 1980). In many cases, short-term A_n-C_a curves are used as a first-step tool for predicting new $[CO_2]$ scenarios. However, not all plants present the same photosynthetic characteristics that determine the A_n-C_a response: g_s, g_m, photochemistry (J_{max}) and Rubisco (V_{cmax}) characteristics may vary across species, thus setting different intrinsic upper limits for A_n (Flexas et al. 2012; Onoda et al. 2017; Nadal and Flexas 2019). Regarding g_m, recent studies have highlighted its variable nature across species: crops, herbaceous and deciduous species tend to have higher g_m than evergreen species among Angiosperms (Gago et al. 2014; Nadal and Flexas 2019). On the other hand, Gymnosperms, Pteridophytes and Bryophytes present relatively low g_m values compared to Angiosperms (Tosens et al. 2016; Veromann-Jürgenson et al. 2017; Carriquí et al. 2019b; Gago et al. 2019). Therefore, species that are more limited by CO_2 diffusion than others at current [CO_2] conditions could potentially be more favored in a future scenario with rising [CO_2], due to partial decrease in diffusive limitations (Niinemets et al. 2011; Flexas et al. 2014). Species with thick, robust leaves, and a high investment in structural components (such as temperate evergreens), have relatively low g_m (0.05–0.15 mol CO_2 m^{-2} s^{-1}); at this g_m range, a rise in atmospheric [CO_2] to 780 ppm would result in a 25% enhancement in A_n relative to their deciduous counterparts (Niinemets et al. 2011). Indeed, the A_n enhancement under increased [CO_2] depends on the photosynthetic capacity of each species. Figure 3.2a illustrates the differential response for each A_n scenario: low A_n plants (with low g_m, g_s, V_{cmax} and J_{max}) would be more favoured than plants with high A_n (Flexas et al. 2014). However, this simulation does not consider the possible photosynthetic down-regulation at high [CO_2]. In this regard, Fig. 3.2b shows the potential effect of CO_2 acclimation on A_n, not only of g_m, but also of g_s, V_{cmax} and J_{max}. Nonetheless, the degree of g_m and photosynthetic acclimation could also depend on species and/or plant type (Zhu et al. 2012), so more research is needed in this area for studies aiming to

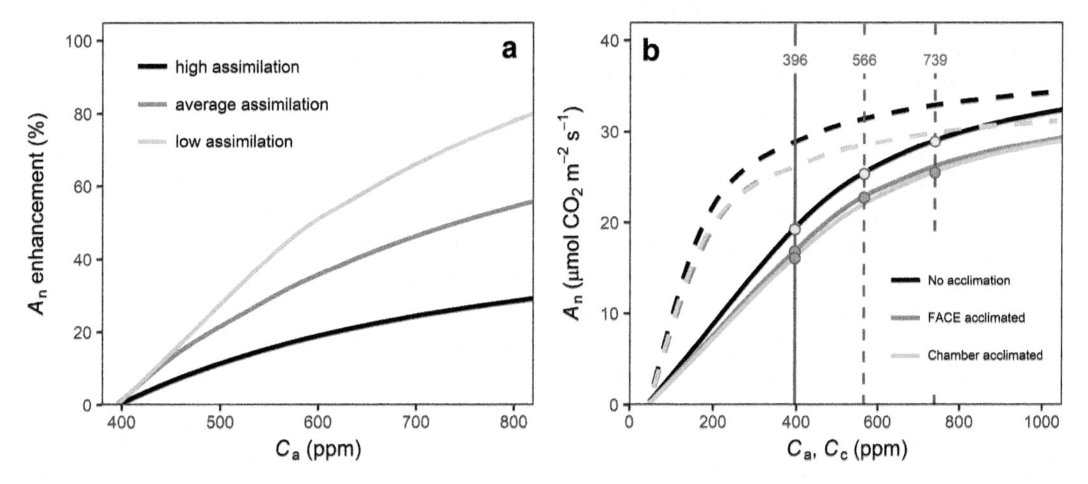

Fig. 3.2. Illustration of the response of net CO_2 assimilation (A_n) to increases of atmospheric [CO_2] (C_a). The differential response to CO_2 across species is illustrated in (**a**) as A_n enhancement (%) relative to A_n at C_a = 396 ppm. Low assimilation species are proportionally more favored by increases in C_a, whereas A_n enhancement in high assimilation species is lower (no acclimation was considered in this approach; the overall A_n enhancement would possibly be lower). The parameters for 'average' were V_{cmax} = 122 µmol CO_2 m^{-2} s^{-1}, J_{max} = 158 µmol e^- m^{-2} s^{-1}, g_s = 0.16 mol CO_2 m^{-2} s^{-1}, g_m = 0.19 mol CO_2 m^{-2} s^{-1}, R_d = 1.25 µmol CO_2 m^{-2} s^{-1} (A_n at 396 ppm = 19 µmol CO_2 m^{-2} s^{-1}); for 'low', V_{cmax} = 50 µmol CO_2 m^{-2} s^{-1}, J_{max} = 70 µmol e^- m^{-2} s^{-1}, g_s = 0.05 mol CO_2 m^{-2} s^{-1}, g_m = 0.07 mol CO_2 m^{-2} s^{-1}, R_d = 0.5 µmol CO_2 m^{-2} s^{-1} (A_n at 396 ppm = 7 µmol CO_2 m^{-2} s^{-1}); and for 'high', V_{cmax} = 175 µmol CO_2 m^{-2} s^{-1}, J_{max} = 220 µmol e^- m^{-2} s^{-1}, g_s = 0.45 mol CO_2 m^{-2} s^{-1}, g_m = 0.55 mol CO_2 m^{-2} s^{-1}, R_d = 2.5 µmol CO_2 m^{-2} s^{-1} (A_n at 396 ppm = 34 µmol CO_2 m^{-2} s^{-1}). The acclimation effect to rising C_a is shown in (**b**); note that the relative increase of A_n is lower when accounting for acclimation to elevated [CO_2]; this photosynthetic down-regulation also explains the observed lower A_n when measured at common C_a. Dashed lines represent the A_n-C_c relationship, whereas continuous lines represent A_n-C_a. Modelling was performed using the FvCB model (Farquhar et al. 1980) accounting for finite g_m (Niinemets et al. 2011). Initial values for each parameter for the 'no acclimation' simulation in (**b**) were the mean values of the g_m studies under elevated [CO_2] listed in Table 3.1 (same as 'average assimilation' in (**a**)). Mean relative change across studies for each parameter was used to account for acclimation to either FACE (mean C_a = 566 ppm) or chamber conditions (mean C_a = 739 ppm). Rubisco kinetics from Bernacchi et al. (2002) were used in all cases.

properly model future changes in primary production considering differences in plant strategies (Ali et al. 2013).

4. The Importance of Accounting for g_m

Since the late 2000s, mesophyll conductance has been considered one of the major factors constraining photosynthesis (Flexas et al. 2008, 2012; Warren 2008a). g_m is now considered essential for properly estimating photosynthetic parameters, especially V_{cmax} and J_{max} (Ethier and Livingston 2004), and also as a key target for photosynthesis improvement (Flexas et al. 2016). However, most studies considering elevated [CO_2] have not accounted for g_m, even when aiming to report V_{cmax} and J_{max} and their possible acclimation to growth [CO_2] (Long et al. 2004; Ainsworth and Rogers 2007). Singsaas et al. (2003) and Crous et al. (2013) pointed out the importance of considering g_m for proper estimation of V_{cmax} and J_{max} in elevated [CO_2] studies. Figure 3.3 shows the agreement between C_i- and C_c-based V_{cmax} and J_{max} estimates in long-term CO_2 studies: C_i-based modelling results in an underestimation of V_{cmax} compared to a C_c-based approach, especially at high V_{cmax} values (Fig. 3.3a), but has weaker effects on J_{max} estimates (Fig. 3.3b). The use of either C_i- or C_c-based parameters generally does not translate into a different effect of elevated [CO_2] on either V_{cmax} nor J_{max}, but it does affect the *absolute* values of these parameters and the V_{cmax}:J_{max} ratio. These errors could have repercussions when aiming to accurately predict ecosystem-scale productivity. Indeed, Sun et al. (2014) reported a 16% correction in modelling the CO_2 fertilization effect for global gross primary production (GPP) when considering g_m and thus C_c-based biochemical parameters. H. Wang et al. (2017) also found a better agreement in Earth System Models with observed and predicted GPP using C_c/C_a instead of C_i/C_a. Overall, these studies highlight the importance of considering g_m in global-scale car-

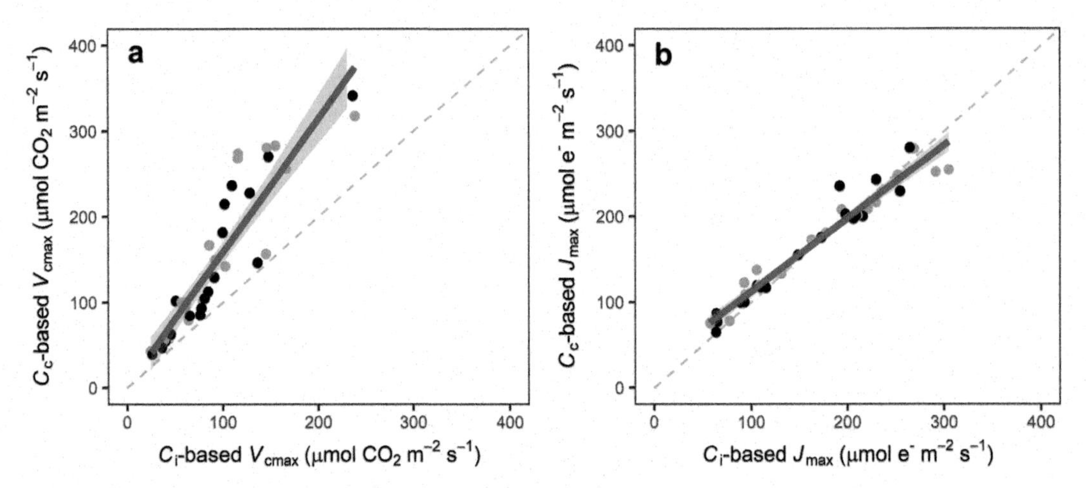

Fig. 3.3. Comparison between C_i- and C_c-based biochemical parameters: maximum velocity of Rubisco carboxylation (V_{cmax}) (**a**) and maximum electron transport rate (J_{max}) (**b**) estimated from A-C_i or A-C_c curves. Ambient and elevated [CO_2] growth conditions are represented by grey and black points, respectively. No differences among [CO_2] growth conditions were detected in any case (ANCOVA; p-value >0.05). Dashed line represents the 1:1 relationship, whereas red line and grey area correspond to a linear regression fit and the 95% confidence interval, respectively. $R^2 = 0.83$, p-value <0.0001, slope = 1.56 for V_{cmax}; $R^2 = 0.95$, p-value <0.0001, slope = 0.86 for J_{max}. Data are from the following references: Singsaas et al. (2003), Possell and Hewitt (2009), Crous et al. (2013), Singh et al. (2013), Aranjuelo et al. (2015), Kitao et al. (2015), DaMatta et al. (2016), Singh and Reddy (2016, 2018), G. Xu et al. (2016), Pérez-Romero et al. (2018), Ruiz-Vera et al. (2017), Sharwood et al. (2017) and Cai et al. (2018).

bon modelling for predicting future climate change scenarios and ecosystem responses (Ali et al. 2013; Rogers et al. 2017).

III. Climate Change: Temperature

A. *Short Term Response*

1. *Photosynthesis Response to Temperature*

Photosynthesis is a biochemical process; as such, it responds to temperature. However, as for CO_2, the responses to short- and long-term changes in temperature are quite different: organisms respond to *relative* changes in temperature depending on their optimal growth temperature (Falk et al. 1996). Long-term changes in growth temperature usually involve a high or low temperature acclimation, which consists of regulation of enzyme expression and lipid membrane stability, among other processes (Yamori et al. 2014). On the other hand, short-term temperature responses of photosynthetic components are directly linked with enzyme activity: electron transport rate and Rubisco carboxylation rate show an enzyme-like thermal response (June et al. 2004; Galmés et al. 2015). These responses partly account for the overall short-term temperature response of A_n (Sage and Kubien 2007; Yamori et al. 2014). However, other processes such as stomatal conductance and respiration are also involved (Atkin and Tjoelker 2003). Nonetheless, little is known about the response of g_m to temperature – and it could also be key for correctly estimating the response of V_{cmax} and J_{max}. The response of g_m to temperature has received considerable attention recently, and its implications for the overall response of net CO_2 assimilation to temperature have been highlighted as key for our understanding of photosynthesis and further photosynthetic modelling (Flexas and Díaz-Espejo 2015; Shrestha et al. 2019).

2. *Mesophyll Conductance Response to Temperature*

Overall g_m Response: Species- and Genotype-Specific

The first reports on g_m responses to temperature from the early 2000s were conflicting. On the one hand, Bernacchi et al. (2002) showed an increase of g_m at higher temperatures (maximum values at 35–37.5 °C) and a steep decrease at 40 °C in tobacco grown at 25–18 °C, and thus a greater g_m limitation at high temperature. On the other hand, Pons and Welschen (2003) and Warren and Dreyer (2006) reported no g_m decline at high temperature (35–38 °C) in the tropical tree *Eperua grandiflora* and the temperate *Quercus canariensis*, respectively. Scafaro et al. (2011) also showed that species of the genus *Oryza* presented the highest g_m at 40 °C. These early studies already suggested a possible species-specific response of g_m to temperature (Bunce 2008). In one of the few field studies including g_m responses to temperature, photosynthetic down-regulation in olive during high summer temperatures was related to decreasing g_m at 30–40 °C (Díaz-Espejo et al. 2007). The g_m decline at high temperatures could be related to growth temperature: indeed, g_m is a key factor in limiting photosynthesis at 35–40 °C in cool-grown plants (Yamori et al. 2006a; Silim et al. 2010; Xue et al. 2016). Nonetheless, differences in g_m thermal responses appear to be more strongly affected by species- or even genotype-specific differences rather than temperature acclimation (Warren 2008b; Dillaway and Kruger 2010; Sakata et al. 2015). In a comprehensive nine species study, von Caemmerer and Evans (2015) highlighted strong differences across species with no clear pattern: some of them displayed a continuous g_m increase from 15 to 40 °C, whereas others showed almost no change. Moreover, g_m and g_s showed no coordination in their response to temperature. Recently, Shrestha et al. (2019) reported

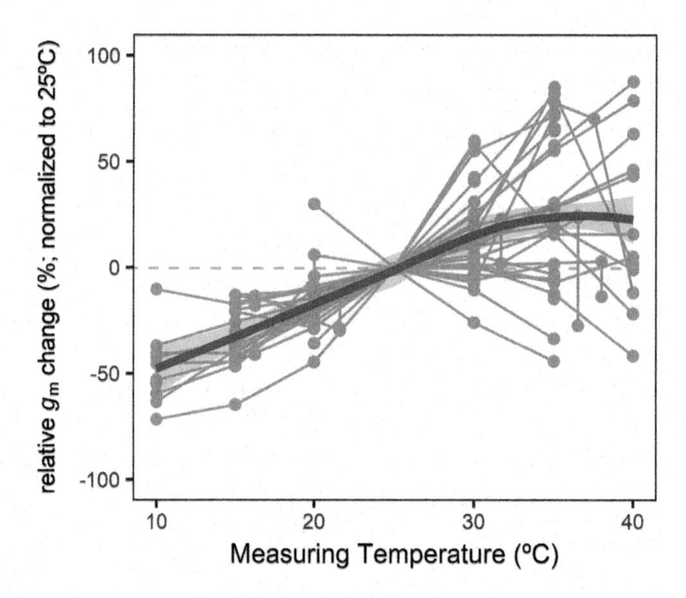

Fig. 3.4. Mesophyll conductance to CO_2 diffusion (g_m) response to short-term (minutes to hours) changes in temperature. Connected points represent the relative g_m change (using g_m at 25 °C as the reference value) of a species for a given treatment and study. Red line and grey area correspond to a local regression fit using the LOESS method and its 95% confidence interval, respectively ('ggplot2'; Wickham 2016). Data are from the following references: Bernacchi et al. (2002), Warren and Dreyer (2006), Yamori et al. (2006a), Díaz-Espejo et al. 2007, Bunce (2008), Velikova et al. (2009), Silim et al. (2010), Fares et al. (2011), Scafaro et al. (2011), Sakata et al. (2015), von Caemmerer and Evans (2015), Qiu et al. (2017), Ubierna et al. (2017) and Greer (2018).

that g_m temperature responses vary at the intra-specific level: the temperature sensitivity of g_m differed among soybean genotypes and leaf age. Reports on g_m short-term responses to temperature are displayed in Fig. 3.4. Overall, the differences across species are clear, especially at the high end of the temperature response (25–40 °C). Little difference occurs from 10 to 25 °C, where g_m decreases below its optimum value by 35–70%, whereas from 25 to 40 °C, g_m may show all variety of responses, from −50% to increases up to 100%. Interestingly, the only report to date regarding g_m thermal responses in C_4 species showed a very similar response to C_3 plants (Ubierna et al. 2017). However, it is worth noting that some of the observed variation across species and genotypes may be due to method discrepancies in g_m estimates. Recently, Sonawane and Cousins (2019) showed that the carbon and oxygen isotope discrimination methods may yield different g_m values at high temperature due to uncertain assumptions regarding isotope equilibrium and H_2O fractionation, which may arise from different temperature-associated pH effects at the sites of CO_2 transport.

The differential response of g_m across species needs to be taken into account for proper estimation of V_{cmax} and J_{max} and their response to temperature (Bernacchi et al. 2002; Warren and Dreyer 2006; Warren 2008b; Díaz-Espejo 2013; von Caemmerer and Evans 2015), as already predicted by theoretical modelling (Juurola et al. 2005) and confirmed by *in vitro* and *in vivo* Rubisco kinetics comparisons (Galmés et al. 2016).

Mechanistic Basis of g_m Response

The response of g_m to short-term changes in temperature can help elucidate its possible mechanistic basis. Despite strong evidence supporting the role of ultrastructure and protein activity determining g_m (Flexas et al. 2018a), the complex interplay between these factors is still to be elucidated. Bernacchi et al. (2002) highlighted that the response of g_m to temperature could be a protein-facilitated process, as its Q_{10} (2.2) was higher than the Q_{10} for CO_2 diffusion in water (1.25). Later studies also suggested the active role of proteins in the g_m thermal response in species such as *Quercus canariensis* (Warren and Dreyer 2006) and *Spinacia oleracea* (Yamori et al. 2006a) based on the discrepancy in Q_{10} commented above. However, some species show a constant g_m over the 30–35 °C span, suggesting that g_m is not solely determined by protein activity, but is a complex process that involves both passive and active CO_2 transport (Warren 2008b). Further modelling may contribute to disentangling the contributions of the different phases determining g_m. Scafaro et al. (2011) investigated the possible role of cell ultrastructure in three *Oryza* species, although T_{cw} did not appear to affect the g_m response to temperature – suggesting a preeminent role of membrane components determining the g_m dependence of temperature. Splitting these components into aqueous and cell membrane/wall phases (g_{liq} and g_{mem}, respectively) allows to distinguish different temperature responses between species. Walker et al. (2013) attributed the relatively lower g_m temperature dependency of *Arabidopsis thaliana* compared to *Nicotiana tabacum* to g_m being mostly determined by g_{liq} instead of g_{mem} in *A. thaliana*, implying a higher presence and/or temperature sensitivity of CO_2 transport-related proteins in the warm-adapted *N. tabacum*. This key role of g_{mem} in the short-term temperature response of g_m was also highlighted by von Caemmerer and Evans (2015): although leaf anatomy (represented by S_c/S) sets the upper limit of g_m across species, membrane activation energy seems to determine the *relative* g_m response. Flexas and Díaz-Espejo (2015) suggested that a progressive decrease of S_c/S due to chloroplast movement or cell shrinkage at elevated temperature could explain a g_m decline, although Shrestha et al. (2019) did not find changes in S_c/S that explained g_m changes to temperature. ABA-mediated processes have also been proposed as a mechanism of g_m thermal responses (Qiu et al. 2017). However, due to the difficult nature of estimating short-term responses of g_m, some experimental reports may not accurately meet the theoretical framework: indeed, the high Q_{10} reported for several species exceeds the Q_{10} of carbonic anhydrase- and aquaporin-facilitated CO_2 transport (Tholen et al. 2012). This could partially arise from modified carbonic anhydrase activity and kinetic properties with varying pH and temperature (Boyd et al. 2015; Sonawane and Cousins 2019). In addition, the g_m response to high temperature may also be confounded with concomitant increases in vapour pressure deficit and transpiration (and thus lower water potential). A recent study discerned both effects and found that the 'net' g_m decrease at high temperature partially arises from leaf water deficit (Y. Li et al. 2019).

B. Long Term Impact of Temperature

1. Photosynthesis Acclimation to Growth Temperature

Temperature is a determining factor in the distribution of plants across different environments. Thus, general temperature increases due to the global warming will modify species distribution, as the capacity to cope with this acclimation depends on different factors. One of the most important factors is photosynthesis, as it is a really temperature-sensitive process (Yamori et al. 2014; Vico et al. 2019). Whereas most plants show considerable capacity to modify their physiological and biochemical traits in order to adjust their A_n to growth temperatures, this capacity has been reported to be closely linked to the species photosynthetic pathway (C_3, C_4 and CAM; Sage and Kubien 2007; Yamori et al. 2014). On one hand, plants grown at low temperature present higher A_n at those temperatures generally due to an increased amount of photosynthetic enzymes such as Rubisco -in order to compensate for the decreased enzymatic activity-, by producing isoforms with better performance at low temperatures (Yamori et al. 2006b), and by the maintenance of greater membrane fluidity (Yamori et al. 2014). On the other hand, plants grown at high temperatures usually show increased A_n due to greater heat tolerance of thylakoid membranes and photosynthetic enzymes, the production of different isoforms of Rubisco activase and respiration acclimation, among others (Yamori et al. 2014). However, in C_3 plants g_m has not been traditionally considered between the main mechanisms that have been proposed to explain A_n acclimation to long-term changes in growth temperature, despite its key role under short-term temperature responses.

2. *Mesophyll Conductance Response to Growth Temperature*

No Clear Effect on g_m

The g_m response to growth temperature – together with other photosynthetic parameters – was first studied in experiments applying relatively wide temperature ranges (over a 10–30 °C span). Yamori et al. (2006a, b) reported the first results regarding g_m acclimation to temperature – combined with its short-term dependence – in spinach plants grown at 15 and 30 °C. Low temperature-grown plants showed a reduction in g_m compared to high temperature-grown plants when measured at 25–35 °C, whereas no differences were observed when measured at 10–20 ° C and g_m significantly limited photosynthesis at higher temperature in cool-grown spinach. Moreover, the conductance ratio (g_m/g_s) was greater in cool plants across all temperatures, mostly due to a reduction in g_s compared to plants grown at 30 °C. This study highlighted the potential role of g_m in constraining photosynthesis at high temperatures, especially in cold-acclimated plants. Fares et al. (2011) also reported a strong decrease in g_m, together with A_n, at 35 °C in *Populus x euramericana*; nonetheless, this effect could be reversed by 'recovering' these plants at 25 °C for 2 weeks. However, *Populus balsamifera* showed the opposite trend: in plants grown at 15 and 25 °C, diffusive limitations (both g_s and g_m) were higher at lower temperatures (Silim et al. 2010). Moreover, some species may be affected by both low and high temperatures. g_m was the dominant limiting factor for photosynthesis at both low (10 °C) and high (30 °C) temperatures in the alpine plant *Mecanopsis horridula* (Zhang 2010). Nonetheless, other similar studies with alpine plants yielded no g_m differences in high vs low altitude populations (Sakata et al. 2015). Figure 3.5 synthesizes the disparity in the g_m response to long-term temperature changes. Although in some cases there are strong increases in g_m (by 50–100%), others show either no response or small decreases in g_m (by 25–50%). Although no clear pattern is observed, most of the large g_m increases occur in species grown at 12–22 °C compared to when they are acclimated at 25–35 °C.

Possible Mechanistic Basis

There are very few reports regarding the possible mechanisms determining changes in g_m to growth temperature, and no clear pattern that can be generalized for C_3 plants. The only studies in which deeper analyses have been carried out are in species adapted to extreme temperatures, which are not representative of most terrestrial plants: in two Antarctic vascular species with a very mild temperature treatment (Sáez et al. 2018a; see below), and in the thermophilic tree *Ziziphus spina-christi* (Zait et al. 2019). In the case of *Z. spina-christi*, plants were acclimated and measured at four growth temperatures (34/28 °C –the optimum growth tempera-

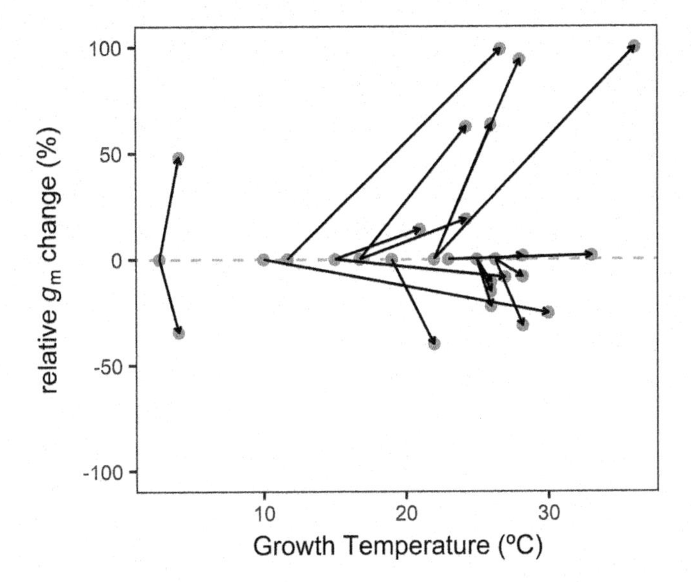

Fig. 3.5. Long-term temperature effect on mesophyll conductance to CO_2 diffusion (g_m). Each vector represents the relative g_m change with increasing growth temperature, taking the g_m value at the lowest temperature as reference within each species/condition/study. g_m measurements were performed at 25 ± 2 °C in all cases regardless of growth temperature. Data are from the following references: Yamori et al. (2006a), Bunce (2008), Silim et al. (2010), Fares et al. (2011), Crous et al. (2013), Lewis et al. (2015), Xue et al. (2016), Cai et al. (2018) and Sáez et al. (2018a).

ture-, 28/22 °C, 22/16 °C, and 16/10 °C). In this species, both LMA and leaf thickness increased, and the fraction of intercellular air space decreased with increasing growth temperature, correlating with a hyperbolical decrease in g_m with decreasing growth temperature. They conclude that the analyzed mesophyll traits alone did not explain g_m acclimation to temperature, but pointed towards a potential effect of T_{cw}. It is worth mentioning that it is possible that the effects of temperature observed in g_m could be, to some extent, artefactual (as detailed in previous sections). This could be related to the different optimum temperatures for A_n, g_m and mitochondrial respiration. Since the respiration optimum temperature for C_3 plants is higher (e.g., just below the temperature at which heat inactivation of enzymes occurs; above 45 °C; Yamori et al. 2014) than for A_n and g_m, an increase in the recycling of (photo) respired CO_2 in the chloroplast stroma could explain the lower (potentially apparent) g_m reported for several species (Tholen et al. 2012).

3. *Temperature Response of g_m in a Climate Change Context*

Small Temperature Increases (1–2 °C) Coupled with Elevated [CO_2]

The g_m response described in the previous section referred to large changes in growth temperature (>10 °C). However, in the context of climate change, the global temperature is expected to experience a more modest increase (1–4 °C; Allen et al. 2018). Very few studies have considered the possible effects of this climate change scenario on g_m. Notably, the photosynthetic response – including changes in g_m – to long-term changes in growing temperature was recently investigated in two Antarctic vascular species, *Colobanthus quitensis* and *Deschampsia antarctica*, under field conditions using open top chambers (OTCs) which simulated a 2 °C increase over a 3-years period (Sáez et al. 2018a). Interestingly, g_m responded differently in each species: whereas g_m was unaffected in *D. antarctica*, g_m in *C. quitensis* showed an increase under the OTC treatment, which was related to a greater proximity of chloroplasts to the cell wall and a higher exposure of mesophyll cells to the intercellular space (S_m/S). This distinct response was later confirmed under laboratory conditions in the 5–10 °C range (Sáez et al. 2018b), although g_m in *D. antarctica* was more sensitive in the 10–16 °C range than in *C. quitensis*, where g_m increase was relatively smaller. As polar and alpine habitats are some of the most sensitive ecosystems under current climate change scenarios (Dolezal et al. 2016; Turner et al. 2016), more studies are needed in plants from extremely cold habitats to confirm the potential role of g_m in photosynthetic responses to temperature increases.

However, climate change scenarios also need to take into account warming with concomitant increases in atmospheric [CO_2]. There are very few reports of g_m responses to both elevated [CO_2] and temperature (in a 1–4 °C range). Salazar-Parra et al. (2012) and Martínez-Lüscher et al. (2015) studied the response of *Vitis vinifera* to a combined treatment of high CO_2 (700 ppm) and temperature (+4 °C). In both cases, no clear trend in g_m was observed, although g_m decreased in some of the developmental stages. However, these studies could not disentangle if there was an interactive effect between [CO_2] and temperature. In the ancient gymnosperm *Wollemia nobilis*, Lewis et al. (2015) did show a significant temperature by [CO_2] interaction effect on g_m, where the increase of g_m with higher temperature at ambient [CO_2] was not reflected at 650 ppm. The interaction effect on g_m may also depend on the plant developmental stage, as reported in two rice cultivars (Cai et al. 2018). On the other hand, Xue et al. (2016) did not find any synergistic effect of warming and CO_2 in *Glycine max* despite applying greater growth temperature ranges (22–36 °C).

IV. Climate Change: Water

A. *Mesophyll Conductance Declines Under Water Stress*

Precipitation regimes and drought episodes will likely be affected by climate change (Trenberth 2011). Water is already the most important limiting factor for food production, and therefore the improvement of crop water use efficiency (*WUE*) and, possibly, net CO_2 assimilation is envisaged as a key challenge for agriculture and plant physiology in a climate change scenario (Loreto and Centritto 2008; Gago et al. 2014). Hence, understanding photosynthetic behavior under water stress is key for possible crop improvement (Nadal and Flexas 2019), as shown by recent studies in photosynthesis and *WUE* engineering (Yang et al. 2016; Lehmeier et al. 2017). g_m is a key component underlying photosynthetic decreases under water stress (Flexas et al. 2004; Galmés et al. 2007; Galle et al. 2009, 2011; Nadal and Flexas 2019) and of A_n and intrinsic water use efficiency enhancement (WUE_i; Flexas et al. 2016). The overall response of g_m to water stress has been recently reviewed (see Nadal and Flexas 2018). g_m declines under moderate to severe water stress and imposes a relatively high limitation on A_n (25–50% across species). Reductions in g_m are hypothesized to be due to the involvement of aquaporin and carbonic anhydrase activity (Pérez-Martin et al. 2014) and to the ABA-mediated coordination with stomatal conductance decline (Mizokami et al. 2015). Recently, it has been described that decreases in S_c/S can also contribute to g_m decline under drought (Ouyang et al. 2017; Han et al. 2019). In most species, a leaf water potential (Ψ_{leaf}) decline of 1–2 MPa implies a significant decrease in g_m (Galmés et al. 2007; Galle et al. 2011; Varone et al. 2012, 2016; Silva et al. 2015). In other species, such as olive, g_m decreases only occur under very negative Ψ_{leaf} due to the species' high drought tolerance (Pérez-Martín et al. 2014; Díaz-Espejo et al. 2018). Despite the strong effect of water stress on g_m, some studies have reported decreased mesophyll limitations after several drought cycles, indicating an acclimation of g_m to some extent (Galle et al. 2011; Han et al. 2016).

Mesophyll conductance is related to other diffusion-related parameters such as g_s (Flexas et al. 2013a) and leaf hydraulic conductance (K_{leaf}; Flexas et al. 2013b), possibly by sharing a common path with water across the leaf mesophyll (Xiong et al. 2017). Recently, the coordination between these different factors – g_m, g_s and K_{leaf} – has received attention in an effort to elucidate the underlying mechanisms of photosynthetic decline under water stress (Dewar et al. 2018; Flexas et al. 2018b). In a recent comprehensive study, X. Wang et al. (2018) detailed the coordination between g_s, g_m and K_{leaf} in rice, highlighting the latter as the factor triggering the decrease of g_s and g_m under drought. However, other species -with great intervarietal diversity- show a delayed response of K_{leaf} and g_m upon water stress, which in some cases may result in enhanced WUE_i (Pou et al. 2012; Théroux-Rancourt et al. 2015), highlighting the possible different strategies across species.

B. *Water Stress Under Increased [CO_2]*

Net CO_2 assimilation decreases under water stress are usually ameliorated under elevated CO_2 conditions (Albert et al. 2011; Song and Huang 2014; Sekhar et al. 2017), although some studies show non-additive effects and a strong predominance of drought (Meng et al. 2013). However, very few studies have evaluated the possible interactive effects on g_m between elevated [CO_2] and water stress conditions. The overall trend towards lower g_m under elevated [CO_2] conditions could add to the well-known down-regulation of g_m under water stress, thus increasing [CO_2] diffusion limitations under both conditions. To the best of our knowledge, only Salazar-Parra et al. (2012) have monitored g_m under drought, elevated [CO_2] and temperature (a combined treatment) in grapevine: CO_2 diffusion strongly limited photosynthesis during water stress through reductions in both g_m and g_s, but the g_m reduction was ameliorated under elevated [CO_2] and temperature. This result contrasts with the overall response of g_m to high [CO_2], suggesting a possible interactive effect on g_m. Ge et al. (2012) showed a greater effect of water stress than elevated [CO_2] on C_c-based estimates of V_{cmax} and J_{max} in the boreal grass *Phalaris arundinacea*, highlighting the importance of considering g_m for unveiling changes on biochemical components. Salinity poses a very similar stress to plants as does drought (Chaves et al. 2009), although the effect of salinity under elevated [CO_2] has been less studied. Similar to drought, only one study has reported the combined effect of salinity and high [CO_2] on g_m, showing a similar degree of amelioration of g_m decrease under elevated [CO_2] in the halophyte *Salicornia ramosissima* (Pérez-Romero et al. 2018). More studies are needed to explore how g_m responds to both water and salt stress together with elevated [CO_2], as the combination of these stresses is expected to be more frequent in the future (Loreto and Centritto 2008).

V. Climate Change: Others

A. *Ozone*

Ozone (O_3) is a phytotoxic air pollutant widely considered to be a risk factor for vegetation and crop yields (Fowler et al. 1999; Matyssek and Innes 1999). Its formation occurs predominantly in highly populated and industrialized areas, and tropospheric [O_3] has more than doubled in the last century in mid-latitudes of the Northern Hemisphere. Ozone concentrations are predicted to keep increasing in the future, albeit with strong regional differences (Hartmann et al. 2013). One of the pernicious effects of O_3 on plant yield is a decrease in A_n (Reich 1987; Dizengremel 2001; Matyssek and Sandermann 2003; Yue et al. 2017), since ozone enters leaves through the stomata and causes oxidative stress. The O_3-related decrease in A_n is commonly correlated with reductions in g_s or decreases in V_{cmax} (Gao et al. 2016; Li et al. 2017). A recent global meta-analysis reported 16 and 29% decreases in g_s when plants where grown at China's current (42 ppb) or an elevated (102 ppb) [O_3] compared to plants grown in filtered clean air (Wittig et al. 2007; Li et al. 2017). However, most studies do not account for g_m. Recent studies reported that high ozone led to reductions in g_m in several species (*Betula pendula*, *Fagus crenata*, *Phaseolus vulgaris*, *Populus deltoides*, *Quercus ilex* and *Quercus pubescens*), with the exception of *Fagus sylvatica*, where g_m remained unaffected (Eichelmann et al. 2004; Velikova et al. 2005; Flowers et al. 2007; Warren et al. 2007; Watanabe et al. 2018; Xu et al. 2019). Moreover, Y. Xu et al. (2019) quantified the limitations to A_n in poplar species, revealing that decreased A_n tin plants exposed to elevated [O_3] was mainly due to mesophyll limitations, and, to a lesser extent, biochemical limitations.

Although the mechanistic bases for this general decline in g_m due to elevated [O_3] are not yet well defined, different potential causes have been identified. Early studies reported that pronounced decreases in A_n were mediated through a collapse of leaf mesophyll cells (Matyssek et al. 1991). Paoletti et al. (2009) reported increases in T_{cw}, and alterations in chloroplast shape, size and location due to elevated [O_3], pointing to their possible effects on g_m. Gao et al. (2016) also reported changes in mesophyll ultrastructure, observing that elevated [O_3] increased foliar damage and induced the degradation of cell walls and chloroplasts, two key traits in setting g_m. Moreover, Eichelmann et al. (2004) pointed to a negative effect of elevated [O_3] on g_m through the modification of aquaporins involved in CO_2 diffusion, thus increasing the mesophyll diffusion resistance.

B. Nutrients

Leaf nutrient content (N and P, among others) is a major factor determining leaf photosynthesis (Wright et al. 2004). N is correlated not only with protein (mainly Rubisco) and chlorophyll levels, but it also influences g_m, possibly through increasing chloroplast number and/or size, thus affecting S_c/S (Onoda et al. 2017). Nutrient availability has been recently highlighted as a potential factor in determining the long-term plant response to elevated [CO_2] (Reich et al. 2018), and thus understanding the effect of nutrient status on g_m under high CO_2 is necessary to fully understand photosynthetic response under both elevated [CO_2] and varying nutrient availability. In this regard, the possible interaction between nutrient status and atmospheric [CO_2] on g_m has been investigated in several species. Most reports show no significant interaction for the studied nutrients (N, P and K), indicating a greater effect of nutrient availability than [CO_2]: g_m responds strongly to nutrient status and shows either no response (Singh et al. 2013; Singh and Reddy 2018) or a minimal decrease (Singh and Reddy 2016; Ruiz-Vera et al. 2017) to elevated [CO_2]. Only Kitao et al. (2015) reported a significant interaction effect between N and [CO_2] in *Betula platyphylla*, where g_m did not decrease at high N content at elevated [CO_2]. Notably, mesophyll limitations were greater under strong K deficiency despite elevated [CO_2] (Singh and Reddy 2018), indicating a possible shift from stomata to mesophyll CO_2 diffusion constraining CO_2 assimilation under those conditions.

VI. Outlook

An increasing amount of evidence highlights the importance of estimating g_m and its mechanistic basis to reveal the effects of increased temperature, atmospheric CO_2 concentration and water stress (among other factors) on photosynthesis, and therefore on future ecosystem behavior and crop yields. Including g_m in modelling increases the accuracy of predictions of photosynthesis and primary production under elevated [CO_2] scenarios (Sun et al. 2014). Moreover, current g_m and C_c-based estimations predict a possible different response of plant groups, where currently diffusion-limited species may be more favoured under rising [CO_2]. In this sense, there is a need to extend the studies of the effects of climate change on g_m beyond angiosperms: changes in g_m in gymnosperms, ferns and mosses have been scarcely considered in climate change studies, despite their potential different responses (Flexas et al. 2014). Further modelling efforts are needed as well for proper estimations of g_m responses to climate change factors and for providing insights into the mechanistic basis of how g_m is altered by short-term and long-term exposure to rising [CO_2], temperatures, water stress and other factors (Flexas et al. 2018a; Sonawane and Cousins 2019). Future research on the variable nature of g_m in response to a dynamic environment is key for a more complete understanding and prediction of photosynthesis response in a climate change scenario.

Acknowledgements

MN was supported by a predoctoral fellowship BES-2015-072578 from the Ministerio de Economía y Competitividad (Spain) co-financed by the European Social Fund. MC was supported by a predoctoral fellowship FPI/1700/2014 from the Conselleria d'Educació, Cultura i Universitats (Govern de les Illes Balears) and the European Social Fund. Research of MN and JF on mesophyll conductance is supported by the project CTM2014-53902-C2-1-P from the Ministerio de Economía y Competitividad (Spain) and the European Regional Development Fund.

References

Ainsworth EA, Rogers A (2007) The response of photosynthesis and stomatal conductance to rising [CO_2]: mechanisms and environmental interactions. Plant Cell Environ 30:258–270

Albert KR, Ro-Poulsen H, Mikkelsen TN, Michelsen A, van der Linden L, Beier C (2011) Interactive effects of elevated CO_2, warming, and drought on photosynthesis of *Deschampsia flexuosa* in a temperate heath ecosystem. J Exp Bot 62:4253–4266

Ali AA, Medlyn BE, Crous KY, Reich PB (2013) A trait-based ecosystem model suggests that long-term responsiveness to rising atmospheric CO_2 concentration is greater in slow-growing than fast-growing plants. Funct Ecol 27:1011–1022

Allen MR, Dube OP, Solecki W, Aragón-Durand F, Cramer W, Humphreys S, Kainuma M, Kala J, …, Zickfeld K (2018). Framing and context. In: Masson-Delmotte V, Zhai P, Pörtner HO, Roberts D, Skea J, Shukla PR, Pirani A, Moufouma-Okia W, …, Waterfield T (eds) Global Warming of 1.5°C. IPCC., in press

Aranjuelo I, Cabrera-Bosquet L, Morcuende R, Avice JC, Nogués S, Araus JL, Martínez-Carrasco R, Pérez P (2011) Does ear C sink strength contribute to overcoming photosynthetic acclimation of wheat plants exposed to elevated CO_2? J Exp Bot 62:3957–3969

Aranjuelo I, Sanz-Sáez A, Jáuregui I, Irigoyen JJ, Araus JL, Sánchez-Díaz M, Erice G (2013) Harvest index, a parameter conditioning responsiveness of wheat plants to elevated CO_2. J Exp Bot 64:1879–1892

Aranjuelo I, Tcherkez G, Jauregui I, Gilard F, Ancín M, Fernández-San Millán A, Larraya L, Veramendi J, …, Farran I (2015) Alternation by thioredoxin f over-expression of primary carbon metabolism and its response to elevated CO_2 in tobacco (*Nicotiana tabacum* L.). Environ Exp Bot 118: 40–48

Atkin OK, Tjoelker MG (2003) Thermal acclimation and the dynamic response of plant respiration to temperature. Trends Plant Sci 8:343–351

Barbour MM, Evans JR, Simonin KA, von Caemmerer S (2016) Online CO_2 and H_2O oxygen isotope fractionation allows estimation of mesophyll conductance in C_4 plants, and reveals that mesophyll conductance decreases as leaves age in both C_4 and C_3 plants. New Phytol 210:875–889

Bellasio C, Beerling DJ, Griffiths H (2016) Deriving C_4 photosynthetic parameters from combined gas exchange and chlorophyll fluorescence using an excel tool: theory and practice. Plant Cell Environ 39:1164–1179

Berghuijs HNC, Yin X, Ho QT, Driever SM, Retta MA, Nicolai BM, Struik PC (2018) Mesophyll conductance and reaction-diffusion models for CO_2 transport in C_3 leaves; needs, opportunities and challenges. Plant Sci 252:62–75

Bernacchi CJ, Portis AR, Nakano H, von Caemmerer S, Long SP (2002) Temperature response of mesophyll conductance. Implications for the determination of rubisco enzyme kinetics and for limitations to photosynthesis in vivo. Plant Physiol 130:1992–1998

Bernacchi CJ, Morgan PB, Ort DR, Long SP (2005) The growth of soybean under free air [CO_2] enrichment (FACE) stimulates photosynthesis while decreasing in vivo rubisco capacity. Planta 220:434–446

Bockers M, Capková V, Tichá I, Scäfer C (1997) Growth at high CO_2 affects the chloroplast number but not the photosynthetic efficiency of photoautotrophic *Marchantia polymorpha* culture cells. Plant Cell Tiss Org 48:103–110

Boyd RA, Gandin A, Cousins AB (2015) Temperature responses of C_4 photosynthesis: biochemical analysis of rubisco, phosphoenolpyruvate carboxylase, and carbonic anhydrase in *Setaria viridis*. Plant Physiol 169:1850–1861

Brodribb TJ, McAdam SAM (2013) Unique responsiveness of angiosperm stomata to elevated CO_2 explained by calcium signalling. PLoS One 8:e82057

Brodribb TJ, McAdam SAM, Jordan GJ, Field TS (2009) Evolution of stomatal responsiveness to CO_2 and optimization of water-use efficiency among land plants. New Phytol 183:839–847

Bunce JA (2008) Acclimation of photosynthesis to temperature in *Arabidopsis thaliana* and *Brassica oleracea*. Photosynthetica 46:517–524

Bunce JA (2009) Use of the response of photosynthesis to oxygen to estimate mesophyll conductance to carbon dioxide in water-stressed soybean leaves. Plant Cell Environ 32:875–881

Cai C, Li G, Yang H, Yang J, Liu H, Struik PC, Luo W, Yin X, …, Guo X (2018) Do all leaf photosynthesis parameters of rice acclimate to elevated CO_2, elevated temperature, and their combination, in FACE environments? Glob Change Biol 24:1685–1707

Carriquí M, Cabrera HM, Conesa MÀ, Coopman RE, Douthe C, Gago J, Gallé A, Galmés J, …, Flexas J (2015) Diffusional limitations explain the lower photosynthetic capacity of ferns as compared with angiosperms in a common garden study. Plant Cell Environ 38:448–460

Carriquí M, Douthe C, Molins A, Flexas J (2019a) Leaf anatomy does not explain apparent short-term responses of mesophyll conductance to light and CO_2 in tobacco. Physiol Plantarum 165:604–618

Carriquí M, Roig-Oliver M, Brodribb TJ, Coopman R, Gill W, Mark K, Niinemets U, Perera-Castro AV, …, Flexas J (2019b) Anatomical constraints to non-stomatal diffusion conductance and photosynthesis in lycophytes and bryophytes. New Phytol 222 1256–1270

Chaves MM, Flexas J, Pinheiro C (2009) Photosynthesis under drought and salt stress: regulation mechanisms from whole plant to cell. Ann Bot 103:551–560

Crous KY, Quentin AG, Lin YS, Medlyn BE, Williams DG, Barton CVM, Ellsworth DS (2013) Photosynthesis of temperate *Eucalyptus globulus* trees outside their native range has limited adjustment to elevated CO_2 and climate warming. Glob Change Biol 19:3790–3807

DaMatta FM, Godoy AG, Menezes-Silva PE, Martins SCV, Sanglard LMVP, Morais LE, Torre-Neto A, Ghini R (2016) Sustained enhancement of photosynthesis in coffee trees grown under free-air CO_2 enrichment conditions: disentangling the contributions of stomatal, mesophyll, and biochemical limitations. J Exp Bot 67:341–352

Dewar R, Mauranen A, Mäkelä A, Hölttä T, Medlyn B, Vesala T (2018) New insights into the covariation of stomatal, mesophyll and hydraulic conductances from optimization models incorporating nonstomatal limitations to photosynthesis. New Phytol 217:571–585

Díaz-Espejo A (2013) New challenges in modelling photosynthesis: temperature dependencies of rubisco kinetics. Plant Cell Environ 36:2104–2107

Díaz-Espejo A, Nicolás E, Fernández JE (2007) Seasonal evolution of diffusional limitations and photosynthetic capacity in olive under drought. Plant Cell Environ 30:922–933

Díaz-Espejo A, Fernández JE, Torres-Ruiz JM, Rodriguez-Dominguez CM, Perez-Martin A, Hernandez-Santana V (2018) The olive tree under water stress: fitting the pieces of response mechanisms in the crop performance puzzle. In: García-Tejero IF, Duran-Zuazo VH (eds) Water Scarcity and Sustainable Agriculture in Semiarid Environment. Elsevier, London, pp 439–479

Dillaway DN, Kruger EL (2010) Thermal acclimation of photosynthesis: a comparison of boreal and temperate tree species along a latitudinal transect. Plant Cell Environ 33:888–899

Dizengremel P (2001) Effects of ozone on the carbon metabolism of forest trees. Plant Physiol Biochem 39:729–742

Dolezal J, Dvorsky M, Kopecky M, Liancourt P, Hiiesalu I, Macek M, Altman J, Chlumska Z, …, Schweingruber F (2016) Vegetation dynamics at the upper elevational limit of vascular plants in Himalaya. Sci Rep 6: 24881

Douthe C, Dreyer E, Epron D, Warren CR (2011) Mesophyll conductance to CO_2, assessed from online TDL-AS records of (CO_2)-C-13 discrimination, displays small but significant short-term responses to CO_2 and irradiance in eucalyptus seedlings. J Exp Bot 62:5335–5346

Douthe C, Dreyer E, Brendel O, Warren CR (2012) Is mesophyll conductance to CO_2 in leaves of three eucalyptus species sensitive to short-term changes of irradiance under ambient as well as low O_2? Funct Plant Biol 39:435–448

Earles JM, Théroux-Rancourt G, Roddy AB, Gilbert ME, McElrone AJ, Brodersen CR (2018) Beyond porosity: 3D leaf intercellular airspace traits that impact mesophyll conductance. Plant Physiol 178:148–162

Eichelmann H, Oja V, Rasulov B, Padu E, Bichele I, Pettai H, Möls T, Kasparova I, …, Laisk A (2004) Photosynthetic parameters of birch (*Betula pendula* Roth) leaves growing in normal and in CO_2- and O_3-enriched atmospheres. Plant Cell Environ 27: 479–495

Ethier GJ, Livingston NJ (2004) On the need to incorporate sensitivity to CO_2 transfer conductance into the Farquhar-von Caemmerer-Berry leaf photosynthesis model. Plant Cell Environ 27:137–153

Evans JR, von Caemmerer S (1996) Carbon dioxide diffusion inside leaves. Plant Physiol 110:339–346

Evans JR, Sharkey TD, Berry JA, Farquhar GD (1986) Carbon isotope discrimination measured concurrently with gas exchange to investigate CO_2 diffusion in leaves of higher plants. Aust J Plant Physiol 13:281–292

Evans JR, von Caemmerer S, Setchell BA, Hudson GS (1994) The relationship between CO_2 transfer conductance and leaf anatomy in transgenic tobacco with a reduced content of rubisco. Aust J Plant Physiol 21:475–495

Evans JR, Kaldenhoff R, Genty B, Terashima I (2009) Resistances along the CO_2 diffusion pathway inside leaves. J Exp Bot 60:2235–2248

Falk S, Maxwell DP, Laudenbach DE, Huner NPA (1996) Photosynthetic adjustment to temperature. In: Baker NR (ed) Photosynthesis and the Environment. Advances in Photosynthesis and Respiration, vol 5. Springer, Dordrecht, pp 367–385

Fares S, Mahmood T, Liu S, Loreto F, Centritto M (2011) Influence of growth temperature and measuring temperature on isoprene emission, diffusive limitations of photosynthesis and respiration in hybrid poplars. Atmos Environ 45:155–161

Farquhar GD, von Caemmerer, Berry JA (1980) A biochemical model of photosynthetic CO_2 assimilation in leaves of C_3 species. Planta 149:78–90

Flexas J, Díaz-Espejo A (2015) Interspecific differences in temperature response of mesophyll conductance: food for thought on its origin and regulation. Plant Cell Environ 38:625–628

Flexas J, Bota J, Loreto F, Cornic G, Sharkey TD (2004) Diffusive and metabolic limitations to photosynthesis under drought and salinity in C_3 plants. Plant Biol 6:269–279

Flexas J, Ribas-Carbó M, Hanson DT, Bota J, Otto B, Cifre J, McDowell N, Medrano H, Kaldenhoff R (2006) Tobacco aquaporin NtAQP1 is involved in mesophyll conductance to CO_2 *in vivo*. Plant J 48:427–439

Flexas J, Díaz-Espejo A, Galmés J, Kaldenhoff R, Medrano H, Ribas-Carbó M (2007) Rapid variations of mesophyll conductance in response to changes in CO_2 concentration around leaves. Plant Cell Environ 30:1284–1298

Flexas J, Ribas-Carbó M, Diaz-Espejo A, Galmés J, Medrano H (2008) Mesophyll conductance to CO_2: current knowledge and future prospects. Plant Cell Environ 31:602–621

Flexas J, Barbour MM, Brendel O, Cabrera HM, Carriquí M, Díaz-Espejo A, Douthe C, Dreyer E, …, Warren CR (2012) Mesophyll diffusion conductance to CO_2: an unappreciated central player in photosynthesis. Plant Sci 193–194: 70–84

Flexas J, Niinemets Ü, Gallé A, Barbour MM, Centritto M, Diaz-Espejo A, Cyril C, Jeroni J, …, Medrano H (2013a) Diffusional conductances to CO_2 as a target for increasing photosynthesis and photosynthetic water-use efficiency. Photosynth Res 117: 45–59

Flexas J, Scoffoni C, Gago J, Sack L (2013b) Leaf mesophyll conductance and leaf hydraulic conductance: an introduction to their measurement and coordination. J Exp Bot 64:3965–3981

Flexas J, Carriquí M, Coopman RE, Gago J, Galmés J, Martorell S, Morales F, Diaz-Espejo A (2014) Stomatal and mesophyll conductances to CO_2 in different plant groups: underrated factors for predicting leaf photosynthesis responses to climate change? Plant Sci 226:41–48

Flexas J, Díaz-Espejo A, Conesa MA, Coopman RE, Douthe C, Gago J, Gallé A, Galmés J, …, Niinemets Ü (2016) Mesophyll conductance to CO_2 and rubisco as targets for improving intrinsic water use efficiency in C_3 plants. Plant Cell Environ 39:965–982

Flexas J, Cano FJ, Carriquí M, Coopman RE, Mizokami Y, Tholen D, Xiong D (2018a) CO_2 diffusion inside photosynthetic organs. In: Adams W III, Terashima I (eds) The Leaf: A Platform for Performing Photosynthesis. Advances in Photosynthesis and Respiration (Including Bioenergy and Related Processes), vol 44. Springer, Cham, pp 163–208

Flexas J, Carriquí M, Nadal M (2018b) Gas exchange and hydraulics during drought in crops: who drives whom? J Exp Bot 69:3791–3795

Flowers MD, Fiscus EL, Burkey KO, Booker FL, Dubois J-JB (2007) Photosynthesis, chlorophyll fluorescence, and yield of snap bean (*Phaseolus vulgaris* L.) genotypes differing in sensitivity to ozone. Environ Exp Bot 61:190–198

Fowler D, Cape IN, Coyle M, Flechard C, Kuylenstierna J, Hicks K, Derwent D, Johnson C, Stevenson D (1999) The global exposure of forests to air pollutants. Water Air Soil Pollut 116:5–32

Gago J, Douthe C, Florez-Sarasa I, Escalona JM, Galmes J, Fernie AR, Flexas J, Medrano H (2014) Opportunities for improving leaf water use efficiency under climate change conditions. Plant Sci 226:108–119

Gago J, Daloso DM, Figueroa CM, Flexas J, Fernie AR, Nikoloski Z (2016) Relationships of leaf net photosynthesis, stomatal conductance, and mesophyll conductance to primary metabolism: a multispecies meta-analysis approach. Plant Physiol 171:1–15

Gago J, Carriquí M, Nadal M, Clemente-Moreno MJ, Coopman RE, Fernie AR, Flexas J (2019) Photosynthesis optimized across land plant phylogeny. Trends Plant Sci 24:947–958

Galle A, Florez-Sarasa I, Tomas M, Pou A, Medrano H, Ribas-Carbo M, Flexas J (2009) The role of mesophyll conductance during water stress and recovery in tobacco (*Nicotiana sylvestris*): acclimation or limitation? J Exp Bot 60:2379–2390

Galle A, Florez-Sarasa I, El Aououad H, Flexas J (2011) The Mediterranean evergreen *Quercus ilex* and the semideciduous *Cistus albidus* differ in their leaf gas exchange regulation and acclimation to repeated drought and re-watering cycles. J Exp Bot 62:5207–5216

Galmés J, Medrano H, Flexas J (2006) Acclimation of rubisco specificity factor to drought in tobacco: discrepancies between *in vitro* and *in vivo* estimations. J Exp Bot 57:3659–3667

Galmés J, Medrano H, Flexas J (2007) Photosynthetic limitations in response to water stress and recovery in Mediterranean plants with different growth forms. New Phytol 175:81–93

Galmés J, Kapralov MV, Copolovici LO, Hermida-Carrera C, Niinemets Ü (2015) Temperature responses of the rubisco máximum carboxylase activity across domains of life: phylogenetic signals, trade-offs, and importance for carbon gain. Photos Res 123:183–201

Galmés J, Hermida-Carrera C, Laanisto L, Niinemets Ü (2016) A compendium of temperature responses of rubisco kinetic traits: variability among and within photosynthetic groups and impacts on photosynthesis modeling. J Exp Bot 67:5067–5091

Gao F, Calatayud V, García-Breijo F, Reig-Armiñana J, Feng ZZ (2016) Effects of elevated ozone on physiological, anatomical and ultrastructural characteristics of four common urban tree species in China. Ecol Indic 67:367–379

Ge ZM, Zhou X, Kellomäki S, Zhang C, Peltola H, Martikainen PJ, Wang KY (2012) Acclimation of photosynthesis in a boreal grass (Phalaris arundinacea L.) under different temperature, CO2, and soil water regimes. Photosynthetica 50(1): 141-151

Grassi G, Magnani F (2005) Stomatal, mesophyll conductance and biochemical limitations to photosynthesis as affected by drought and leaf ontogeny in ash and oak trees. Plant Cell Environ 28:834–849

Greer DH (2018) Photosynthetic responses to CO_2 at different leaf temperatures in leaves of apple trees (*Malus domestica*) grown in orchard conditions with different levels of soil nitrogen. Environ Exp Bot 155:56–65

Gu L, Sun Y (2014) Artefactual responses of mesophyll conductance to CO_2 and irradiance estimated with the variable *J* and online isotope discrimination methods. Plant Cell Environ 37:1231–1249

Han JM, Meng HF, Wang SY, Jiang CD, Liu F, Zhang WF, Zhang YL (2016) Variability of mesophyll conductance and its relationship with water use efficiency in cotton leaves under drought pretreatment. J Plant Physiol 194:61–71

Han J, Lei Z, Zhang Y, Yi X, Zhang W, Zhang Y (2019) Drought-introduced variability of mesophyll conductance in *Gossypium* and its relationship with leaf anatomy. Physiol Plantarum 166:873–887

Hanba YT, Shibasaka M, Hayashi Y, Hayakawa T, Kasamo K, Terashima I, Katsuhara M (2004) Overexpression of the barley aquaporin HvPIP2;1 increases internal CO_2 conductance and CO_2 assimilation in the leaves of transgenic rice plants. Plant Cell Physiol 45:521–529

Harley PC, Loreto F, Di Marco G, Sharket TD (1992) Theoretical considerations when estimating the mesophyll conductance to CO_2 flux by analysis of the response of photosynthesis to CO_2. Plant Physiol 98:1429–1436

Hartmann DL, Klein Tank AMG, Rusticucci M, Alexander LV, Brönnimann S, Charabi Y, Dentener F, Dlugokencky E, …, Zhai PM (2013) Observations: atmosphere and surface. In: Stocker TF, Qin D, Plattner GK, Tignor M, Allen SK, Boschung J, Nauels A, Xia Y, …, Midgley PM (eds). Climate Change 2013: The Physical Science Basis. Cambridge/New York, IPCC/Cambridge University Press

Hassiotou F, Ludwig M, Renton M, Veneklaas EJ, Evans JR (2009) Influence of leaf dry mass oer area, CO_2, and irradiance on mesophyll conductance in sclerophylls. J Exp Bot 60:2303–2314

Ho QT, Berghuijs HNC, Watté R, Verboven P, Herremans E, Yin X, Retta MA, Aernouts B, … Nicolai BM (2016) Three-dimensional microscale modelling of CO_2 transport and light propagation in tomato leaves enlightens photosynthesis. Plant Cell Environ 39: 50–51

June T, Evans JR, Farquhar GD (2004) A simple new equation for the reversible temperature dependence of photosynthetic electron transport: a study on soybean leaf. Funct Plant Biol 31:275–283

Juurola E, Aalto T, Thum T, Vesala T, Hari P (2005) Temperature dependence of leaf-level CO_2 fixation: revising biochemical coefficients through analysis of leaf three-dimensional structure. New Phytol 166:205–215

Kitao M, Yazaki K, Kitaoka S, Fukatsu E, Tobita H, Komatsu M, Maruyama Y, Koike T (2015) Mesophyll conductance in leaves of Japanese white birch (*Betula platyphylla* var. *japonica*) seedlings grown under elevated CO_2 concentration and low N availability. Physiol Plantarum 155:435–445

Kolbe AR, Cousins AB (2018) Mesophyll conductance in *Zea mays* responds transiently to CO_2 availability: implications for transpiration efficiency in C_4 crops. New Phytol 217:1463–1474

Lehmeier C, Pajor R, Lundgren MR, Mathers A, Sloan J, Bauch M, Mitchell A, Bellasio C, … Fleming AJ (2017) Cell density and airspace patterning in the leaf can be manipulated to increase leaf photosynthetic capacity. Plant J 92: 981–994

Lewis JD, Phillips NG, Logan BA, Smith RA, Aranjuelo I, Clarke S, Offord CA, Frith, A, …, Tissue DT (2015) Rising temperature may negate the stimulatory effect of rising CO_2 on growth and physiology of Wollemi pine (*Wollemia nobilis*). Funct Plant Biol 42: 836–850

Li P, Feng ZZ, Calatayud V, Yuan XY, Xu YS, Paoletti E (2017) A meta-analysis on growth, physiological, and biochemical responses of woody species to ground-level ozone highlights the role of plant functional types. Plant Cell Environ 40:2369–2380

Li Y, Song X, Li S, Salter WT, Barbour MM (2019) The role of leaf water potential in the temperature response of mesophyll conductance. New Phytol. https://doi.org/10.1111/nph.16214

Long SP, Ainsworth EA, Rogers A, Ort DR (2004) Rising atmosphere carbon dioxide: plants FACE the future. Annu Rev Plant Biol 55:591

Loreto F, Centritto M (2008) Leaf carbon assimilation in a water-limited world. Plant Biosyst 142:154–161

Loreto F, Harley PC, Di Marco G, Sharkey TD (1992) Estimation of mesophyll conductance to CO_2 flux by three different methods. Plant Physiol 98:1437–1443

Maherali H, Reid CD, Polley HW, Johnson HB, Jackson RB (2002) Stomatal acclimation over a subambient to elevated CO_2 gradient in a C_3/C_4 grassland. Plant Cell Environ 25:557–566

Majeau N, Coleman JR (1996) Effect of CO_2 concentration on carbonic anhydrase and ribulose-1,5-bisphosphate carboxylase/oxygenase expression in pea. Plant Physiol 112:569–574

Martínez-Lüscher J, Morales F, Sánchez-Díaz M, Delrot S, Aguirreolea J, Gomès E, Pascual I (2015) Climate change conditions (elevated CO_2 and temperature) and UV-B radiation affect grapevine (*Vitis vinifera* cv. Tempranillo) leaf carbon assimilation, altering fruit ripening rates. Plant Sci 236:168–176

Martins SCV, Galmés J, Molins A, DaMatta FM (2013) Improving the estimation of mesophyll conductance to CO_2: on the role of electron transport rate correction and respiration. J Exp Bot 64:3285–3298

Matyssek R, Innes JL (1999) Ozone- a risk factor for forest trees and forests in Europe? Water Air Soil Pollut 116:199–226

Matyssek R, Sandermann H (2003) Impact of ozone on trees: an ecophysiological perspective. Prog Bot 64:349–404

Matyssek R, Günthardt-Goerg MS, Keller T, Scheidegger C (1991) Impairment of the gas exchange and structure in birch leaves (*Betula pendula*) caused by low ozone concentrations. Trees 5:5–13

Medlyn BE, Barton CVM, Broadmeadow MSJ, Ceulemans R, De Angelis P, Forstreuter M, Freeman M, Jackson SB, …, Jarvis PG (2001) Stomatal conductance of forest species after long-term exposure to elevated CO_2 concentration: a synthesis. New Phytol 149: 247–264

Meng GT, Li GX, He LP, Chai Y, Kong JJ, Lei YB (2013) Combined effects of CO_2 enrichment and drought stress on growth and energetic properties in the seedlings of a potential bioenergy crop *Jatropha curcas*. J Plant Growth Regul 32:542–550

Mizokami Y, Noguchi K, Kojima M, Sakakaibara H, Terashima I (2015) Mesophyll conductance decreases in the wild type but not in an ABA-deficient mutant (*aba1*) of *Nicotiana plumbaginifolia* under drought conditions. Plant Cell Environ 38:388–398

Mizokami Y, Noguchi K, Kojima M, Sakakibara H, Terashima I (2019a) Effects of instantaneous and growth CO_2 levels, and ABA on stomatal and mesophyll conductances. Plant Cell Environ 42:1257–1269

Mizokami Y, Sugiura D, Watanbe CKA, Betsuyaku E, Inada N, Terashima I (2019b) Elevated CO_2-induced changes in mesophyll conductance and anatomical traits in wild type and carbohydrate-metabolism mutants of Arabidopsis. J Exp Bot 70:4807–4818

Nadal M, Flexas J (2018) Mesophyll conductance to CO_2 diffusion: effects of drought and opportunities for improvement. In: García-Tejero IF, Duran-Zuazo VH (eds) Water Scarcity and Sustainable Agriculture in Semiarid Environment. Elsevier, London, pp 403–438

Nadal M, Flexas J (2019) Variation in photosynthetic characteristics with growth form in a water-limited scenario: implications for assimilation rates and water use efficiency in crops. Agr Water Manage 216:457–472

Niinemets Ü, Reichstein M (2003) Controls on the emission of plant volatiles through stomata: differential sensitivity of emission rates to stomatal closure explained. J Geophys Res 108:4208

Niinemets Ü, Díaz-Espejo A, Flexas J, Galmés J, Warren CR (2009) Role of mesophyll diffusion conductance in constraining potential photosynthetic productivity in the field. J Exp Bot 60:2249–2270

Niinemets Ü, Flexas J, Peñuelas J (2011) Evergreens favored by higher responsiveness to increased CO_2. Trends Ecol Evol 26:136–142

Ogée J, Wingate L, Genty B (2018) Estimating mesophyll conductance from measurements of $C^{18}OO$ photosynthetic discrimination and carbonic anhydrase activity. Plant Physiol 178:728–752

Oksanen E, Riikonen J, Kaakinen S, Holopainen T, Vapaavuori E (2005) Structural characteristics and chemical composition of birch (*Betula pendula*) leaves are modified by increasing CO_2 and ozone. Glob Change Biol 11:732–748

Onoda Y, Wright IJ, Evans JR, Hikosaka K, Kitajima K, Niinmets Ü, Poorter H, Tosens T, …, Westoby M (2017) Physiological and structural tradeoffs underlying the leaf economics spectrum. New Phytol 214: 1447–1463

Osborn HL, Alonso-Cantabrana H, Sharwood RE, Covshoff S, Evans JR, Furbank RT, von Caemmerer S (2017) Effects of reduced carbonic anhydrase activity on CO_2 assimilation rates in *Setaria viridis*: a transgenic analysis. J Exp Bot 68:299–310

Ouyang W, Struik PC, Yin X, Yang J (2017) Stomatal conductance, mesophyll conductance, and transpiration efficiency in relation to leaf anatomy in rice and wheat genotypes under drought. J Exp Bot 68:5191–5205

Overdieck D (1989) The effects of preindustrial and predicted future atmospheric CO_2 concentration on *Lyonia mariana* L.D. Don. Funct Ecol 3:569–576

Paoletti E, Contran N, Bernasconi P, Gunthardt-Goerg MS, Vollenweider P (2009) Structural and physiological responses to ozone in manna ash (*Fraxinus ornus* L.) leaves of seedlings and mature trees under controlled and ambient conditions. Sci Total Environ 407:1631–1643

Pérez-Martin A, Michelazzo C, Torres-Ruiz JM, Flexas J, Fernández JE, Sebastiani L, Diaz-Espejo A (2014) Regulation of photosynthesis and stomatal and mesophyll conductance under water stress and recovery in olive trees: correlation with gene expression of carbonic anhydrase and aquaporins. J Exp Bot 65:3143–3156

Pérez-Romero JA, Idaszkin YL, Barcia-Piedras JM, Duarte B, Redondo-Gómez S, Caçador I, Mateos-Naranjo E (2018) Disentangling the effect of atmospheric CO_2 enrichment on the halophyte *Salicornia ramosissima* J. woods physiological performance under optimal and suboptimal saline conditions. Plant Physiol Bioch 127:617–629

Pons TL, Welschen RAM (2003) Midday depression of net photosynthesis in the tropical rainforest tree *Eperua grandiflora*: contributions of stomatal and internal conductances, respiration and rubisco. Tree Physiol 23:937–947

Pons TL, Flexas J, von Caemmerer S, Evans JR, Genty B, Ribas-Carbo M, Brugnoli E (2009) Estimating mesophyll conductance to CO_2: methodology, potential errors, and recommendations. J Exp Bot 60:2217–2234

Poorter H, Niinemets Ü, Poorter L, Wright IJ, Villar R (2009) Causes and consequences of variation in leaf mass per area (LMA): a meta-analysis. New Phytol 182:565–588

Porter MA, Grodzinski B (1984) Acclimation to CO_2 in bean. Carbonic anhydrase and ribulose bisphosphate carboxylase. Plant Physiol 74:413–416

Possell M, Hewitt CN (2009) Gas exchange and photosynthetic performance of the tropical tree *Acacia nigrescens* when grown in different CO_2 concentrations. Planta 229:837–846

Pou A, Medrano H, Tomàs M, Martorell S, Ribas-Carbó M, Flexas J (2012) Anisohydric behaviour in grapevines results in better performance under moderate water stress and recovery than isohydric behaviour. Plant Soil 359:335–349

Qiu C, Ethier G, Pepin S, Dubé P, Desjardins Y, Grosselin A (2017) Persistent negative temperature response of mesophyll conductance in red raspberry (*Rubus idaeus* L.) leaves under both high and low vapour pressure deficits: a role for abscisic acid? Plant Cell Environ 40:1940–1959

Reich PB (1987) Quantifying plant response to ozone: a unifying theory. Tree Physiol 3:63–91

Reich PB, Hobbie SE, Lee TD, Pastore MA (2018) Unexpected reversal of C_3 versus C_4 grass response to elevated CO_2 during a 20-year field experiment. Science 360:317–320

Robertson EJ, Leech RM (1995) Significant changes in cell and chloroplast development in young wheat leaves (*Triticum aestivum* cv Hereward) grown in elevated CO_2. Plant Physiol 107:63–71

Rogers A, Medlyn BE, Dukes JS, Bonan G, von Caemmerer S, Dietze MC, Kattge J, Leakey ADB, … Zaehle S (2017) A roadmap for improving the representation of photosynthesis in earth system models. New Phytol 213 22–42

Roumet C, Bel MP, Sonie L, Jardon F, Roy J (1996) Growth response of grasses to elevated CO_2: a physiological plurispecific analysis. New Phytol 133:595–603

Ruiz-Vera UM, De Souza AP, Long SP, Ort DR (2017) The role of sink strength and nitrogen availability in the down-regulation of photosynthetic capacity in field-grown *Nicotiana tabacum* L. at elevated CO_2 concentration. Front Plant Sci 8:998

Sáez PL, Cavieres LA, Galmés J, Gil-Pelegrín E, Peguero-Pina JJ, Sancho-Knapik D, Vivas M, Sanhueza C, … Bravo LA (2018a) *In situ* warming in the Antarctic: effects on growth and photosynthesis in Antarctic vascular plants. New Phytol 218: 1406–1418

Sáez PL, Galmés J, Ramírez CF, Poblete L, Rivera BK, Cavieres LA, Clemente-Morenob MJ, Flexasb J, … Bravo LA (2018b) Mesophyll conductance to CO_2 is the most significant limitation to photosynthesis at different temperatures and water availabilities in Antarctic vascular species. Env Exp Bot 156: 279–287

Sage RF, Kubien DS (2007) The temperature response of C_3 and C_4 photosynthesis. Plant Cell Environ 30:1086–1106

Sakata T, Nakano T, Kachi N (2015) Effects of internal conductance and rubisco on the optimum temperature for leaf photosynthesis in *Fallopia japonica* growing at different altitudes. Ecol Res 30:163–171

Salazar-Parra C, Aguirreolea J, Sánchez-Díaz M, Irigoyen JJ, Morales F (2012) Photosynthetic response of Tempranillo grapevine to climate change scenarios. Ann Appl Biol 161:277–292

Saxe H, Ellsworth DS, Heath J (1998) Tree and forest functioning in an enriched CO_2 atmosphere. New Phytol 139:395–436

Scafaro AP, von Caemmerer S, Evans JR, Atwell BJ (2011) Temperature response of mesophyll conductance in cultivated and wild *Oryza* species with contrasting mesophyll cell wall thickness. Plant Cell Environ 34:1999–2008

Sekhar KM, Reddy KS, Reddy AR (2017) Amelioration of drought-induced negative responses by elevated CO_2 in field grown short rotation coppice mulberry (*Morus* spp.), a potential bio-energy tree crop. Photosynth Res 132:151–164

Sharkey TD, Bernacchi CJ, Farquhar GD, Singsaas (2007) Fitting photosynthetic carbon dioxide response curves for C_3 leaves. Plant Cell Environ 30:1035–1040

Sharwood RE, Crous KY, Whitney SM, Ellsworth DS, Ghannoum O (2017) Linking photosynthesis and leaf N allocation under future elevated CO_2 and climate warming in *Eucalyptus globulus*. J Exp Bot 68:1157–1167

Shrestha A, Song X, Barbour MM (2019) The temperature response of mesophyll conductance, and its copmonent conductance, varies between species and genotypes. Photosynth Res 141:65–82

Sicher RC, Kremer DF, Rodermel SR (1994) Photosynthetic acclimation to elevated CO_2 occurs in transformed tobacco with decreased ribulose-1,5-bisphosphate carboxylase/oxygenase content. Plant Physiol 104:409–415

Silim SN, Ryan N, Kubien DS (2010) Temperature responses of photosynthesis and respiration in *Populus balsamifera* L.: acclimation versus adaptation. Photosynth Res 104:19–30

Silva EN, Silveira JAG, Ribeiro RV, Vieira SA (2015) Photoprotective function of energy dissipation by thermal processes and photorespiratory mechanisms in *Jatropha curcas* plants during different intensities of drought and after recovery. Env Exp Bot 110:36–45

Singh SK, Reddy VR (2016) Method of mesophyll conductance estimation: its impact on key biochemical parameters and photosynthetic limitations in phosphorus-stressed soybean across CO_2. Physiol Plantarum 157:234–254

Singh SK, Reddy VR (2018) Co-regulation of photosynthetic processes under potassium deficiency across CO_2 levels in soybean: mechanisms of limitations and adaptations. Photosynth Res 137:183–200

Singh SK, Badgujar G, Reddy VR, Fleisher DH, Bunce JA (2013) Carbon dioxide diffusion across stomata and mesophyll and photo-biochemical processes as affected by growth CO_2 and phosphorus nutrition in cotton. J Plant Physiol 170:801–813

Singsaas EL, Ort DR, DeLucia EH (2003) Elevated CO_2 effects on mesophyll conductance and its consequences for interpreting photosynthetic physiology. Plant Cell Environ 27:41–50

Smith RA, Lewis JD, Ghannoum O, Tissue DT (2012) Leaf structural responses to pre-industrial, current and elevated atmospheric [CO_2] and temperature affect leaf function in *Eucalyptus sideroxylon*. Funct Plant Biol 39:285–296

Sonawane BV, Cousins AB (2019) Uncertainties and limitations of using carbon-13a¡ and oxygen-18 leaf isotope exchange to estimate the temperature response of mesophyll CO_2 conductance in C_3 plants. New Phytol 222:122–131

Song Y, Huang B (2014) Differential effectiveness of doubling ambient atmospheric CO_2 concentration mitigating adverse effects of drought, heat, and combined stress in Kentucky bluegrass. J Amer Soc Hort Sci 139:364–373

Sorrentino G, Haworth M, Wahbi S, Mahmood T, Zuomin S, Centritto M (2016) Abscisic acid induces rapid reductions in mesophyll conductance to carbon dioxide. PLoS One 11:e0148554

Sugiura D, Watanabe CKA, Betsuyaku E, Terashima I (2017) Sink-source balance and down-regulation of photosynthesis in *Raphanus sativus*: effects of grafting, N and CO_2. Plant Cell Physiol 58:2043–2056

Sun Y, Gu L, Dickinson RE, Norby RJ, Pallardy SG, Hoffman FM (2014) Impact of mesophyll diffusion on estimated global land CO_2 fertilization. PNAS 111:15774–15779

Tazoe Y, von Caemmerer S, Badger MR, Evans JR (2009) Light and CO_2 do not affect the mesophyll conductance to CO_2 diffusion in wheat leaves. J Exp Bot 60:2291–2301

Teng N, Wang J, Chen T, Wu X, Wang Y, Lin J (2006) Elevated CO_2 induces physiological, biochemical and structural changes in leaves of *Arabidopsis thaliana*. New Phytol 172:92–103

Terashima I, Hanba YT, Tholen D, Niinemets Ü (2011) Leaf functional anatomy in relation to photosynthesis. Plant Physiol 155:108–116

Théroux-Rancourt G, Éthier G, Pepin S (2015) Greater efficiency of water use in poplar clones having a delayed response of mesophyll conductance to drought. Tree Physiol 35:172–184

Tholen D, Ethier G, Genty B, Pepin S, Zhu XG (2012) Variable mesophyll conductance revisited: theoretical background and experimental implications. Plant Cell Environ 35:2087–2103

Tomàs M, Flexas J, Copolovici L, Galmés J, Hallik L, Medrano H, Ribas-Carbó M, Tosens T, …, Niinemets Ü (2013) Importance of leaf anatomy in determining mesophyll diffusion conductance to CO_2 across species: quantitative limitations and scaling up by models. J Exp Bot 64 2269–2281

Tosens T, Nishida K, Gago J, Coopman RE, Cabrero HM, Carriquí M, Laanisto L, Morales L, …, Flexas J (2016) The photosynthetic capacity in 35 ferns and fern allies: mesophyll CO_2 diffusion as a key trait. New Phytol 209 1576–1590

Trenberth KE (2011) Changes in precipitation with climate change. Clim Res 47:123–138

Turner J, Lu H, White I, King JC, Phillips T, Hosking JS, Bracegirdle TJ, Marshall GJ, … Deb P (2016) Absence of 21st century warming on Antarctic peninsula consistent with natural variability. Nature 535: 411–415

Ubierna N, Gandin A, Boyd RA, Cousins AB (2017) Temperature response of mesophyll conductance in three C_4 species calculated with two methods: ^{18}O discrimination and *in vitro* V_{pmax}. New Phytol 214:66–80

Ubierna N, Gandin A, Cousins AB (2018) The response of mesophyll conductance to short-term variation in CO_2 in the C_4 plants *Setaria viridis* and *Zea mays*. J Exp Bot 69:1159–1170

Uprety DC, Dwivedi N, Mohan R, Paswan G (2001) Effect of elevated CO_2 concentration on leaf structure of *Brassica juncea* under water stress. Biol Plantarum 44:149–152

Varone L, Ribas-Carbo M, Cardona C, Gallé A, Medrano H, Gratani L, Flexas J (2012) Stomatal and non-stomatal limitations to photosynthesis in seedlings and saplings of Mediterranean species pre-conditioned and aged in nurseries: different response to water stress. Env Exp Bot 75:235–247

Varone L, Vitale M, Catoni R, Gratani (2016) Physiological differences of five holm oak (*Quercus ilex* L.) ecotypes growing under common growth conditions were related to native local climate. Plant Spec Biol 31:196–210

Velikova V, Tsonev T, Pinelli P, Alessio GA, Loreto F (2005) Localized ozone fumigation system for studying ozone effects on photosynthesis, respiration, electron transport rate and isoprene emission in field-grown Mediterranean oak species. Tree Physiol 25:1523–1532

Velikova V, Tsonev T, Barta C, Centritto M, Koleva D, Stefanova M, Busheva M, Loreto F (2009) BVOC emissions, photosynthetic characteristics and changes in chloroplast ultrastructure of *Platanus orientalis* L. exposed to elevated CO_2 and high temperature. Environ Pollut 157:2629–2637

Veromann-Jürgenson LL, Tosens T, Laanisto L, Niinemets Ü (2017) Extremely thick cell walls and low mesophyll conductance: welcome to the world of ancient living! J Exp Bot 68:1639–1653

Vico G, Way DA, Hurry V, Manzoni S (2019) Can leaf photosynthesis acclimate to rising and more variable temperatures? Plant Cell Environ 42:1913–1928

von Caemmerer S, Evans JR (1991) Determination of the average partial pressure of CO_2 in chloroplasts from leaves of several C_3 plants. Aust J Plant Physiol 18:287–305

von Caemmerer S, Evans JR (2015) Temperature responses of mesophyll conductance differ greatly between species. Plant Cell Environ 38:629–637

von Cammerer S (2013) Steady-state models of photosynthesis. Plant Cell Environ 36:1617–1630

Walker B, Ariza LS, Kaines S, Badger MR, Cousins AB (2013) Temperature response of *in vivo* rubisco kinetics and mesophyll conductance in *Arabidopsis thaliana*: comparisons to *Nicotiana tabacum*. Plant Cell Environ 36:2108–2119

Wang Z, Anderson OR, Griffin KL (2004) Chloroplast numbers, mitochondrion numbers and carbon assimilation physiology of *Nicotiana sylvestris* as affected by CO_2 concentration. Environ Exp Bot 51:21–31

Wang H, Prentice IC, Keenan TF, Davis TW, Wright IJ, Cornwell WK, Evans BJ, Peng C (2017) Towards a universal model for carbon dioxide uptake by plants. Nat Plants 3:734–741

Wang X, Du T, Huang J, Peng S, Xiong D (2018) Leaf hydraulic vulnerability triggers the decline in stomatal and mesophyll conductance during drought in rice. J Exp Bot 69:4033–4045

Warren CR (2008a) Stand aside stomata, another actor deserves Centre stage: the forgotten role of the internal conductance to CO_2 transfer. J Exp Bot 59:1475–1487

Warren CR (2008b) Does growth temperature affect the temperature responses of photosynthesis and internal conductance to CO_2? A test with *Eucalyptus regnans*. Tree Physiol 28:11–19

Warren CR, Dreyer E (2006) Temperature response of photosynthesis and internal conductance to CO_2: results from two independent approaches. J Exp Bot 57:3057–3067

Warren CR, Löw M, Matyssek R, Tausz M (2007) Internal conductance to CO_2 transfer of adult *Fagus sylvatica*: variation between sun and shade leaves and due to free-air ozone fumigation. Environ Exp Bot 59:130–138

Watanabe M, Kaminaki Y, Mori M, Okabe S, Arakawa I, Kinose Y, Nakaba S, Izuta T (2018) Mesophyll conductance to CO_2 in leaves of Siebold's beech (*Fagus crenata*) seedlings under elevated ozone. J Plant Res 131:907–914

Wickham H (2016) ggplot2: elegant graphics for data analysis. Springer, New York. ISBN. isbn:978-3-319-24277-4

Wittig VE, Ainsworth EA, Long SP (2007) To what extent do current and projected increases in surface ozone affect photosynthesis and stomatal conductance of trees? A meta-analytic review of the last 3 decades of experiments. Plant Cell Environ 30:1150–1162

Wright IJ, Reich PB, Westoby M, Ackerly DD, Baruch Z, Bongers F, Cavender-Bares J, Chapin T, … Villar R (2004) The worldwide leaf economics spectrum. Nature 428: 821–827

Xiong D, Liu X, Liu L, Douthe C, Li Y, Peng S, Huang J (2015) Rapid responses of mesophyll conductance to changes of CO_2 concentration, temperature and irradiance are affected by N supplements in rice. Plant Cell Environ 38:2541–2550

Xiong D, Flexas J, Yu T, Peng S, Huang J (2017) Leaf anatomy mediates coordination of leaf hydraulic conductance and mesophyll conductance to CO_2 in *Oryza*. New Phytol 213:572–583

Xu CY, Salih A, Ghannoum O, Tissue DT (2012) Leaf structure characteristics are less important than leaf chemical properties in determining the response of leaf mass per area and photosynthesis of *Eucalyptus saligna* to industrial-age changes in [CO_2] and temperature. J Exp Bot 63:5829–5841

Xu G, Singh SK, Reddy VR, Barnaby JY, Sicher RC, Li T (2016) Soybean grown under elevated CO_2 benefits more under low temperature than high temperature stress: varying response of photosynthetic limitations, leaf metabolites, growth, and seed yield. J Plant Physiol 205:20–32

Xu Y, Feng Z, Shang B, Dai L, Uddling J, Tarvainen L (2019) Mesophyll conductance limitation of photosynthesis in poplar under elevated ozone. Science Total Environ 657:136–145

Xue W, Otieno D, Ko J, Werner C, Tenhunen J (2016) Conditional variations in temperature response of photosynthesis, mesophyll and stomatal control of water use in rice and winter wheat. Field Crop Res 199:77–88

Yamori W, Noguchi K, Hanba YT, Terashima I (2006a) Effects of internal conductance on the temperature dependence of the photosynthetic rate in spinach leaves from contrasting growth temperatures. Plant Cell Physiol 47:1069–1080

Yamori W, Suzuki K, Noguchi K, Nakai M, Terashima I (2006b) Effects of rubisco kinetics and rubisco activation state on the temperature dependence of the photosynthetic rate in spinach leaves from contrasting growth temperatures. Plant Cell Environ 29:1659–1670

Yamori W, Hikosaka K, Way DA (2014) Temperature response of photosynthesis in C_3, C_4 and CAM plants: temperature acclimation and temperature adaptation. Photosynth Res 119:101–117

Yang Z, Liu J, Tischer SV, Christmann A, Windisch W, Schnyder H, Grill E (2016) Leveraging abscisic acid receptors for efficient water use in *Arabidopsis*. Proc Natl Acad Sci 113:6791–6796

Yin X, Struik PC, Romero P, Harbinson J, Evers JB, van der Putten PEL, Vos J (2009) Using combined measurements of gas exchange and chlorophyll fluorescence to estimate parameters of a biochemical C_3 photosyntthesis model: a critical appraisal and a new integrated approach applied to leaves in a wheat (*Triticum aestivum*) canopy. Plant Cell Environ 32: 448–464

Yue X, Unger N, Harper K, Xia XG, Liao H, Zhu T, Xiao JF, Feng ZZ, Li J (2017) Ozone and haze pollution weakens net primary productivity in China. Atmos Chem Phys 17:6073–6089

Zait Y, Shtein I, Schwartz A (2019) Long-term acclimation to drought, salinity and temperature in the thermophilic tree *Ziziphus spina-christi*: revealing different tradeoffs between mesophyll and stomatal conductance. Tree Physiol 39:701–716

Zhang SB (2010) Temperature acclimation of photosynthesis in *Mecanopsis horridula* var. *racemose* Prain. Bot Stud 51:457–464

Zhu C, Ziska L, Zhu J, Zeng Q, Xie Z, Tang H, Jia X, Hasegawa T (2012) The temporal and species dynamics of photosynthetic acclimation in flag leaves of rice (*Oryza sativa*) and wheat (*Triticum aestivum*) under elevated carbon dioxide. Physiol Plantarum 145:395–405

Chapter 4

Photosynthetic Acclimation to Temperature and CO_2: The Role of Leaf Nitrogen

André G. Duarte, Mirindi E. Dusenge, Sarah McDonald, Kristyn Bennett, Karen Lemon, Julianne Radford
Department of Biology, University of Western Ontario, London, ON, Canada

Danielle A. Way*
Department of Biology, University of Western Ontario, London, ON, Canada

Nicholas School of the Environment, Duke University, Durham, NC, USA

Terrestrial Ecosystem Science & Technology Group, Environmental & Climate Sciences Department, Brookhaven National Laboratory, Upton, NY, USA

Summary

Rising CO_2 and increasing temperatures affect photosynthesis directly, but can also impact net CO_2 assimilation rates by altering leaf nitrogen. We review the effects of high CO_2 and warming on photosynthesis, including photosynthetic acclimation, focusing on the role of leaf nitrogen. While elevated CO_2 often leads to a decrease in leaf nitrogen and an associated down-regulation of photosynthesis, there is growing evidence from field experiments that photosynthetic down-regulation may be avoided over decadal timescales if plants increase their access to soil nitrogen. In an analysis of 59 species grown at elevated CO_2, we found no relationship between the increase in growth CO_2 concentration and changes in leaf nitrogen content, implying that decreases in leaf nitrogen concentration and leaf mass per area largely offset each other. However, there was a positive relationship between changes in leaf nitrogen

*Author for correspondence, e-mail: dway4@uwo.ca

K. M. Becklin et al. (eds.), *Photosynthesis, Respiration, and Climate Change*, Advances in Photosynthesis and Respiration 48, https://doi.org/10.1007/978-3-030-64926-5_4

content and the degree of photosynthetic down-regulation in response to elevated CO_2 in C_3, but not C_4, species. We also show that across 48 species, elevated growth temperatures did not have a consistent effect on leaf nitrogen content. While warming-induced changes in leaf nitrogen content were positively correlated with changes in photosynthesis in woody species, there was no relationship between shifts in leaf nitrogen content and shifts in photosynthesis in herbaceous C_3 species. Our results highlight the importance of understanding how leaf nitrogen dynamics change in response to climate change, and the general coupling of leaf nitrogen status and photosynthetic performance under both elevated CO_2 and temperature conditions.

I. Introduction

Since the Industrial Revolution, atmospheric CO_2 concentrations have risen from ~275 ppm (Joos and Spahni 2008) to about 410 ppm (Dlugokencky and Tans 2018), an almost 50% increase. The rise in CO_2 and other greenhouse gases has increased global mean temperatures by ~1 °C compared to the period from 1880 to 1900 (NOAA 2019), with some regions, such as Alaska, already being 2.5–3.3 °C warmer now than their mean twentieth century average temperatures (NOAA 2019). As CO_2 concentrations continue to rise, potentially as high as 1000 ppm by the year 2100 (Ciais et al. 2013), global temperatures will increase even further. Currently, global mean annual air temperatures are predicted to warm 1–3.7 °C by the end of the century under a business as usual scenario (Ciais et al. 2013), which would mean increased temperatures of over 8 °C in high latitude regions (Collins et al. 2013).

Photosynthesis plays a key role in the global carbon cycle, with terrestrial vegetation absorbing ~123 Gt C each year from the air (Ciais et al. 2013). This large carbon flux helps determine atmospheric CO_2 concentrations, and thus the rate of future climate change. Photosynthesis also dictates the availability of carbon that can be used for plant growth, thereby underlying crop yield and food security (Long et al. 2006; Zhu et al. 2010). Given the importance of photosynthesis on a global scale for both climate and agriculture, it is crucial to understand how photosynthesis will be altered by ongoing climate change, particularly rising CO_2 and temperatures.

The ability of photosynthesis to acclimate to different growth conditions, including variation in temperature and CO_2 concentration, has been studied for decades (e.g., Pearcy 1977; Mooney et al. 1978; Berry and Björkman 1980; Sage et al. 1989; Long and Drake 1991). Despite this long history, we still lack a general framework for predicting how photosynthesis will respond to the expected increases in CO_2 concentration and temperature over the coming decades (Dusenge et al. 2019). Critically, photosynthetic rates are often correlated with leaf nitrogen status (Evans 1989; Makino et al. 1994; Wright et al. 2004), and acclimation to both CO_2 and warming is often linked to changes in leaf nitrogen concentration (Tjoelker et al. 1998; Ainsworth and Long 2005). The question of how nitrogen availability will affect plant responses to rising temperatures and CO_2 concentrations has important implications for our projections of future productivity and vegetation-atmosphere interactions (Hungate et al. 2003; Reich et al. 2006; Reich and Hobbie 2013; Wieder et al. 2015). Here, we therefore review the current state of our knowledge of photosynthetic acclimation responses to both increasing growth temperatures and atmospheric CO_2 concentrations in C_3 plants, focusing on the role of leaf and plant nitrogen status.

II. Modeling the CO_2 and Temperature Dependence of Photosynthesis

In the Farquhar, Berry and von Caemmerer (1980) biochemical model of photosynthesis, net CO_2 assimilation rates are limited by either Rubisco (ribulose-1,5-bisphosphate carboxylase/oxygenase) limitations, RuBP (ribulose-1,5-bisphosphate) regeneration limitations, or triose phosphate utilization (TPU) limitations. Depending on the leaf internal CO_2 concentration, light intensity, and temperature, different biochemical limitations will limit net photosynthesis (Farquhar et al. 1980; Sage and Kubien 2007; Busch and Sage 2017). At low CO_2 concentrations, Rubisco limitations prevail because there is insufficient CO_2 to saturate Rubisco carboxylation rates, but Rubisco oxygenation rates are high (i.e., photorespiration), assuming that the supply of RuBP is saturating. Thus, photosynthesis is limited by the maximum carboxylation rate of Rubisco (V_{cmax}). But as CO_2 substrate availability increases at higher CO_2 concentrations, photorespiration is increasingly suppressed. Under these conditions, either the ability of electron transport to support the regeneration of RuBP or other reactions within the Calvin Cycle itself become limiting for photosynthesis, such that photosynthesis is limited by the maximum electron transport rate (J_{max}) (Farquhar et al. 1980). At very high CO_2 concentrations, the ability to convert triose phosphates into sucrose and starch, and thereby to release inorganic phosphate, puts an upper ceiling on the net photosynthetic rate (Sharkey 1985; Sage 1990; Sage 1994).

A similar switch between the biochemical limitations of photosynthesis occurs as leaf temperature varies at current CO_2 concentrations. At low leaf temperatures, TPU limitations usually constrain photosynthesis (Sage and Sharkey 1987; Labate and Leegood 1988; Yang et al. 2016), partly because TPU-limited photosynthesis is highly temperature-sensitive (Jolliffe and Tregunna 1973; Cen and Sage 2005; Yang et al. 2016). This often transitions to a Rubisco limitation at moderately low temperatures, whereas at higher leaf temperatures, either RuBP regeneration or Rubisco limitations typically prevail (Sage and Kubien 2007; Busch and Sage 2017).

III. Photosynthetic Acclimation to Rising CO_2

Increased CO_2 concentrations stimulate instantaneous rates of photosynthesis, since CO_2 is the substrate for Rubisco, and suppress photorespiration (Farquhar et al. 1980). This short-term response of photosynthesis to elevated CO_2 is well-established, and led to initial predictions that rising atmospheric CO_2 concentrations would stimulate leaf carbon fixation rates in plants, producing a CO_2 fertilization effect (Drake et al. 1997).

The first experiments to test this idea were performed with potted plants in growth chambers and glasshouses, where CO_2 could be controlled. These studies found that although high CO_2 concentrations did indeed increase photosynthetic rates at first, photosynthesis in the high CO_2-grown plants decreased over time (Clough et al. 1981; DeLucia et al. 1985; Cure and Acock 1986; Arp 1991). This CO_2 acclimation, or photosynthetic down-regulation, was stronger and more rapid in plants grown in small pots, with low nitrogen availability, or with a high source-sink ratio (Clough et al. 1981; Cure and Acock 1986; Curtis and Wang 1998; Jablonski et al. 2002). In plants with a large sink capacity, extra carbohydrates resulting from enhanced photosynthetic rates at high CO_2 could be used to build new tissues or be consumed by other carbon sinks (e.g., root exudation), and photosynthetic down-regulation was weak or absent (Herold 1980; Campbell et al. 1988). But when sink strength was limited, whether by low nutrient availability, restricted rooting volume, or manipulation of source-sink ratios, non-structural carbohydrates accumulated in the leaves of plants exposed to high CO_2 (DeLucia et al. 1985; Arp 1991). This build-up of leaf carbohydrates indicates a source-sink imbalance in carbon at the whole plant level, which induces feedback inhibition (Herold 1980; DeLucia et al. 1985). Increases in leaf sucrose under elevated CO_2 are thought to induce a hexokinase signaling pathway that suppresses Rubisco transcription, leading to a down-regulation of Rubisco concentrations (Moore et al. 1999), with Rubisco concentration decreases of up to 60% reported (Sage et al. 1989; Rowland-Bamford et al. 1991). Since Rubisco is the most abundant enzyme in C_3 leaves, accounting for up to 28% of leaf nitrogen (Evans 1989) and 55% of soluble leaf protein (Campbell et al. 1988), reductions in Rubisco concentrations result in strong reductions in leaf N.

The results from these studies highlighted that the N status of the plant needed to be taken into account when predicting photosynthetic responses to rising CO_2. The question therefore turned to whether photosynthetic down-regulation would occur in natural ecosystems, where rooting volume is essentially unlimited and fluxes of soil nitrogen are large. In both field chambers (such as open-top chambers) and Free Air CO_2 Enrichment (FACE) studies, strong stimulations of photosynthesis are seen when a high CO_2 treatment is initially imposed (Bowes 1991; Ainsworth and Rogers 2007). In some field studies, this initial stimulation of photosynthesis was maintained for months or years (Radin et al. 1987; Ziska et al. 1990; Bader et al. 2010). However, even when there are no artificial limitations on plant nitrogen uptake, acclimation to elevated CO_2 often occurs. In almost all CO_2 field experiments, some degree of photosynthetic acclimation has been found (Leuzinger et al. 2011; Wang et al. 2012; Warren et al. 2015). In most of these cases, high CO_2 does enhance carbon uptake, just not to the extent expected without down-regulation (Ainsworth and Rogers 2007; Leakey et al. 2009). These higher photosynthetic rates increase productivity in most FACE experiments, often over multiple years (Herrick and Thomas 2001; Ainsworth and Long 2005; Luo et al. 2006; Nowak et al. 2004; Norby et al. 2005; Liberloo et al. 2009; McCarthy et al. 2010; Way et al. 2010; Aranjuelo et al. 2011, 2011; Dawes et al. 2011; Zak et al. 2011; Ellsworth et al. 2012; Wang et al. 2012; Terrer et al. 2016). But in some FACE experiments, strong down-regulation of photosynthesis in high CO_2 plots led to either very little or even no CO_2 fertilization after only a few years of study (Luo et al. 2004; Dukes et al. 2005; Adair et al. 2009; Bader et al. 2009; Norby et al. 2010; Leuzinger et al. 2011; Newingham et al. 2013; Warren et al. 2015; Ellsworth et al. 2017).

In cases where there is a weak or nonexistent CO_2 fertilization effect, nutrient limitations have been proposed to constrain the response to CO_2 (Luo et al. 2004; Dukes et al. 2005; Finzi et al. 2006; Norby et al. 2010; Ellsworth et al. 2017). The Progressive Nitrogen Limitation (PNL) theory posits that soil nitrogen becomes an increasingly limiting nutrient for ecosystems as nitrogen is sequestered in plant tissue and soil organic matter, reducing the ability of plants to fully respond to elevated CO_2 (Luo et al. 2004, 2006). The role of nitrogen availability in dictating the strength of CO_2 acclimation in field experiments can be directly evaluated by looking at studies that combined CO_2 and fertilizer treatments. Over a ten-year FACE experiment, Ainsworth et al. (2003) found that photosynthetic capacity was reduced 23% in *Lolium perrene* grown at elevated CO_2 with low N, but only by 12% in the high N, high CO_2 plots, confirming the role of nitrogen availability in reducing photosynthetic down-regulation. But at the Duke FACE experiment, photosynthetic capacity of *Pinus taeda* was largely unchanged by elevated CO_2 and nitrogen fertilization over 10 years (Paschalis et al. 2017), even though annual woody growth at the site was increased 47% in the high CO_2, high nitrogen plots (Oren et al., 2001). Similar results were found in the EUROFACE experiment, where poplar species were grown under elevated CO_2, either with or without supplemental fertilizer. High CO_2 had no impact on poplar V_{cmax}, though it increased J_{max} by 15%, and there was little difference between photosynthetic traits in the fertilized and non-fertilized trees in the high CO_2 treatment, while biomass production was stimulated up to 29% (Liberloo et al. 2007). Fertilization also had little impact on photosynthetic responses to high CO_2 in rice, where even under the low nitrogen treatment, there was sufficient nitrogen to support growth under elevated CO_2 (Seneweera et al. 2002). A recent analysis of long-term FACE studies (7–11 years in duration) found no evidence that the impact of elevated CO_2 on aboveground net primary productivity declined over time, and instead found that the impact of elevated CO_2 on leaf nitrogen concentrations diminished over time (Feng et al. 2015).

How could plants grown at elevated CO_2 maintain high leaf nitrogen concentrations over time and minimize photosynthetic acclimation to CO_2? Elevated CO_2 can increase belowground carbon allocation, thereby increasing plant nitrogen uptake and helping maintain positive CO_2 effects on productivity over time (Drake et al. 2011). At both the Duke and Oak Ridge National Laboratory FACE sites, as well as the Rhinelander FACE site where nitrogen limited tree growth, elevated CO_2 increased nitrogen uptake (Finzi et al. 2007), likely through a combination of enhanced fine root production (Norby et al. 2004) and greater allocation of carbon to soil microbes and mycorrhizae (Finzi et al. 2007). Similarly, nitrogen uptake and nitrogen mineralization rates increased in elevated CO_2 plots in a grassland system, but these effects were primarily related to improved soil moisture conditions caused by a decrease in transpiration rates in high CO_2 plants (Hungate et al. 1996). However, neither root growth nor root activity was altered by elevated CO_2 in an alpine site, indicating that belowground traits may be less sensitive to rising CO_2 in late successional sites (Handa et al. 2008). While PNL focuses explicitly on nitrogen, and nitrogen is likely to limit the CO_2 responsiveness of vegetation for temperate and boreal ecosystems, in tropical soils, phosphorous limitations are likely to be more common (Goll et al. 2012; Ellsworth et al. 2017).

Both early pot studies and more recent field work clearly show that how rising CO_2 affects leaf nitrogen status is closely tied to how rising CO_2 affects photosynthesis. While the impact of elevated CO_2 on leaf nitrogen has been analyzed in previous studies (e.g., Ainsworth and Long 2005 for FACE studies), the relationship between high CO_2-induced changes in leaf nitrogen and high CO_2-induced effects on photosynthesis has not been evaluated, to our knowledge. We therefore searched Web of Science for studies that grew plants at both ambient and elevated CO_2 concentrations and then measured both leaf nitrogen and net CO_2 assimilation rates at a common CO_2 concentration (A_{common}, as an index of the degree of down-regulation of photosynthesis). Data were extracted from the text, figures, tables or supplemental material, with data in figures quantified using DataThief III (version 1.5, www.datathief.org). This search generated 45 studies (Table 4.1) that encompassed 59 species and 75 contrasting data points: 40 contrasts for C_3 herbaceous species, 7 for C_4 species, 17 for deciduous woody species, and 11 for evergreen woody species. Twenty-one of the studies measured field-grown plants and 24 studies pot-grown plants, resulting in 42 contrasts from fields studies and 33 contrasts from pot studies. Leaf nitrogen and A_{common} were all expressed on, and converted to when necessary, a leaf area basis for direct comparison of the two variables. The relationship between changes in leaf nitrogen content and the increase in growth CO_2 concentration, and the relationship between changes in A_{common} in response to changes in leaf nitrogen content in high

Table 4.1. Studies used for the elevated CO_2 analysis

Species	PFT	Study	Type of study
Acacia smallii	Deciduous woody	Polley et al. (1997)	Pot
Acer rubrum	Deciduous woody	Bauer et al. (2001)	Field
Achillea millefolium	C_3 herbaceous	Lee et al. (2001)	Field
Agropyron repens	C_3 herbaceous	Lee et al. (2001)	Field
Agrostis capillaris	C_3 herbaceous	Davey et al. (1999)	Field
Alnus glutinosa	Deciduous woody	Vogel and Curtis (1995)	Pot
Amorpha canescens	C_3 herbaceous	Lee et al. (2001)	Field
Andropogon gerardii	C_4 herbaceous	Lee et al. (2001)	Field
Anemone cylindrica	C_3 herbaceous	Lee et al. (2001)	Field
Avena barbata	C_3 herbaceous	Jackson et al. (1995)	Field
Betula alleghaniensis	Deciduous woody	Bauer et al. (2001)	Field
Betula papyrifera	Deciduous woody	Cao et al. (2007), Tjoelker et al. (1998)	Pot
Brassica oleracea	C_3 herbaceous	Sage et al. (1989)	Pot
Bromus inermis	C_3 herbaceous	Lee et al. (2001)	Field
Chenopodium album	C_3 herbaceous	Sage et al. (1989)	Pot
Eucalyptus saligna	Evergreen woody	Ghannoum et al. (2010)	Pot
Eucalyptus sideroxylon	Evergreen woody	Ghannoum et al. (2010)	Pot
Fagus sylvatica	Deciduous woody	Liozon et al. (2000)	Pot
Fargesia rufa Yi	C_3 herbaceous	Li et al. (2013)	Pot
Geum reptans	C_3 herbaceous	Körner and Diemer (1994)	Pot
Glycine max	C_3 herbaceous	Ainsworth et al. (2006), Rosenthal et al. (2014), Sims et al. (1998)	Field/pot
Gossypium hirsutum	C_3 herbaceous	Wong (1979)	Pot
Koeleria cristata	C_3 herbaceous	Lee et al. (2001)	Field
Larix laricina	Deciduous woody	Tjoelker et al. (1998)	Pot
Lespedeza capitata	C_3 herbaceous	Lee et al. (2001)	Field
Liquidambar styraciflua	Deciduous woody	Herrick and Thomas (2001)	Field
Lolium perenne	C_3 herbaceous	Davey et al. (1999), Rogers et al. (1998), von Caemmerer et al. (2001)	Field
Lupinus albus	C_3 herbaceous	Campbell and Sage (2006)	Pot
Lupinus perennis	C_3 herbaceous	Lee et al. (2001)	Field
Medicago sativa	C_3 herbaceous	Aranjuelo et al. (2009), Sanz-Sáez et al. (2010)	Pot
Panicum bisulcatum	C_3 herbaceous	Pinto et al. (2014)	Pot
Panicum milioides	C_3 herbaceous	Pinto et al. (2014)	Pot
Paspalum dilatatum	C_4 herbaceous	von Caemmerer et al. (2001)	Field
Petalostemum villosum	C_3 herbaceous	Lee et al. (2001)	Field
Phaseolus vulgaris	C_3 herbaceous	Araya et al. (2008), Cowling and Sage (1998), Jifon and Wolfe (2002)	Pot
Picea abies	Evergreen woody	Hättenschwiler and Körner (1996), Roberntz and Stockfors (1998), Urban et al. (2012)	Field/pot
Picea mariana	Evergreen woody	Tjoelker et al. (1998)	Pot
Pinus radiata	Evergreen woody	Griffin et al. (2000), Turnbull et al. (1998)	Field
Pinus strobus	Evergreen woody	Bauer et al. (2001)	Field
Pinus sylvestris	Evergreen woody	Jach and Ceulemans (2000)	Field
Pinus taeda	Evergreen woody	Ellsworth et al. (2012)	Field
Populus x euramericana	Deciduous woody	Calfapietra et al. (2005)	Field
Populus grandidentata	Deciduous woody	Curtis and Teeri (1992)	Field
Quercus geminata	Deciduous woody	Hymus et al. (2002)	Field
Quercus ilex	Deciduous woody	Staudt et al. (2001)	Pot

(continued)

Table 4.1. (continued)

Species	PFT	Study	Type of study
Quercus myrtifolia	Deciduous woody	Hymus et al. (2002)	Field
Quercus rubra	Deciduous woody	Bauer et al. (2001), Cavender-Bares et al. (2000)	Pot
Ranunculus glacialis	C_3 herbaceous	Körner and Diemer (1994)	Pot
Schizachyrium scoparium	C_4 herbaceous	Lee et al. (2001)	Field
Shorea leprosula	Deciduous woody	Leakey et al. (2002)	Pot
Solanum tuberosum	C_3 herbaceous	Sage et al. (1989)	Pot
Solidago rigida	C_3 herbaceous	Lee et al. (2001)	Field
Sorghastrum nutans	C_4 herbaceous	Lee et al. (2001)	Field
Sorghum bicolor	C_3 herbaceous	Watling et al. (2000)	Pot
Trifolium repens	C_3 herbaceous	Davey et al. (1999), von Caemmerer et al. (2001)	Field
Trifolium subterraneum	C_3 herbaceous	von Caemmerer et al. (2001)	Field
Triticum aestivum L	C_3 herbaceous	Del Pozo et al. (2007), Gutiérrez et al. (2013)	Field
Triticum durum	C_3 herbaceous	Aranjuelo et al. (2011, 2011)	Field
Zea mays	C_4 herbaceous	Kim et al. (2006), Pinto et al. (2014), Wong (1979)	Pot

PFT indicates plant functional type; type of study indicates whether plants were grown in pots or in the field

CO_2-grown plants were analyzed using linear regression models, considering the interaction between the dependent variable and PFT. All analyses were performed in R (R Core Team, 2014, version 3.4.3).

Across the entire data set, the increase in growth CO_2 was 291 ± 117 ppm; exposure to these elevated CO_2 conditions reduced A_{common} by 1.06 ± 1.35μmol m^{-2} s^{-1} and decreased leaf nitrogen content by 0.17 ± 0.10 g m^{-2} (all means ± standard deviations). We found no evidence for a greater decrease in leaf nitrogen content in plants grown at higher CO_2 conditions ($p = 0.72$; Fig. 4.1). Across the PFTs, we observed no change in leaf nitrogen as growth CO_2 increased in C_3 herbaceous ($p = 0.11$), C_4 herbaceous ($p = 0.38$), deciduous woody ($p = 0.06$), or evergreen woody ($p = 0.06$) species. It must be noted that this result does not necessarily mean that leaf nitrogen concentration does not decrease more strongly under higher CO_2 concentrations. Leaf nitrogen content (the variable evaluated here) is the product of leaf nitrogen concentration and leaf mass per area (LMA); leaf nitrogen concentration decreases while LMA increases under high CO_2 (Yin 2002), and the contrasting responses of these variables helps explain our results, including the trend towards increasing leaf nitrogen content with increasing CO_2 concentration seen in deciduous woody species (Fig. 4.1). We did find a positive relationship between the change in leaf nitrogen content and the change in A_{common} ($p = 0.006$) in plants grown in elevated CO_2, confirming that when leaf nitrogen content decreases under elevated CO_2, photosynthesis also tends to decrease (Fig. 4.2). Changes in leaf nitrogen content positively correlated with changes in A_{common} in the three C_3 PFTs (C_3 herbaceous ($p < 0.001$), deciduous woody ($p = 0.006$), and evergreen woody ($p = 0.04$) species). This was not the case in C_4 plants ($p = 0.40$), which may reflect their CO_2-concentrating mechanism, or may reflect the small range we observed for changes in leaf nitrogen content in the C_4 data set.

Our results, and the large body of work discussed above, highlight that the availability and acquisition of nitrogen clearly play a role in the degree of photosynthetic acclimation that occurs in response to elevated

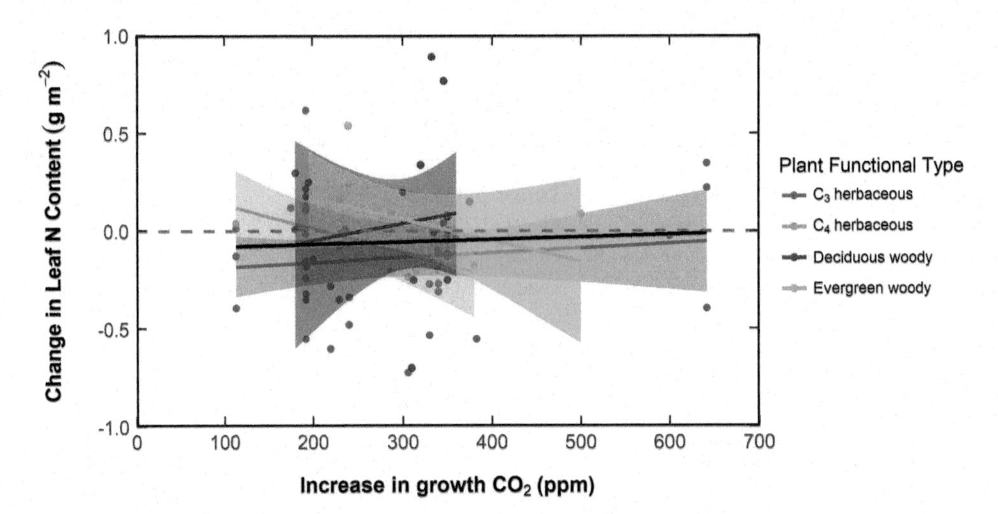

Fig. 4.1. Higher growth CO_2 concentrations do not result in larger decreases in leaf nitrogen content. Symbols and lines indicate data and regression lines from different plant functional types: the entire data set (black line, $p = 0.72$, $R^2 = 0.002$), C_3 herbaceous species (blue points and line, $p = 0.11$, $R^2 = 0.014$), C_4 species (yellow points and line, $p = 0.38$, $R^2 = 0.63$), deciduous woody species (red points and line, $p = 0.06$, $R^2 = 0.018$) and evergreen woody species (green points and line, $p = 0.06$, $R^2 = 0.14$). Shaded regions represent 95% confidence intervals for each plant functional type

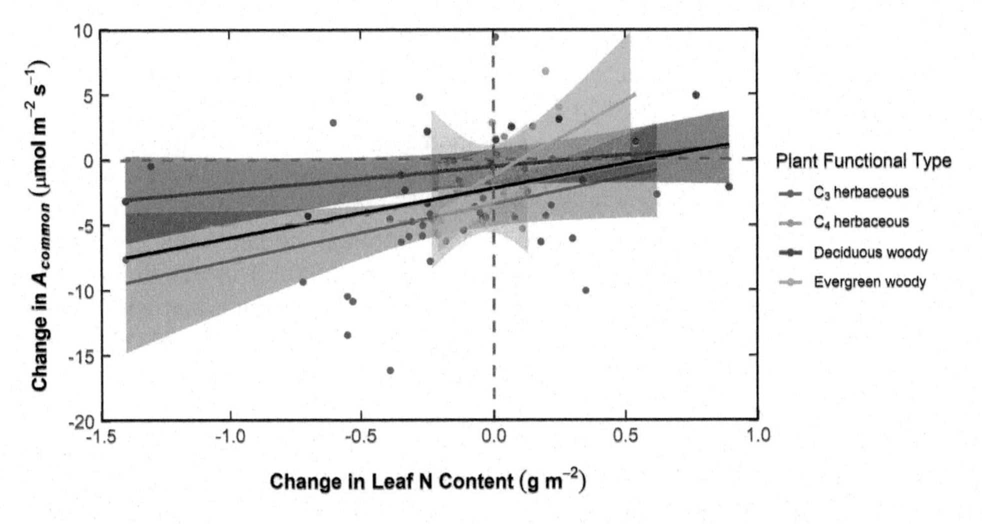

Fig. 4.2. Changes in leaf nitrogen content in response to elevated CO_2 are positively correlated with the degree to which elevated CO_2 concentrations alter photosynthesis measured at a common CO_2 concentration (A_{common}). Symbols and lines indicate data and regression lines from different plant functional types: the entire data set (black line, $p = 0.006$, $R^2 = 0.12$), C_3 herbaceous species (blue points and line, $p < 0.001$, $R^2 = 0.10$), C_4 species (yellow points and line, $p = 0.40$, $R^2 = 0.06$), deciduous woody species (red points and line, $p = 0.006$, $R^2 = 0.14$) and evergreen woody species (green points and line, $p = 0.04$, $R^2 = 0.45$). Shaded regions represent 95% confidence intervals for each plant functional type

CO_2 growth conditions. However, elevated CO_2 can impact plant nitrogen dynamics not only through indirect effects on root growth and allocation patterns, but also directly through its effect on photorespiration. Photorespiration helps control both carbon and nitrogen cycling in plants: NH_3 is needed to make glycine, and ammonium is released when serine is produced (Dusenge et al. 2019). Conditions that suppress photorespiration, such as rising CO_2 concentrations, should therefore also decrease plant nitrogen demand. Indeed, photorespiration rates, which are set by the ratio CO_2/O_2, are correlated with nitrogen uptake capacity, and more specifically, nitrate uptake capacity (Rachmilevitch et al. 2004; Bloom et al. 2010, 2012, 2014; Rubio-Asensio and Bloom, 2017; Busch et al. 2018; Wujeska-Klause et al. 2019), as photorespiratory malate export from the chloroplast provides energy in the form of NADH for nitrate reduction (Bloom 2015). In a meta-analysis, plants grown under elevated CO_2 had a 16% lower nitrate uptake capacity than ambient CO_2-grown plants, though CO_2 had no impact on ammonium uptake capacity (Cheng et al. 2012). These physiological shifts altered soil nitrogen pools in elevated CO_2, leading to a 27% increase in soil nitrate concentrations, but an 8% depletion of soil ammonium concentrations (Cheng et al. 2012). Given that many plants species show a preference for taking up either nitrate or ammonium (Boudsocq et al. 2012; Britto and Kronzucker 2013), reductions in nitrogen uptake in species that preferentially assimilate nitrate could predispose them to stronger photosynthetic down-regulation under elevated CO_2 than ammonium-preferring species.

Elevated CO_2 can also affect plant nitrogen availability through its impact on mycorrhizae. The higher photosynthetic rates in plants grown at elevated CO_2 can lead to a greater allocation of carbohydrates to mycorrhizae (Sanders et al. 1998; Drigo et al. 2010). Since mycorrhizal hyphae are thinner and longer than plant fine roots (Smith and Read 2008), increased mycorrhizal growth under elevated CO_2 should increase plant nitrogen acquisition and help reduce photosynthetic acclimation. But not all mycorrhizae are equal with regard to rising CO_2 concentrations. Plants that interact with ectomycorrhizal fungi show strong increases in biomass under elevated CO_2, while those with an arbuscular mycorrhizal symbiosis have little growth response (Terrer et al. 2016). The increased sink strength of ectomycorrhizal plant species under high CO_2, and their greater ease of acquiring nitrogen, translates into a lower degree of photosynthetic down-regulation than in arbuscular mycorrhizal plant species (Terrer et al. 2018).

IV. Photosynthetic Acclimation to Temperature

Net CO_2 assimilation rates follow a unimodal response to leaf temperature, and the impact of higher temperatures on net photosynthesis is dependent on both the initial leaf temperature and the degree of warming (Berry and Björkman 1980; Hikosaka et al. 2006; Sage and Kubien 2007; Way and Yamori, 2014; Yamori et al. 2014). Warming stimulates net photosynthesis when leaf temperatures are below the photosynthetic thermal optimum, as the activity of photosynthetic enzymes (and therefore photosynthetic capacity, i.e., V_{cmax} and J_{max}) increase exponentially up to a temperature near 40–45 °C (Jordan and Ogren 1984; Harley and Tenhunen 1991). For C_3 species, the photosynthetic thermal optimum is usually 25–30 °C (Yamori et al. 2014): at leaf temperatures above this optimum, respiration and photorespiration rates increase faster than photosynthetic capacity (Long 1991), so the total effect is a decline in net photosynthesis. But even within the processes that control photosynthetic capacity, the thermal responses of different components differ. RuBP-regeneration-limited photosynthesis is more temperature-sensitive than Rubisco-limited photosynthesis (Kirschbaum and Farquhar 1984; Hikosaka et al. 2006). Mesophyll conductance, which describes CO_2 movement from the intercellular airspace into the chloroplast, usually increases with temperature (Bernacchi et al. 2002; Evans and von Caemmerer 2013; von

Caemmerer and Evans 2015), though it can also be relatively temperature-insensitive (von Caemmerer and Evans 2015) or even decrease at higher leaf temperatures (Way et al. 2019). If high temperatures correlate with high vapor pressure deficit, then stomatal conductance will also decrease as leaf temperature increases, restricting CO_2 diffusion into the leaf and further suppressing net photosynthetic rates (Lin et al. 2012).

While these biochemical and stomatal responses explain how short-term changes in leaf temperature impact photosynthesis, acclimation to a new growth temperature alters the shape of this initial temperature response curve (Berry and Björkman 1980; Hikosaka et al. 2006; Yamori et al. 2014, Way and Yamori, 2014; Smith and Dukes 2017). The exact mechanisms underlying this shift are less well characterized than those leading to photosynthetic down-regulation to CO_2, and differ somewhat between plants acclimated to warmer versus cooler thermal regimes. A change in growth temperature usually alters the phospholipid membrane composition of leaf tissues, to maintain an appropriate membrane fluidity: low temperatures increase the unsaturation of fatty acids (Wilson and Crawford 1974; Berry and Raison 1981), while the opposite is true under high temperatures (Pearcy 1978; Kunst et al. 1989). Respiration is reduced in plants grown at warmer temperatures (Atkin and Tjoelker 2003; Slot and Kitajima 2015), and this respiratory acclimation is responsible for some of the change in the response of net photosynthesis to temperature (Way and Yamori, 2014). Different versions of enzymes or different concentrations of photosynthetic proteins can also be produced when growth temperature is altered (Badger et al. 1982; Strand et al. 1999; Stitt and Hurry 2002; Yamori et al. 2005). For example, Rubisco activase has two isozymes, one of which is more heat tolerant than the other, and high growth temperatures in some species can induce the expression of the more thermally tolerant enzyme (Crafts-Brandner et al. 1997; Law and Crafts-Brandner 2001). Given the differing thermal sensitivities of RuBP-regeneration-limited and Rubisco-limited photosynthesis, the ratio of J_{max} to V_{cmax} should be higher in cool-grown plants to optimally allocate nitrogen investment between these processes (Hikosaka 1997). Indeed, a recent global meta-analysis of plants measured during different parts of the year found that the ratio of J_{max} to V_{cmax} was negatively related to recent growth temperatures (Kumarathunge et al., 2019). Additionally, leaves that develop at a cool temperature are usually thicker, with a higher leaf mass per area, than those from a warmer growth temperature (Loveys et al., 2002; Atkin et al. 2006; Way and Sage 2008a). Plants grown under different temperatures therefore usually have dissimilar photosynthetic rates when measured under a common set of environmental conditions.

The biochemical, physiological, and anatomical changes that occur in a new growth temperature often improve the photosynthetic performance of the plant, but this is not always the case. Across a broad range of studies, the photosynthetic thermal optimum increases with warmer growth temperatures by 0.55–0.62 °C per 1 °C rise in temperature (Way and Yamori, 2014; Yamori et al., 2014; Kumarathunge et al., 2019). This shift in the thermal optimum is more strongly related to changes in photosynthetic capacity than in stomatal conductance, and the thermal optima for both V_{cmax} and J_{max} were positively related to growth temperature across a study of 141 species (Kumarathunge et al., 2019). But a shift in the thermal optimum does not necessarily mean that the net photosynthetic rate achieved in the new growth environment (A_{growth}), or at the thermal optimum (A_{opt}), is comparable between different thermal regimes. In 68% of experiments, A_{growth} is similar, or even higher, in plants grown at a warmer temperature in comparison to the control treatment (Way and Yamori, 2014). But this means that in a third of the cases examined, warming suppresses net CO_2 assimilation rates at the new growth temperature (Way and Yamori, 2014). Although it has been known for decades that increased temperatures do not always improve leaf carbon uptake (Berry and Björkman 1980), more recent work has shown that acclimation to warmer temperatures can actually reduce carbon uptake in the new thermal environment. This scenario, termed detractive adjustments by Way and Yamori (2014), implies that perhaps net photosynthesis is not the variable being optimized in warmer temperatures.

Many of the changes that occur during photosynthetic thermal acclimation are correlated with leaf nitrogen. Acclimation to low temperatures is strongly linked to increases in leaf nitrogen and photosynthetic enzyme concentrations (Hüner et al. 1993; Martindale and Leegood 1997; Stitt and Hurry 2002). Cold conditions suppress enzyme activities (Arcus et al. 2016), so by increasing enzyme concentrations, leaf metabolic rates are increased and the direct effect of low temperatures on individual enzymes is offset. Most photosynthetic enzymes show an increased capacity in cold-grown plants, including Rubisco (Holaday et al. 1992; Hurry et al. 1995; Martindale and Leegood 1997), sucrose synthase (Crespi et al. 1991) and sucrose-phosphate synthase (Holaday et al. 1992; Martindale and Leegood 1997). Because sucrose synthesis is particular sensitive to cold (i.e., TPU limitations are prevalent at low temperatures; Stitt and Grosse 1988), acclimation to cool conditions increases the transcription, concentration and activity of cytosolic fructose-1,6-bisphosphatase and sucrose phosphate synthase (Strand et al. 1997; Strand et al. 1999; Hurry et al. 2000), key enzymes in sucrose synthesis, to a larger extent than the increase in Rubisco activity.

The link between leaf nitrogen and photosynthetic acclimation to warming is less clear. In plants where warming alters photosynthesis, there is usually a corresponding change in leaf nitrogen (e.g., Tjoelker et al. 1999; Lewis et al. 2004; Way and Sage 2008a). But higher growth temperatures can lead to either a decrease in leaf nitrogen concentrations (Dwyer et al. 2007; Drake et al. 2015) or an increase (Lewis et al. 2004; Yamaguchi et al. 2016).

The question of whether plants with higher leaf nitrogen, whether due to variation among species or to fertilization, might differ in their capacity for photosynthetic thermal acclimation from plants with low leaf nitrogen has received relatively little attention. Martindale and Leegood (1997) showed that nitrogen deficiency inhibited the ability of plants to acclimate photosynthesis to low temperatures, since cold acclimation requires increased enzyme production and concentrations. But in a rapidly warming climate, we need a better understanding of how leaf nitrogen concentrations will respond to increasing growth temperatures, and whether this helps explain how photosynthesis is affected by warming. This becomes even more important when we note that warming stimulates net nitrogen mineralization rates by 46%, increasing nitrogen availability for plants (Rustad et al. 2001). There is therefore the possibility of strong coupling between plant and soil nitrogen dynamics in elevated temperatures, similar to those discussed above for high CO_2 concentrations, but the consequences of this for plant carbon uptake are poorly understood.

To evaluate how changes in growth temperature affect leaf nitrogen content, and how changes in leaf nitrogen could affect thermal acclimation of photosynthesis to warming, we searched Web of Science for studies that investigated the effect of growing plants at two or more temperatures and included both measurements of net CO_2 assimilation rates at the growth temperature (A_{growth}) and leaf nitrogen. Data were extracted directly from the main text, supplemental material, or from figures using DataThief III (version 1.5, www.datathief.org) from studies in controlled environments (i.e., greenhouses and growth chambers) where growth temperature could be clearly delineated, using studies from Dusenge et al. (2019) as a starting point. From the 23 studies we identified (Table 4.2), A_{growth} and leaf nitrogen content

Table 4.2. Studies used for the elevated growth temperature analysis

Species	PFT	Study
Acer saccharum	Deciduous woody	Carter and Cavaleri (2018)
Anoperus glandulosus	Evergreen woody	Scafaro et al. (2017)
Atherosperma moschatum	Evergreen woody	Scafaro et al. (2017)
Betula alleghaniensis	Deciduous woody	Gunderson et al. (2010)
Betula nana	Deciduous woody	Heskel et al. (2014)
Castanopsermun austral	Evergreen woody	Scafaro et al. (2017)
Corymbia calophylla	Evergreen woody	Aspinwall et al. (2017)
Cryptocarya mackinnoniana	Evergreen woody	Scafaro et al. (2017)
Cucumis sativus	C_3 herbaceous	Yamori et al. (2010)
Eriophorum vaginatum	Evergreen woody	Heskel et al. (2014)
Eucalyptus globulus	Evergreen woody	Crous et al. (2013)
Eucalyptus grandis	Evergreen woody	Drake et al. (2015)
Eucalyptus regnans	Evergreen woody	Scafaro et al. (2017)
Eucalyptus saligna	Evergreen woody	Ghannoum et al. (2010); Xu et al. (2012)
Eucalyptus sideroxylon	Evergreen woody	Ghannoum et al. (2010); Smith et al. (2012)
Eucalyptus tereticornis	Evergreen woody	Aspinwall et al. (2016); Drake et al. (2015)
Eucryphia lucida	Evergreen woody	Scafaro et al. (2017)
Hedycarya angustifolia	Evergreen woody	Scafaro et al. (2017)
Hevea brasiliensis	Evergreen woody	Kositsup et al. (2009)
Liquidambar styraciflua	Deciduous woody	Gunderson et al. (2010)
Nicotiana tabacum	C_3 herbaceous	Yamori et al. (2010)
Oryza sativa	C_3 herbaceous	Nagai and Makino (2009); Yamori et al. (2010)
Phaseolus vulgaris	C_3 herbaceous	Cowling and Sage (1998)
Picea abies	Evergreen woody	Kroner and Way (2016)
Picea crassifolia	Evergreen woody	Zhang et al. (2018)
Picea koraiensis	Evergreen woody	Zhang et al. (2015)
Picea likiangensis	Evergreen woody	Zhang et al. (2015)
Picea mariana	Evergreen woody	Way and Sage (2008a, b)
Picea meyeri	Evergreen woody	Zhang et al. (2015)
Picea wilsonii	Evergreen woody	Zhang et al. (2018)
Populus grandidentata	Deciduous woody	Gunderson et al. (2010)
Quercus ilex	Evergreen woody	Mediavilla et al. (2016)
Quercus serrata	Deciduous woody	Yamagushi et al. (2016)
Quercus suber	Evergreen woody	Mediavilla et al. (2016)
Qurercus rubra	Deciduous woody	Gunderson et al. (2010)
Rubus chamaemorus	C_3 herbaceous	Heskel et al. (2014)
Secale cereale	C_3 herbaceous	Yamori et al. (2010)
Solanum lycopersicum	C_3 herbaceous	Yamori et al. (2010)
Solanum tuberosum	C_3 herbaceous	Yamori et al. (2010)
Spinacia olearacea	C_3 herbaceous	Yamori et al. (2010)
Stipa krylovii	C_3 herbaceous	Chi et al. (2013)
Synima cordierorum	Evergreen woody	Scafaro et al. (2017)
Syzygium sayeri	Evergreen woody	Scafaro et al. (2017)
Tasmannia lanceolata	Evergreen woody	Scafaro et al. (2017)
Tilia americana	Deciduous woody	Carter and Cavaleri (2018)
Triticum aestivum	C_3 herbaceous	Nagai and Makino (2009), Yamori et al. (2010)
Tritocosecale sp.	C_3 herbaceous	Yamori et al. (2010)
Vicia faba	C_3 herbaceous	Yamori et al. (2010)

PFT indicates plant functional type

were collected from 48 species, generating 122 contrasts for three plant functional types: 31 contrasts for C_3 herbaceous species, 25 for deciduous woody species, and 66 for evergreen woody species. Leaf nitrogen content and A_{growth} were expressed, and converted when necessary, as g m^{-2} and μmol CO_2 m^{-2} s^{-1}, respectively. The relationship between changes in leaf nitrogen content in response to increases in growth temperature, as well as the relationship between changes in A_{growth} in response to changes in leaf nitrogen content were analyzed using linear regression models, considering the interaction between the dependent variable and PFT. All analyses were performed in R (R Core Team, 2014, version 3.4.3).

Within our data set, increases in growth temperature had no significant impact on leaf nitrogen content overall ($p = 0.89$), or in any of the three plant functional types (C_3 herbaceous species, $p = 0.20$; deciduous trees, $p = 0.08$; evergreen trees, $p = 0.29$; Fig. 4.3). However, there was a positive relationship between the effect of warming on leaf nitrogen content and the effect of warming on net CO_2 assimilation rates at the growth temperature in deciduous ($p = 0.02$) and evergreen trees ($p = 0.0002$; Fig. 4.4). In contrast, changes in leaf nitrogen content that occur in response to warming had little impact on A_{growth} in C_3 herbaceous species ($p = 0.98$) (Fig. 4.4).

We thus find little evidence for a directional response of leaf nitrogen content in response to warming, at least in controlled studies, where plants are usually fertilized and have high soil nitrogen. When elevated temperatures do alter leaf nitrogen content in woody species, these changes in leaf nitrogen status translate into changes in CO_2 assimilation rates as may be expected. However, in herbaceous species, photosynthesis tended to be stimulated by warming, regardless of whether leaf nitrogen content increased or decreased. This could be explained if species that had reduced leaf nitrogen in warmer conditions reallocated to enhance nitrogen use efficiency (i.e., exhibited qualitative changes in how nitrogen was partitioned), while species with increases in leaf nitrogen simply increased their overall

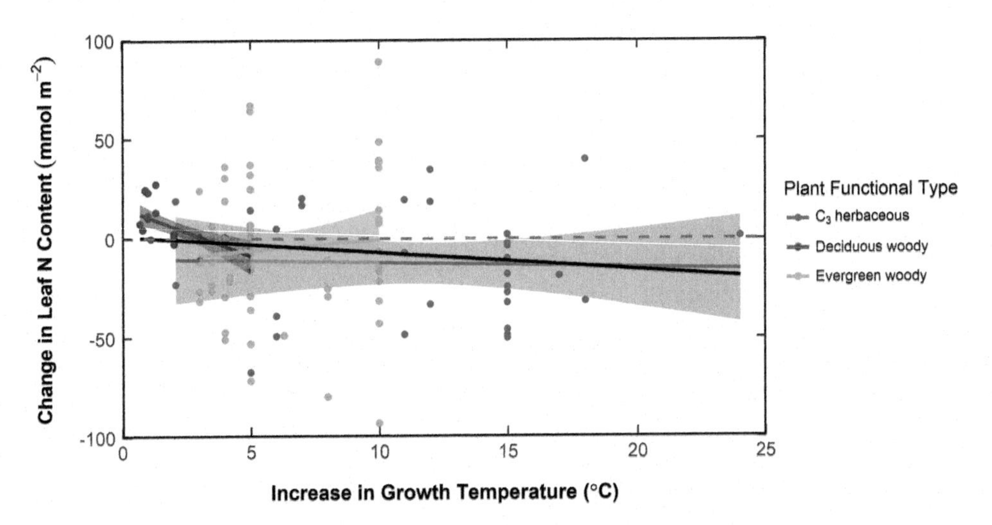

Fig. 4.3. Warming has no consistent effect on leaf nitrogen content. Symbols and lines indicate data and regression lines from different plant functional types: the entire data set (black line, $p = 0.89$, $R^2 = 0.018$), C_3 herbaceous species (blue points and line, $p = 0.20$, $R^2 = 0.001$), deciduous woody species (red points and line, $p = 0.08$, $R^2 = 0.37$) and evergreen woody species (green points and line, $p = 0.29$, $R^2 = 0.008$). Shaded regions represent 95% confidence intervals for each plant functional type

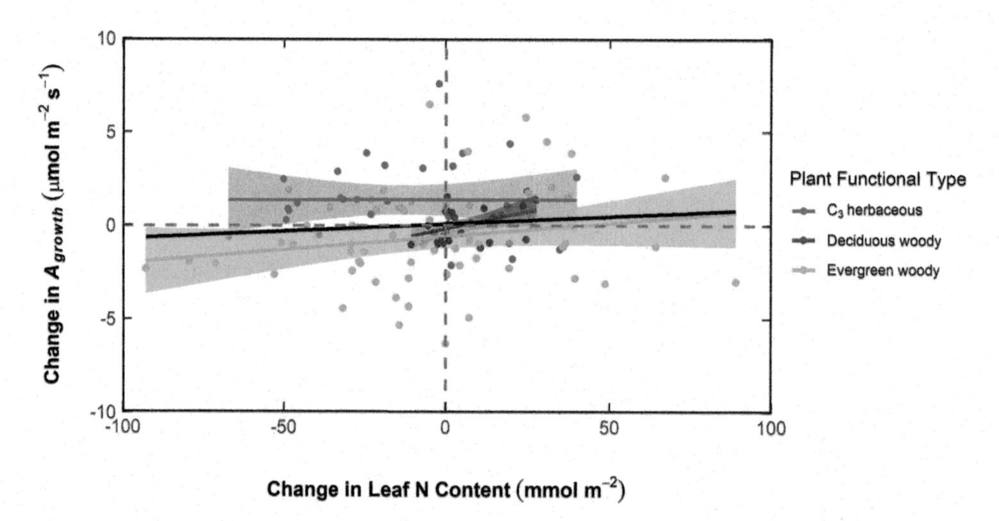

Fig. 4.4. Changes in leaf nitrogen content in response to warming are positively correlated with the degree to which elevated growth temperatures alter photosynthesis in the new growth temperature (A_{growth}) for woody species, but not herbaceous C_3 plants. Symbols and lines indicate data and regression lines from different plant functional types: the entire data set (black line, $p = 0.28$, $R^2 = 0.008$), C_3 herbaceous species (blue points and line, $p = 0.98$, $R^2 = 0.001$), deciduous woody species (red points and line, $p = 0.02$, $R^2 = 0.18$) and evergreen woody species (green points and line, $p = 0.002$, $R^2 = 0.033$). Shaded regions represent 95% confidence intervals for each plant functional type

investment in photosynthesis (i.e., made a quantitative changes in investing N) (Way and Sage 2008b). While our analyses cannot address these ideas directly, the uncoupling between changes in leaf nitrogen content and changes in photosynthetic performance in C_3 herbaceous species deserves further study.

V. Conclusions

Early studies found strong photosynthetic down-regulation and declines in leaf nitrogen under elevated CO_2, however recent work has highlighted the need to consider the ability of field-grown plants to maintain their nitrogen status (and therefore their CO_2 uptake rates) through changes in root allocation, nitrogen uptake capacity and symbiotic relationships. Overall, we found that higher growth CO_2 concentrations did not induce stronger reductions in leaf nitrogen content, although CO_2-induced changes in leaf nitrogen content were positively correlated with the degree of photosynthetic down-regulation expressed at high CO_2. The impacts of warming on leaf nitrogen content were variable, with no discernable pattern. While warming-induced increases in leaf nitrogen content correlate with improved photosynthetic performance in woody species, they show little correlation with the impacts of warming on photosynthesis in herbaceous C_3 species.

Acknowledgements

DAW acknowledges support from the Natural Sciences and Engineering Research Council of Canada in the form of a Discovery Grant, as well as support for a Visiting Fellowship from the Research School of Biology at the Australian National University.

References[1]

Adair EC, Reich PB, Hobbie SE, Knops JMH (2009) Interactive effects of time, CO_2, N, and diversity on total belowground carbon allocation and ecosystem carbon storage in a grassland community. Ecosystems 12:1037–1052

Ainsworth EA, Long SP (2005) What have we learned from 15 years of free-air CO2 enrichment (FACE)? A meta-analytic review of the responses of photosynthesis, canopy properties and plant production to rising CO_2. New Phytol 165:351–372

Ainsworth EA, Rogers A (2007) The response of photosynthesis and stomatal conductance to rising [CO_2]: mechanisms and environmental interactions. Plant Cell Environ 30:258–270

Ainsworth EA, Davey PA, Hymus GJ, Osborne CP, Rogers A, Blum H, Nösberger J, Long SP (2003) Is stimulation of leaf photosynthesis by elevated carbon dioxide concentration maintained in the long term? A test with *Lolium perenne* grown for 10 years at two nitrogen fertilization levels under free air CO_2 enrichment (FACE). Plant Cell Environ 26:705–714

§Ainsworth EA, Rogers A, Leakey AD, Heady LE, Gibon Y, Stitt M, Schurr U (2006) Does elevated atmospheric [CO_2] alter diurnal C uptake and the balance of C and N metabolites in growing and fully expanded soybean leaves? J Exp Bot 58: 579–591

§Aranjuelo I, Irigoyen JJ, Nogués S, Sánchez-Díaz M (2009) Elevated CO_2 and water-availability effect on gas exchange and nodule development in N_2-fixing alfalfa plants. Environ Exp Bot 65: 18–26

Aranjuelo I, Ebbets AL, Evans RD, Tissue DT, Nogués S, Van Gestel N, Payton P, Ebbert V, … Smith SD (2011) Maintenance of C sinks sustains enhanced CO_2 assimilation during long-term exposure to elevated [CO_2] in Mojave Desert shrubs. Oecologia 167: 339–354

§Aranjuelo I, Cabrera-Bosquet L, Morcuende R, Avice JC, Nogués S, Araus JL, Martínez-Carrasco R, Pérez P (2011) Does ear C sink strength contribute to overcoming photosynthetic acclimation of wheat plants exposed to elevated CO_2? J Exp Bot 62: 3957–3969

§Araya T, Noguchi KO, Terashima I (2008) Manipulation of light and CO_2 environments of the primary leaves of bean (*Phaseolus vulgaris* L.) affects photosynthesis in both the primary and the first trifoliate leaves: involvement of systemic regulation. Plant Cell Environ 31: 50–61

[1]Those marked with * were used in the CO_2 meta-analysis; those marked with § were used in the temperature meta-analysis.

Arcus VL, Prentice EJ, Hobbs JK, Mulholland AJ, Van der Kamp MW, Pudney CR, Parker EJ, Schipper LA (2016) On the temperature dependence of enzyme-catalyzed rates. Biochemistry 55:1681–1688

Arp WJ (1991) Effects of source-sink relations on photosynthetic acclimation to elevated CO_2. Plant Cell Environ 14:869–875

*Aspinwall MJ, Drake JE, Campany C, Vårhammar A, Ghannoum O, Tissue DT, Reich PB, Tjoelker MG (2016) Convergent acclimation of leaf photosynthesis and respiration to prevailing ambient temperatures under current and warmer climates in *Eucalyptus tereticornis*. New Phytol 212: 354–367

*Aspinwall MJ, Vårhammar A, Blackman CJ, Tjoelker MG, Ahrens C, Byrne M, Tissue DT, Rymer PD (2017) Adaptation and acclimation both influence photosynthetic and respiratory temperature responses in *Corymbia calophylla*. Tree Physiol 37: 1095–1112

Atkin OK, Tjoelker MG (2003) Thermal acclimation and the dynamic response of plant respiration to temperature. Trends Plant Sci 8:343–351

Atkin OK, Loveys BR, Atkinson LJ, Pons TL (2006) Phenotypic plasticity and growth temperature: understasdning interspecific variability. J Exp Bot 57:267–281

Bader M, Hiltbrunner E, Körner C (2009) Fine root responses of mature deciduous forest trees to free air carbon dioxide enrichment (FACE). Funct Ecol 23:913–921

Bader MK-F, Siegwolf R, Körner C (2010) Sustained enhancement of photosynthesis in mature deciduous forest trees after 8 years of free air CO_2 enrichment. Planta 232:1115–1125

Badger MR, Björkman O, Armond PA (1982) An analysis of photosynthetic response and adaptation to temperature in higher-plants – temperature acclimation in the desert evergreen *Nerium oleander* L. Plant Cell Environ 5:85–99

§Bauer GA, Berntson GM, Bazzaz FA (2001) Regenerating temperate forests under elevated CO_2 and nitrogen deposition: comparing biochemical and stomatal limitation of photosynthesis. New Phytol 152: 249–266

Bernacchi CJ, Portis AR, Nakano H, von Caemmerer S, Long SP (2002) Temperature response of mesophyll conductance. Implications for the determination of rubisco enzyme kinetics and for limitations to photosynthesis *in vivo*. Plant Physiol 130:1992–1998

Berry J, Björkman O (1980) Photosynthetic response and adaptation to temperature in higher plants. Annu Rev Ecol Syst 31:491–543

Berry JA, Raison JK (1981) Responses of macrophytes to temperature. In: Lange OL, Nobel PS, Osmond

CB, Zeigler H (eds) Physiological Plant Ecology. 1. Responses to the Physical Environment. Springer, Berlin, pp 277–338

Bloom AJ (2015) The increasing importance of distinguishing among plant nitrogen sources. Curr Opin Plant Biol 25:10–16

Bloom AJ, Burger M, Rubio Asensio JS, Cousins AB (2010) Carbon dioxide enrichment inhibits nitrate assimilation in wheat and *Arabidopsis*. Science 328:899–903

Bloom AJ, Asensio JSR, Randall L, Rachmilevitch S, Cousins AB, Carlisle EA (2012) CO_2 enrichment inhibits shoot nitrate assimilation in C_3 but not C_4 plants and slows growth under nitrate in C_3 plants. Ecology 93:355–367

Bloom AJ, Burger MA, Kimball BJ, Pinter JP (2014) Nitrate assimilation is inhibited by elevated CO_2 in field-grown wheat. Nat Clim Chang 4:477–480

Boudsocq S, Niboyet A, Lata JC, Raynoud X, Loeuille N, Mathieu J, Blouin M, Abbadie L, Barot S (2012) Plant preference for ammonium versus nitrate: a neglected determinant of ecosystem functioning? Am Nat 180:60–69

Bowes G (1991) Growth at elevaetd CO_2 – photosynthetic responses mediated through rubisco. Plant Cell Environ 14:795–806

Britto DT, Kronzucker HJ (2013) Ecological significance and complexity of N-source preference in plants. Ann Bot 112:957–963

Busch FA, Sage RF (2017) The sensitivity of photosynthesis to O_2 and CO_2 concentration identifies strong rubisco control above the thermal optimum. New Phytol 213:1036–1051

Busch FA, Sage RF, Farquhar GD (2018) Plants increase CO_2 uptake by assimilating nitrogen via the photorespiratory pathway. Nat Plants 4:46–54

§Calfapietra C, Tulva I, Eensalu E, Perez M, De Angelis P, Scarascia-Mugnozza G, Kull O (2005) Canopy profiles of photosynthetic parameters under elevated CO_2 and N fertilization in a poplar plantation. Environ Pollut 137: 525–535

§Campbell CD, Sage RF (2006) Interactions between the effects of atmospheric CO_2 content and P nutrition on photosynthesis in white lupin (*Lupinus albus* L.). Plant Cell Environ 29: 844–853

Campbell WJ, Allen LH, Bowes G (1988) Effects of CO_2 concentration on Rubisco activity, amount and photosynthesis in soybean leaves. Plant Physiol 88:1310–1316

§Cao B, Dang QL, Zhang S (2007) Relationship between photosynthesis and leaf nitrogen concentration in ambient and elevated [CO_2] in white birch seedlings. Tree Physiol 27: 891–899

*Carter KR, Cavaleri M (2018) Within-canopy experimental leaf warming induces photosynthetic decline instead of acclimation in two northern hardwood species . Front For Glob Chang 1, Article 11

§Cavender-Bares J, Potts M, Zacharias E, Bazzaz FA (2000) Consequences of CO_2 and light interactions for leaf phenology, growth, and senescence in *Quercus rubra*. Glob Chang Biol 6: 877–887

Cen YP, Sage RF (2005) The regulation of rubisco activity in response to variation in temperature and atmospheric CO_2 partial pressure in sweet potato. Plant Physiol 139:979–990

Cheng L, Booker FL, Tu C, Burkey KO, Zhou L, Shew HD, Rufty TW, Hu S (2012) Arbuscular mycorrhizal fungi increase organic carbon decomposition under elevated CO_2. Science 337:1084–1087

*Chi Y, Xu M, Shen R, Yang Q, Huang, B, Wan, S (2013). Acclimation of foliar respiration and photosynthesis in response to experimental warming in a temperate steppe in northern China. PLoS One, 8, e56482

Ciais P, Sabine C, Bala G, Bopp L, Brovkin V, Canadell J, Chhabra A Jones C (2013) Carbon and other biogeochemical cycles. In: Stocker TF, Qin D, Plattner G-K, Tignor M, Allen SK, Boschung J, Nauels A, Xia Y, …, Midgley PM (eds) Climate Change 2013: The Physical Science Basis. Contribution of Working Group I to the Fifth Assessment Report of the Intergovernamental Panel on Climate Change, Cambridge University Press, Cambridge, UK

Clough JM, Peet MM, Kramer PJ (1981) Effects of high atmospheric CO_2 and sink size on rates of photosynthesis of a soybean cultivar. Plant Physiol 67:1007–1010

Collins M, Knutti R, Arblaster J, Dufresne J-L, Fichefet T, Friedlingstein P, Gao X, Shongwe M (2013) Long-term climate change: projections, commitments and irreversibility. In: Stocker TF, Qin D, Plattner G-K, Tignor M, Allen SK, Boschung J, Nauels A, Xia Y, …, Midgley PM (eds) Climate Change 2013: The Physical Science Basis. Contribution of Working Group I to the Fifth Assessment Report of the Intergovernmental Panel on Climate Change, Cambridge University Press , Cambridge, UK

Core Team R (2014) R: a language and environment for statistical computing. In: R foundation for statistical computing. Austria, Vienna. http://www.R--project.org/

*§Cowling SA, Sage RF (1998) Interactive effects of low atmospheric CO_2 and elevated temperature on growth, photosynthesis and respiration in *Phaseolus vulgaris*. Plant Cell Environ 21: 427–435

Crafts-Brandner SJ, van de Loo FJ, Salvucci ME (1997) The two forms of ribulose-1,5-bisphosphate carboxylase/oxygenase activase differ in sensitivity to elevated temperature. Plant Physiol 114:439–444

Crespi MD, Zabaleta ZJ, Pontis HG, Salerno GL (1991) Sucrose synthase expression during cold acclimation in wheat. Plant Physiol 96:887–891

*Crous KY, Quentin AG, Lin YS, Medlyn BE, Williams DG, Barton CV, Ellsworth DS (2013) Photosynthesis of temperate *Eucalyptus globulus* trees outside their native range has limited adjustment to elevated CO_2 and climate warming. Glob Chang Biol 19: 3790–3807

Cure JD, Acock B (1986) Crop responses to carbon dioxide doubling: a literature survey. Agric For Meterol 38:127–145

§Curtis PS, Teeri JA (1992) Seasonal responses of leaf gas exchange to elevated carbon dioxide in *Populus grandidentata*. Can J For Res 22: 1320–1325

Curtis PS, Wang X (1998) A meta-analysis of elevated CO_2 effects on woody plant mass, form, and physiology. Oecologia 113:299–313

§Davey PA, Parsons AJ, Atkinson L, Wadge K, Long SP (1999) Does photosynthetic acclimation to elevated CO_2 increase photosynthetic nitrogen-use efficiency? A study of three native UK grassland species in open-top chambers. Funct Ecol 13: 21–28

Dawes MA, Hättenschwiler S, Bebi P, Hagedorn F, Handa IT, Körner C, Rixen C (2011) Species-specific tree growth responses to 9 years of CO_2 enrichment at the alpine treeline. J Ecol 99:383–394

§Del Pozo A, Pérez P, Gutiérrez D, Alonso A, Morcuende R, Martínez-Carrasco R (2007) Gas exchange acclimation to elevated CO_2 in upper-sunlit and lower-shaded canopy leaves in relation to nitrogen acquisition and partitioning in wheat grown in field chambers. Environ Exp Bot 59: 371–380

DeLucia EH, Sasek TW, Strain BR (1985) Photosynthetic inhibition after long-term exposure to elevated levels of atmospheric carbon dioxide. Photosynth Res 7:175–184

Dlugokencky E, Tans P (2018) Trends in atmospheric carbon dioxide, National Oceanic & Atmospheric Administration, Earth System Research Laboratory (NOAA/ESRL). http://www.esrl.noaa.gov/gmd/ccgg/trends/global.html. Accessed 20 Feb 2019

Drake BG, Gonzalez-Meler MA, Long SP (1997) More efficient plants: a consequence of rising atmospheric CO_2? Annu Rev Plant Physiol Plant Mol Biol 48:609–639

Drake JE, Gallet-Budynek A, Hofmockel KS, Bernhardt ES, Billings SA, Jackson RB, Johnsen KS, Moore DJ (2011) Increases in the flux of carbon belowground stimulate nitrogen uptake and sustain the long-term enhancement of forest productivity under elevated CO_2. Ecol Lett 14:349–357

*Drake JE, Aspinwall MJ, Pfautsch S, Rymer PD, Reich PB, Smith RA, Crous KY, Tissue DT, …, Tjoelker MG (2015) The capacity to cope with climate warming declines from temperate to tropical latitudes in two widely distributed *Eucalyptus* species. Glob Chang Biol 21: 459–472

Drigo B, Pijl AS, Duyts H, Kielak AM, Gamper HA, Houtekamer MJ, Boschker HT, Kowalchuk GA (2010) Shifting carbon flow from roots into associated microbial communities in response to elevated atmospheric CO_2. Proc Natl Acad Sci U S A 107:10938–10942

Dukes JS, Chiariello NR, Cleland EE, Moore LA, Shaw MR, Thayer S, Tobeck T, Mooney HA, …, Field CB (2005) Responses of grassland production to single and multiple global environmental changes. PLoS Biol 3: 1829–1837

Dusenge ME, Duarte AG, Way DA (2019) Plant carbon metabolism and climate change: elevated CO_2 and temperature impacts on photosynthesis, photorespiration and respiration. New Phytol 221:32–49

Dwyer SA, Ghannoum O, Nicotra A, von Caemmerer S (2007) High temperature acclimation of C_4 photosynthesis is linked to changes in photosynthetic biochemistry. Plant Cell Environ 30:53–66

§Ellsworth DS, Thomas R, Crous KY, Palmroth S, Ward E, Maier C, DeLucia E, Oren R (2012) Elevated CO_2 affects photosynthetic responses in canopy pine and subcanopy deciduous trees over 10 years: a synthesis from Duke FACE. Glob Chang Biol 18: 223–242

Ellsworth DS, Anderson IC, Crous KY, Cooke J, Drake JE, Gherlenda AN, Gimeno TE, Macdonald CA et al (2017) Elevated CO_2 does not increase eucalypt forest productivity on a low-phosphorus soil. Nat Clim Chang 7:279–282

Evans JR (1989) Photosynthesis and nitrogen relationships in leaves of C_3 plants. Oecologia 78:9–19

Evans JR, von Caemmerer S (2013) Temperature response of carbon isotope discrimination and mesophyll conductance in tobacco. Plant Cell Environ 36:745–756

Farquhar GD, von Caemmerer S, Berry JA (1980) A biochemical model of photosynthetic CO_2 assimilation in leaves of C_3 species. Planta 149:78–90

Feng Z, Rütting T, Pleijel H, Wallin G, Reich PB, Kammann CI, Newton PCD, Kobayashi K et al (2015) Constraints to nitrogen acquisition of terrestrial plants under elevated CO_2. Glob Chang Biol 21:3152–3168

Finzi AC, Moore DJ, DeLucia EH, Lichter J, Hofmockel KS, Jackson RB, Kim HS, Pippen JS (2006) Progressive nitrogen limitation of ecosystem processes under elevated CO_2 in a warm-temperate forest. Ecology 87:15–25

Finzi AC, Norby RJ, Calfapietra C, Gallet-Budynek A, Gielen B, Holmes WE, Hoosbeek MR, Ledford J (2007) Increases in nitrogen uptake rather than nitrogen-use efficiency support higher rates of temperate forest productivity under elevated CO_2. Proc Natl Acad Sci U S A 104:14014–14019

*§Ghannoum O, Phillips NG, Sears MA, Logan BA, Lewis JD, Conroy JP, Tissue DT (2010) Photosynthetic responses of two eucalypts to industrial-age changes in atmospheric [CO_2] and temperature. Plant Cell Environ 33: 1671–1681

Goll DS, Brovkin V, Parida BR, Reick CH, Kattge J, Reich PB, Van Bodegom PM, Niinemets Ü (2012) Nutrient limitation reduces land carbon uptake in simulations with a model of combined carbon, nitrogen and phosphorus cycling. Biogeosciences 9:3547–3569

§Griffin KL, Tissue DT, Turnbull MH, Whitehead D (2000) The onset of photosynthetic acclimation to elevated CO_2 partial pressure in field-grown *Pinus radiata* D. Don. After 4 years. Plant Cell Environ 23:1089–1098

*Gunderson CA, O'Hara KH, Campion CM, Walker AV, Edwards NT (2010) Thermal plasticity of photosynthesis: the role of acclimation in forest responses to a warming climate. Glob Chang Biol 16: 2272–2286

§Gutiérrez D, Morcuende R, Del Pozo A, Martínez-Carrasco R, Pérez P (2013) Involvement of nitrogen and cytokinins in photosynthetic acclimation to elevated CO_2 of spring wheat. J Plant Physiol 170: 1337–1343

Handa IT, Hagedorn F, Hättenschwiler S (2008) No stimulation in root production in response to 4 years of *in situ* CO_2 enrichment at the Swiss treeline. Funct Ecol 22:348–358

Harley PC, Tenhunen JD (1991) Modelling the photosynthetic response of C_3 leaves to environmental factors. In: Boote KJ, Loomis RS (eds) Modelling Crop Photosynthesis: From Biochemistry to Canopy. CSSA, Madison, pp 17–39

§Hättenschwiler S, Körner. (1996) System-level adjustments to elevated CO_2 in model spruce ecosystems. Glob Chang Biol 2: 377–387

Herold A (1980) Regulation of photosynthesis by sink activity – the missing link. New Phytol 86:131–144

§Herrick JD, Thomas RB (2001) No photosynthetic down-regulation in sweetgum trees (*Liquidambar styraciflua* L.) after three years of CO_2 enrichment at the Duke Forest FACE experiment. Plant Cell Environ 24: 53–64

*Heskel MA, Bitterman D, Atkin OK, Turnbull MH, Griffin KL (2014) Seasonality of foliar respiration in two dominant plant species from the Arctic tundra: response to long-term warming and short-term temperature variability. Funct Plant Biol, 41: 287–300

Hikosaka K (1997) Modelling optimal temperature acclimation of the photosynthetic apparatus in C_3 plants with respect to nitrogen use. Ann Bot 80:721–730

Hikosaka K, Ishikawa K, Borjigidai A, Muller O, Onoda Y (2006) Temperature acclimation of photosynthesis: mechanisms involved in the changes in temperature dependence of photosynthetic rate. J Exp Bot 57:291–302

Holaday AS, Martindale W, Alred R, Brooks A, Leegood RC (1992) Changes in activities of enzymes of carbon metabolism in leaves during exposure to low temperature. Plant Physiol 98:1105–1114

Hüner NPA, Öquist G, Hurry VM, Krol M, Falk S, Griffith M (1993) Photosynthesis, photoinhibition and low temperature acclimation in cold tolerant plants. Photosynth Res 37:19–39

Hungate BA, Chapin FS III, Zhong H, Holland EA, Field CB (1996) Stimulation of grassland nitrogen cycling under carbon dioxide enrichment. Oecologia 109:149–153

Hungate BA, Dukes JS, Shaw MR, Luo Y, Field CB (2003) Nitrogen and climate change. Science 302:1512–1513

Hurry VM, Keerberg O, Pärnik T, Gardeström P, Öquist G (1995) Cold hardening results in increased activity of enzymes involved in carbon metabolism in leaves of winter rye (*Secale cereal* L.). Planta 195:554–562

Hurry V, Strand A, Furbank R, Stitt M (2000) The role of inorganic phosphate in the development of freezing tolerance and the acclimatization of photosynthesis to low temperature is revealed by the *pho* mutants of *Arabidopsis thaliana*. Plant J 24:383–396

§Hymus GJ, Snead TG, Johnson DP, Hungate BA, Drake BG (2002) Acclimation of photosynthesis and respiration to elevated atmospheric CO_2 in two scrub oaks. Glob Chang Biol 8: 317–328

Jablonski LM, Wang X, Curtis PS (2002) Plant reproduction under elevated CO_2 conditions: a meta-analysis of reports on 79 crop and wild species. New Phytol 156:9–26

§Jach ME, Ceulemans R (2000) Effects of season, needle age and elevated atmospheric CO_2 on photosynthesis in scots pine (*Pinus sylvestris*). Tree Physiol 20: 145–157

§Jackson RB, Luo Y, Cardon ZG, Sala OE, Field CB, Mooney HA (1995) Photosynthesis, growth and density for the dominant species in a CO_2-enriched grassland. J Biogeogr 1: 221–225

§Jifon JL, Wolfe DW (2002) Photosynthetic acclimation to elevated CO_2 in *Phaseolus vulgaris* L. is altered by growth response to nitrogen supply. Glob Chang Biol 8: 1018–1027

Jolliffe PA, Tregunna EB (1973) Environmental regulation of the oxygen effect on apparent photosynthesis in wheat. Can J Bot 51:841–853

Joos F, Spahni R (2008) Rates of change in natural and anthropogenic radiative forcing over the past 20,000 years. Proc Natl Acad Sci U S A 105:1425–1430

Jordan DB, Ogren WL (1984) The CO_2/O_2 specificity of ribulose 1,5-*bis*phosphate carboxylase/oxygenase. Dependence on ribulose bisphosphate concentration, pH and temperature. Planta 161:308–313

§Kim SH, Sicher RC, Bae H, Gitz DC, Baker JT, Timlin DJ, Reddy VR (2006) Canopy photosynthesis, evapotranspiration, leaf nitrogen, and transcription profiles of maize in response to CO_2 enrichment. Glob Chang Biol 12: 588–600

Kirschbaum MUF, Farquhar GD (1984) Temperature dependence of whole leaf photosynthesis in *Eucalyptus pauciflora* Sieb. ex Spreng. Aust J Plant Physiol 11:519–538

§Körner C, Diemer M (1994) Evidence that plants from high altitudes retain their greater photosynthetic efficiency under elevated CO_2. Funct Ecol 1: 58–68

*Kositsup B, Montpied P, Kasemsap P, Thaler P, Améglio T, Dreyer E (2009) Photosynthetic capacity and temperature responses of photosynthesis of rubber trees (*Hevea brasiliensis* Müll. Arg.) acclimate to changes in ambient temperatures. Trees 23: 357–365

*Kroner Y, Way DA (2016) Carbon fluxes acclimate more strongly to elevated growth temperatures than to elevated CO_2 concentrations in a northern conifer. Glob Chang Biol 22: 2913–2928

Kumarathunge D, Medlyn BE, Drake JE, Tjoelker MG, Aspinwall MJ, Battaglia M, Cano FJ, Carter K, Cavaleri MA, Cernusak L, Chambers JQ, Crous KY, De Kauwe MG, Dillaway DN, Dreyer E, Ellsworth DS, Ghannoum O, Han Q, Hikosaka K, Jensen AM, Kelly JWG, Kruger EL, Mercado LM, Onoda Y, Reich PB, Rogers A, Slot M, Smith NG, Tarvainen L, Tissue DT, Togashi HF, Tribuzy ES, Uddling J, Vårhammar A, Wallin G, Warren JM, Way DA (2019) Acclimation and adaptation components of the temperature dependence of plant photosynthesis at the global scale. New Phytol 222:768–784

Kunst L, Browse J, Somerville C (1989) Enhanced thermal tolerance in a mutant of Arabidopsis deficient in palmitic acid unsaturation. Plant Physiol 91:401–408

Labate CA, Leegood RC (1988) Limitation of photosynthesis by changes in temperature. Planta 173:519–527

Law RD, Crafts-Brandner SJ (2001) High temperature stress increases the expression of wheat ribulose-1,5-bisphosphate carboxylase/oxygenase activase protein. Arch Biochem Biophys 386:261–267

§Leakey AD, Press MC, Scholes JD, Watling JR (2002) Relative enhancement of photosynthesis and growth at elevated CO_2 is greater under sunflecks than uniform irradiance in a tropical rain forest tree seedling. Plant Cell Environ 25: 1701–1714

Leakey ADB, Ainsworth EA, Bernacchi CJ, Rogers A, Long SP, Ort DR (2009) Elevated CO_2 effects on plant carbon, nitrogen, and water relations: six important lessons from FACE. J Exp Bot 60:2859–2876

§Lee TD, Tjoelker MG, Ellsworth DS, Reich PB (2001) Leaf gas exchange responses of 13 prairie grassland species to elevated CO_2 and increased nitrogen supply. New Phytol 150: 405–418

Leuzinger S, Luo Y, Beier C, Dieleman W, Vicca S, Körner C (2011) Do global change experiments overestimate impacts on terrestrial ecosystems? Trends Ecol Evol 26:236–241

Lewis JD, Lucash M, Olszyk DM, Tingey DT (2004) Relationships between needle nitrogen concentration and photosynthetic responses of Douglas-fir seedlings to elevated CO_2 and temperature. New Phytol 162:355–364

§Li Y, Zhang Y, Zhang X, Korpelainen H, Berninger F, Li C (2013) Effects of elevated CO_2 and temperature on photosynthesis and leaf traits of an understory dwarf bamboo in subalpine forest zone, China. Physiol Plant 148: 261–272

Liberloo M, Tulva I, Raïm O, Kull O, Ceulemans R (2007) Photosynthetic stimulation under long-term CO_2 enrichment and fertilization is sustained across a closed *Populus* canopy profile (EUROFACE). New Phytol 173:537–549

Liberloo M, Lukac M, Calfapietra C, Hoosbeek MR, Gielen B, Miglietta F, Scarascia-Mugnozza GE, Ceulemans R (2009) Coppicing shifts CO_2 stimulation of poplar productivity to above-ground pools: a synthesis of leaf to stand level results from the POP/EUROFACE experiment. New Phytol 182:331–346

Lin Y-S, Medlyn BE, Ellsworth DS (2012) Temperature responses of leaf net photosynthesis: the role of component processes. Tree Physiol 32:219–231

§Liozon R, Badeck FW, Genty B, Meyer S, Saugier B (2000) Leaf photosynthetic characteristics of beech (*Fagus sylvatica*) saplings during three years of exposure to elevated CO_2 concentration. Tree Physiol 20: 239–247

Long SP (1991) Modification of the response of photosynthetic productivity to rising temperature by atmospheric CO_2 concentrations: has its importance been underestimated? Plant Cell Environ 14:729–739

Long SP, Drake BG (1991) Effect of the long-term elevation of CO_2 concentration in the field on the quantum yield of photosynthesis of the C_3 sedge, *Scirpus olneyi*. Plant Physiol 96:221–226

Long SP, Zhu X-G, Naidu SL, Ort DR (2006) Can improvement in photosynthesis increase crop yields? Plant Cell Environ 29:315–330

Loveys BR, Sheurwater I, Pons TL, Fitter A, Atkin OK (2002) Growth temperature influences the underlying components of relative growth rate: an investigation using inherently fast- and slow-growing plant species. Plant Cell Environ 25:975–988

Luo Y, Su BO, Currie WS, Dukes JS, Finzi A, Hartwig U, Hungate B, Pataki DE (2004) Progressive nitrogen limitation of ecosystem responses to rising atmospheric carbon dioxide. Bioscience 54:731–739

Luo Y, Hui D, Zhang D (2006) Elevated CO_2 stimulates net accumulations of carbon and nitrogen in land ecosystems: a meta-analysis. Ecology 87:53–63

Makino A, Nakano H, Mae T (1994) Responses of ribulose-1,5-bisphosphate carboxylase, cytochrome-f, and sucrose synthesis enzymes in rice leaves to leaf nitrogen and their relationships to photosynthesis. Plant Physiol 105:173–179

Martindale W, Leegood RC (1997) Acclimation of photosynthesis to low temperature in *Spinacia oleracea* L. II Effects of nitrogen supply. J Exp Bot 48:1873–1880

McCarthy HR, Oren R, Johnsen KH, Gallet-Budynek A, Pritchard SG, Cook CW, LaDeau SL, Jackson RB, Finzi AC (2010) Re-assessment of plant carbon dynamics at the Duke free-air CO2 enrichment site: interactions of atmospheric [CO_2] with nitrogen and water availability over stand development. New Phytol 185:514–528

*Mediavilla S, González-Zurdo P, Babiano J, Escudero A (2016) Responses of photosynthetic parameters to differences in winter temperatures throughout a temperature gradient in two evergreen tree species. Eur J For Res 135: 871–883

Mooney HA, Björkman O, Collatz GJ (1978) Photosynthetic acclimation to temperature in the desert shurb, *Larrea divaricata*. 1. Carbon dioxide exchange characteristics of intact leaves. Plant Physiol 61:406–410

Moore BD, Cheng SH, Sims D, Seemann JR (1999) The biochemical and molecular basis for photosynthetic acclimation to elevated atmospheric CO_2. Plant Cell Environ 22:567–582

*Nagai T, Makino A (2009) Differences between rice and wheat in temperature responses of photosynthesis and plant growth. Plant Cell Physiol 50: 744–755

Newingham BA, Vanier CH, Charlet TN, Ogle K, Smith SD, Nowak RS (2013) No cumulative effect of ten years of elevated CO_2 on perennial plant biomass components in the Mojave Desert. Glob Chang Biol 19:2168–2181

NOAA (2019) National Centers for Environmental Information, State of the Climate: global climate report for annualn.d.. https://www.ncdc.noaa.gov/sotc/global/201813. Accessed 20 Feb 2019

Norby RJ, Ledford J, Reilly CD, Miller NE, O'Neill EG (2004) Fine-root production dominates response of a deciduous forest to atmospheric CO_2 enrichment. Proc Nat Acad Sci U S A 101:9689–9693

Norby RJ, DeLucia EH, Gielen B, Calfapietra C, Giardina CP, King JS, Ledford J, De Angelis P (2005) Forest response to elevated CO_2 is conserved across a broad range of productivity. Proc Natl Acad Sci 102:18052–18056

Norby RJ, Warren JM, Iversen CM, Medlyn BE, McMurtrie RE (2010) CO_2 enhancement of forest productivity constrained by limited nitrogen availability. Proc Nat Acad Sci U S A 107:19368–19373

Nowak RS, Ellsworth DS, Smith SD (2004) Functional responses of plants to elevated atmospheric CO_2 – do photosynthetic and productivity data from FACE experiments support early predictions? New Phytol 162:253–280

Oren R, Ellsworth DS, Johnsen KH, Phillips N, Ewers BE, Maier C, Schäfer KV, McCarthy H, Hendrey G, McNulty SG, Katul GG (2001) Soil fertility limits carbon sequestration by forest ecosystems in a CO_2 -enriched atmosphere. Nature 411:469–472

Paschalis A, Katul GG, Fatichi S, Palmroth S, Way D (2017) On the variability of the ecosystem responsc to elevated atmospheric CO_2 across spatial and temporal scales at the Duke Forest FACE experiment. Agric For Meteorol 232:367–383

Pearcy RW (1977) Acclimation of photosynthetic and respiratory carbon dioxide exchange to growth temperature in *Atriplex lentiformis* (Torr.) wats. Plant Physiol 59:795–799

Pearcy RW (1978) Effect of growth temperature on the fatty acid composition of the leaf lipids in *Atriplex lentiformis* (Torr.) wats. Plant Physiol 61:484–486

§Pinto H, Sharwood RE, Tissue DT, Ghannoum O (2014) Photosynthesis of C_3, C_3–C_4, and C_4 grasses at glacial CO_2. J Exp Bot 65: 3669–3681

§Polley HW, Johnson HB, Mayeux HS (1997) Leaf physiology, production, water use, and nitrogen dynamics of the grassland invader *Acacia smallii* at elevated CO_2 concentrations. Tree Physiol 17: 89–96

Rachmilevitch S, Cousins AB, Bloom AJ (2004) Nitrate assimilation in plant shoots depends on photorespiration. Proc Nat Acad Sci U S A 101:11506–11510

Radin JW, Kimball BA, Hendrix DL, Mauney JR (1987) Photosynthesis of cotton plants exposed to elevated levels of carbon dioxide in the field. Photosynth Res 12:191–203

Reich PB, Hobbie SE (2013) Decade-long soil nitrogen constraint on the CO_2 fertilization of plant biomass. Nat Clim Chang 3:278–282

Reich PB, Hungate BA, Luo Y (2006) Carbon-nitrogen interactions in terrestrial ecosystems in response to rising atmospheric carbon dioxide. Annu Rev Ecol Evol Syst 37:611–636

§Roberntz P, Stockfors JA (1998) Effects of elevated CO_2 concentration and nutrition on net photosynthesis, stomatal conductance and needle respiration of field-grown Norway spruce trees. Tree Physiol 18: 233–241

§Rogers A, Fischer BU, Bryant J, Frehner M, Blum H, Raines CA, Long SP (1998) Acclimation of photosynthesis to elevated CO_2 under low-nitrogen nutrition is affected by the capacity for assimilate utilization. Perennial ryegrass under free-air CO2 enrichment. Plant Physiol 118: 683–689

§Rosenthal DM, Ruiz-Vera UM, Siebers MH, Gray SB, Bernacchi CJ, Ort DR (2014) Biochemical acclimation, stomatal limitation and precipitation patterns underlie decreases in photosynthetic stimulation of soybean (*Glycine max*) at elevated [CO_2] and temperatures under fully open air field conditions. Plant Sci 226: 136–146

Rowland-Bamford AJ, Baker JT, Allen LH Jr, Bowes G (1991) Acclimation of rice to changing atmospheric carbon dioxide concentrations. Plant Cell Environ 14:577–583

Rubio-Asensio JS, Bloom AJ (2017) Inorganic nitrogen form: a major player in wheat and *Arabidopsis* responses to elevated CO_2. J Exp Bot 68:2611–2625

Rustad L, Campbell J, Marion G, Norby R, Mitchell M, Hartley A, Cornelissen J, Gurevitch J (2001) A meta-analysis of the response of soil respiration, net nitrogen mineralization, and aboveground plant growth to experimental ecosystem warming. Oecologia 126:543–562

Sage RF (1990) A model describing the regulation of ribulose-1,5-*bis*phosphate carboxylase, electron transport, and triose phosphate use in response to light intensity and CO_2 in C_3 plants. Plant Physiol 94:1728–1734

Sage RF (1994) Acclimation of photosynthesis to increasing atmospheric CO_2: the gas exchange perspective. Photosynth Res 39:351–368

Sage RF, Kubien DS (2007) The temperature response of C_3 and C_4 photosynthesis. Plant Cell Environ 30:1086–1106

Sage RF, Sharkey TD (1987) The effect of temperature on the occurrence of O_2 and CO_2 insensitive photosynthesis in field grown plants. Plant Physiol 84:658–664

Sage RF, Sharkey TD, Seemann JR (1989) Acclimation of photosynthesis to elevated CO_2 in 5 C_3 species. Plant Physiol 89:590–596

Sanders IR, Streitwolf-Engel R, van der Heijden MGA, Boller T, Wiemken A (1998) Increased allocation to external hyphae of arbuscular mycorrhizal fungi under CO_2 enrichment. Oecologia 117:496–503

§Sanz-Sáez Á, Erice G, Aranjuelo I, Nogués S, Irigoyen JJ, Sánchez-Díaz M (2010) Photosynthetic down-regulation under elevated CO_2 exposure can be prevented by nitrogen supply in nodulated alfalfa. J Plant Physiol 167: 1558–1565

*Scafaro AP, Xiang S, Long BM, Bahar NHA, Weerasinghe LK, Creek D, Evans JR, Reich PB, Atkin OK (2017) Strong thermal acclimation of photosynthesis in tropical and temperate wet-forest tree species: the importance of altered Rubisco content. Glob Chang Biol 23: 2783–2800

Seneweera SP, Conroy JP, Ishimaru K, Ghannoum O, Okada M, Lieffering M, Kim HY, Kobayoshi K (2002) Changes in source-sink relations during development influence photosynthetic acclimation of rice to free air CO_2 enrichment (FACE). Funct Plant Biol 29:945–953

Sharkey TD (1985) Photosynthesis in intact leaves of C_3 plants: physics, physiology, and rate limitations. Bot Rev 51:53–105

§Sims DA, Luo Y, Seemann JR (1998) Comparison of photosynthetic acclimation to elevated CO_2 and limited nitrogen supply in soybean. Plant Cell Environ 21: 945–952

Slot M, Kitajima K (2015) General patterns of acclimation of leaf respiration to elevated temperatures across biomes and plant types. Oecologia 177:885–900

Smith NG, Dukes JS (2017) Short-term acclimation to warmer temperatures accelerates leaf carbon exchange processes across plant types. Glob Chang Biol 23:4840–4853

Smith SE, Read DJ (2008) Mycorrhizal Symbiosis, 2nd edn. Academic, San Diego

*Smith RA, Lewis JD, Ghannoum O, Tissue DT (2012) Leaf structural responses to pre-industrial, current and elevated atmospheric [CO_2] and temperature affect leaf function in *Eucalyptus sideroxylon*. Funct Plant Biol 39: 285–296

§Staudt M, Joffre R, Rambal S, Kesselmeier J (2001) Effect of elevated CO_2 on monoterpene emission of young *Quercus ilex* trees and its relation to structural and ecophysiological parameters. Tree Physiol 21: 437–445

Stitt M, Grosse H (1988) Interactions between sucrose synthesis and CO_2 fixation IV. Temperature-dependent adjustment of the relation between sucrose synthesis and CO_2 fixation. J Plant Physiol 133:392–400

Stitt M, Hurry V (2002) A plant for all seasons: alterations in photosynthetic carbon metabolism during cold acclimation in *Arabidopsis*. Curr Opin Plant Biol 5:199–206

Strand Å, Hurry V, Gustafsson P, Gardestrom P (1997) Development of *Arabidopsis thaliana* leaves at low temperatures releases the suppression of photosynthesis and photosynthetic gene expression despite the accumulation of soluble carbohydrates. Plant J 12:605–614

Strand Å, Hurry V, Henkes S, Hüner N, Gustafsson P, Gardestrom P, Stitt M (1999) Acclimation of *Arabidopsis* leaves developing at low temperatures. Increasing cytoplasmic volume accompanies increased activities of enzymes in the Calvin Cycle and in the sucrose-biosynthesis pathway. Plant Physiol 119:1387–1397

Terrer C, Vicca S, Hungate BA, Phillips RP, Prentice IC (2016) Mycorrhizal association as a primary control of the CO_2 fertilization effect. Science 353:72–74

Terrer C, Vicca S, Stocker BD, Hungate BA, Phillips RP, Recih PB, Finzi AC, Prentice IC (2018) Ecosystem responses to elevated CO_2 governed by plant-soil interactions and the cost of nitrogen acquisition. New Phytol 217:507–522

§Tjoelker MG, Oleksyn J, Reich PB (1998) Seedlings of five boreal tree species differ in acclimation of net photosynthesis to elevated CO_2 and temperature. Tree Physiol 18: 715–726

Tjoelker MG, Oleksyn J, Reich PB (1999) Acclimation of respiration to temperature and CO_2 in seedlings of boreal tree species in relation to plant size and relative growth rate. Glob Chang Biol 5:679–691

§Turnbull MH, Tissue DT, Griffin KL, Rogers GN, Whitehead D (1998) Photosynthetic acclimation to long-term exposure to elevated CO_2 concentration in *Pinus radiata* D. Don. Is related to age of needles. Plant Cell Environ 21: 1019–1028

§Urban O, Hrstka M, Zitová M, Holišová P, Šprtová M, Klem K, Calfapietra C, De Angelis P, Marek MV (2012) Effect of season, needle age and elevated CO_2 concentration on photosynthesis and Rubisco acclimation in *Picea abies*. Plant Physiol Biochem 58: 135–141

§Vogel CS, Curtis PS (1995) Leaf gas exchange and nitrogen dynamics of N_2-fixing, field-grown *Alnus glutinosa* under elevated atmospheric CO_2. Glob Chang Biol 1: 55–61

von Caemmerer S, Evans JR (2015) Temperature responses of mesophyll conductance differ greatly between species. Plant Cell Environ 38:629–637

§von Caemmerer S, Ghannoum O, Conroy JP, Clark H, Newton PC (2001) Photosynthetic responses of temperate species to free air CO_2 enrichment (FACE) in a grazed New Zealand pasture. Funct Plant Biol 28: 439–450

Wang D, Heckathorn SA, Wang X, Philpott SM (2012) A meta-analysis of plant physiological and growth responses to temperature and elevated CO_2. Oecologia 169:1–13

Warren JM, Jensen AM, Medlyn BE, Norby RJ, Tissue DT (2015) Carbon dioxide stimulation of photosynthesis in *Liquidambar styraciflua* is not sustained during a 12-year field experiment. AoB Plants 7:plu074

§Watling JR, Press MC, Quick WP (2000) Elevated CO_2 induces biochemical and ultrastructural changes in leaves of the C_4 cereal sorghum. Plant Physiol 123:1143–1152

*Way DA, Sage RF (2008a) Elevated growth temperatures reduce the carbon gain of black spruce *Picea mariana* (Mill.) B.S.P. Glob Chang Biol 14: 624–636

*Way DA, Sage RF (2008b) Thermal acclimation of photosynthesis in black spruce (*Picea mariana* (Mill.) B.S.P.). Plant Cell Environ 31:1250–1262

Way DA, Yamori W (2014) Thermal acclimation of photosynthesis: on the importance of adjusting definitions and accounting for thermal acclimation of respiration. Photosynth Res 119:89–100

Way DA, LaDeau SL, McCarthy HR, Clark JS, Oren R, Finzi AC, Jackson RB (2010) Greater seed production in elevated CO_2 is not accompanied by reduced seed quality in *Pinus taeda* L. Glob Chang Biol 16:1046–1056

Way DA, Aspinwall MJ, Drake J, Crous K, Campany C, Ghannoum O, Tissue D, Tjoelker MG (2019) Light respiration responses to warming in field-grown trees: a comparison of the thermal sensitivity of the Kok and Laisk methods. New Phytol 222:132–143

Wieder WR, Cleveland CC, Smith WK, Todd-Brown K (2015) Future productivity and carbon storage limited by terrestrial nutrient availability. Nat Geosci 8:441–444

Wilson JM, Crawford RMM (1974) The acclimatization of plants to chilling temperatures in relation to the fatty-acid composition of leaf polar lipids. New Phytol 73:805–820

Wong SC (1979) Elevated atmospheric partial pressure of CO_2 and plant growth. Oecologia 44:68–74

Wright IJ, Reich PB, Westoby M, Ackerly DD, Baruch Z, Bongers F, Cavender-Bares J, Flexas J (2004) The worldwide leaf economic spectrum. Nature 428:821–827

Wujeska-Klause A, Crous KY, Ghannoum O, Ellsworth DS (2019) Lower photorespiration in elevated CO_2 reduces leaf N concentrations in mature Eucalyptus trees in the field. Glob Chang Biol. https://doi.org/10.1111/14555

*Xu CY, Salih A, Ghannoum O, Tissue DT (2012) Leaf structural characteristics are less important than leaf chemical properties in determining the response of leaf mass per area and photosynthesis of *Eucalyptus saligna* to industrial-age changes in [CO_2] and temperature. J Exp Bot, 63: 5829–5841

*Yamaguchi DP, Nakaji T, Hiura T, Hikosaka K (2016) Effects of seasonal change and experimental warming on the temperature dependence of photosynthesis in the canopy leaves of *Quercus serrata*. Tree Physiol 36: 1283–1295

Yamori W, Noguchi K, Terashima I (2005) Temperature acclimation of photosynthesis in spinach leaves: analyses of photosynthetic components and temperature dependencies of photosynthetic partial reactions. Plant Cell Environ 28:536–547

*Yamori W, Noguchi K, Hikosaka K, Terashima I (2010) Phenotypic plasticity in photosynthetic temperature acclimation among crop species with different cold tolerances. Plant Physiol 152: 388–399

Yamori W, Hikosaka K, Way DA (2014) Temperature response of photosynthesis in C3, C4 and CAM plants: acclimation and adaptation. Photosynth Res 119:101–117

Yang JT, Preiser AL, Li Z, Weise SE, Sharkey TD (2016) Triose phosphate use limitation of photosynthesis: short-term and long-term effects. Plan Theory 243:687–698

Yin X (2002) Responses of leaf nitrogen concentration and specific leaf area to atmospheric CO_2 enrichment: a retrospective synthesis across 62 species. Glob Chang Biol 8:631–642

Zak DR, Pregitzer KS, Kubiske ME, Burton AJ (2011) Forest productivity under elevated CO_2 and O_3: positive feedbacks to soil N cycling sustain decade-long net primary productivity enhancement by CO_2. Ecol Lett 14:1220–1226

*Zhang XW, Wang JR, Ji MF, Milne RI, Wang MH, Liu JQ, Shi S, Yang SL, Zhao CM (2015). Higher thermal acclimation potential of respiration but not photosynthesis in two alpine *Picea* taxa in contrast to two lowland congeners. PLoS One 10: e0123248

*Zhang X, Chen, L, Wang, J, Wang, M, Yang, S, Zhao, C (2018) Photosynthetic acclimation to long-term high temperature and soil drought stress in two spruce species (*Picea crassifolia* and *P. wilsonii*) used for afforestation. J For Res 29: 363–372

Zhu X-G, Long SP, Ort DR (2010) Improving photosynthetic efficiency for greater yield. Annu Rev Plant Biol 61:235–261

Ziska LH, Drake BG, Chamberlain S (1990) Long-term photosynthetic response in single leaves of a C_3 and C_4 salt marsh species grown at elevated atmospheric CO_2 in situ. Oecologia 83:469–471

Chapter 5

Trichome Responses to Elevated Atmospheric CO_2 of the Future

James M. Fischer*
Carl R. Woese Institute for Genomic Biology, University of Illinois at Urbana Champaign, Urbana, IL, USA

Department of Ecology and Evolutionary Biology, University of Kansas, Lawrence, KS, USA

and

Joy K. Ward
College of Arts and Sciences, Department of Biology, Case Western Reserve University, Cleveland, OH, USA

Department of Ecology and Evolutionary Biology, University of Kansas, Lawrence, KS, USA

*Author for correspondence, e-mail: jf047@illinois.edu

K. M. Becklin et al. (eds.), *Photosynthesis, Respiration, and Climate Change*, Advances in Photosynthesis and Respiration 48, https://doi.org/10.1007/978-3-030-64926-5_5

Summary

Leaf hairs (trichomes) are small and rigid epidermal structures that serve in herbivore defense, temperature regulation, boundary layer fortification, and UV-B protection, and can even act as mechanosensory switches indicating insect herbivore presence (Zhou et al. 2016; Xiao et al. 2017). As such, leaf trichomes have impacts on overall plant physiology, photosynthetic efficiency, fitness, and plant-environment interactions (Calo et al. 2006; Løe et al. 2007; Sletvold et al. 2010; Nguyen et al. 2016; Xiao et al. 2017; Imboden et al. 2018; Kergunteuil et al. 2018; Kim 2019). Elevated [CO_2] affects plants across multiple scales, ranging from the molecular and physiological levels to the ecosystem level (Masle 2000; Bidart-Bouzat et al. 2005; Teng et al., 2006; Medeiros and Ward 2013; Becklin et al. 2016; Dong et al. 2018). It has been noted that leaf trichomes shift in density on the leaf surface when grown at elevated [CO_2] in a number of species (Masle 2000; Bidart-Bouzat et al. 2005; Lake and Wade 2009; Karowe and Grubb 2011; Guo et al. 2013, 2014). Elevated [CO_2] *decreases* trichome densities by as much as 60% in some species (wheat, *Arabidopsis*) and increases densities by as much as 57% in other species (*Brassica rapa, Medicago truncatula*) (Karowe and Grubb 2011; Guo et al. 2013, 2014). However, the responses of trichomes to elevated [CO_2] remain critically understudied, and little is known about the molecular/developmental mechanisms driving these responses. As trichomes are physiologically important structures for numerous food crop species (e.g., wheat, soybean) and critical ecological species, it is imperative that further research be dedicated to understanding the implications of shifting trichome densities in an elevated [CO_2] environment of the future. Here we review the ecological roles of trichomes, describe what is known about elevated [CO_2]-trichome phenotypic responses, provide a relevant background of trichome genetics, and present potential molecular/developmental mechanisms behind elevated [CO_2]-trichome responses. Finally, we propose future research directions to stimulate future elevated [CO_2]-trichome research in order to uncover the mechanisms that control altered trichome densities at elevated [CO_2].

I. Introduction

The atmospheric concentration of carbon dioxide ([CO_2]) is currently approximately 410 ppm, having risen from a pre-industrial value (approximately 200 years ago) of 270 ppm (NOAA 2016). Atmospheric [CO_2] is projected to reach 800–1000 ppm by the end of this century if CO_2 emissions are not curtailed (NOAA 2016). Elevated [CO_2] affects plants across multiple scales, ranging from the ecosystem level to the physiological and molecular levels, whereby elevated [CO_2] can alter leaf gas exchange, plant nutritional quality, growth, development, and production of secondary defense compounds (Masle 2000; Bidart-Bouzat et al. 2005; Teng et al. 2006; Medeiros and Ward 2013; Becklin et al. 2016; Dong et al. 2018). Elevated [CO_2] may also influence the production and density of micro-structures on the leaf surface—such as stomata (pores required for gas exchange) and trichomes (leaf hairs) (Masle 2000; Bidart-Bouzat 2004; Bidart-Bouzat et al. 2005; Lake and Wade 2009; Karowe and Grubb 2011; Haus et al. 2018). Such changes can have major effects on plant carbon assimilation, water use, and defense. Stomata densities (number of stomata normalized by leaf area) and/or indices (percentage of stomata to epidermal cells) shift in a large number of plant species grown at elevated [CO_2] and under other climatic shifts (Woodward and Kelly 1995; Teng et al. 2006) (See chapter "Stomatal Responses to Climate Change"). Wide-scale attention has been paid to understanding the mechanisms that control stomatal density shifts at elevated [CO_2], ranging from changes in cuticular wax composition to mobile signals transported through the phloem (Gray et al. 2000; Lake et al. 2001; Ainsworth and Rogers 2007; Lake and Woodward 2008; Haus et al. 2018; See Xu et al. 2016b for an excellent review on stomata and elevated [CO_2]).

Trichomes have also been shown to shift in density at elevated [CO_2], along with stomata densities (Bidart-Bouzat et al. 2005; Lake and Wade 2009). However, altered trichome production at elevated [CO_2] and the mechanism(s) controlling these responses remain critically understudied. Generally, the ecological and physiological roles of trichomes have been well characterized, as has the genetic and developmental pathways behind trichome initiation (Pesch and Hülskamp 2009; Balkunde et al. 2010; Bickford 2016; Xiao et al. 2017). This past work provides an excellent foundation for trichome research that is aimed at understanding the implications of elevated [CO_2] on trichome production and provides great insights for enhancing our ability to elucidate the molecular mechanisms driving trichome density shifts at elevated [CO_2].

In a review of the literature, we found that elevated [CO_2] reduces trichome densities by as much as 60% in some species (e.g., wheat) and promotes trichome density increases as high as 57% in other species (e.g., *Brassica rapa*) (Masle 2000; Karowe and Grubb 2011; Guo et al. 2013; Guo et al. 2014). These altered trichome densities, that may deviate from locally adaptive trichome densities, could potentially alter plant fitness, herbivory damage, water use efficiency (WUE), fungal infection rates, and photosynthetic efficiency (Manetas 2003; Calo et al. 2006; Løe et al. 2007; Sletvold et al. 2010; Kergunteuil et al. 2018; Kim 2019). Therefore, trichome density shifts may have large implications for the responses of native species and food crops (e.g., wheat, soybean) to future global climate change, and further research will be necessary for understanding the underlying mechanisms of shifting trichome densities in an elevated [CO_2] future.

In this review we describe (1) the multiple roles of trichomes, (2) trichome responses to elevated [CO_2], (3) molecular mechanisms of trichome production, (4) potential mechanisms for shifted trichome densities at elevated [CO_2], and (5) directions for future research. We have culled information on numerous trichome genes, and potential connections between elevated [CO_2], trichome densities/patterning, and other elevated [CO_2]-responsive physiological/developmental processes as the basis for this review. Moreover, the purpose of this review is to unite multiple fields in an effort to reveal a framework of potential molecular mechanisms that control the responses of trichome productivity to rising [CO_2].

II. The Multiple Roles of Trichomes

Trichomes appear in a host of different forms, ranging from single-cell rigid hairs, as in *Arabidopsis thaliana* (Fig. 5.1a) to

Abbreviations: CPC – CAPRICE; ET – Ethylene; FDH – FIDDLEHEAD; GA – Gibberellic acid; GFP – Green fluorescent protein; GL1 – GLABROUS1; GL2 – GLABROUS2; GL3 – GLABROUS3; GO – gene ontology; HIC – HIGH CARBON DIOXIDE; IAA – Indole-3-acetic acid; JA – Jasmonic acid; JAZ – JASMONATE-ZIM DOMAIN; KCS – Ketoacyl-CoA synthase; miR156 – MicroRNA156; PAP1 – PRODUCTION OF ANTHOCYANIN PIGMENT 1; PS II – Photosystem II; SA – Salicylic acid; SIM – SIAMESE; SlIAA15 – *Solanum lycopersicum* indole-3-acetic acid 15; SPL – SQUAMOSA PROMOTER BINDING-LIKE; TCL1 – TRICHOMELESS1; TRY – TRYPTYCHON; TTG1 – TRANSPARENT TESTA GLABRA 1; UPL3 – UBIQUITIN PROTEIN LIGASE 3; VLCFA – Very long chain fatty acids; WER – WEREWOLF; WUE – Water use efficiency

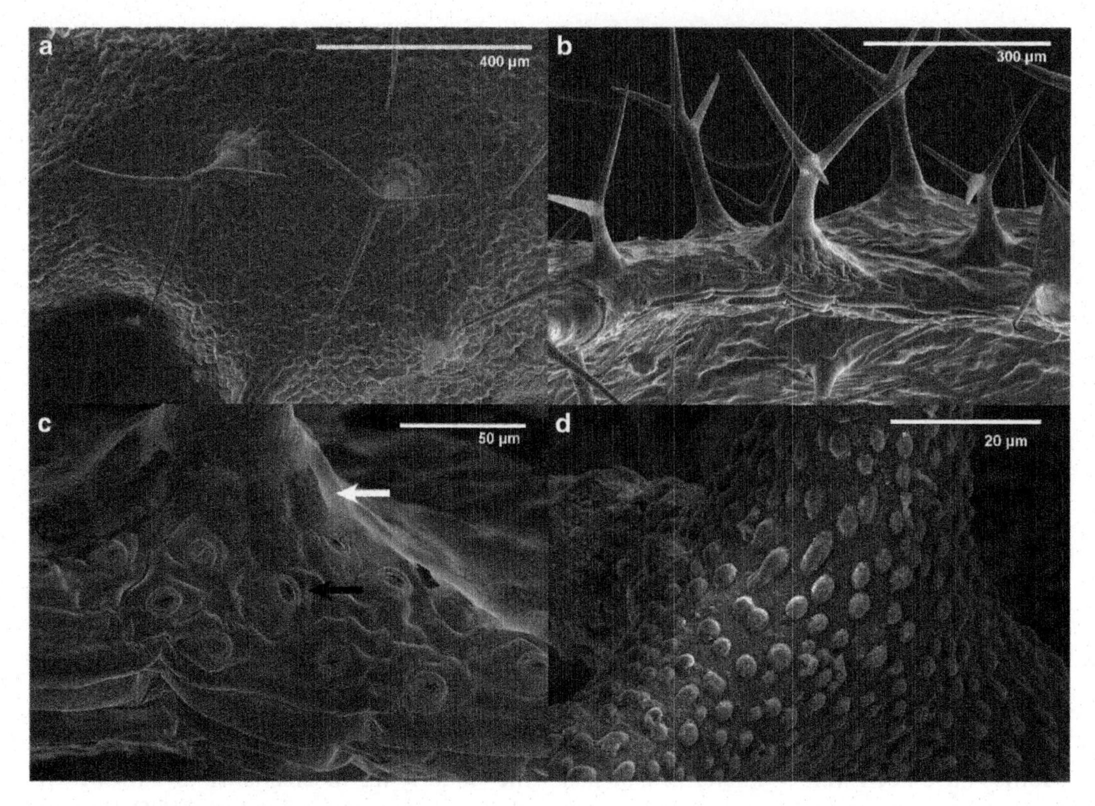

Fig. 5.1. Trichomes across the leaf surface of *Arabidopsis thaliana*. Trichomes are hair-like structures (<<1 mm) that are important in physiology, defense, and overall plant survival. (**a**, **b**): Trichomes across the adult leaf surface. (**c**): Close up of trichome base. White arrow indicates a trichome socket cell and the black arrow indicates a stomate. (**d**): Close up of trichome papillae at the meeting point of three branches. (Credits: Fischer, J.M., Rosa-Molinar, E, Kilcrease, J.)

fluid-secreting (glandular), multicellular structures such as those found in tomato (*Solanum lycopersicum*). With highly varied functions and structural morphologies, trichomes can be found on all aerial organs of the plant, across the majority of plant species (Werker 2000). In this review, we primarily focus on leaf trichomes since they have the largest implications for plant water use and carbon assimilation. Leaf trichomes have impacts on overall plant physiology, photosynthetic efficiency, fitness, and plant-environment interactions (Xiao et al. 2017). More specifically, leaf trichomes act in herbivore defense, temperature regulation, boundary layer fortification, and UV-B protection. Further, there have been shown to be trade-offs between trichome production and fitness/performance, and elevated [CO_2] may potentially offset this balance (Hanley et al. 2007; Sletvold et al. 2010). Below we provide a more thorough description of the many roles that trichomes play in plant physiology, defense, and overall survival.

A. Herbivore Defense

One primary physiological role of non-glandular trichomes is mechanical defense of the leaf. Sharp branch tips on trichomes (Fig. 5.1b) deter large herbivores and can injure smaller insects or trap them until starvation or predation occurs (Johnson 1975; Krimmel and Pearse 2013). In glandular trichomes, defensive capabilities such as those present on *Solanum lycopersicum* (cultivated tomato) rely on noxious compounds produced from the tip of the trichome as a means of chemical defense against insect herbivores (Glas et al. 2012; Bergau et al. 2015). Trichomes have also recently been shown to act as mechanosensory switches whereby a large insect herbivore can buckle a trichome, triggering a shift in leaf pH in the area surrounding the trichome, potentially initiating secondary metabolite defense mechanisms (Zhou et al. 2016).

A lack or reduced level of trichome-based herbivore defenses may be detrimental to plant survivorship, which is especially important when considering potential shifts in trichome densities at elevated [CO_2]. In natural *Arabidopsis lyrata* populations from across Sweden, glabrous (lacking trichomes) individuals consistently exhibited increased herbivore damage compared to trichome-producing individuals (Løe et al. 2007). The difference in herbivory was even more pronounced for mixed populations with both glabrous and trichome-producing individuals, whereby there was selection for higher trichome densities in mixed populations (Løe et al. 2007).

B. Temperature Regulation

Trichomes aid in temperature regulation on the leaf surface, which primarily occurs through reflectance of solar radiation. Pubescent lines of *Encelia farinosa*, for example, were shown to regulate leaf temperature through trichome-mediated light reflectance, whereas glabrous *Encelia frutescens* maintained leaf temperature through heat-dissipating transpiration (Ehleringer 1988). Although *E. farinosa* and *E. frutescens* modulate leaf temperature through two different physiological mechanisms, leaf temperatures were found to be highly similar between the two species—signifying that trichome production is one adaptive pathway that can replace other physiological responses (Ehleringer 1988). There is also wide intraspecific variation in trichome densities, and this density variation can lead to differences in the mechanisms used for leaf temperature modulation. For example, within *E. farinose,* a wide variation in trichome densities exists whereby less pubescent *E. farinosa* individuals increase transpiration to modulate leaf temperatures, while more pubescent individuals reflect excess light via denser trichome mats (Ehleringer and Björkman 1978; Ehleringer and Mooney 1978; Bickford 2016).

Conversely, extremely trichome-dense leaves may experience increased overheating. For example, dense pubescence in *Euphorbia wallichii*, native to the Himalaya-Hengduan Mountains, absorbed high amounts of solar radiation during times of peak radiation, leading to reduced heat loss and an increased likelihood of overheating (Peng et al. 2015). Others have hypothesized that the connection between dense pubescence and increased leaf temperature may be adaptive to pubescent arctic species, but this hypothesis requires further investigation (Peng et al. 2015).

C. Boundary Layer Fortification

In addition to regulating temperature and thereby water loss, trichomes also reduce water loss through transpiration via fortification of the boundary layer between the leaf surface and the ambient outside air (Konrad et al. 2015). The thickness of the boundary layer, fortified by trichome presence, scales with resistance to water diffusion (Amada et al. 2017). The contributions of trichomes to improved water use efficiency (WUE; carbon gain per water loss) vary significantly among species. In *Metrosideros polymorpha*, only 9% of resistance to water flux can be attributed to trichome resistance (Amada et al. 2017). Conversely, *S. lycopersicum* trichome densities are significantly positively correlated with WUE and significantly negatively correlated with stomatal conductance (g_s) (Galdon-Armero et al. 2018). Further, the ratio of trichomes to stomata (T/S) on the leaf surface also positively correlates with WUE in tomato (Galdon-Armero et al.

2018). T/S, therefore, may be viewed as a measurement to interconnect stomatal densities, trichome densities, and gas exchange (Galdon-Armero et al. 2018). In addition to T/S, whole leaf characteristics also play a role in determining the magnitude to which trichomes can control water loss. Small leaves subjected to low wind speeds, for example, had high contributions of trichome layers to water flux resistance (Schreuder et al. 2001). Through modelling, the relationship between leaf size and shape, wind speed and the effects of trichome characteristics and densities is useful for determining whether trichomes play a significant role in boundary layer resistance in a given scenario, and understanding these relationships may prove important in a changing climate (Schreuder et al. 2001).

D. UV-B & Photosystem II Protection

Trichomes protect the leaf photosystems from excess light via a papillae coating on the trichome surface (Fig. 5.1d) and through phenolic compounds, which individually lower internal leaf temperatures and prevent UV-B damage to photosystem II (PS II) (Johnson 1975; Ehleringer and Björkman 1978; Manetas 2003). When trichomes were experimentally removed from photosynthetic stems of *Verbascum speciosum*, hairless stems exhibited significant decreases in photosystem II yield (23% decrease overall) and electron transport rates (14% decrease overall) when exposed to UV-B (Manetas 2003). Increased UV-B exposure, in conjunction with lowered trichome densities, has other physiological effects in addition to impaired photosynthesis. For example, *Arabidopsis* mutants with reduced or absent trichomes exhibited increased effects of UV-B, such as shifts in flowering time due to the impacts of stress on plant development (Yan et al. 2012). Furthermore, a trichome mutant (*gl1*) with reduced trichome densities displayed a 62% delay in flowering time when exposed to elevated UV-B, while the control (Col-0) and a trichome over-producing mutant (*try82*) displayed only 41% and 35% delays in flowering time, respectively (Yan et al. 2012). Thus, altered leaf trichome densities not only shift the micromorphology of the leaf, but can influence the entire development of the plant under high levels of stress.

E. Environmental Adaptation

Maintaining an optimal trichome density for the enhancement of plant fitness may involve a fine balance between maximizing trichome density to inhibit herbivory and alleviate the impacts of environmental stressors (e.g., UV-B), while simultaneously preventing the impacts of overproduction of trichomes including fungal infections and excessive energy expenditure (Fig. 5.2) (Calo et al. 2006; Løe et al. 2007; Sletvold et al. 2010; Nguyen et al. 2016; Imboden et al. 2018; Kergunteuil et al. 2018; Kim 2019). Such a balance between too few and too many trichomes is important to consider with regards to shifting trichome densities at elevated [CO_2]. Trichomes of maize, *Brachypodium distachyon*, and barley act as entry points for *Fusarium* infection, with fungal hyphae gaining entry into the leaf through trichome base socket cells (Fig. 5.1c) (Peraldi et al. 2011; Nguyen et al. 2016; Imboden et al. 2018). In addition, a glabrous *Arabidopsis* trichome mutant (*gl1*) showed increased resistance to the necrotrophic fungus *Botrytis cinereal* compared to wildtype, while a trichome overproducing mutant (*try*) was more susceptible to fungal infection (Calo et al. 2006). Therefore, excessive trichome density may not be entirely beneficial, as increased trichome densities carry increased risk of detrimental fungal infections, and possibly other yet unknown outcomes.

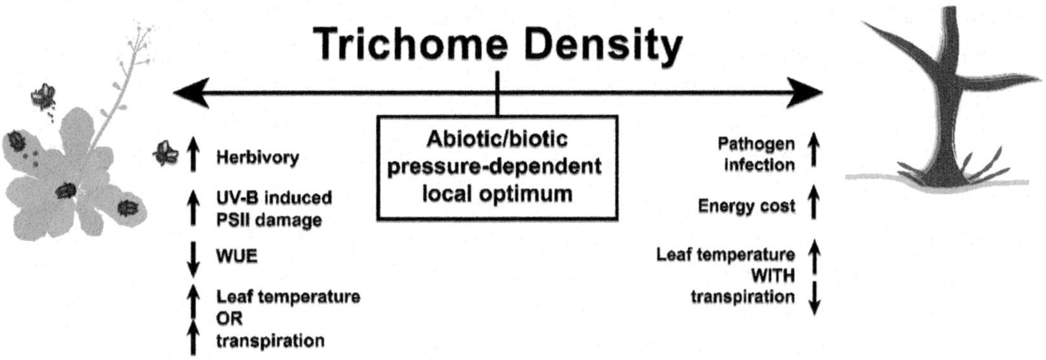

Fig. 5.2. There may exist a trichome density optimum that depends on the abiotic/biotic pressures of an individual plant within a specific environment. With reductions in leaf trichome densities, a plant may be subjected to increased intensity of herbivore damage, increased UV-B damage, decreased WUE, and leaf temperature/transpiration. Conversely, an increase in trichome densities may lead to increased pathogen infection, higher energy diversions to trichome growth, and either an increase in leaf temperature or a decrease in transpiration. (Figure by J.M. Fischer using publications from Mauricio 1998; Calo et al. 2006; Hanley et al. 2007; Løe et al. 2007; Sletvold et al. 2010; Peraldi et al. 2011; Nguyen et al. 2016; Imboden et al. 2018; Kergunteuil et al. 2018, Kim 2019)

Trichomes can be energetically costly to produce, whereby trichome-dense lines of *A. lyrata* grown in the absence of herbivory exhibited decreased fitness and performance compared to plants exposed to herbivory where trichomes were beneficial (Sletvold et al. 2010). A similar observation was made in *A. thaliana* whereby trichome density was found to be significantly negatively correlated with fruit number, suggesting fitness implications as a result of excessive trichome production (Mauricio 1998). Across an elevational gradient in the Swiss Alps, plant communities located at the lowest and highest elevations were the most trichome dense, while mid-elevation communities had the lowest overall trichome densities (Kergunteuil et al. 2018). This observation may be explained by the fact that high elevation sites are adapted to higher UV-B exposure, while low elevation sites may be subjected to higher herbivory pressure (Kergunteuil et al. 2018). Thus, mid-elevation communities in this case may have experienced both relaxed herbivory and UV stress that could have contributed to lower energetic costs devoted to trichome production (Kergunteuil et al. 2018). Therefore, the fitness/performance costs of trichome production is context dependent, influenced by plant species, micro-habitat, and environmental pressures, and likely the interaction of these factors (Hanley et al. 2007; Sletvold et al. 2010).

Taken together, the above studies indicate that trichomes play physiological and ecological roles across many species, and that selective pressures may "fine-tune" levels of trichome densities to match local abiotic/biotic pressures (e.g., risk of fungal infection, UV-B levels, herbivory) (Løe et al. 2007; Kergunteuil et al. 2018). Therefore, we must understand and better predict how trichome densities may be altered in response to future climate change, including rising atmospheric [CO_2]. Since many major crop species rely on trichome-dense leaves to reduce environmental stress, a shift in trichome density in future global climate scenarios could pose a challenge to crop productivity and food production. In more natural ecological settings, if trichome densities are shifted outside of adaptive realms, a plant may be more susceptible to overheating, herbivory, excessive energy investment, or increased fungal infection. In addition to shifting trichome densities, elevated [CO_2] may reduce stomatal densities and stomatal apertures, which can reduce the cooling ability of the leaf through transpiration limitation, further contributing to potential heat stress on leaves through an interactive effect through alterations on both micro-structures (Engineer et al. 2016).

The potential impacts of altered trichome densities on crop and native species physiology, survival, and yield in future environments are poorly understood and therefore we cannot fully predict plant responses in the future without this understanding. Future work must address the potential for elevated [CO_2] to disrupt past adaptive trichome densities of a variety of species, and how such disruptions may affect plant physiology, defense, and overall fitness. While the roles of trichomes are well known, as are the consequences of trichome density shifts—through leaf manipulation and mutant analyses—we know little about the effects of elevated [CO_2] on trichome initiation and densities and the mechanisms that control these responses. Therefore, we take this opportunity to summarize the current state of this field and describe how to motivate research to move this field forward in order to improve predictions of the impacts of rising [CO_2] on future plants that will be growing and adapting to elevated [CO_2] environments on a global scale.

III. Trichome Responses to Elevated [CO_2]

At elevated [CO_2] (720–900 ppm [CO_2]), trichome density responses occur across a spectrum, with some species exhibiting significant decreases in trichome density while others show significantly higher trichome densities when compared to plants grown in ambient [CO_2] (350–400 ppm [CO_2]) (Fig. 5.3) (Masle 2000; Bidart-Bouzat 2004; Bidart-Bouzat et al. 2005; Lake and Wade 2009; Karowe and Grubb 2011; Guo et al. 2013; Guo et al. 2014). Responses of leaf trichome densities in plants grown at elevated [CO_2] have been characterized primarily in the model species *Arabidopsis thaliana*. *Arabidopsis* lines Cvi and Col-0 experience significant trichome density *decreases* in response to elevated [CO_2] (approximately 34.5% and 51.2% decreases, respectively) (Masle 2000; Bidart-Bouzat 2004; Bidart-Bouzat et al. 2005; Lake and Wade 2009). The other lines examined were found to be non-responsive (Ler, Zn-0, Edi, Wu-0, Can-0) in trichome densities at elevated [CO_2] (Lake, JA and Wade, RN unpublished, 2009; Lake and Wade 2009). Similar to *Arabidopsis*, wheat (*Triticum aestivum* L.) shows intraspecific variation in trichome density responses to elevated [CO_2] (900 ppm [CO_2]) (Masle 2000). Trichome densities of the Hartog cultivar decreased significantly at elevated [CO_2], with an approximate 60% decrease compared to plants grown at ambient conditions. However, it is worth noting

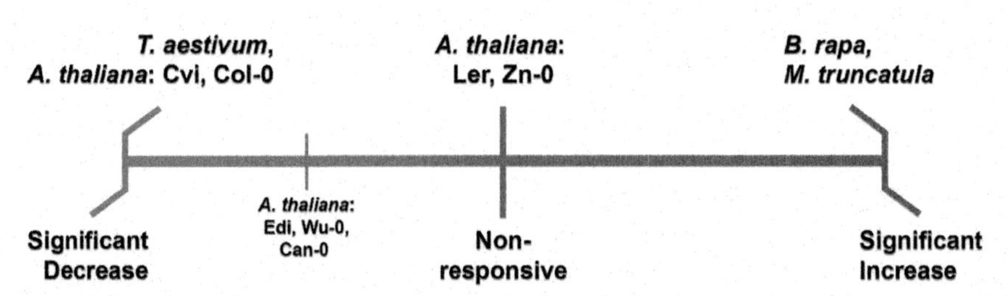

Fig. 5.3. Trichome density shifts among *Brassica rapa*, *Medicago truncatula*, *Triticum aestivum*, and *Arabidopsis thaliana* ecotypes grown at elevated [CO_2] (720–900 ppm [CO_2]) compared to current [CO_2] (350–400 ppm [CO_2]) (Figure based on work by Masle 2000; Bidart-Bouzat 2004; Bidart-Bouzat et al. 2005; Lake and Wade 2009; Karowe and Grubb 2011; Guo et al. 2013, 2014)

that this significant trichome density response was only observed in *non-vernalized* Hartog plants. Furthermore, the decrease response observed in wheat was line dependent, with no response observed in the Birch cultivar (Masle 2000). Therefore, the effects of elevated [CO_2] on trichome density are far reaching, including important crop species, with the significance of the trichome density decrease partially dependent upon other environmental factors.

Along with the above examples of trichome reductions, some species exhibited significant *increases* in trichome densities at elevated [CO_2]. For example, elevated [CO_2] increases trichome densities in *Brassica rapa* (approximate 57% increase, 744 ppm [CO_2]) and *Medicago truncatula* (up to an approximate 42% increase; 750 ppm [CO_2]) (Karowe and Grubb 2011; Guo et al. 2013; Guo et al. 2014). Elevated [CO_2] increased the densities of non-glandular trichomes on both young and mature *M. truncatula* (cv. Jemalong) leaves (approximately 26% and 13% density increase respectively), but [CO_2] enrichment increased glandular trichomes only on *mature* leaves (approximate 42% increase) (Guo et al. 2014). As such, the four species studied up to this point in regard to the impacts of [CO_2] enrichment on trichome densities—*Arabidopsis*, wheat, *B. rapa*, and *M. truncatula*—all have shown a degree of intraspecific variation. Furthermore, the trichome density responses of these species at elevated [CO_2] also show variation dependent upon either trichome types, leaf stages, and/or environmental factors experienced by the plants. Overall, there is too little research to determine whether increases or decreases are most common among and within species; more phenotypic data is needed.

As described above, in the limited work available, elevated [CO_2] has substantial effects on trichome densities among many species, but the molecular mechanisms underlying shifts in trichome densities at elevated [CO_2] remain unclear. Fortunately, a wealth of molecular information concerning the initiation and patterning of trichomes is available and will be key for understanding trichome responses to elevated [CO_2]. Next in this review, we cover the relevant molecular pathways of trichome production that may pertain to elevated [CO_2], and describe how these pathways are affected by various environmental perturbations (e.g., drought). We then describe potential molecular mechanisms driving responses of trichome densities to [CO_2] enrichment, utilizing the strong foundation of knowledge on the molecular/developmental basis of trichome initiation and production.

IV. Molecular Mechanisms of Trichome Initiation and Patterning

Prior to discussing the potential mechanisms that may underly altered trichome production at elevated [CO_2], we first provide a brief introduction on trichome genetics as is relevant to understanding the impacts of elevated [CO_2] (for more details see excellent work by Yang and Ye 2012, and Pattanaik et al. 2014 for reviews of this field; see Pesch et al. 2015 for a relevant update to this field). We will focus on the core genetic mechanisms of trichome production that are most likely to be relevant to elevated [CO_2] effects on trichome densities. The molecular pathways driving trichome initiation in the leaf have been well-elucidated over the past 30 years and trichomes have long served as models for single-cell development and initiation, providing an excellent molecular/developmental foundation for understanding how elevated [CO_2] may intercept trichome pathways (Hülskamp 2004).

A. *Trichome Initiation*

Here we outline the trichome initiation pathway, beginning with young protodermal cells in the developing leaf primordia, leading up to the point at which the trichome cell begins to protrude from the leaf. Trichome initiation begins with accumulation of transcription factors GLABROUS1 (GL1) and GLABROUS3 (GL3) in the protodermal cells of early developing leaves (Larkin et al. 1993; Payne et al. 2000; Balkunde et al. 2010). Trichome initiation, and the subsequent resulting trichome pattern, are *de novo*, in that all young protodermal cells on the leaf surface are potentially able to enter into the trichome initiation pathway, ubiquitously expressing *GL1* and *GL3* (Schnittger et al. 1999; Larkin et al. 2003; Zhang et al. 2003; Pesch and Hülskamp 2009). Later in leaf development, *GL1* and *GL3* expression is restricted to trichome initials—cells that have entered the early stages of the trichome morphogenesis pathway (Schnittger et al. 1999; Zhang et al. 2003; Kryvych et al. 2008). Due to stochastic variation in expression among protodermal cells, some cells have increased concentrations of the trichome initiation positive regulators (Larkin et al. 2003). These cells will potentially commit to the trichome pathway, and subsequently inhibit neighboring cells from entering the trichome pathway (Larkin et al. 2003; Hülskamp 2004). GL1-GL3 binding creates the primary building block for the trichome multimer (Fig. 5.4a), while the physical GL1-GL3 protein interaction is stabilized by the binding of TRANSPARENT TESTA GLABRA1 (TTG1) to GL3 to create a multimeric complex (Fig. 5.4b, c.1) (Payne et al. 2000; Larkin et al. 2003; Hülskamp 2004; Ramsay and Glover 2005; Pesch and Hülskamp 2009; Yoshida et al. 2009; Balkunde et al. 2010; Tsuji and Coe 2013). The GL1/GL3/TTG1 multimer is necessary for the activation of downstream trichome morphogenesis genes, such as those controlling the cell expansion necessary to create the distinct protuberance of the trichome from the leaf (Fig. 5.4c.1) (Szymanski and Marks 1998; Walker et al. 2000; Johnson et al. 2002; Hülskamp 2004; Wang and Chen 2008; Balkunde et al. 2010; Yang and Ye 2012). A cell is locked into the trichome fate once endoreduplication begins, controlled by the cyclin-dependent protein kinase inhibitor *SIAMESE* (*SIM*), the transcription of which is regulated by the GL1/GL3/TTG1 multimer (Walker et al. 2000; Yang and Ye 2012). Furthermore, the GL1/GL3/TTG1 multimer promotes expression of homeodomain transcription factor *GLABROUS2* (*GL2*), mutants of which show defects in post-initiation trichome structure (Johnson et al. 2002; Ohashi et al. 2002; Hülskamp 2004; Wang and Chen 2008; Balkunde et al. 2010).

Incongruous with older literature, the protein complex controlling trichome initiation is not strictly a trimer complex (e.g., GL1-GL3-TTG1), and many organizations of the complex are possible (Fig. 5.4b) (Pesch et al. 2015; Zhang 2018). The GL1-binding domain of GL3 also binds proteins controlling anthocyanin biosynthesis (PRODUCTION OF ANTHOCYANIN PIGMENT 1 [PAP1]) and root hair initiation (WEREWOLF [WER]) (Zhang et al. 2003; Ramsay and Glover 2005). Furthermore, proteins related to phytohormone signal transduction (JAZs, DELLAs) and proteasome protein degradation (UBIQUITIN PROTEIN LIGASE 3 [UPL3]) heterodimerize with GL3 at the C-terminal domain (Qi et al. 2011; Patra et al. 2013; Qi et al. 2014). GL3 homodimerization coupled with the range of GL3-heterodimerizing myb proteins results in numerous potential organizations of the multimer complex (Fig. 5.4b) (Payne et al. 2000; Zhang et al. 2003; Ramsay and Glover 2005; Qi et al. 2011; Patra et al. 2013; Qi et al. 2014; Pesch et al. 2015; Zhang 2018). Such numerous organizations of the multimer allow for a large swathe of different gene targets, and the multimer promotes different biological processes dependent upon multimer composition. Later in this review, we delve into how elevated [CO_2] may

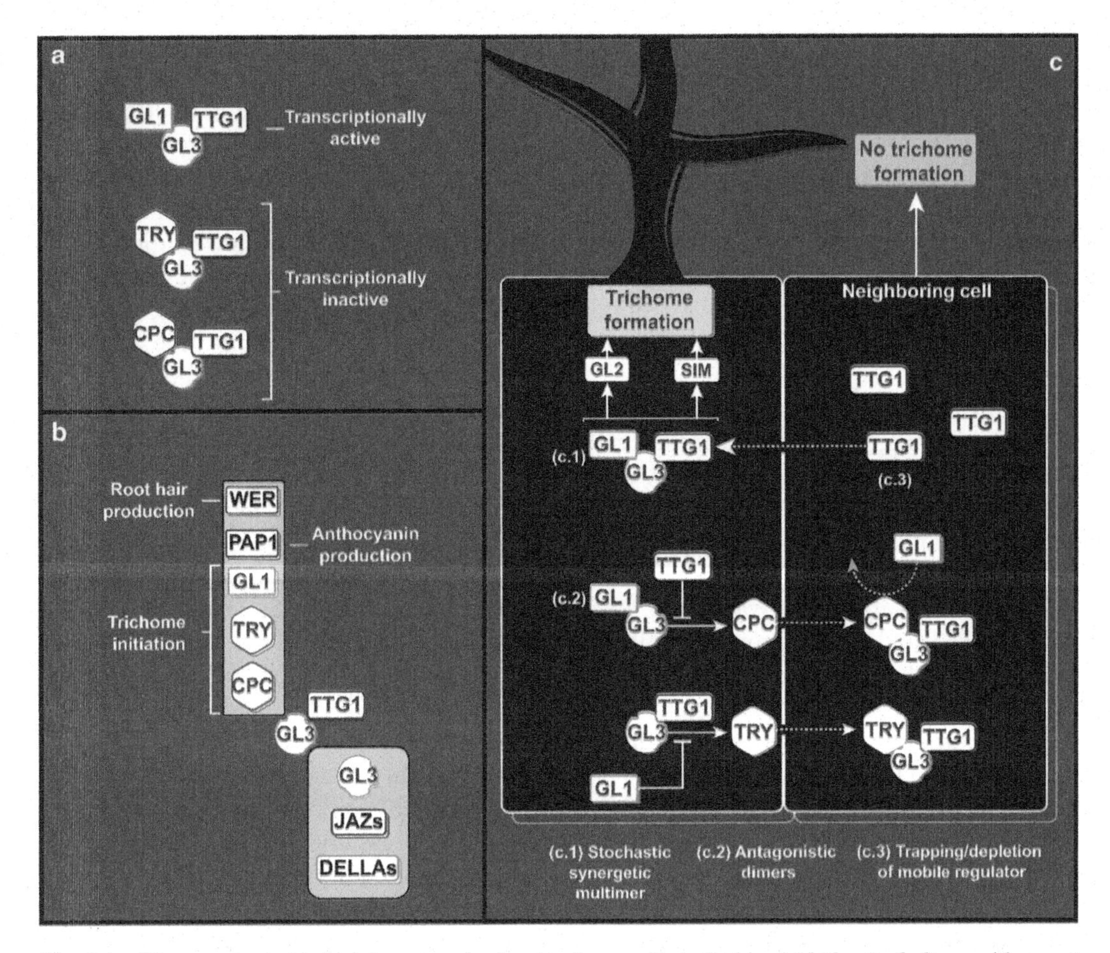

Fig. 5.4. The genetics behind trichome production has been well studied in *Arabidopsis thaliana* with recent advances in understanding formation of the multimeric trichome protein complex. (**a**) A transcriptionally active complex contains GL1, GL3, and potential accessory protein TTG1. The negative regulators TRY and CPC block the formation of this complex, forming transcriptionally inactive complexes. (**b**) GL3 binds a range of proteins across the three GL3 protein binding domains, leading to a high number of potential multimer complex organizations. (**c**) The GL1-GL3-TTG1 multimer promotes *GL2* and *SIM* transcription and in turn GL2 and SIM promote trichome initiation (**c**.1). TTG1 is cell-to-cell mobile and is trapped in cells with stochastically increased concentrations of GL3, which depletes TTG1 from surrounding cells (**c**.3). The *mobile* negative regulators TRY and CPC are promoted by antagonistic dimers formed from members of the synergetic multimer. TRY and CPC enter surrounding cells, prevent the formation of the synergetic multimer, and inhibit trichome initiation (**c**.2). (Figure created by J.M. Fischer utilizing work by Wada et al. 1997; Hülskamp et al. 1994; Szymanski and Marks 1998; Payne et al. 2000; Walker et al. 2000; Johnson et al. 2002; Larkin et al. 2003; Zhang et al. 2003; Hülskamp 2004; Ramsay and Glover 2005; Bouyer et al. 2008; Wang and Chen 2008; Pesch and Hülskamp 2009; Yoshida et al. 2009; Balkunde et al. 2010; Qi et al. 2011; Yang and Ye 2012; Patra et al. 2013; Tsuji and Coe 2013; Qi et al. 2014; Wang and Chen 2014; Pesch et al. 2015; Zhang 2018)

impact overlapping pathways (e.g., trichome initiation and anthocyanin biosynthesis), affecting organization of the multimer with subsequent effects on trichome initiation and other pathways utilizing components of the multimer.

B. *Negative Regulation and Patterning*

A trichome patterning mechanism is intertwined intimately with the trichome initiation pathway, as there must presumably be a defined distribution of trichomes *across* the leaf (or stem) in order to allow adaptive trichome functionality (Hülskamp 2004). Furthermore, as trichomes form across the leaf surface, they are less often found directly next to one another than a random pattern model would predict (Larkin et al. 2003). Any impacts of elevated [CO_2] on this patterning mechanism could potentially alter trichome functioning, therefore we briefly delve into the mechanics of trichome patterning below.

In a mechanism presumably to reduce trichome clumping, trichome precursor cells produce mobile regulators, TRIPTYCHON (TRY) and CAPRICE (CPC), to prevent neighboring cells from entering the trichome precursor state (Hülskamp et al. 1994; Wada et al. 1997; Wang and Chen 2014). TRY and CPC proteins exit the source cell and enter surrounding protodermal cells (Fig. 5.4c.2) (Wang and Chen 2014). Thoroughly compromised protodermal cells then fail to enter into trichome initiation, as negative regulators compete with GL1 for binding to GL3 that prevents the formation of the trimeric complex (Hülskamp 2004). TRY/CPC are R3-MYB proteins lacking DNA binding domains, and therefore replacement of GL1 with TRY or CPC in the multimeric complex creates an inactive complex incapable of DNA-binding (Fig. 5.4a) (Hülskamp 2004; Wang and Chen 2014). As such, *GL2* and *SIM* transcription—and the transition from protodermal cell to trichome precursor—are prevented by deactivation of the GL1/GL3/TTG1 multimer (Payne et al. 2000; Wang and Chen 2014).

Arabidopsis trichome formation is initiated in the protodermal cells of developing leaves, following the intracellular stochastic buildup of a multimeric protein complex. This complex has often been proposed as a three-protein complex (For excellent reviews see: Ramsay and Glover 2005; Pesch and Hülskamp 2009; Balkunde et al. 2010), producing negative regulators that block the formation of the trimer. However, research now indicates a more complex system of competition among an array of complexes, both multimeric and dimeric (Pesch et al. 2015; Zhang 2018). Throughout this review, a mixed model theory will be used, in accordance with the most recent molecular trichome studies. The newest dimer concept will be used to address the production of trichome negative regulators, but this concept does not address positive regulator production. Therefore, the trimer/multimer concept will be used for positive regulator production, as current data indicates.

Antagonistic dimer complexes, composed of members of the multimer, promote either *TRY* or *CPC*, depending on the protein composition of the promoting dimer (Hülskamp 2004; Pesch et al. 2015; Zhang 2018). The GL1-GL3 dimer promotes *CPC*, while the formation of the GL1-GL3 dimer is blocked by TTG1. TTG1 binds with GL3 (TTG1-GL3) to promote *TRY*, while the formation of this dimer is blocked by GL1. Therefore, GL1 is capable of downregulating *TRY* expression, while TTG1 is capable of downregulating *CPC* (Fig. 5.4c.2) (Hülskamp 2004; Pesch et al. 2015; Zhang 2018).

Further contributing to the patterning mechanism, TTG1 is a mobile protein, and when bound to the non-mobile GL1-GL3 complex, becomes trapped in the cell in which this complex is contained (Fig. 5.4c.3) (Bouyer et al. 2008; Balkunde et al. 2010). Therefore, TTG1 is more likely to become trapped in cells with stochastically higher amounts of GL3, depleting TTG1 in the surrounding cells. This trapping-depletion mechanism adds further stochasticity to trichome initiation, dependent upon the stochasticity of initial GL3 concentrations in the protodermal cells (Bouyer et al. 2008; Balkunde et al. 2010).

C. Phytohormones and Trichome Initiation

Trichome densities are in part determined by a subset of intertwined endogenous phytohormones, and as discussed later, concentrations of these hormones are affected at elevated [CO_2] (Traw and Bergelson 2003; Maes et al. 2008; Pattanaik et al. 2014). Jasmonic acid (JA) and gibberellic acid (GA) act in concert to promote the transcriptional functions of the GL1 and GL3 proteins. GL1 and GL3 DNA-binding sites are bound by antagonistic JAZ and DELLA proteins, blocking GL1/GL3 transcriptional functions. These transcription-blocking JAZ and DELLA proteins degrade following JA and GA application, respectively (Qi et al. 2014). Additionally, JA directly regulates 'default' *GL3* expression, positively controlling trichome densities in wild-type plants (Yoshida et al. 2009). Therefore, JA and GA act synergistically to promote trichome initiation and wild-type trichome densities.

The positive regulatory actions of JA/GA are tempered via salicylic acid (SA). SA applied exogenously to the leaf decreases trichome formation (Traw and Bergelson 2003). In support of this, trichome density of leaves post JA application increased 31.4%, but trichome densities decreased 0.9% with application of JA and SA in tandem (*Arabidopsis* ecotype Ler) (Traw and Bergelson 2003). Similarly, GA application alone increased trichome number 72%, while GA application coupled with SA application increased trichome number by only 29.6% in *Arabidopsis* (Traw and Bergelson 2003). The effect of phytohormones on trichome density is dependent upon leaf stage; *GL1* and *GL3* were highly induced by GA and JA in leaf seven of *Arabidopsis*, but much less so in leaf three (Maes et al. 2008). Finally, GA and JA affect negative regulators of trichome initiation as well. *TRY* expression is significantly reduced with application of JA and GA in *Arabidopsis*, while *CPC* expression is significantly induced via JA and GA (Maes et al. 2008). Therefore, wild-type trichome density is in part determined by a balance between GA, JA, and SA in developing leaves that is currently not well understood. As discussed later, elevated [CO_2] affects levels of these phytohormones, and it is tempting to predict a role of phytohormone shifts in the responses of trichome densities to elevated [CO_2].

D. *Root Hair-Trichome Pleiotropy*

Root hair initiation—which is known to be altered at elevated [CO_2]—and trichome initiation share core genetic machinery (*TTG1*, *GL3*, *GL2*, etc.), save for *GL1*. In lieu of GL1, WER binds to GL3 in the protein multimer controlling root hair initiation (WER-GL3-TTG1) (Schiefelbein 2003). Despite these overlaps, this genetic machinery plays opposite roles in the initiation of the two structures (Schiefelbein 2003). Namely, positive regulators of trichome initiation (e.g., *GL2*, *GL3*) serve as negative regulators of root hair initiation, and vice versa (e.g., negative trichome regulator *TRY* is a positive regulator of root hair initiation).

Niu et al. found that positive regulators of root hair initiation *CPC* and *TRY* were upregulated in the root at elevated [CO_2], while negative regulators of root hair initiation *TTG1*, *GL2*, and *GL3* were downregulated in the root (Niu et al. 2011). Such shifts in core root hair genes lead to increased root hair numbers at elevated [CO_2] (Yue et al. 2009; Niu et al. 2011). As such, although observed in a separate organ of the plant, there is evidence for elevated [CO_2] affecting core trichome machinery (*CPC*, *TRY*, *TTG1*, *GL2*, *GL3*), which may have implications for trichome production as well.

E. *Molecular Mechanisms for Environmental Perturbations of Trichome Production*

While the molecular mechanisms driving trichome density shifts at elevated [CO_2] remain unknown, the mechanisms driving trichome density shifts under several other environmental perturbations have been better described. Drought and excess UV-B have been shown to increase trichome densities in leaves that are initiated subsequent to application of these stressors, in both *Arabidopsis* and *Caragana korshinskii* (Peashrub) (Traw and Bergelson 2003; Yan et al. 2012; Ning et al. 2016). Across experimental field sites in the Loess Plateau (China), drought stressed *C. korshinskii* plants showed an upregulation in core trichome genes *GL2* and *TTG1* (positive regulators), as well as an upregulation in phytohormone GA (Ning et al. 2016). Furthermore, *CPC*—a core negative regulator of trichome initiation—was downregulated in drought-stressed *C. korshinskii* (Ning et al. 2016). Similarly, the core trichome gene *GL3* was shown to be upregulated in UV-B responsive *Arabidopsis* mutants exposed to excess UV-B, which in turn increased trichome densities (Yan et al. 2012).

Clearly, there is indication in the literature that shifts in trichome densities in response to environmental perturbations can be traced back to core trichome genes. In the sections that follow, we will use these known molecular mechanisms that produce altered trichome production under well-studied stressors (e.g., drought) to provide a foundation for better understanding the possible mechanisms that may be driving altered trichome production at elevated [CO_2]. Being that this field is relatively new, we propose paths for short-term and long-term investigations in the hopes that more attention is paid to this critical research area in the near future.

V. Potential Mechanisms for Altered Trichome Densities at Elevated [CO_2]

While the underlying mechanisms driving altered trichome density at elevated [CO_2] are currently unknown, the molecular trichome literature provides a trove of potential mechanisms for explaining the impacts of elevated [CO_2] on trichome production. In this review, we describe potential mechanisms driving shifts in trichome densities at elevated [CO_2] and outline potential research avenues that would further distinguish these possible mechanisms. We outline known elevated [CO_2] responses outside of trichome initiation, and how these responses may be intertwined with trichome initiation, either in developmental signaling or elevated [CO_2] responsive-pathway overlaps. By combining information from numerous publications, we determined overlaps between known elevated [CO_2]-responsive pathways and the trichome initiation pathway (Fig. 5.5a offers a general overview of how elevated [CO_2] connects to overarching physiological/developmental categories, and Fig. 5.5b offers a closer look at the genes within these physiological and developmental categories and potential pleiotropies). This was conducted using numerous literature sources to identify pleiotropies and overlaps in these pathways and constructed using a database of trichome genes (keyword-coded citation database available upon request), and displayed with the R package GOplot (Lolle et al. 1997; Tsuji and Coe 2013; Yan et al. 2014; Walter et al. 2015; Matias-Hernandez et al. 2016). We use the information shown in Fig. 5.5—such as the connection between anthocyanin genes (e.g., *PAP1*), trichome genes (e.g., *GL3*), and elevated [CO_2]—to formulate the potential mechanisms we highlight below (Fig. 5.5b). A majority of our hypotheses concern overlap between trichome initiation and other pathways, both at the whole-plant and leaf-levels. As shown in Fig. 5.5, there are numerous connections between anthocyanin, flowering, stomata production, and trichome production; a number of these overlapping pathways are known to be elevated [CO_2]-responsive. We draw from the growing elevated [CO_2] literature to extrapolate possible mechanisms from known elevated [CO_2]-responsive pathways (e.g., stomata).

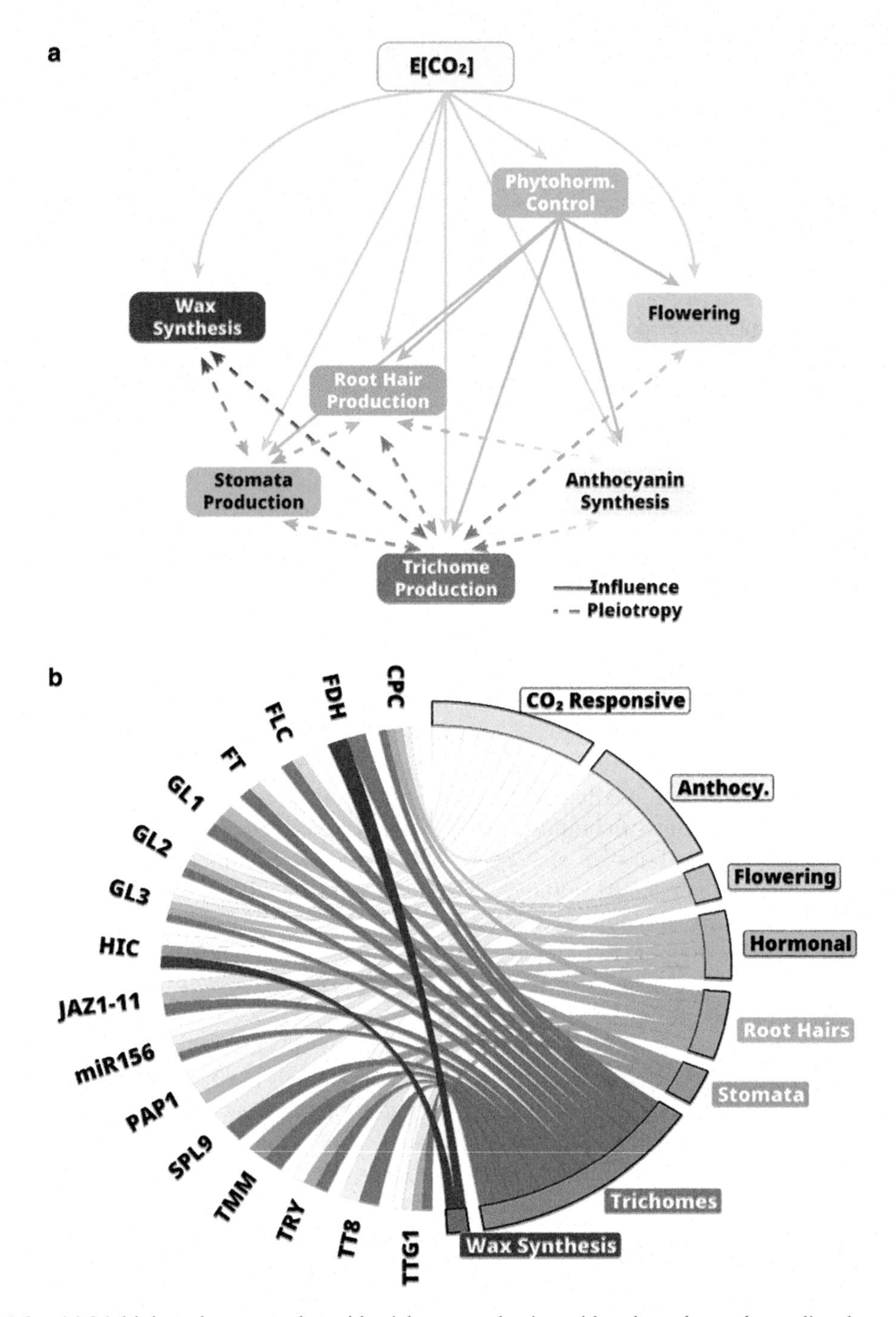

Fig. 5.5. (**a**) Multiple pathways overlap with trichome production, either through a pathway directly controlling trichome production, or through pleiotropy. Dashed lines indicate pleiotropy between the trichome pathway and another pathway, while solid lines indicate control over the connected pathway. (**b**) Pleiotropy of various genes controlling the initiation of trichomes. On the left-hand side are genes related to trichome initiation, while in the boxes on the right-hand side are biological roles the genes play. "Elevated [CO_2] responsive" refers to genes that have been found to respond to elevated [CO_2] within any organ or time point of plant growth. Figure created using R package GOplot. (The figure was created by J.M. Fischer, based on results from Lolle et al. 1997; Wada et al. 1997; Szymanski and Marks 1998; Yephremov 1999; Gray et al. 2000; Payne et al. 2000; Walker et al. 2000; Bird and Gray 2002; Johnson et al. 2002; Ohashi et al. 2002; Schiefelbein 2003; Traw and Bergelson 2003; Zhang et al. 2003; Li et al. 2008; Maes et al. 2008; Springer et al. 2008; Yoshida et al. 2009; Yu et al. 2010; Gou et al. 2011; Niu et al. 2011; Qi et al. 2011; Yan et al. 2012; May et al. 2013; Tsuji and Coe 2013; Qi et al. 2014; Shi and Xie 2014; Wang and Chen 2014; Xue et al. 2014; Yan et al. 2014; Pesch et al. 2015; Walter et al. 2015; Wang et al. 2015; Engineer et al. 2016; Hegebarth et al. 2016; Matias-Hernandez et al. 2016)

A. *Mechanism: Phytohormone Concentration Shifts*

Trichome density shifts at elevated [CO_2] may be tied into a larger [CO_2]-response cascade, rather than a direct mechanism targeting trichome density (Fig. 5.6a). Phytohormones influence most plant physiological processes and are known to shift in concentration at elevated [CO_2] (Ivakov and Persson 2013; Matias-Hernandez et al. 2016). In 2016, Sun et al. briefly acknowledged phytohormone concentration shifts as a possible mechanism behind trichome density shifts at elevated [CO_2] (Sun et al. 2016). SA concentration was shown to increase by 70% and JA concentration decreased by 35% in plants grown at elevated [CO_2] (750 ppm [CO_2]) (Zavala et al. 2012; Sun et al. 2013). As SA is a potent inhibitor of trichome initiation and JA is a positive regulator of trichome initiation, the observed shifts at elevated [CO_2] could potentially decrease *GL3* transcription and thus trichome initiation itself (Fig. 5.6a) (Teng et al. 2006; Zavala et al. 2012; Sun et al. 2013; Sun et al. 2016). Yet, further work is required to determine if *GL3* is downregulated at elevated [CO_2], and whether phytohormones might be the mediators of such a response (Teng et al. 2006; Sun et al. 2013; Sun et al. 2016). On the contrary, GA is a required positive regulator of *GL3* and *GL1* transcription, and is *upregulated* (55% increase) at elevated [CO_2] (700 ppm [CO_2]) (Chien and Sussex 1996; Perazza et al. 1998; Traw and Bergelson 2003; Teng et al. 2006; Tian et al. 2016). Increased GA may fail to rescue trichome production at elevated [CO_2], as GA requires the presence of elevated [CO_2]-attenuated JA for trichome initiation (Qi et al. 2014). Finally, as JA and GA affect the expression of core trichome negative regulators differentially (i.e., JA and GA upregulate *CPC* but downregulate *TRY*), further understanding of the shifts in complex ratios of phytohormones and core trichome genes at elevated [CO_2] is needed (Maes et al. 2008).

Phytohormone concentration shifts at elevated [CO_2] are also thought to be tied to increased carbohydrate reserves, and therefore altered trichome density may be potentially traced upstream to whole-plant carbohydrate status (Teng et al. 2006). To test the role of phytohormones, we recommend experiments to measure phytohormone ratios at elevated [CO_2] (e.g., SA:JA:GA) with concurrent measures of gene expression in core trichome genes and other potentially related factors such as plant carbohydrate status). Furthermore, more work is needed to better understand how elevated [CO_2] effects on phytohormones and trichomes may alter herbivory. Although elevated [CO_2] increases the defense hormone SA, the reduction of JA and ethylene (ET) at elevated [CO_2] decreases the time it takes for aphids to reach the phloem, increasing aphid population abundance and adult aphid mass (Sun et al. 2013; Sun et al. 2016). As such, further study of how the combined effects of elevated [CO_2] on phytohormones *and* defensive trichome densities affect herbivore damage will be important.

B. *Mechanism: Cuticle and Signal Transmission*

Alternative to changes in the trichome-altering signal itself (e.g., mechanism A, Phytohormone concentration shifts), elevated [CO_2] could alter the *efficiency* to which a trichome-influencing signal spreads across the leaf (Fig. 5.6c). Since trichome initiation occurs within a specific timeframe, the speed or distance to which a trichome signal can reach surrounding competent protodermal cells may in part determine the final outcome of trichome density across a leaf (Lloyd et al. 1994; Larkin et al. 2003). The uppermost wax layer of the leaf cuticle may act as a conduit for signals which alter trichome and stomata density, and changes in cuticle wax composition are known to shift the densities of leaf structures, including trichomes and stomata (Gray et al. 2000;

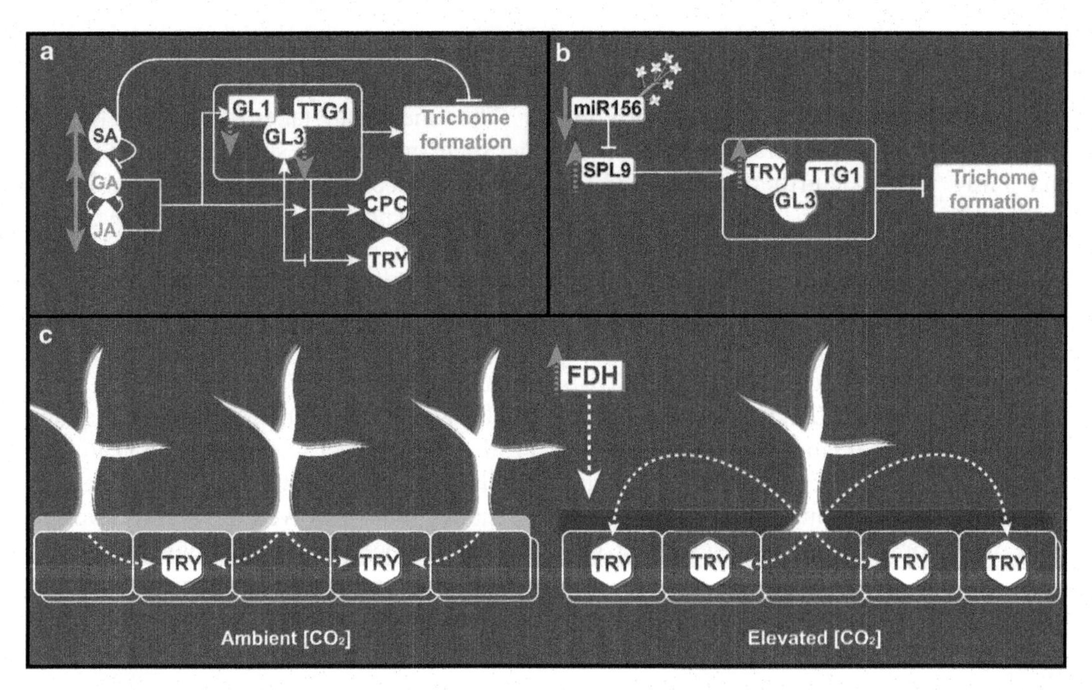

Fig. 5.6. Here we use trichome density decreases as the outcome of each potential mechanism for efficiency, and present potential mechanisms behind shifts in trichome densities under elevated [CO_2]. Solid lines indicate known connections/responses pulled from the literature; dashed lines indicate predicted connections/responses. Arrows either pointing up or down next to genes/proteins indicates an up or down regulation, respectively, at elevated [CO_2]. (**a**) Shifting concentrations of phytohormones alter transcription of key trichome genes, decreasing trichome numbers (Based on work from Teng et al. 2006; Zavala et al. 2012; Sun et al. 2013; Sun et al. 2016). (**b**) Flowering regulators *miR156* and *SPL9* respond at elevated [CO_2] in a manner that may increase transcription of core negative trichome regulator TRY, decreasing trichome numbers as TRY increases at elevated [CO_2] (Based on work from Van Der Kooij and De Kok 1996; Springer and Ward 2007; Yu et al. 2010; May et al. 2013; Ioannidi et al. 2016; Matias-Hernandez et al. 2016; Sun et al. 2016; Xu et al. 2016a). (**c**) Similar to the relationship between wax gene *HIC* and stomatal densities at elevated [CO_2], *FDH* is a wax gene known to affect trichome densities. Here we hypothesize that if *FDH* is elevated [CO_2]-responsive similar to the *FDH* relative *HIC*, shifts in wax constituency may cause differential signal transmission across cuticular wax at elevated [CO_2]. Although there are numerous possibilities for wax-mobile signals, in this example we show TRY as the signal with elevated [CO_2] increasing the distance TRY can travel across the cuticle. (Based on work from Yephremov 1999; Gray et al. 2000; Bird and Gray 2002; Morohashi and Grotewold 2009; Hegebarth et al. 2016; Haus et al. 2018). Figure from J.M. Fischer

Bird and Gray 2002). *HIGH CARBON DIOXIDE* (*HIC*), primarily expressed in guard cells, contributes to cuticle wax constituency through biosynthesis of long chain fatty acids (VLCFA) (Engineer et al. 2016). In *HIC Arabidopsis* mutants grown at elevated [CO_2], stomatal densities significantly increased, compared to the non-responsive wild-type background (ecotype C24) (Gray et al. 2000). Bird and Gray (2002) offer two hypotheses for how *HIC* prevents increases in wildtype stomatal density at elevated [CO_2]: VLCFAs temper stomatal densities by either (1) aiding in the diffusion of an unidentified stomata inhibitor across the surface of the leaf, or (2) directly inhibiting stomatal initiation itself (Bird and Gray 2002). Trichome production partially relies on a gene highly homologous to HIC—*FIDDLEHEAD* (*FDH*/*KCS10*), also a member of the 21-gene 3-ketoacyl-CoA synthase (KCS) family —with both genes involved in the synthesis of VLCFA elongases (Lolle et al. 1997; Yephremov 1999; Gray et al.

2000; Joubes et al. 2008). *FDH* mutation in *Arabidopsis* decreases leaf trichome densities approximately 50% compared to wild-type (ecotype Ler) (Yephremov 1999; Bird and Gray 2002). Furthermore, *FDH* is expressed primarily in trichomes and is required in the early stages of trichome formation, with core trichome proteins GL1 and GL3 targeting the *FDH* promotor (Gray et al. 2000; Bird and Gray 2002; Morohashi and Grotewold 2009; Hegebarth et al. 2016). Due to the role of *FDH* in trichome initiation, and the close relationship between *FDH* and *HIC*, it is tempting to infer that *FDH* and wax constituency might play a role in alterations of trichome initiation at elevated [CO_2] (Fig. 5.6c). If this mechanism were in fact driving trichome density shifts at elevated [CO_2], further research may reveal exciting insights into the role of cuticle wax as a signaling mechanism. Furthermore, as the primary role of the waxy cuticle is to reduce water loss through the leaf epidermis, elevated [CO_2] may affect water loss through both shifts in cuticle wax composition *and* trichome densities (Riederer and Schreiber 2001). Therefore, more work is required to connect wax composition, trichome densities, and stomatal densities at elevated [CO_2], and how these possibly intertwined responses may interact to affect leaf and whole-plant water loss.

C. *Mechanism: Flowering and Trichome Pleiotropy*

The following potential mechanism for shifting trichome densities at elevated [CO_2] takes a more whole-plant approach to the potential molecular mechanism underlying altered trichome densities at elevated [CO_2]. Here we consider a mechanism where trichome density shifts may be responding to alterations in whole-plant developmental programs that are sensitive to elevated [CO_2] (Fig. 5.6b). The following potential mechanism is derived from known overlaps between floral initiation and trichome initiation, along with well-characterized responses of floral initiation to elevated [CO_2].

The time a plant takes to reach the flowering stage is known to shift at elevated [CO_2], and both accelerations towards floral initiation and delays of floral initiation have been observed in numerous species at elevated [CO_2] (Springer and Ward 2007; Springer et al. 2008). Trichome initiation and floral initiation overlap in several key pathways, including *microRNA156* (*miR156*) (Yu et al. 2010; Xue et al. 2014; Matias-Hernandez et al. 2016; Xu et al. 2016a). *miR156* is a small non-coding RNA, responsible for silencing genes involved in several developmental processes including flowering. *miR156* is highest in young *Arabidopsis* seedlings and declines as plants age and approaches flowering, during which time levels of *miR156*-silenced *SQUAMOSA PROMOTER BINDING-LIKES* (*SPLs*) increase (Yu et al. 2010). *mir156*-regulated *SPLs (*e.g., *SPL9*) are responsible for the transition to flowering, and relevant to this review, upregulate the negative trichome regulators *TRY* and *TRICHOMELESS1* (*TCL1*) (Yu et al. 2010). Furthermore, SPL4 and SPL5 recruit TTG1, disrupting the formation of the multimer promoter complex and reducing *GL2* transcription (Ioannidi et al. 2016).

miR156 and associated *SPLs* not only control the floral transition, but also play a large role in trichome initiation, and therefore it is tempting to consider that trichome responses at elevated [CO_2] are an indirect result of over-riding developmental pathways that are commonly sensitive to elevated [CO_2] (e.g., flowering time). Elevated [CO_2] has been shown to affect *miR156* and *SPLs* transcription, but the effect on trichome initiation is wholly unknown. *miR156* expression in *Arabidopsis* (Col-0) rosette tissue decreases at elevated [CO_2], and Col-0 tends to flower earlier at elevated [CO_2], which may be attributable to an accelerated decrease in *miR156* at elevated [CO_2] (Van Der Kooij and De Kok 1996;

Springer and Ward 2007; May et al. 2013). In addition to phytohormone shifts, Sun et al. briefly proposed *miR156* as potentially playing a role in shifting trichome densities at elevated [CO_2], although no further work has been published to date (Sun et al. 2016). As such, an accelerated decline of *miR156* at elevated [CO_2] could potentially increase *SPLs*, *TRY*, and *TCL1* transcription in late rosette leaves, disrupting formation of the multimer complex (Fig. 5.6b) (Van Der Kooij and De Kok 1996; Springer and Ward 2007; Yu et al. 2010; May et al. 2013; Ioannidi et al. 2016; Matias-Hernandez et al. 2016; Sun et al. 2016; Xu et al. 2016a). If this mechanism is in fact relevant, the interconnection between developmental speed (e.g., time to flower) and trichome production would be an important direction for elevated CO_2 studies in the future. As such, we recommend extensive phenotyping of the trichome response across all stages of development in *Arabidopsis* to test for a potential correlation between the time in the overall life cycle that that a leaf is initiated (e.g., early in the plant life cycle versus just prior to flowering) and the trichome response of a given leaf to elevated [CO_2]. Furthermore, we recommend analysis of trichome genes across whole-plant developmental at elevated [CO_2] to test for differences in trichome gene expression that correlate with shifts in *miR156* expression.

VI. Other Potential Mechanisms

A. Anthocyanin and Trichome Pleiotropy

Leaf-level stress responses could incidentally affect trichome initiation at elevated [CO_2] through pleiotropic overlaps with the trichome production pathway. Anthocyanin response to elevated [CO_2] may decrease trichome densities through interactions between the trichome and anthocyanin pathways. Anthocyanin production is controlled by a complex similar to that of the GL1-GL3-TTG1 complex, with an anthocyanin specific MYB protein, PAP1, in lieu of the trichome-specific GL1; furthermore, numerous other trichome genes are pleiotropic with anthocyanin production (e.g., *miR156*, *CPC*, *GL2*) (Zhang et al. 2003; Maes et al. 2008; Zhu et al. 2009; Gou et al. 2011; Shi and Xie 2014; Pesch et al. 2015; Wang et al. 2015). *PAP1* expression and anthocyanin concentrations increase in *Arabidopsis* (Col-0) at elevated [CO_2] (Li et al. 2008; Takatani et al. 2014). Therefore, protein stoichiometries between the trichome and anthocyanin pathways could be upended at elevated [CO_2] as PAP1 concentrations increase. Shared components between the two pathways, such as the GL3-TTG1 dimer, could potentially be bound by increased PAP1 concentrations at elevated [CO_2] and removed from the available protein pool (Zhang et al. 2003; Li et al. 2008; Takatani et al. 2014). PAP1-GL3-TTG1 multimers may result in reduced GL3-TTG1 dimers available for GL1 binding and trichome initiation.

Conversely, *PAP1* over-expressors (*pap1-D*) do not display noticeable disruptions in trichome production; but otherwise there is little information regarding potential trade-offs between anthocyanin production and trichome initiation and therefore this area requires further investigation (Tohge et al. 2005). Testing for this potential mechanism would require measurement (e.g., a series of pulldown experiments) of ratios between anthocyanin-related multimers (e.g., PAP1-GL3-TTG1, PAP1-GL3) compared to trichome-related multimers (e.g., GL1-GL3-TTG1, GL1-GL3). We therefore recommend future experiments investigate protein behavior at elevated [CO_2] with an emphasis on trichome and anthocyanin multimer protein components.

B. Effects of Elevated [CO_2] on Cell Division

We also briefly consider that elevated [CO_2] may alter the *time frame* in which trichomes can initiate during leaf development, in turn affecting final trichome density through shifts strictly in leaf development. The basal area of the developing leaf is the primary site of trichome initiation and is the main portion of the leaf expressing the core trichome genes (Balkunde et al. 2010). Elevated [CO_2] increases the rate of cell division in the basal portion of the leaf (shown in *Populus*) and can decrease the time a cell spends in the division and elongation zones (shown in wheat) (Masle 2000; Ferris et al. 2001). As mentioned previously, trichome initiation can only occur within a specific time frame (Larkin et al. 2003). Therefore, it is tempting to think that there could be a relationship between basal cell division rate and trichome density. Yet, this relationship remains untested. Furthermore, the effects of elevated [CO_2] on cell division, trichome gene expression, and other related traits (e.g., anthocyanin) are almost wholly unrealized, and more work is required to determine whether increased basal cell division rates could affect trichome densities at elevated [CO_2].

VII. Directions for Future Research

Teasing apart the relationships between elevated [CO_2] and trichomes is still an almost wholly unrealized field of study. Future work (Table 5.1) must research the relationships between elevated [CO_2] and trichomes, as trichomes play important physiological and defense roles in crop and wild species. Therefore, we recommend further development of specific trichome-elevated [CO_2] research tracts that relate to the specific mechanisms proposed above, as well as new approaches for the overall field as described below.

A. *Full-Leaf Trichome Patterning*

Extensive phenotyping is still needed to further understand the genetic underpinnings of the elevated [CO_2] trichome response. In addition to trichome density as is commonly measured, we recommend a complete phenotyping of trichome *patterning* across the entire leaf surface in response to elevated [CO_2]. Also, past work generally addressed

Table 5.1. Avenues of proposed research for future advances in the elevated [CO_2]-trichome field (see text for more details)

Future research avenues	Related resources
Mobile signal from mature-to-developing leaves	Haus et al. (2018)
Cuticle wax composition at elevated [CO_2]	
Proteomics of developing leaves at elevated [CO_2]; core trichome protein ratios	Zhang (2018)
Single cell analyses across time; how elevated [CO_2]-accelerated basal cellular division/development may affect trichome densities	Masle (2000), Ferris et al. (2001))
RNA-sequencing analyses throughout leaf development for plants grown in ambient and elevated [CO_2]	
Effects of elevated [CO_2] on phytohormone ratios (JA:GA:SA) in a single leaf	Traw and Bergelson (2003), Yoshida et al. (2009)
Ecophysiological impacts of trichome shifts at elevated [CO_2], including:	Sun et al. (2016)
WUE	
Photosynthetic efficiency	
UV-B reflectance	
Herbivory	
Leaf temperature	
Effects on herbivore communities	
Effects on intraspecific and interspecific competition	
Crop output	

elevated [CO_2]-trichome density shifts in small subsections of the leaf, but work has yet to be done on whole-leaf trichome densities with attention paid to how elevated [CO_2] affects trichome patterning (Bidart-Bouzat et al. 2005; Lake and Wade 2009). As many trichome genes (e.g., *TRY*, *CPC*, *TTG1*) play dual roles in trichome initiation and trichome patterning, an elevated [CO_2] effect on trichome initiation through core genes may affect both trichome initiation and trichome patterning (Digiuni et al. 2008; Balkunde et al. 2010; Zhang 2018). As a majority of the potential mechanisms we highlighted above (Fig. 5.6) involve potential shifts in expression of core trichome-patterning genes elevated [CO_2] could theoretically affect both trichome density *and* patterning. For example, mechanism A would lead to trichome density shifts through a potential change in *GL3* and *GL1*, and *CPC* expression (Fig. 5.6a); mechanism C on the other hand would drive alterations in *TRY* and *TTG1* expression (Fig. 5.6b). In both of these examples, patterning would be affected, as the genes shifting in expression belong to both the trichome density and trichome patterning pathways. Therefore, more work is needed to investigate the effects of elevated [CO_2] on trichome patterning, and whether shifts in trichome densities are tethered to shifts in leaf trichome patterning.

To test if there is alteration in the quality of the trichome initiation signal itself, we believe future experiments should not only observe signal quantity (e.g., phytohormone concentrations, core gene expression), but also measure qualities of the signal (e.g., protein location). For example, we encourage the use of green fluorescent protein (GFP) to observe if elevated [CO_2] may affect localization of trichome proteins, such as mobile negative regulators TRY and CPC. Measuring the dispersion distance of negative regulators from the source trichome, with new advances in molecular plant microscopy, would be an interesting method for detecting the influence of elevated [CO_2] on trichome initiation and patterning.

B. Differently Aged Leaves

We recommend more analyses of trichome responses at the leaf level across various stages of plant development in order to determine whether altered whole-plant development could play a role in the trichome density responses to elevated [CO_2]. In previous elevated [CO_2]-trichome research, only mature adult leaves have been analyzed and therefore there are gaps in our understanding, such as how trichome densities of different leaf stages (e.g., juvenile versus adult leaf) respond to elevated [CO_2] (Bidart-Bouzat et al. 2005; Lake and Wade 2009). To fully understand potential mechanisms of trichome shifts at elevated [CO_2], further and more detailed phenotyping is necessary. Additionally, related research only investigates a single leaf pair from the rosette, therefore we recommend future research phenotype elevated [CO_2]-trichome density shifts across the spectrum of juvenile to adult leaf transition, with a focus on *Arabidopsis thaliana*. (Bidart-Bouzat et al. 2005; Lake and Wade 2009).

In the responses of stomatal densities to elevated [CO_2]—which are similar to the phenotypic responses of trichome densities at elevated [CO_2]—there is precedence for a signaling system from mature to developing leaves. At elevated [CO_2], a mobile signal from mature to developing leaves may alter stomatal densities of developing leaves according to the environment under which the mature leaf developed (Lake et al. 2001; Haus et al. 2018). Coupe et al. (2006) hypothesized that the mobile signal may be related to sugar and hormonal signaling, and some evidence may indicate microRNAs as a potential signal (Coupe et al. 2006). Similar to the potential roles of carbohydrates in the mobile signal altering stomatal densities, Guo et al. attributes the elevated [CO_2]-increased *M. truncatula* trichome densities to elevated primary metabolism at elevated [CO_2], as the increased pool of resources can be routed to structural defenses (Guo et al. 2014). On the contrary, *Arabidopsis* line Col-0 also sharply increases primary metabolism at elevated [CO_2], but *decreases* trichome densities with increased [CO_2] (Bidart-Bouzat et al. 2005; Teng et al. 2006; Lake and Wade 2009; Noguchi et al. 2015). Therefore, more work needs to be done on the connection between primary metabolism, elevated [CO_2], and trichome production across species. Furthermore, we recommend future research utilize the dual-compartment CO_2 chambers from Haus et al. 2018 to investigate whether the trichome density-altering signal is mobile from mature to developing leaves, similar to the signal altering stomatal densities at elevated [CO_2] (Haus et al. 2018).

C. *Ecological and Physiological Implications*

The implications of elevated [CO_2] induced trichome density shifts is almost wholly unrealized. As mentioned in the subsection "Trichome density and environmental adaptation", a disruption of trichome densities from the locally fine-tuned levels of trichome densities could, hypothetically, alter herbivory damage, WUE, fungal infection rates, and photosynthetic efficiency, as well as overall fitness. We recommend thorough investigations into the impacts of elevated [CO_2] on the potential consequences of trichome density shifts, such as fungal infection rates, across lines that differ in their trichome responses to elevated [CO_2]. As most of our potential mechanisms described above indicate trade-offs between trichomes and other physiologically important functions, we believe it is important to utilize lines in which, for example, anthocyanin concentrations and trichome densities both shift at elevated [CO_2], compared with lines that only respond in one of these traits (and control lines that do not respond in either trait). We currently understand very little regarding the potential implications of shifted trichome densities at elevated [CO_2], but further investigations are critical since shifting trichome densities could scale up to affect plant community composition and plant-herbivore interactions.

In summary, trichomes play physiologically important roles in herbivore defense, temperature regulation, boundary layer fortification, and UV-B protection, which can impact overall plant physiology, photosynthetic efficiency, fitness, and plant-environment interactions (Xiao et al. 2017). Studies have shown effects of elevated [CO_2] on trichome density, with significant increases in some species (e.g., *B. rapa*) and significant decreases in others (e.g., wheat) (Masle 2000; Karowe and Grubb 2011; Guo et al. 2013; Guo et al. 2014). Yet the molecular mechanisms driving trichome density shifts at elevated [CO_2] remain elusive. The potential mechanisms we highlighted above included broad responses to elevated [CO_2] affecting trichome densities (mechanism A. phytohormone shifts), leaf-scale changes affecting trichome initiation signaling (mechanism B. cuticle and signal transmission) and overlaps between elevated [CO_2]-responsive pathways and the trichome pathway (mechanism C. flowering and trichome pleiotropy). Further work must be done on the relationship between trichome production and elevated [CO_2], and the potential molecular mechanisms that drive trichome density shifts at elevated [CO_2], as described above. We recommend further trichome phenotyping of leaves at elevated [CO_2], and expansive genotyping of responsive and non-responsive individuals to further elucidate potential mechanisms. An understanding of these molecular mechanisms is crucial to predicting how wild and crop species may respond to a changing climate and will be a factor for helping us to best improve crop performance in preparation for future climate change.

Acknowledgments

Supported by IOS NSF grant to JKW. JMF was supporting by the Garden Club of America Corliss Knapp Engle Scholarship in Horticulture. The authors would like to acknowledge the University of Kansas Microscopy and Imaging Laboratory (KUMAI), Dr. Eduardo Rosa-Molinar, and Dr. Jim Kilcrease for help with trichome imaging via SEM.

References

Ainsworth EA, Rogers A (2007) The response of photosynthesis and stomatal conductance to rising [CO_2]: mechanisms and environmental interactions. Plant Cell Environ 30:258–270

Amada G, Onoda Y, Ichie T, Kitayama K (2017) Influence of leaf trichomes on boundary layer conductance and gas-exchange characteristics in *Metrosideros polymorpha* (Myrtaceae). Biotropica 49:482–492

Balkunde R, Pesch M, Hülskamp M (2010) Trichome patterning in *Arabidopsis thaliana*: from genetic to molecular models. Curr Top Dev Biol 91:299–321

Becklin KM, Anderson JT, Gerhart LM, Wadgymar SM, Wessinger CA, Ward JK (2016) Examining plant physiological responses to climate change through an evolutionary lens. Plant Physiol 172:635–649

Bergau N, Bennewitz S, Syrowatka F, Hause G, Tissier A (2015) The development of type VI glandular trichomes in the cultivated tomato *Solanum lycopersicum* and a related wild species *S habrochaites*. BMC Plant Biol:15

Bickford CP (2016) Ecophysiology of leaf trichomes. Funct Plant Biol 43:807–814

Bidart-Bouzat MG (2004) Herbivory modifies the lifetime fitness response of *Arabidopsis thaliana* to elevated CO_2. Ecology 85:297–303

Bidart-Bouzat M, Mithen R, Berenbaum M (2005) Elevated CO_2 influences herbivory-induced defense responses of *Arabidopsis thaliana*. Oecologia 145:415–424

Bird SM, Gray JE (2002) Signals from the cuticle affect epidermal cell differentiation. New Phytol 157:9–23

Bouyer D, Geier F, Kragler F, Schnittger A, Pesch M, Wester K, Balkunde R, Timmer J, … Hülskamp M (2008) Two-dimensional patterning by a trapping/depletion mechanism: the role of *TTG1* and *GL3* in *Arabidopsis* trichome formation. PLoS Biol 6:1166–1177

Calo L, García I, Gotor C, Romero LC (2006) Leaf hairs influence phytopathogenic fungus infection and confer an increased resistance when expressing a Trichoderma alpha-1,3-glucanase. J Exp Bot 57:3911–3920

Chien JC, Sussex IM (1996) Differential regulation of trichome formation on the adaxial and abaxial leaf surfaces by gibberellins and photoperiod in *Arabidopsis thaliana* (L.) Heynh. Plant Physiol 111:1321–1328

Coupe SA, Palmer BG, Lake JA, Overy SA, Oxborough K, Woodward FI, Gray JE, Quick WP (2006) Systemic signalling of environmental cues in *Arabidopsis* leaves. J Exp Bot 57:329–341

Digiuni S, Schellmann S, Geier F, Greese B, Pesch M, Wester K, Dartan B, Mach V, … Hülskamp M (2008) A competitive complex formation mechanism underlies trichome patterning on *Arabidopsis* leaves. Mol Syst Biol 4:1–11

Dong J, Gruda N, Lam SK, Li X, Duan Z (2018) Effects of elevated CO_2 on nutritional quality of vegetables: a review. Front Plant Sci 9

Ehleringer JR (1988) Comparative ecophysiology of *Encelia farinosa* and *Encelia frutescens*. I Energy Balance Consider Oecol 76:553–561

Ehleringer J, Björkman O (1978) Pubescence and leaf spectral characteristics in a desert shrub, *Encelia farinosa*. Oecologia 36:151–162

Ehleringer JR, Mooney HA (1978) Leaf hairs: effects on physiological activity and adaptive value to a desert shrub. Oecologia 37:183–200

Engineer C, Hashimoto-Sugimoto M, Negi J, Israelsson-Nordström M, Azoulay-Shemer T, Wouter-Jan R, Iba K, Schroeder J (2016) CO_2 sensing and CO_2 regulation of stomatal conductance: advances and open questions. Trends Plant Sci 21:16–30

Ferris R, Sabatti M, Miglietta F, Mills RF, Taylor G (2001) Leaf area is stimulated in *Populus* by free air CO_2 enrichment (POPFACE), through increased cell expansion and production. Plant Cell Environ 24:305–315

Galdon-Armero J, Fullana-Pericas M, Mulet PA, Conesa MA, Martin C, Galmes J (2018) The ratio of trichomes to stomata is associated with water use efficiency in *Solanum lycopersicum* (tomato). Plant J 96:607–619

Glas JJ, Schimmel BCJ, Alba JM, Escobar-Bravo R, Schuurink RC, Kant MR (2012) Plant glandular trichomes as targets for breeding or engineering of resistance to herbivores. Int J Mol Sci 13:17077–17103

Gou J, Felippes FF, Liu C, Weigel D, Wang J (2011) Negative regulation of anthocyanin biosynthesis in *Arabidopsis* by a *miR156*-targeted *SPL* transcription factor. Plant Cell 23:1512–1522

Gray J, Holroyd G, van der Lee F, Bahrami A, Sijmons P, Woodward F, Schuch W, Hetherington A (2000) The *HIC* signalling pathway links CO_2 perception to stomatal development. Nature 408:713–716

Guo H, Sun Y, Li Y, Liu X, Zhang W, Ge F (2013) Elevated CO_2 decreases the response of the ethylene signaling pathway in *Medicago truncatula* and increases the abundance of the pea aphid. New Phytol 201:279–291

Guo H, Sun Y, Li Y, Liu X, Wang P, Zhu-Salzman K, Ge F (2014) Elevated CO_2 alters the feeding behavior of the pea aphid by modifying the physical and chemical resistance of *Medicago truncatula*. Plant Cell Environ 37:2158–2168

Hanley ME, Lamont BB, Fairbanks MM, Rafferty CM (2007) Plant structural traits and their role in anti-herbivore defence. Perspect Plant Ecol Evol System 8:157–178

Haus MJ, Li M, Chitwood DH, Jacobs TW (2018) Long-distance and trans-generational stomatal patterning by CO_2 across *Arabidopsis* organs. Front Plant Sci 9

Hegebarth D, Buschhaus C, Wu M, Bird D, Jetter R (2016) The composition of surface wax on trichomes of *Arabidopsis thaliana* differs from wax on other epidermal cells. Plant J 88:762–774

Hülskamp M (2004) Plant trichomes: a model for cell differentiation. Mol Cell Biol 5:471–480

Hülskamp M, Misera S, Jurgens G (1994) Genetic dissection of trichome cell development in *Arabidopsis*. Cell 76:555–556

Imboden L, Afton D, Trail F (2018) Surface interactions of *Fusarium graminearum* on barley. Mol Plant Pathol 19:1332–1342

Ioannidi E, Rigas S, Tsitsekian D, Daras G, Alatzas A, Makris A, Tanou G, Argiriou A, … Kanellis AK (2016) Trichome patterning control involves TTG1 interaction with SPL transcription factors. Plant Mol Biol 92: 675–687

Ivakov A, Persson S (2013) Plant cell shape: modulators and measurements. Front Plant Sci 4:439

Johnson HB (1975) Plant pubescence: An ecological perspective. Bot Rev 41:233–258

Johnson CS, Kolevski B, Smyth DR (2002) *TRANSPARENT TESTA GLABRA2*, a Trichome and seed coat development gene of *Arabidopsis*, encodes a WRKY transcription factor. Plant Cell 14:1359–1375

Joubes J, Raffaele S, Bourdenx B, Garcia C, Laroche-Traineau J, Moreau P, Domergue F, Lessire R (2008) The VLCFA elongase gene family in *Arabidopsis thaliana*: phylogenetic analysis, 3D modelling, and expression profiling. Plant Mol Biol 67:547–566

Karowe D, Grubb C (2011) Elevated CO_2 increases constitutive phenolics and trichomes, but decreases inducibility of phenolics in *Brassica rapa* (Brassicaceae). J Chem Ecol 37:1332–1340

Kergunteuil A, Descombes P, Glauser G, Pellissier L, Rasmann S (2018) Plant physical and chemical defence variation along elevation gradients: a functional trait-based approach. Oecologia 187:561–571

Kim KW (2019) Plant trichomes as microbial habitats and infection sites. Eur J Plant Pathol 154:157–169

Konrad W, Burkhardt J, Ebner M, Roth-Nebelsick A (2015) Leaf pubescence as a possibility to increase water use efficiency by promoting condensation. Ecohydrology 8:480–492

Krimmel BA, Pearse IS (2013) Sticky plant traps insects to enhance indirect defence. Ecol Lett 16:219–224

Kryvych S, Nikiforova V, Herzog M, Perazza D, Fisahn J (2008) Gene expression profile of the different stages of *Arabidopsis thaliana* trichome development on the single cell level. Plant Physiol Biochem 46:160–173

Lake JA, Wade RN (2009) Plant-pathogen interactions and elevated CO_2: morphological changes in favour of pathogens. J Exp Bot 60:3123–3131

Lake J, Woodward F (2008) Response of stomatal numbers to CO_2 and humidity: control by transpiration rate and abscisic acid. New Phytol 179:397–404

Lake JA, Quick WP, Beerling DJ, Woodward FI (2001) Signals from mature to new leaves. Nature 411

Larkin JC, Oppenheimer DG, Pollock S, Marks MD (1993) Arabidopsis GLABROUS1 gene requires downstream sequences for function. Plant Cell 5:1739–1748

Larkin JC, Brown ML, Schiefelbein J (2003) How do cells know what they want to be when they grow up? Lessons from epidermal patterning in *Arabidopsis*. Annu Rev Plant Biol 54:403–430

Li P, Ainsworth EA, Leakey ADB, Ulanov A, Lozovaya V, Ort DR, Bohnert HJ (2008) *Arabidopsis* transcript and metabolite profiles: ecotype-specific responses to open-air elevated [CO_2]. Plant Cell Environ:1673–1687

Lloyd AM, Schena M, Walbot V, Davis RW (1994) Epidermal cell fate determination in *Arabidopsis*: patterns defined by a steroid-inducible regulator. Science 266:436–439

Løe G, Toräng P, Gaudeul M, Ågren J (2007) Trichome production and spatiotemporal variation in herbivory in the perennial herb *Arabidopsis lyrata*. Oikos 116:134–142

Lolle SJ, Berlyn GP, Engstrom EM, Krolikowski KA, Reiter W, Pruitt RE (1997) Developmental regulation of cell interactions in the *Arabidopsis fiddlehead-1* mutant: a role for the epidermal cell wall and cuticle. Dev Biol 189:311–321

Maes L, Inzé D, Goossens A (2008) Functional specialization of the TRANSPARENT TESTA GLABRA1 network allows differential hormonal control of laminal and marginal trichome initiation in *Arabidopsis* rosette leaves. Plant Physiol 148:1453–1464

Manetas Y (2003) The importance of being hairy: the adverse effects of hair removal on stem photosynthesis of *Verbascum speciosum* are due to solar UV-B radiation. New Phytol 158:503–508

Masle J (2000) The effects of elevated CO_2 concentrations on cell division rates, growth patterns, and blade anatomy in young wheat plants are modulated by factors related to leaf position, vernalization, and genotype. Plant Physiol 122:1399–1413

Matias-Hernandez L, Aguilar-Jaramillo AE, Cigliano RA, Sanseverino W, Pelaz S (2016) Flowering and trichome development share hormonal and transcription factor regulation. J Exp Bot 67:1209–1219

Mauricio R (1998) Costs of resistance to natural enemies in field populations of the annual plant *Arabidopsis thaliana*. Am Soc Natural 151:20–28

May P, Liao W, Wu Y, Shuai B, McCombie W, Zhang M, Liu Q (2013) The effects of carbon dioxide and temperature on microRNA expression in *Arabidopsis* development. Nat Commun 4:2145

Medeiros JS, Ward JK (2013) Increasing atmospheric [CO_2] from glacial through future levels affects drought tolerance via impacts on leaves, xylem, and their integrated function. New Phytol 199:738–748

Morohashi K, Grotewold E (2009) A systems approach reveals regulatory circuitry for *Arabidopsis* trichome initiation by the GL3 and GL1 selectors. PLoS Genet 5

Nguyen TTX, Dehne H, Steiner U (2016) Maize leaf trichomes represent an entry point of infection for *Fusarium* species. Fungal Biol 120:895–903

Ning P, Wang J, Zhou Y, Gao L, Wang J, Gong C (2016) Adaptional evolution of trichome in *Caragana korshinskii* to natural drought stress on the Loess Plateau, China. Ecol Evol 6:3786–3795

Niu Y, Jin C, Jin G, Zhou Q, Lin X, Tang C, Zhang Y (2011) Auxin modulates the enhanced development of root hairs in *Arabidopsis thaliana* (L.) Heynh. Under elevated CO_2. Plant Cell Environ 34:1304–1317

NOAA (2016). https://www.esrl.noaa.gov/gmd/ccgg/trends/. Accessed 30 Sept 2018

Noguchi K, Watanabe CK, Terashima I (2015) Effects of elevated atmospheric CO_2 on primary metabolite levels in *Arabidopsis thaliana* Col-0 leaves: an examination of metabolome data. Plant Cell Physiol 56:2069–2078

Ohashi Y, Oka A, Ruberti I, Morelli G, Aoyama T (2002) Entopically additive expression of *GLABRA2* alters the frequency and spacing of trichome initiation. Plant J 29:359–369

Patra B, Pattanaik S, Yuan L (2013) Ubiquitin protein ligase 3 mediates the proteasomal degradation of GLABROUS 3 and ENHANCER OF GLABROUS 3, regulators of trichome development and flavonoid biosynthesis in *Arabidopsis*. Plant J 74:435–447

Pattanaik S, Patra B, Singh SK, Yuan L (2014) An overview of the gene regulatory network controlling trichome development in the model plant, *Arabidopsis*. Front Plant Sci 5

Payne CT, Zhang F, Lloyd AM (2000) *GL3* encodes a bHLH protein that regulates trichome development in *Arabidopsis* through interaction with GL1 and TTG1. Genetics:1349–1362

Peng D, Niu Y, Song B, Chen J, Li Z, Yang Y, Sun H (2015) Wooly and overlapping leaves dampen temperature fluctuations in reproductive organ of an alpine Himalayan forb. J Plant Ecol 8:159–165

Peraldi A, Beccari G, Steed A, Nicholson P (2011) *Brachypodium distachyon*: a new pathosystem to study Fusarium head blight and other *Fusarium* diseases of wheat. BMC Plant Biol 11

Perazza D, Vachon G, Herzog M (1998) Gibberellins promote trichome formation by up-regulating *GLABROUS1* in *Arabidopsis*. Plant Physiol 117:375–383

Pesch M, Hülskamp M (2009) One, two, three…models for trichome patterning in *Arabidopsis*? Curr Opin Plant Biol 12:587–592

Pesch M, Schultheiß I, Klopffleisch K, Uhrig JF, Koegl M, Clemen CS, Simon R, Weidtkamp-Peters S, Hülskamp M (2015) TRANSPARENT TESTA GLABRA1 and GLABRA1 compete for binding to GLABRA3 in *Arabidopsis*. Plant Physiol 168:584–597

Qi T, Song S, Ren Q, Wu D, Huang H, Chen Y, Fan M, Peng W, … Xie D (2011) The jasmonate-ZIM-domain proteins interact with the WD-repeat/bHLH/MYB complexes to regulate jasmonate-mediated anthocyanin accumulation and trichome initiation in *Arabidopsis thaliana*. Plant Cell 23: 1795–1814

Qi T, Huang H, Wu D, Yan J, Qi Y, Song S, Xie D (2014) *Arabidopsis* DELLA and JAZ proteins bind the WD-repeat/bHLH/MYB complex to modulate gibberellin and jasmonate signaling synergy. Plant Cell 26:1118–1133

Ramsay NA, Glover BJ (2005) MYB-bHLH-WD40 protein complex and the evolution of cellular diversity. Trends Plant Sci 10:63–70

Riederer M, Schreiber L (2001) Protecting against water loss: analysis of the barrier properties of plant cuticles. J Exp Bot 52:2023–2032

Schiefelbein J (2003) Cell-fate specification in the epidermis: a common patterning mechanism in the root and shoot. Curr Opin Plant Biol 6:74–78

Schnittger A, Folkers U, Schwab B, Jurgens G, Hülskamp M (1999) Generation of a spacing pattern: the role of *TRIPTYCHON* in trichome patterning in *Arabidopsis*. Plant Cell 11:1105–1116

Schreuder MDJ, Brewer CA, Heine C (2001) Modelled influences of non-exchanging trichomes on leaf boundary layers and gas exchange. J Theor Biol 210:23–32

Shi M, Xie D (2014) Biosynthesis and metabolic engineering of anthocyanins in *Arabidopsis thaliana*. Recent Pat Biotechnol 8:47–60

Sletvold N, Huttunen P, Handley R, Karkkainen K, Agren J (2010) Cost of trichome production and resistance to a specialist insect herbivore in *Arabidopsis lyrata*. Evol Ecol 24:1307–1319

Springer CJ, Ward JK (2007) Flowering time and elevated atmospheric CO_2. New Phytol 176:243–255

Springer CJ, Orozco RA, Kelly JK, Ward JK (2008) Elevated CO_2 influences the expression of floral-initiation genes in *Arabidopsis thaliana*. New Phytol 178:63–67

Sun Y, Guo H, Zhu-Salzman K, Ge F (2013) Elevated CO_2 increases the abundance of the peach aphid on *Arabidopsis* by reducing jasmonic acid defenses. Plant Sci 210:128–140

Sun Y, Guo H, Ge F (2016) Plant-aphid interactions under elevated CO_2: some cues from aphid feeding behavior. Front Plant Sci 7:1–10

Szymanski DB, Marks MD (1998) *GLABROUS1* overexpression and *TRYPTYCHON* alter the cell cycle and trichome cell fate in *Arabidopsis*. Plant Cell 10:2047–2062

Takatani N, Ito T, Kiba T, Mori M, Miyamoto T, Maeda S, Omata T (2014) Effects of high CO_2 on growth and metabolism of *Arabidopsis* seedlings during growth with a constantly limited supply of nitrogen. Plant Cell Physiol 55:281–292

Teng N, Wang J, Tong C, Wu X, Wang Y, Lin J (2006) Elevated CO_2 induces physiological, biochemical and structural changes in leaves of *Arabidopsis thaliana*. New Phytol 172:92–103

Tian H, Qi T, Li Y, Wang C, Ren C, Song S, Huang H (2016) Regulation of the WD-repeat/bHLH/MYB complex by gibberellin and jasmonate. Plant Signal Behav 11

Tohge T, Nishiyama Y, Hirai MY, Yano M, Nakajima J, Awazuhara M, Inoue E, Takahashi H, … Saito K (2005) Functional genomics by integrated analysis of metabolome and transcriptome of *Arabidopsis* plants over-expressing an MYB transcription factor. Plant J 42: 218–235

Traw MB, Bergelson J (2003) Interactive effects of jasmonic acid, salicylic acid, and gibberellin on induction of trichomes in *Arabidopsis*. Plant Physiol 133:1367–1375

Tsuji J, Coe L (2013) The *glabra1* mutation affects the stomatal patterning of *Arabidopsis thaliana* rosette leaves. Bios 84:92–97

Van Der Kooij TAW, De Kok LJ (1996) Impact on elevated CO_2 on growth and development of *Arabidopsis thaliana* L. Phyton 36:173–184

Wada T, Tachibana T, Shimura Y, Okada K (1997) Epidermal cell differentiation in *Arabidopsis* determined by a *Myb* homolog, *CPC*. Science 277:1113–1116

Walker JD, Oppenheimer DG, Concienne J, Larkin JC (2000) *SIAMESE*, a gene controlling the endoreduplication cell cycle in *Arabidopis thaliana* trichomes. Development 127:3931–3940

Walter W, Sanchez-Cabo F, Ricote M (2015) GOplot: an R package for visually combining expression data with functional analysis. Bioinformatics 31:2912–2914

Wang S, Chen J (2008) Arabidopsis transient expression analysis reveals that activation of *GLABRA2* may require concurrent binding of GLABRA1 and GLABRA3 to the promoter of *GLABRA2*. Plant Cell Physiol 49:1792–1804

Wang S, Chen J (2014) Regulation of cell fate determination by single-repeat R3 MYB transcription factors in *Arabidopsis*. Front Plant Sci 5

Wang X, Wang, X., Hu, Q., Dai, X., Tian, H., Zheng, K., Wang, X., Mao, T., …, Wang, S. (2015) Characterization of an activation-tagged mutant uncovers a role of *GLABRA2* in anthocyanin biosynthesis in *Arabidopsis*. Plant J: 300-311

Werker E (2000) Trichome diversity and development. Adv Bot Res 31:1–35

Woodward F, Kelly C (1995) The influence of CO_2 concentration on stomatal density. New Phytol 131:311–327

Xiao K, Mao X, Lin Y, Xu H, Zhu Y, Cai Q, Xie H, Zhang J (2017) Trichome, a functional diversity phenotype in plant. Mol Biol 6

Xu M, Hu T, Zhao J, Park M, Earley KW, Wu G, Yang L, Poethig RS (2016a) Developmental functions of *miR156*-regulated *SQUAMOSA PROMOTER BINDING PROTEIN-LIKE* (*SPL*) genes in *Arabidopsis thaliana*. PLoS Genet 12:1–29

Xu Z, Jiang Y, Jia B, Zhou G (2016b) Elevated-CO_2 response of stomata and its dependence on environmental factors. Front Plant Sci 7

Xue X, Zhao B, Chao L, Chen D, Cui W, Mao Y, Wang LJ, Chen X (2014) Interaction between two timing microRNAs controls trichome distribution in *Arabidopsis*. PLoS Genet 10

Yan A, Pan J, An L, Gan Y, Feng H (2012) The responses of trichome mutants to enhanced ultraviolet-B radiation in *Arabidopsis thaliana*. J Photochem Photobiol B Biol 113:29–35

Yan L, Cheng X, Jia R, Qin Q, Guan L, Du H, Hou S (2014) New phenotypic characteristics of three *tmm* alleles in *Arabidopsis thaliana*. Plant Cell Rep 33:719–731

Yang C, Ye Z (2012) Trichomes as models for studying plant cell differentiation. Cell Mol Life Sci 70:1937–1948

Yephremov A (1999) Characterization of the *FIDDLEHEAD* gene of *Arabidopsis* reveals a link between adhesion response and cell differentiation. Plant Cell 11:2187–2202

Yoshida Y, Sano R, Wada T, Takabayashi J, Okada K (2009) Jasmonic acid control of *GLABRA3* links inducible defense and trichome patterning in *Arabidopsis*. Development 136:1039–1048

Yu N, Cai W, Wang S, Shan C, Wang L, Chen X (2010) Temporal control of trichome distribution by *microRNA156*-targeted *SPL* genes in *Arabidopsis thaliana*. Plant Cell 22:2322–2335

Yue W, Shao-Ting D, Ling-Ling L, Li-Dong H, Ping F, Xian-Yong L, Yong-Song Z, Hai-Long W (2009) Effect of CO_2 elevated on root growth and its relationship with indole acetic acid and ehtylene in tomato seedlings. Pedosphere 19:570–576

Zavala JA, Nabity PD, DeLucia EH (2012) An emerging understanding of mechanisms governing insect herbivory under elevated CO_2. Annu Rev Entomol 58:79–97

Zhang B (2018) Evolutionary analysis of MYBs-bHLH-WD40 complexes formation and their functional relationship in Planta. (Doctoral Dissertation) Retrieved from https://www.kupsubuni-koelnde/8367/1/Bipei%20Zhang-PhD%20Dissertationpdf

Zhang F, Gonzalez A, Zhao M, Payne CT, Lloyd A (2003) A network of redundant bHLH proteins functions in all TTG1-dependent pathways of *Arabidopsis*. Development 130:4859–4869

Zhou LH, Liu SB, Wang PF, Lu TJ, Xu F, Genin GM, Pickard BG (2016) The *Arabidopsis* trichome is an active mechanosensory switch. Plant Cell Environ 40:611–621

Zhu H, Fitzsimmons K, Khandelwal A, Kranz RG (2009) CPC, a single-repeat R3 MYB, is a negative regulator of anthocyanin biosynthesis in Arabidopsis. Epigenet Plant Dev 2:790–802

Part III

Population- and Community-Level Responses of Photosynthesis and Respiration to Climate Change

Chapter 6

Intraspecific Variation in Plant Responses to Atmospheric CO_2, Temperature, and Water Availability

Michael J. Aspinwall*
School of Forestry and Wildlife Sciences, Auburn University, Auburn, AL, USA

Thomas E. Juenger
Department of Integrative Biology, University of Texas at Austin, Austin, TX, USA

Paul D. Rymer and David T. Tissue
Hawkesbury Institute for the Environment, Western Sydney University, Penrith, NSW, Australia

and

Alexis Rodgers
Department of Biology, University of North Florida, Jacksonville, FL, USA

*Author for correspondence, e-mail: aspinwall@auburn.edu

K. M. Becklin et al. (eds.), *Photosynthesis, Respiration, and Climate Change*, Advances in Photosynthesis and Respiration 48, https://doi.org/10.1007/978-3-030-64926-5_6

Summary

In this chapter we compile data on intraspecific variation in plant reproductive, growth, and physiological responses to changes in atmospheric CO_2, temperature, and water availability. In total, we extracted data from 71 studies comprising a total of 79 species representing all major growth forms, functional groups, and biomes. Cumulatively, these studies examined responses to environmental change in 1154 genotypes. We used these data to examine: (1) the extent to which natural populations and genotypes within species vary in their response to increasing CO_2, warmer temperatures, and reduced water availability, and (2) whether intraspecific variation in these responses differs among growth forms, functional groups, biomes, and the phenotypic trait. In general, genotypes or populations of many species showed a wide range of responses to elevated CO_2, warming, and reduced water availability. However, probability values (p-values) for genotype-by-environment interaction terms (usually from analysis of variance) varied from <0.0001 to >0.90 depending upon the study design (and species), the environmental factor, and the scale of the trait. More studies reported significant intraspecific variation in plant responses to increasing temperature and decreasing water availability than intraspecific variation in plant responses to increasing CO_2. Thus, warmer and drier conditions may be more likely to result in evolutionary changes within species than increasing CO_2 alone. We also find that intraspecific variation in plant responses to environmental change is generally higher for reproductive and growth traits than for leaf-scale physiological traits. Even so, moderate intraspecific variation in physiological responses could result in substantial variation in growth and reproductive responses among genotypes. We conclude by discussing our understanding of genetic features that influence genotype-by-environment interactions. We go on to identify future research directions for advancing our understanding of the causes and consequences of intraspecific variation in plant responses to global change.

I. Introduction

Fossil fuel combustion and land use change are contributing to alterations in global climate. Recently (2002–2011), the concentration of atmospheric CO_2 ($[CO_2]$) has increased at an average rate of 2.0 ppm year^{-1}, and barring substantial reductions in emissions, atmospheric CO_2 may exceed 900 ppm by the end of the twenty-first century (Collins et al. 2013). Rising atmospheric $[CO_2]$ has contributed to climate warming, and global mean surface temperatures could increase 3–4 °C by the middle of the century (Collins et al. 2013). Climate warming has also been implicated in more frequent and extreme precipitation and drought events (Groisman et al. 2005; Hansen et al. 2012), and net declines in soil moisture in many, but not all, regions. Atmospheric CO_2, temperature, and water availability alone and in combination strongly influence plant growth and function. An important question is how plant populations and species will respond to these rapid changes in climate.

There are four potential ways that plant populations can respond to climate change: extinction, migration, plasticity (or acclimation), and adaptation (Aitken et al. 2008; Nicotra et al. 2010). For some populations, extinction may occur due to small population sizes, low recruitment, poor tolerance of environmental change, and compounding effects of disease or insect attack (Thomas et al. 2004; Feeley and Silman 2009). Migration is possible for some populations, and much effort has been put towards predicting migration of plant populations and changes in species distributions (e.g., Higgins et al. 2003; Thuiller et al. 2008). Yet plants are sessile, and migration may be difficult for species that have limited dispersal ability, encounter barriers to migration, are slow to reproduce, or are competitively excluded from suitable habitat outside their current range. For many plant populations, short-term plasticity or acclimation, and long-term adaptation, are the most likely routes for persistence under climate change. 'Phenotypic plasticity' is defined as the ability of an individual to adjust its growth, morphology, or physiology in response to environmental change (Bradshaw 1965). 'Acclimation' is a form of phenotypic plasticity and is more succinctly defined as a reversible physiological adjustment (Berry and Björkman 1980; Atkin and Tjoelker 2003; Way and Yamori 2014). Evolution by natural selection leads to populations that are locally adapted to a wide range of conditions, with adaptive genetic variation often found within and among populations (Ahrens et al. 2019). Evolutionary and adaptive responses to climate change will depend upon both constitutive adaptive genetic variation in traits related to fitness *and* intraspecific variation in response to environmental change; i.e., genetic variation in plasticity or acclimation. "Standing" genetic variation within populations enables evolutionary responses to climate change (Jump et al. 2008; Hoffmann and Srgo 2011). Genetic variation in plasticity or acclimation can be determined by examining population, provenance, seedlot or clone responses to environmental change, broadly referred to as genotype × environment interactions (G × E). 'Genotypes' are typically selected individuals or clones, and populations are made up of many genotypes. Importantly, evolutionary and adaptive responses to climate change can occur when there is selection on plasticity or acclimation, and genotypes or populations vary in plasticity or acclimation (as evidenced by G × E). Indeed, there is evidence that the occurrence of G × E can facilitate the evolution of plasticity (Springate et al. 2011). Overall, G × E or the lack thereof, is often seen as an indicator of species adaptive capacity in the face of climate change (Franks et al. 2014).

G × E can take different forms including plasticity arising from changes in variance among genotypes or populations across environments, and plasticity resulting from genotype or population rank changes across environments. Genotypes or populations may also show variable linear or non-linear responses to continuous environmental variation (Des Marais et al. 2013), which may be important for identifying response thresholds to environmental drivers. The plasticity of a genotype or population is sometimes indexed (0–1 scale) based on the slope of its 'reaction norm' across an environmental gradient (Fig. 6.1; Valladares et al. 2007; Nicotra et al. 2010), where steeper slopes represent higher plasticity. But, in many cases, plasticity can more easily be quantified as the relative change (i.e., percent change) in a phenotypic trait of a given genotype in response to environmental change.

The interpretation and implications of genetic variation in plasticity may depend upon the trait, the direction and magnitude of change relative to the control (or baseline), and the global change factor. For example, when fitness-related traits (survival, reproduction) show evidence of genetic variation in response to global change, there is an indication that species may undergo evolutionary responses to climate change (i.e., change in allele/trait frequencies). On the other hand, biochemical, physiological and growth responses to environmental conditions regulate growth and fitness responses, highlighting the importance of understanding plasticity at different scales. Photosynthesis, respiration, and their associated traits (e.g., stomatal conductance, leaf nitrogen) determine rates of carbon uptake and use and changes in these traits can culminate in increased or decreased growth and reproduction. It is also possible that plasticity or accli-

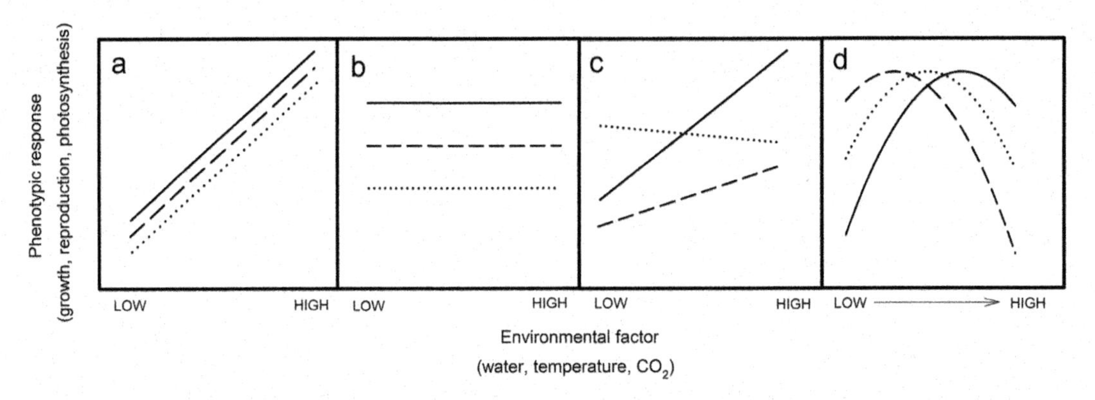

Fig. 6.1. Conceptual figure showing (**a**) plasticity, depicted as a directional phenotypic response to environmental change across three genotypes (shown as different lines), (**b**) adaptation, depicted as genetic differentiation in phenotypic traits across different environmental conditions, (c) genetic variation in plasticity, where different genotypes show contrasting responses to environmental change, and (**d**) genetic variation in plasticity, where genotypes show contrasting non-linear (i.e., threshold) phenotypic responses along an environmental gradient. In panel (**c**) we would predict that all genotypes respond positively to CO_2 (i.e., dotted line is unlikely) such that detection of G × E would be based on differences in the magnitude of the response to increasing CO_2 among genotypes

mation can be adaptive or maladaptive (Ghalambor et al. 2007). For example, genotypes or populations that acclimate to climate warming by maintaining homeostatic rates of respiration may show greater fitness than genotypes or populations that do not fully acclimate. On the other hand, plasticity or acclimation that creates a mismatch between the fitness optima and the mean environment may be maladaptive. Importantly, low plasticity or greater stability in one trait may be the result of greater molecular or biochemical plasticity, and whole-plant traits such as size, leaf area, and rooting depth can moderate leaf-scale responses to CO_2, heat, and drought.

This chapter focuses on the following questions: (1) to what extent do natural populations and genotypes within species vary in their response to variable atmospheric CO_2, temperature, and water availability? and (2) to what extent does intraspecific variation in plant responses to atmospheric CO_2, temperature, and water differ among growth forms, functional groups, biomes, and the scale or function of the phenotypic trait? To answer these questions, we compile and synthesize data from the primary literature on intraspecific variation in plant responses to increasing atmospheric CO_2 and temperature, and reductions in water availability. Our synthesis is focused on naturally occurring genotypes (i.e., no domesticated cultivars or families) and populations. For simplicity, we refer to both as 'genotypes'. We focus on three areas of plant performance: reproduction, growth (height, diameter, leaf area, or biomass), and leaf-scale traits associated with carbon (C) and water fluxes, and N investment (photosynthesis, stomatal conductance, respiration, leaf nitrogen). We searched the literature using Google Scholar with the search terms 'plant', 'evolution', 'plasticity', 'acclimation', 'elevated CO_2', 'population', 'photosynthesis', 'respiration', 'warming', 'temperature', 'drought', 'water availability', 'genotype', and 'genotype-by-environment interaction'. We excluded publications that did not examine our focal traits or used an experimental setup that did not allow for the testing of genotype-by-environment interactions. We also included studies we knew examined reproductive, growth, or physiological responses to CO_2, temperature, and water availability, even if they did not appear in our search. For each study we extracted the probability (p) value for the genotype (or population) × environment (CO_2, temperature, water) interaction term (when reported) which serves as a test for intraspecific variation in plasticity. When genotype means under each experimental treatment were provided in tables we calculated the relative response (percent change) of the least and most responsive (or plastic) genotype in the study. When data were provided in figures we estimated genotype means under each treatment and calculated the relative response of the least and most plastic genotype. These calculations provided an estimate of the range (minimum, maximum) of responses to CO_2, temperature, and water availability within individual species. We used this data to produce boxplots summarizing the distribution (interquartile range), variance (95% confidence intervals), median, and mean response of the least and most responsive genotypes of all species for a given environmental factor (CO_2, temperature, water availability), performance trait, and functional group. In total, we gathered and examined data from 71 studies comprising a total of 79 species representing all major growth forms, functional groups, and biomes. Cumulatively, these studies examined responses to environmental change in 1154 genotypes. We identified important knowledge gaps and future research directions related to intraspecific variation in plant responses to global climate change.

II. Intraspecific Variation in Plant Response to Atmospheric CO_2

For context, we briefly review plant responses to atmospheric CO_2. At the leaf-scale, C_3 plants typically show increased rates of net photosynthesis (A) measured at elevated atmospheric CO_2 (eCO_2), which is the result of increased substrate (i.e., CO_2) availability for Rubisco and reduced photorespiration (Drake et al. 1997, reviewed by Dusenge et al. 2018). Averaged across C_3 species, eCO_2 stimulates A by roughly 30% (Ainsworth and Rogers 2007). PEP carboxylase activity and specialized anatomy allows C_4 plants to concentrate CO_2 at the site of Rubisco carboxylation; thus A is nearly saturated in C_4 species under current atmospheric [CO_2] and these species typically show smaller increases in A at eCO_2 (Ainsworth and Rogers 2007). Increasing atmospheric [CO_2] results in increased intercellular [CO_2], triggering a reduction in stomatal conductance (g_s). Averaged across species and studies, growth at eCO_2 decreases g_s by roughly 20% (Medlyn et al. 2001; Ainsworth and Rogers 2007). Increased A and decreased g_s results in increased water-use efficiency (>50%). Increased A and water-use efficiency at eCO_2 partly explain the increase in plant biomass production at eCO_2 that has been observed across many studies.

In some cases, growth at eCO_2 can lead to photosynthetic down-regulation, which can limit productivity responses to eCO_2. Down-regulation occurs when photosynthate supply exceeds the demand or capacity of carbon (C) sinks (growth, defense, maintenance and respiration). The imbalance is sensed by mesophyll cells, which reduce the amount or activity of photosynthetic enzymes (i.e., Rubisco), often manifested as a decline in leaf N concentrations (Drake et al. 1997; Tissue et al. 1999; Ainsworth and Long 2005; Ainsworth and Rogers 2007).

Respiratory responses to increasing atmospheric CO_2 remain uncertain (Dusenge et al. 2018). Short-term increases in CO_2 do not affect leaf mitochondrial respiration in the dark (R). But long-term effects of growth at eCO_2 on leaf R have been inconsistent, with both decreases (Wullschleger et al. 1992; Griffin et al. 2001a; Bruhn et al. 2007) and increases in R observed in plants grown at eCO_2 (Griffin et al. 2001b).

In total, we identified 28 studies that tested whether natural genotypes within species differ in their response to CO_2 (Table 6.1). Eighteen studies were conducted in growth chambers or greenhouses, seven in open-top chambers, two in a Free-Air CO_2 Enrichment (FACE) facility, and one was carried out in a 'solardome' (Table 6.1). Treatment CO_2 concentrations ranged from 180μmol mol^{-1} to 800μmol mol^{-1} (Table 6.1). We focused on responses to eCO_2 relative to CO_2 treatments closest to ambient. Across studies, the average 'ambient' and elevated treatment CO_2 concentrations were 368 ± 18μmol mol^{-1} and 661 ± 55μmol mol^{-1}, respectively. Among these studies, 29 species were represented; 45% were trees or shrubs (2 broadleaved evergreen tree (BET) species, 6 broadleaved deciduous tree (BDT) species, 1 broadleaved evergreen shrub (BES) species, 6 needled evergreen tree (NET) species), 31% were grasses (1 C_4, 8 C_3), and 21% were forbs (2 perennial, 4 annual). Most species represented subtropical, temperate or boreal biomes, with very few species representing tropical or scrub biomes.

The number of genotypes included in each study varied from 2 to 162 (Table 6.1, average = 14). Combining all studies and species, responses to CO_2 were examined in 521 individual genotypes. Mean trait values of genotypes or populations under each CO_2 treatment were not always reported, particularly if the *p*-value for the genotype × CO_2 interaction was greater than 0.05. In these cases, we were unable to determine the range of responsiveness to CO_2 among genotypes or populations. When data were reported, we calculated responses to elevated [CO_2] of the least and most responsive genotypes as: 100 × (genotype mean at elevated growth [CO_2] – genotype mean at ambient growth [CO_2]) ÷ (genotype mean at ambient growth [CO_2]). These responses are summarized in Table 6.1.

A. Reproductive Responses to Atmospheric CO_2

Eight studies (*n* = 6 species), examined reproductive responses to increasing CO_2. Reproductive responses were quantified as either flower or fruit/seed mass per plant, or seed production (number). A *p*-value <0.05 for the genotype × CO_2 interaction was observed in one species (*Arabidopsis thaliana*, Tonsor and Scheiner 2007; Table 6.1). If *p*-values alone determine the significance of intraspecific variation in response to eCO_2, we would conclude that intraspecific variation in reproductive responses to increasing CO_2 was limited. Even so, there was considerable variation in reproductive responses to increasing CO_2 among genotypes of individual species. The lowest reproductive responses of individual genotypes ranged from −25% to −2%, and the highest reproductive responses of individual genotypes ranged from 29% to 100% (Table 6.1). Across genotypes of all species, the average reproductive responses to increasing CO_2 ranged from −15 ± 11% (standard deviation throughout) to 49 ± 26% (Fig. 6.2a). None of the studies examined reproductive responses to increasing CO_2 in long-lived woody plant species, C_4 grasses, or perennial forbs.

B. Growth and Biomass Responses to Atmospheric CO_2

Twenty studies (*n* = 27 species), quantified growth, biomass, and survival responses to increasing CO_2. A *p*-value <0.05 for the genotype × CO_2 interaction term was observed in four species. Across genotypes of all species, average productivity responses to increasing CO_2 ranged from 5 ± 24% to 73 ± 64% (Fig. 6.2a). Most studies examined responses within BDT species (*n* = 6) or NET species (*n* = 4). In general, a broader range of intraspecific variation in growth responses to increasing CO_2 was observed in BDT species (Fig. 6.3a; 5 ± 29% to 108 ± 91%) than in NET species (7 ± 11% to 57 ± 46%). Most BDT species represented temperate regions and most NET species represented boreal regions. Thus, temperate species showed a slightly larger range of intraspecific variation in growth responses to

Table 6.1. Summary of experimental studies that have examined intraspecific variation in plant responses to variable atmospheric CO_2

Reference	Environment	Species	Functional type	Biome	CO_2 treatments (umol mol^{-1})	Genotypes	Response variable, *P*-value, and range of responses: Reproduction	Growth, biomass, survival	Leaf *A*	Leaf g_s	Leaf *R*	Leaf N
Blackman et al. (2016)	GH	*Eucalyptus camaldulensis*	BET	Subtropical – temperate forest	400,640	14 clones			(P > 0.05, range NA)	(P > 0.05, range NA)		(P > 0.05, range NA)
Aspinwall et al. (2018)	GH	*Eucalyptus camaldulensis*	BET	Subtropical – temperate forest	400,640	17 clones		Total mass (P = 0.47, 0 to 50%)	(P = 0.12, 12 to 43%)	(P = 0.46, −30 to 70%)		Mass-based (P = 0.71, −32 to −1%); area-based (P = 0.65, −20 to 21%)
Aspinwall et al. (2017b)	GH	*Eucalyptus grandis*	BET	Tropical – subtropical forest	400,640	15 provenances					Area-based (P > 0.10, −40 to 96%); mass-based (P > 0.10, −36 to 39%)	
Kubiske et al. (1998)	OTC	*Populus tremuloides*	BDT	Boreal forest	380,720 (at either high or low N conditions)	2 clones			High N (P = NA, −4% to 88%); low N (P = NA, 15% to 67%)			
Wang et al. (2000)	OTC	*Populus tremuloides*	BDT	Boreal forest	370, 720	6 clones		Total mass (P > 0.10, −29 to 94%)	(P < 0.05, 14 to 68%)	(P < 0.10, −51 to −4%)		
Cseke et al. (2009)	FACE	*Populus tremuloides*	BDT	Boreal forest	372,560	2 clones		Stem volume (P = NA, 8 to 50%)	(P = NA, 13 to 38%)			
Bazzaz et al. (1995)	GC	*Betula alleghaniensis*	BDT	Temperate forest	350,600	3 genotypes		Total mass (P < 0.05, 18% to 150%)				

(continued)

Table 6.1. (continued)

							Response variable, *P*-value, and range of responses					
Reference	Environment	Species	Functional type	Biome	CO_2 treatments (umol mol^{-1})	Genotypes	Reproduction	Growth, biomass, survival	Leaf *A*	Leaf g_s	Leaf *R*	Leaf N
Riikonen et al. (2004)	OTC	*Betula pendula*	BDT	Boreal forest	370,700	2 clones		Total mass (P = 0.07, −2 to 40%)				
Riikonen et al. (2005)	OTC	*Betula pendula*	BDT	Boreal forest	370,700	2 clones			(P < 0.01, 25 to 40%)	(P = 0.93, −18 to −26%)		
Mohan et al. (2004)*	GC	*Acer rubrum*	BDT	Temperate forest	180,270, 360,600	3 provenances		(P = 0.01, 0 to 73%)				
Spinnler et al. (2003)	OTC	*Fagus sylvatica*	BDT	Temperate forest	370,570	4 provenances		Total mass (P > 0.05, −17 to 50%) high N only				
Polley et al. (1999)*	GH	*Prosopis glandulosa*	BDT	Scrub	370,700	6 families		Survival following drought (P = NA, 60 to 296%)				
Huang et al. (2015)	GH	*Telopea speciosissima*	BES	Subtropical forest	400,640	2 ecotypes		Total mass (P = 0.45, 27 to 36%)	(P = 0.34, 17 to 21%)	(P = 0.91, −3 to 8%)		
Spinnler et al. (2003)	OTC	*Picea abies*	NET	Boreal forest	370,570	8 provenances		Total mass (P > 0.05, −4 to 85%) high N only				
Townend (1993)	GC	*Picea sitchensis*	NET	Temperate rainforest	350,600	4 clones		Growth rate (P < 0.05, 5% to 16%)				
Bigras and Bertrand (2006)	GH	*Picea mariana*	NET	Boreal forest	370,710	2 provenances		Shoot mass (P = 0.12, range NA)	(P > 0.10, 42 to 53%)		Area-basis (P = NA, −28 to 45%)	Mass-basis (P = 0.19, range NA)
Mycroft et al. (2009)	GH	*Picea glauca*	NET	Boreal forest	370,740	29 genotypes		Total mass (P < 0.01, 23 to 108%)				
Cantin et al. (1996)	GC	*Pinus banksiana*	NET	Temperate – boreal forest	390,700	15 families		Height (P > 0.05, 5 to 20%)	(P > 0.05, 75 to 132%)			

Norton et al. (1999)	Solardome	*Agrostis curtisii*	C4 grass	Temperate grassland	356, 673	10 populations		Shoot mass ($P > 0.05$, −52 to 80%)		
Roumet et al. (1999)	GH	*Bromus erectus*	C3 grass	Temperate grassland	350,700	14 genotypes				Mass-basis ($P > 0.05$, −11 to −1%)
Roumet et al. (2002)	GH	*Bromus erectus*	C3 grass	Temperate grassland	350,700	14 genotypes	Spike mass ($P > 0.05$, −25 to 100%, year 1 only)	($P > 0.05$, −11 to 38%, year 1 only)		
Roumet et al. (2002)	GH	*Dactylis glomerata*	C3 grass	Temperate grassland	350,700	14 genotypes	Spike mass ($P > 0.05$, −25 to 30%, year 1 only)	($P < 0.10$, 1 to 77%, year 1 only)		
Roumet et al. (1999)	GH	*Dactylis glomerata*	C3 grass	Temperate grassland	350,700	14 genotypes				Mass-basis ($P < 0.05$, −23 to −3.6%)
Lüscher and Nösberger (1997)	FACE	*Dactylis glomerata*	C3 grass	Temperate grassland	350,600	9–14 genotypes		Aboveground mass ($P > 0.05$; range NA)		
Lüscher and Nösberger (1997)	FACE	*Lolium perenne*	C3 grass	Temperate grassland	350,600	9–14 genotypes		Aboveground mass ($P > 0.05$; range NA)		
Lüscher and Nösberger (1997)	FACE	*Lolium multiforum*	C3 grass	Temperate grassland	350,600	9–14 genotypes		Aboveground mass ($P > 0.05$; range NA)		
Lüscher and Nösberger (1997)	FACE	*Arrhenatherum elatius*	C3 grass	Temperate grassland	350,600	9–14 genotypes		Aboveground mass ($P > 0.05$; range NA)		
Lüscher and Nösberger (1997)	FACE	*Festuca pratensis*	C3 grass	Temperate grassland	350,600	9–14 genotypes		Aboveground mass ($P > 0.05$; range NA)		

(continued)

Table 6.1. (continued)

							Response variable, *P*-value, and range of responses					
Reference	Environment	Species	Functional type	Biome	CO_2 treatments (umol mol^{-1})	Genotypes	Reproduction	Growth, biomass, survival	Leaf *A*	Leaf g_s	Leaf *R*	Leaf N
Lüscher and Nösberger (1997)	FACE	*Holcus lanatus*	C3 grass	Temperate grassland	350,600	9–14 genotypes		Aboveground mass (P > 0.05; range NA)				
Lüscher and Nösberger (1997)	FACE	*Trisetum flavescens*	C3 grass	Temperate grassland	350,600	9–14 genotypes		Aboveground mass (P > 0.05; range NA)				
Lüscher and Nösberger (1997)	FACE	*Trifolium repens*	Perennial forb	Temperate grassland	350,600	9–14 genotypes		Aboveground mass (P > 0.05; range NA)				
Lüscher and Nösberger (1997)	FACE	*Trifolium pratense*	Perennial forb	Temperate grassland	350,600	9–14 genotypes		Aboveground mass (P > 0.05; range NA)				
Bazzaz et al. (1995)	GC	*Abutilon theophrastri*	Annual forb	Temperate grassland	350,600	8 genotypes	Fruit/seed mass (p = 0.63, −22 to 44%)					
Ward and Strain (1997)	GC	*Arabidopsis thalinana*	Annual forb	Global distribution	200,350,700	6 genotypes	Seeds per plant (P > 0.05, −2 to 39%)	Total mass (P > 0.05, 13 to 41%)				
Tonsor and Scheiner (2007)	GC	*Arabidopsis thalinana*	Annual forb	Global distribution	250,355,530,710	35 genotypes	Flower mass (P < 0.0001, range NA)		(P = 0.98, range NA)			Mass-based (P = 0.39, range NA)
Lau et al. (2007)	GC	*Arabidopsis thalinana*	Annual forb	Global distribution	368,560	162 RILs	Fruit number (P > 0.05, range NA)	Total mass (P > 0.05, range NA)				
Curtis et al. (1994)	OTC	*Raphanus raphanistrum*	Annual forb	Temperate grassland – forest	338, 685	5 families	Seeds per plant (P > 0.05, −6 to 52%)					
Jenkins Klus et al. (2001)	OTC	*Plantago lanceolata*	Annual forb	Temperate grassland	350,710	18 families from 2 populations		Total mass (P > 0.05, 6 to 11% comparing populations)	(P < 0.01, −10 to 150% across all families)	(P = 0.44, range NA)		Mass-basis (P < 0.02, −9 to 42% only 3 families considered)

Wulff and Alexander (1985)	GC	*Plantago lanceolata*	Annual forb	Temperate grassland	350,675	4 families	Seed mass (P = NA, −7% to 29%)	Leaf area (P = NA, 37% to 68%)				
Total studies: 27		Total species: 29				Range = 2 to 162, mean = 14, Total = 521	n = 8 studies (6 species)	n = 20 studies (27 species)	n = 11 studies (8 species)	n = 6 studies (5 species)	n = 2 studies (2 species)	n = 7 studies (6 species)

Information on the experimental infrastructure (*GH* greenhouse, *GC* growth chamber, *OTC* open-top chamber, *FACE* free air CO_2 enrichment), species, functional type (*BET* broadleaved evergreen tree, *BES* broadleaved evergreen shrub, *BDT* broadleaved deciduous tree, *NET* needle-like evergreen tree, etc.), the biome represented by the species, the growth CO_2 concentrations, and the number of genotypes, clones, families, population, or provenances tested is provided for each study. For each parameter (reproduction, growth/biomass/survival, net photosynthesis (A), stomatal conductance (g_s), dark respiration (R), and nitrogen (N)), we provide the p-value for the genotype × CO_2 interaction (if provided in reference, *NA* not available), and the range in responsiveness (plasticity) to eCO_2 among genotypes. Responses to eCO_2 were calculated as: 100 × (genotype mean at highest CO_2 – genotype mean at lowest CO_2) ÷ (genotype mean at lowest CO_2). Thus, a positive percent change indicates a positive response to eCO_2 and a negative percent change represents a negative response to eCO_2. If data for calculating responses to eCO_2 were not available (NA) in the reference or supplementary material, the range in responsiveness was not calculated. Note: in some cases, experiments tested for three-way interactions or specific traits that do not fit perfectly within one of the categories of response variables. Where appropriate, we provide clarification on the tests conducted and the variables measured

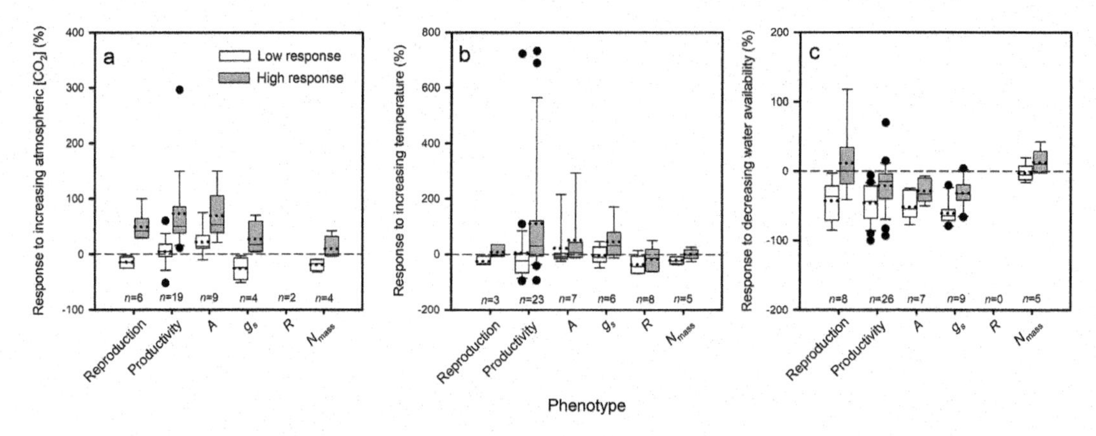

Fig. 6.2. Box-plots of the lowest and highest reproductive, productivity (growth, biomass production) and leaf-scale (*A* net photosynthesis, g_s stomatal conductance, *R* dark respiration, N_{mass} leaf N per unit leaf dry mass) responses (percent change) to (**a**) increasing atmospheric CO_2, (**b**) increasing air temperature, and (**c**) decreasing water availability observed across genotypes of all species and studies shown in Tables 6.1, 6.2 and 6.3. Boxes represent the interquartile range and error bars represent 95% confidence intervals. Solid and dotted lines within each box-plot represent the median and mean responses, respectively. Responses falling outside the 95% confidence intervals are shown as filled circles. The number of data points used in constructing each box-plot pair is shown below each box-plot. Note: box-plots were not constructed if there were less than three data points

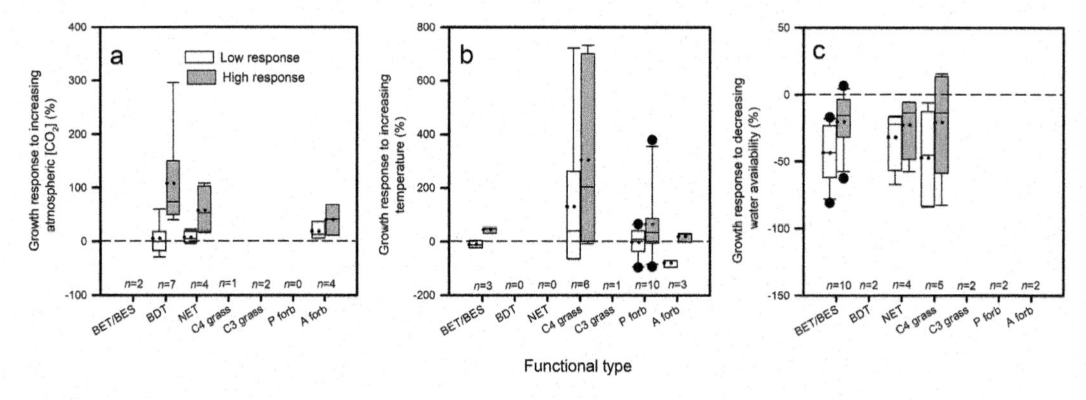

Fig. 6.3. Box-plots comparing the lowest and highest productivity responses (percent change) of genotypes of plants species representing different functional groups (*BET/BES* broadleaved evergreen trees/broadleaved evergreen shrubs, *BDT* broadleaved deciduous tree species, *NET* needled evergreen tree species, *C4* grasses, *C3* grasses, *P forb* perennial forbs, *A forb* annual forbs) to (**a**) increasing atmospheric CO_2, (**b**) increasing air temperature, and (**c**) decreasing water availability. Boxes represent the interquartile range and error bars represent 95% confidence intervals. Solid and dotted lines within each box-plot represent the median and mean responses, respectively. Responses falling outside the 95% confidence intervals are shown as filled circles. The number of data points used in constructing each box-plot pair is shown below each box-plot. Note: box-plots were not constructed if there were less than three data points

increasing CO_2 ($-1 \pm 25\%$ to $63 \pm 42\%$) than boreal species ($0 \pm 17\%$ to $66 \pm 35\%$). Given limited data on grasses and forbs, the extent of intraspecific variation in growth responses to CO_2 was unclear for these groups (Fig. 6.3a).

C. Photosynthetic, Stomatal, and Respiratory Responses to Atmospheric CO_2

Eleven studies ($n = 8$ species) examined intraspecific variation in photosynthetic responses to increasing CO_2. A p-value <0.05

for the genotype × CO_2 interaction term was observed in three species. Across genotypes of all species, the average photosynthetic response to increasing CO_2 ranged from 22 ± 24% to 69 ± 44% (Fig. 6.2a). Boreal tree species were the most well-represented group (6 studies, $n = 4$ species), with average photosynthetic responses to increasing CO_2 among genotypes ranging from 30 ± 25% to 68 ± 35%. It was not possible to compare intraspecific variation in photosynthetic responses to CO_2 among growth forms given that data were limited for grasses and forbs.

Six studies (n = 4 species), examined intraspecific variation in stomatal responses to CO_2. A p-value <0.05 for the genotype × CO_2 interaction term was not observed in any species. All studies that examined stomatal responses to CO_2 were conducted in tree species, with two studies conducted in tropical-subtropical species and boreal species, respectively (Table 6.1). Across genotypes of these species, the average stomatal response to increasing CO_2 ranged from −26 ± 20% to 27 ± 30% (Fig. 6.2a). At least one genotype in each study showed an increase in g_s when grown at elevated CO_2. The extent to which genotypes or populations vary in respiratory responses to CO_2 is unclear given the lack of data (Table 6.1 and Fig. 6.2a). Moreover, intraspecific variation in leaf N (mass-basis) responses to CO_2 have not been widely studied (7 studies, $n = 6$ species). Across genotypes of these species, the average response of leaf N to increasing CO_2 ranged from −19 ± 13% to 10 ± 19% (Fig. 6.2a). Two of these studies reported p-values <0.05 for the interaction term testing for genotypic differences in leaf N responses to CO_2.

III. Intraspecific Variation in Plant Responses to Temperature

The effects of changing temperature on reproduction, growth, and physiology depend upon underlying temperature responses of each process, with growth responses to temperature strongly influenced by temperature responses of A and R. The breadth, temperature optimum, and rate of A at the temperature optimum varies considerably among plant species (Kattge and Knorr 2007; Kumarathunge et al. 2019). Respiration measured at a common temperature varies considerably among species (Atkin et al. 2015), although the shape of the temperature response of R is relatively similar across species (Heskel et al. 2016). There is evidence that species and population differences in temperature responses are associated with temperature adaptation, with warm-origin species and populations sometimes showing higher temperature optimums of A, but lower leaf N and lower rates of R measured at a common temperature compared to cool-origin species and populations (Oleksyn et al. 1998; Bresson et al. 2011).

The effects of changing temperature may be explained by whether the change in growth temperature 'moves' reproduction, growth and physiology up or down the temperature response curve. For example, many warm-origin plants are hypothesized to be operating near or at their optimal temperature such that warming might have negative effects on A, growth, and reproduction. Plants from cooler climates may be operating below their optimal temperature such that warming might increase A, growth, and reproduction. Indeed, there is some support for this hypothesis (Way and Oren 2010; Drake et al. 2015; Crous et al. 2018).

Plants also show considerable capacity for thermal acclimation of leaf physiology, demonstrated by a change in the temperature response of A or R. Typically, acclimation to warmer temperatures results in an increase in the temperature optimum of A and a reduction in R measured at a common temperature, while acclimation to cooler temperatures results in a decrease in the temperature optimum of A and an increase in R measured at a common temperature (Atkin and Tjoelker 2003; Slot and Kitajima 2015; Aspinwall et al. 2016). An important question is whether thermal acclimation capacity varies among genotypes or populations within species.

We identified 15 studies that tested whether natural genotypes or populations within species vary in their response to temperature (Table 6.2). Ten studies were conducted in growth chambers or greenhouses with set temperature treatments (contrasting day/night air temperatures, contrasting mean diel temperatures), four were conducted in common garden experiments at thermally contrasting sites (i.e., spatial differences in mean annual temperature) or at a single site with measurements across cool and warm seasons (i.e., temporal differences in mean daily temperature), and one was carried out in a field warming experiment with an average warming of 3 °C above ambient (Springate and Kover 2014). Across these studies, 33 species were represented, 33% of which were trees or shrubs (3 BET species, 5 BDT species, 1 BES species, 2 NET species), 21% were grasses (6 C_4, 1 C_3), and 45% were forbs (11 perennial, 4 annual). Most species represented tropical-subtropical and temperate biomes, with very few species representing boreal regions. The number of genotypes included in each study varied widely, ranging from 2 to 320 (Table 6.2; average = 14). Cumulatively, these studies examined responses across 499 genotypes. When data were reported, we calculated responses to *increasing* temperature of the least and most responsive genotypes as: 100 × (genotype mean at the warmest temperature – genotype mean at the coolest temperature) ÷ (genotype mean at coolest temperature).

A. Reproductive Responses to Temperature

Four studies ($n = 19$ species) examined intraspecific variation in reproductive responses to temperature. *P*-values <0.05 for the genotype × temperature interaction term were reported in 17 species. The reciprocal transplant study by Ensslin and Fischer (2015) included many of these species, but mean values for reproduction of individuals genotypes at cool and warm sites were not provided so we were unable to calculate the minimum and maximum reproductive response to temperature among genotypes of these species. Data on intraspecific variation in reproductive responses to temperature were only available for three species (Table 6.2), all of which were temperate perennial forbs. Across genotypes of these species, the average reproduction response to increasing temperature ranged from was −26 ± 17% to 9 ± 23% (Fig. 6.2b).

B. Growth and Biomass Responses to Temperature

Nine studies ($n = 26$ species) examined intraspecific variation in growth, biomass, or survival responses to temperature. *P*-values <0.05 for the genotype × temperature interaction term were reported for 20 species (Table 6.2), indicating that intraspecific variation in plant growth responses to temperature is common across many plant species. In total, data on intraspecific variation in growth responses to temperature were reported for 23 species, 19 of which were C_4 grasses or forbs. Across genotypes of these species, the average growth or biomass response to increasing temperature ranged from 22 ± 162% to 117 ± 212% (Fig. 6.2b). A broader range of intraspecific variation in growth responses to increasing temperature was observed in C_4 grass species (131 ± 298% to 305 ± 333%) and annual forb species (−78 ± 14% to 20 ± 17%) than BET and BES species (−10 ± 14% to 45 ± 12%) and perennial forb species (−2 ± 51% to 64 ± 127%, Fig. 6.3b). Data were limited for BDT and NET species, as well as boreal or temperate plant species.

Table 6.2. Summary of experimental studies that have examined intraspecific variation in plant responses to temperature variation or climate warming

							Response variable, *P*-value, and range of responses					
Reference	Environment	Species	Functional type	Biome	Temperature treatments	Genotypes	Reproduction	Growth, biomass, survival	Leaf *A*	Leaf g_s	Leaf *R*	Leaf N
Aspinwall et al. (2017a)	GH	*Corymbia calophylla*	BET	Temperate forest	26/12 & 32/18 °C (day/night T_{air})	4 populations			At 25 °C (P = NA[a], −12 to 44%)	At 25 °C (P = NA[a], −48 to 3%)	Area-basis at 20 °C (P = NA[a], −25 to −4%)	Area-basis (P > 0.05, 16 to 42%)
Drake et al. (2015)	GH	*Eucalyptus tereticornis*	BET	Tropical – temperate forest	21.5/25.0/28.5 °C (mean diel T_{air})	12 provenances		Total mass (P < 0.05, −23% to 55%)	(P < 0.05, −8 to 31%)	(P < 0.05, −12 to 45%)	Area-basis at growth temperature (P = NA, −11 to 51%)	Mass-basis (P < 0.05, −7 to −2%)
Drake et al. (2015)	GH	*Eucalyptus grandis*	BET	Tropical – temperate forest	18.0/21.5/25.0/28.5 °C (mean diel T_{air})	9 provenances		Total mass (P < 0.05) (−12% to 32%)	(P < 0.05, −5 to 6%)	(P < 0.05, −8 to 20%)	Area-basis at growth temperature (P = NA, 14 to 26%)	Mass-basis (P > 0.05, −10 to 8%)
Huang et al. (2015)	GH	*Telopea speciosissima*	BES	Temperate forest	26/16, 30/20 C (day/night T_{air})	2 ecotypes		Total mass (P = 0.02, 4 to 47%)	(P = 0.65, −15 to −9%)	(P = 0.02; −20 to 11%)		
Gunderson et al. (2000)	GC, OTC	*Acer saccharum*	BDT	Temperate forest	27/14, 31/18 C (day/night T_{air})	2 populations			(P > 0.10, range NA)		Area-basis at 25 °C (P = NA, −4 to 0%)	
Lee et al. (2005)	Field (CG)	*Acer rubrum*	BDT	Temperate forest	6 time points varying in 3-day mean T_{air} (range: ~13 to 27 °C)	3 populations					Mass-basis at 24 °C (P > 0.05, −67 to −60%)	Mass-basis (P > 0.30, −37 to 27%)
Weston and Bauerle (2007)	GC	*Acer rubrum*	BDT	Temperate forest	27/25 °C, 33/25 °C (day/night T_{air})	2 genotypes			High O_2 (P > 0.05, −25 to −11%) Low O_2 (P > 0.05, −18 to −9%)			
Silim et al. (2010)	GH	*Populus balsamifera*	BDT	Boreal forest	19/10, 27/16 C (day/night T_{air})	2 origins (2 populations per origin)			At 27 °C, (P = NA, 4 to 7%)	At 27 °C; (P = NA, 46 to 49%)	Area-basis at 25 °C (P > 0.58, range NA)	

(continued)

Table 6.2. (continued)

							Response variable, *P*-value, and range of responses					
Reference	Environment	Species	Functional type	Biome	Temperature treatments	Genotypes	Reproduction	Growth, biomass, survival	Leaf *A*	Leaf g_s	Leaf *R*	Leaf N
Lee et al. (2005)	Field (CG)	*Quercus alba*	BDT	Temperate forest	9 time points varying in 3-day mean T_{air} (range: ~12.5 to 27 °C)	3 populations					Mass-basis at 24 °C (P > 0.05, −68 to −61%)	Mass-basis (P > 0.19, −23 to 0%)
Lee et al. (2005)	Field (CG)	*Quercus rubra*	BDT	Temperate forest	9 time points varying in 3-day T_{air} (range: ~13 to 29 °C)	3 populations					Mass-basis at 24 °C (P > 0.05, −68 to −58%)	Mass-basis (P > 0.61, −30 to −24%)
Benomar et al. (2016)	Field (CG)	*Picea glauca*	NET	Boreal forest	3 sites; 1.5, 2.4, 3.7 °C (mean annual T_{air})	6 provenances		Height (P = 0.14, range NA)	(P = 0.62, range NA)	P = 0.57, range NA)		Mass-basis (P = 0.39, range NA)
Tjoelker et al. (2009)	Field (CG)	*Pinus banksiana*	NET	Temperate – boreal forest	5 time points varying in 3-day mean T_{air} (range: −6.6 to 20.4 °C)	8 provenances					Mass-basis at 5 °C (P < 0.05; −62 to −22%)	
Ensslin and Fischer (2015)	Field (CG)	*Bothriochloa insculpta*	C4 grass	Tropical grassland	18.7 °C, 24.2 °C (mean annual T_{air})	4 maternal families	Reproductive mass (P = 0.02, range NA)	Aboveground mass (P = 0.01, −66 to 122%)				
Ensslin and Fischer (2015)	Field (CG)	*Cymbopogon caesius*	C4 grass	Tropical grassland	18.7 °C, 24.2 °C (mean annual T_{air})	2 maternal families	Reproductive mass (P = 0.02, range NA)	Aboveground mass (P = 0.01, 722 to 733%)				
Ensslin and Fischer (2015)	Field (CG)	*Digitaria velutina*	C4 grass	Tropical grassland	18.7 °C, 24.2 °C (mean annual T_{air})	3 maternal families	Reproductive mass (P = 0.02, range NA)	Aboveground mass (P = 0.01, −63 to −7%)				
Ensslin and Fischer (2015)	Field (CG)	*Heteropogon contortus*	C4 grass	Tropical grassland	18.7 °C, 24.2 °C (mean annual T_{air})	2 maternal families	Reproductive mass (P = 0.02, range NA)	Aboveground mass (P = 0.01, 14 to 288%)				

Ensslin and Fischer (2015)	Field (CG)	*Rhynchelythrum repens*	C4 grass	Tropical grassland	18.7 °C, 24.2 °C (mean annual T_{air})	6 maternal families	Reproductive mass (P = 0.02, range NA)	Aboveground mass (P = 0.01, −65 to 4%)		
Ensslin and Fischer (2015)	Field (CG)	*Sehima nervosum*	C4 grass	Tropical grassland	18.7 °C, 24.2 °C (mean annual T_{air})	2 maternal families	Reproductive mass (P = 0.02, range NA)	Aboveground mass (P = 0.01, 109 to 690%)		
Frei et al. (2014)	GC	*Briza media*	C3 grass	Temperate grassland	12 °C, 16 °C mean diel T_{air}	2 populations		Total mass (P > 0.10, 15 to 25%)		
Frei et al. (2014)	GC	*Trifolium montanum*	Perennial forb	Temperate grassland	12 °C, 16 °C mean diel T_{air}	2 populations	# of flowers (P > 0.05, −6 to 36%)	Total mass (P > 0.10, 48 to 51%)		
Frei et al. (2014)	GC	*Ranunculus bulbosus*	Perennial forb	Temperate forest/ grassland	12 °C, 16 °C mean diel T_{air}	2 populations	# of flowers (P > 0.10, −34 to −6%)	Total mass (P > 0.05, −12 to −5%)		
Molina-Montenegro et al. (2013)	GC	*Taraxacum officinale*	Perennial forb	Temperate grassland	5 °C, 25 °C mean T_{air}	5 populations		Total mass (P = 0.01, 3 to 380%)	At growth T_{air} (P = 0.01, 215 to 294%)	At growth T_{air} (P < 0.001, 20 to 172%)
Li et al. (2016)	GC	*Solidago canadensis*	Perennial forb	Temperate grassland	22/17 °C, 32/27 °C (day/night T_{air})	7 populations		Total mass (P < 0.0001, range NA)		
Kahl et al. (2019)	GH	*Silene vulgaris*	Perennial forb	Temperate grassland	18 °C, 21 °C mean T_{air}	25 populations	Flowers (P < 0.01, −37 to −3%)	Aboveground mass (P = 0.67, −23 to −4%)		
Ensslin and Fischer (2015)	Field (CG)	*Malvastrum coromandelianum*	Perennial forb	Tropical grassland	18.7 °C, 24.2 °C (mean annual T_{air})	3 maternal families	Reproductive mass (P = 0.02, range NA)	Aboveground mass (P = 0.01, −72 to 30%)		
Ensslin and Fischer (2015)	Field (CG)	*Tridax procumbens*	Perennial forb	Tropical grassland	18.7 °C, 24.2 °C (mean annual T_{air})	7 maternal families	Reproductive mass (P = 0.02, range NA)	Aboveground mass (P = 0.01, −38 to 66%)		

(continued)

Table 6.2. (continued)

							Response variable, *P*-value, and range of responses					
Reference	Environment	Species	Functional type	Biome	Temperature treatments	Genotypes	Reproduction	Growth, biomass, survival	Leaf *A*	Leaf g_s	Leaf *R*	Leaf N
Ensslin and Fischer (2015)	Field (CG)	*Emilia discifolia*	Perennial forb	Tropical grassland	18.7 °C, 24.2 °C (mean annual T_{air})	4 maternal families	Reproductive mass (P = 0.02, range NA)	Aboveground mass (P = 0.01, −65 to −40%)				
Ensslin and Fischer (2015)	Field (CG)	*Euphorbia heterophylla*	Perennial forb	Tropical grassland	18.7 °C, 24.2 °C (mean annual T_{air})	5 maternal families	Reproductive mass (P = 0.02, range NA)	Aboveground mass (P = 0.01, −17 to 150%)				
Ensslin and Fischer (2015)	Field (CG)	*Euphorbia hirta*	Perennial forb	Tropical grassland	18.7 °C, 24.2 °C (mean annual T_{air})	3 maternal families	Reproductive mass (P = 0.02, range NA)	Aboveground mass (P = 0.01, −15 to 26%)				
Ensslin and Fischer (2015)	Field (CG)	*Justicia flava*	Perennial forb	Tropical grassland	18.7 °C, 24.2 °C (mean annual T_{air})	3 maternal families	Reproductive mass (P = 0.02, range NA)	Aboveground mass (P = 0.01, −96 to −92%)				
Ensslin and Fischer (2015)	Field (CG)	*Amaranthus hybridus*	Annual forb	Tropical grassland	18.7 °C, 24.2 °C (mean annual T_{air})	8 maternal families	Reproductive mass (P = 0.02, range NA)	Aboveground mass (P = 0.01, −70 to 0%)				
Ensslin and Fischer (2015)	Field (CG)	*Bidens pilosa*	Annual forb	Tropical grassland	18.7 °C, 24.2 °C (mean annual T_{air})	13 maternal families	Reproductive mass (P = 0.02, range NA)	Aboveground mass (P = 0.01, −70 to 28%)				
Ensslin and Fischer (2015)	Field (CG)	*Richardia scabra*	Annual forb	Tropical grassland	18.7 °C, 24.2 °C (mean annual T_{air})	15 maternal families	Reproductive mass (P = 0.02, range NA)	Aboveground mass (P = 0.01, −95 to 31%)				

Springate and Kover (2014)	Field, warming cables	*Arabidopsis thaliana*	Annual forb	Global distribution	Ambient, ambient +2–3 °C	320 near isogenic lines	# of fruits (P = 0.02, range NA)	Rosette diameter (P = 0.25, range NA)				
Total studies: 15		Total species: 33				Range = 2 to 320, mean = 15, Total = 499	n = 4 studies (19 species)	n = 9 studies (26 species)	n = 8 studies (9 species)	n = 6 studies (7 species)	n = 6 studies (9 species)	n = 4 studies (7 species)

Information on the experimental infrastructure (*GH* greenhouse, *GC* growth chamber, *OTC* open-top chamber, *CG* one or more common gardens at thermally contrasting sites, or a single site with repeated measurements over seasons), species, functional type (*BET* broadleaved evergreen tree, *BES* broadleaved evergreen shrub, *BDT* broadleaved deciduous tree, *NET* needle-like evergreen tree, etc.), the biome represented by the species, the growth temperature concentrations (mean diel temperature, mean day/night temperature, seasonal temperature range, mean annual temperature), and the number of genotypes, clones, families, population, or provenances tested is provided for each study. For each parameter (reproduction, growth/biomass/survival, net photosynthesis (*A*), stomatal conductance (g_s), dark respiration (*R*), and nitrogen (N)), we provide the *p*-value for the genotype × temperature interaction (if provided or NA if not available), and the range in responsiveness (plasticity) to temperature among genotypes. Responses to temperature was always calculated as: 100 × (genotype mean at the warmest temperature – genotype mean at the coolest temperature) ÷ (genotype mean at coolest temperature). A positive percent change indicates a positive response to increasing temperatures and a negative percent change represents a negative response to increasing temperatures. If data for calculating responses to temperature were not available (NA) in the reference or supplementary material, the range in responsiveness was not calculated. Note: in some cases, experiments tested for three-way interactions or specific traits that do not fit perfectly within one of the categories of response variables. Where appropriate, we provide clarification on the tests conducted and the variables measured

C. Photosynthetic, Stomatal, and Respiratory Responses to Temperature

Eight studies ($n = 9$ species) examined intraspecific variation in photosynthetic responses to variable growth temperature (Table 6.2). *P*-values <0.05 for the genotype × temperature interaction term was reported (or inferred) for four species; three were BET and BES species from Australia and one was a perennial forb (Molina-Montenegro et al. 2013). Two studies examined responses in boreal species (Silim et al. 2010; Benomar et al. 2016) and both reported little variation in photosynthetic responses to increasing growth temperature among genotypes. Tree and shrub species were the most well-studied groups, with average photosynthetic responses to increasing growth temperature varying among all genotypes from −10 ± 10% to 11 ± 22%. Across genotypes of all species, average photosynthetic responses to increasing temperature ranged from 22 ± 86% to 52 ± 109% (Fig. 6.2b).

Six studies ($n = 6$ species), mostly the same studies that examined photosynthetic responses to temperature, examined g_s responses to variable growth temperature. *P*-values <0.05 for the genotype × temperature interaction term were reported (or inferred) for five species. Across genotypes of all species, average photosynthetic responses to increasing temperature ranged from −4 ± 33% to 50 ± 63% (Fig. 6.2b). Across genotypes of BET and BES species, the average response of g_s to increasing growth temperature ranged from −22 ± 18% to 20 ± 18%. In comparison, two genotypes of *Populus balsamifera* showed similar increases in g_s in response to increasing growth temperature (Table 6.2, Silim et al. 2010). All five genotypes of *Taraxacum officinale* examined by Molina-Montenegro et al. (2013) increased g_s with temperature, yet responses ranged from 20 to 172% (Table 6.2). These data suggest that considerable genetic variation in stomatal responses to temperature exists within species.

Six studies ($n = 9$ species) examined respiratory responses to variable growth temperature. All species were trees; three were BET or BES species, five were BDT species, and one was a NET species. Dark respiration was measured at a common temperature in seven species and measured at prevailing growth temperatures in two species (Table 6.2). Among the species where *R* was measured at a common temperature, two species reported *p*-values <0.05 for the genotype × temperature interaction term (Tjoelker et al. 2009; Aspinwall et al. 2017a). Across all species, the average response of *R* (measured at common temperature) to increasing temperature among genotypes ranged from −49 ± 28% to −34 ± 29% (Fig. 6.2b). Thus, most if not all genotypes of all species acclimated to increasing temperatures by reducing basal rates of *R*, and the average difference in *R* between genotypes was relatively small. The same group of studies (and species therein) also tended to report intraspecific variation in leaf N responses to temperature. In general, leaf N was reported on a mass-basis. A *p*-value <0.05 for the genotype × temperature interaction term was reported in only one species (*Eucalyptus tereticornis*, Drake et al. 2015). In general, leaf N was lower under warmer growth temperatures. Across genotypes of all species, the average response of leaf N to increasing temperature ranged from −21 ± 13% to 2 ± 18% (Fig. 6.2b). Data on intraspecific variation in leaf respiratory and N responses to increasing temperature were not available for genotypes of grasses (C_3 or C_4) or forbs.

IV. Intraspecific Variation in Plant Responses to Water Availability

Among all abiotic factors, water availability effects on plants have been the most well-studied. Water availability, largely determined by precipitation (as well as atmospheric demand and soil texture), is a major factor shaping evolutionary adaptation. At the plant-scale, cell division and elongation are turgor-related processes, so reductions in water availability can quickly slow plant growth (Hsiao 1973). Reductions in turgor also cause stomata to gradually close, which limits diffusion of CO_2 into leaves and C fixation. Medium to long-term drought conditions can lead to photosynthetic impairment, and C limitation can lead to reduced *R* (Mitchell et al. 2013; Drake et al. 2017). The effects of water limitation on leaf N can be complex. Water limitation can reduce N uptake and thus leaf N per unit mass (N_{mass}, He and Dijkstra 2014). However, some plants acclimate to water limitation by increasing leaf N per unit surface area (N_{area}) which presumably aids C fixation when water limitation reduces g_s and CO_2 diffusion into the leaf (e.g., Weih et al. 2011). The effects of water limitation on reproduction may be complex. In some plants, particularly annual species, water limitation can lead to earlier reproduction. However, persistent drought can limit C available for reproduction, which is considered more expensive than vegetative growth. The rate at which water becomes limiting, the severity of the limitation, and the length of the limitation all determine the magnitude of the effects of water limitation on plants.

We identified 33 studies that tested whether natural genotypes or populations within species vary in their response to water availability (Table 6.3). Eighteen studies were conducted in growth chambers or greenhouses, often with two treatments: control (well-watered) and water limitation. Six studies were conducted in common garden experiments at two or more sites differing in seasonal or mean annual precipitation. Nine studies were conducted in a single common garden experiment with two or more rainfall (or irrigation) manipulation treatments (Table 6.3). In total, 30 species were represented, 63% of which were trees or shrubs (9 BET species, 3 BES species, 2 BDT species, 5 NET species), 17% were grasses (3 C_4, 2 C_3), and 20% were forbs (4 perennial, 2 annual). Nearly all these species were representative of temperate or Mediterranean/scrub biomes. The number of genotypes included in each study varied from 2 to 26 (Table 6.3; average = 6). Cumulatively, these studies examined responses across 230 genotypes. When data were reported, we calculated responses to *reduced* water availability as: 100 × (genotype mean in driest treatment – genotype mean in wettest treatment) ÷ (genotype mean in wettest treatment) (Table 6.3).

A. *Reproductive Responses to Water Availability*

Eleven studies (n = 9 species) examined intraspecific variation in reproductive responses to water availability. Reproductive responses were quantified differently among studies and species, but included: flowering (yes or no), flower production (number of flowers), seed production (number), flowering propensity (proportion of flowering plants), and belowground bud production (asexual reproduction). *P*-values <0.05 for the genotype × water availability interaction term for reproduction were reported for two species; *Artemisia californica* (Pratt and Mooney 2013) and *Silene vulgaris* (Kahl et al. 2019). Most genotypes of most species reduced reproduction under reduced water availability, although responses ranged from strongly negative to strongly positive in some species (Table 6.2, *Avena barbata*; Maherali et al. 2010, *Arabidopsis lyrata*; Paccard et al. 2014). Across genotypes of all species, the average response of reproduction to reduced water availability ranged from −43 ± 28% to 12 ± 49% (Fig. 6.2c). Almost all studies were carried out with herbaceous species.

Table 6.3. Summary of experimental studies that have examined intraspecific variation in plant responses to water availability

							Response variable, *P*-value, and range of responses					
Reference	Environment	Species	Functional type	Biome	Treatments	Genotypes	Reproduction	Growth, biomass, survival	Leaf *A*	Leaf g_s	Leaf *R*	Leaf N
McKeirnan et al. (2015)	GH	*Eucalyptus globulus*	BET	Temperate forest	2 treatments (well-watered, drought=50% of previous 3-day evaporation)	4 populations		Aboveground mass (P > 0.05; −23 to −4%)				Mass-basis (P = NA; −5 to 11%)
Guarnaschelli et al. (2003)†	GH	*Eucalyptus globulus*	BET	Temperate forest	3 treatments (control = daily water, moderate= water every 6 days, severe= water every 9 days)	3 provenances		Height (P = 0.02; −29 to −8%)				
McLean et al. (2014)	Field (CG)	*Eucalyptus tricarpa*	BET	Temperate forest	2 sites (472, 840 mm annually)	9 populations		Height (P < 0.001; −37 to 6%)				Mass-basis (P > 0.05; −16 to −2%) Area-basis (P > 0.05; −4 to 17%)
McKiernan et al. (2016)	GH	*Eucalyptus viminalis*	BET	Temperate forest	2 treatments (well-watered, drought = 50% of previous 3 day evaporation)	4 populations		(P > 0.05; −17 to −4%)				Mass-basis (P = NA; −4 to 15%)
Baquedano et al. (2008)	Field (RM)	*Quercus coccifera*	BET	Mediterranean	2 treatments (based on soil water potential; control = −0.05 MPa, stress = −0.4 MPa)	2 populations			(P > 0.05; −62 to −41%)	(P > 0.05; −68 to −66%)		
Gimeno et al. (2009)	GH	*Quercus ilex*	BET	Mediterranean/ scrub	2 treatments (control = 15 to 20% VWC; drought=6 to 10% VWC	6 populations			(P = 0.20; −77 to −33%)	(P = 0.28; range NA)		
Arend et al. (2011)	Field (RM)	*Quercus petraea*	BET	Temperate forest	2 treatments (within growing season; control = 15 to 25% VWC, drought = 5 to 15% VWC)	4 provenances		Shoot mass (P < 0.001; −51 to −38%)				
Arend et al. (2011)	Field (RM)	*Quercus pubescens*	BET	Temperate forest	2 treatments (within growing season; control = 15 to 25% VWC, drought = 5 to 15% VWC)	4 provenances		Shoot mass (P < 0.001; −45 to −14%)				
Arend et al. (2011)	Field (RM)	*Quercus robur*	BET	Temperate forest	2 treatments (within growing season; control = 15 to 25% VWC, drought = 5 to 15% VWC)	4 provenances		Shoot mass (P < 0.001; −62 to −32%)				

Ramirez-Valiente et al. (2010)	Field (CG)	*Quercus suber*	BET	Scrub	1 site, 2 years (year 1 rainfall = 469 mm, year 2 rainfall = 645 mm)	5 ecotypes		Annual growth (P = 0.04; −81 to −63%)			Mass-basis (P = 0.35; range NA)
Pratt and Mooney (2013)	Field (RM)	*Artemisia californica*	BES	Mediterranean/ scrub	2 treatments (high = ambient rainfall +70 mm, low = ambient rainfall)	5 populations	Flower production (P = 0.01; −73 to −16%)	Canopy volume (P = 0.05; −66 to −32%)			(P = 0.43; range NA)
Grant et al. (2005)	Field (CG)	*Cistus albidus*	BES	Mediterranean/ scrub	2 treatments (control = 275 mm, drought=65 mm)	2 populations		Branch length (P < 0.01; −43 to −24%)		(P > 0.05; −50 to −30%)	
Lazaro-Nogal et al. (2015)	GH	*Senna candolleana*	BES	Mediterranean/ scrub	2 treatments (moist = field capacity; drought=22% of field capacity)	4 populations		Height (P < 0.05; −22 to −16%)	(P < 0.05; −51 to −17%)	(P < 0.05; −65 to −38%)	
Rose et al. (2009)	Field (RM)	*Fagus sylvatica*	BDT	Temperate forest	3 treatments (control = 40% moisture content, moderate = −20%, moisture content, severe = 10% moisture content)	2 provenances		Total biomass (P > 0.05; both −44%)			Mass-basis (P > 0.05; −8 to −2%)
Lei et al. (2006)	GH	*Populus przewalskii*	BDT	Temperate forest	3 treatments (100%, 50%, and 25% field capacity)	2 populations		Total mass (P < 0.001; −90 to −49%)			
Lopez et al. (2009)†	GC	*Pinus canariensis*	NET	Temperate/ subtropical forest	3 treatments (control, −1 MPa water potential, −1.5 MPa water potential)	5 provenances		Total mass (P > 0.05; −19 to −7%)			
Baquedano et al. (2008)	Field (RM)	*Pinus halepensis*	NET	Mediterranean/ scrub	2 treatments (control = −0.05 MPa soil water potential, stress = −0.4 MPa)	2			(P > 0.05; −54 to −50%)	(P > 0.05; −69 to −53%)	
Lamy et al. (2013)	Field (CG)	*Pinus pinaster*	NET	Mediterranean forest	2 sites (wet = 800 mm annually, dry = 452 mm annually)	6 populations		Height (P < 0.001; −67 to −58%)			
Kerr et al. (2015)	Field (CG)	*Pinus ponderosa*	NET	Temperate forest	2 treatments (control = 50% of previous 3 day evaporation, Drought = 10% of previous 3 day evaporation)	2 populations		Final shoot mass (P < 0.05; −25 to −21%)			

(continued)

Table 6.3. (continued)

							Response variable, *P*-value, and range of responses					
Reference	Environment	Species	Functional type	Biome	Treatments	Genotypes	Reproduction	Growth, biomass, survival	Leaf *A*	Leaf g_s	Leaf *R*	Leaf N
Cregg (1994)	GH	*Pinus ponderosa*	NET	Temperate forest	2 treatments (control = 3 events per week, drought = series of sequential dry downs)	9 provenances		Diameter (P > 0.05; −16 to −6%), days until 50% mortality after withholding water (P > 0.05; 16 to 49%)				
Bansal et al. (2015)†	Field (CG)	*Pseudotsuga menziesii*	NET	Temperate forest	3 sites (cool/wet = 1635 mm, moderate = 1525 mm, warm/dry = 540 mm annually)	7 ecotypes (regions)				Minimum transpiration (P < 0.001; −44 to −25%)		
Avolio and Smith (2013)†	GH	*Andropogon gerardii*	C4 grass	Temperate grassland	3 treatments (low = 4.9 mm, average = 12.3 mm, high = 19.7 mm per week)	5 genotypes	Below ground buds (P = 0.98; range NA)	Total mass (P < 0.01; −45 to −35%)	(P = 0.64; range NA)	(P = 0.01; −57 to −33%)		
Barney et al. (2009)†	GH	*Panicum virgatum*	C4 grass	Temperate grassland	4 treatments (control = 20–35% VWC, flooded, drought=5% VWC, severe drought= < 5%VWC)	2 ecotypes		(P < 0.01; −84 to −83%)	(P > 0.05; −66 to −43%)	(P > 0.05; −63 to −36%)		
Hartman et al. (2012)	Field (RM)	*Panicum virgatum*	C4 grass	Temperate grassland	3 treatments (470, 626, 782 mm annually)	3 ecotypes	Flowering tillers (P > 0.05; range NA)	Aboveground mass (P > 0.05; range NA)	(P > 0.05; range NA)	(P > 0.05; range NA)		
O'Keefe et al. (2013)	Field (RM)	*Panicum virgatum*	C4 grass	Temperate grassland	2 treatments (ambient = 21 mm/ 6 days; altered = 42 mm/12 days)	3 populations	% reproductive mass (P > 0.05; −3 to 5%)	Aboveground mass (P > 0.05; −6 to 15%)	July (P > 0.05; −8 to 20%) September (P > 0.05; −41 to 0%)	July (P > 0.05; −2 to 3%) September (P > 0.05; −39 to 5%)		

Aspinwall et al. (2017c)†	Field (RM)	*Panicum virgatum*	C4 grass	Temperate grassland	2 sites, 4 treatments per site Site 1: 249, 438, 698, 910 mm (March–October) Site 2: 226, 412, 599, 883 mm (March–October)	9 genotypes		Aboveground mass (site × precipitation × genotype, P < 0.05; site 1: −85 to 7%); site 2: −80 to 15%)		
Loreti and Oesterheld (1996)	GH	*Paspalum dilatatum*	C4 grass	Tropical grassland	2 treatments (control = 8 to 11% VWC, drought=2 to 5% VWC, exlcuding flooding treatment)	3 populations		Total mass (P = 0.80; −19 to −14%)		
Sherrard et al. (2009)	GH	*Avena barbata*	C3 grass	Temperate grassland	2 treatments (wet = 31% VWC, dry = <5% VWC)	26 RILs	Seed production (P > 0.05; range NA)	Aboveground mass (P > 0.05; range NA)	(P > 0.05; range NA)	(P > 0.05; range NA)
Maherali et al. (2010)	GH	*Avena barbata*	C3 grass	Temperate grassland	2 treatments (wet = 31% VWC, dry = <5% VWC)	26 RILs	Seed production (P = 0.60, −63 to 43%)		Pre-reproduction (P = 0.92, −42 to 13%) Reproduction (P = 0.03,-35 to 171%)	Pre-reproduction (P = 0.87, −79 to −22%) Reproduction (P = 0.40, −76 to −13%)
Volaire (1995)	Field (RM)	*Dactylis glomerata*	C3 grass	Temperate grassland	2 treatments (control = weekly irrigation, drought=no summer irrigation)	5 populations		Aboveground mass (P = NA; −100 to −93%)		
Gianoli (2004)	Field (RM)	*Convolvulus arvensis*	Perennial forb	Semi arid to temperate forest	2 treatments (well- watered every 2 days, watered every 4 days)	2 populations	Seed number (P > 0.05; −42 to −41%)	Aboveground mass (P > 0.05; −26 to −18%)		
Gianoli and Gonzalez-Teuber (2005)	GH	*Convolvulus chilensis*	Perennial forb	Semi-arid to temperate forest	2 treatments (control = water every 3–4 days, low water=water every 9–10 days)	3 populations		Total mass (P > 0.05, range NA)		
Kahl et al. (2019)†	GH	*Silene vulgaris*	Perennial forb	Temperate grassland	3 treatments (high = 90 mm, medium = 75 mm, low = 65 mm)	25 populations	Flowers (P < 0.001, −31 to 9%)	Total mass (P = 0.84, −27 to 70%)		
Li et al. (2016)	GC	*Solidago canadensis*	Perennial forb	Temperate grassland	2 treatments (75–80% FC, 20–25% FC)	9 populations		Total mass (P > 0.05, range NA)		

(continued)

Table 6.3. (continued)

Reference	Environment	Species	Functional type	Biome	Treatments	Genotypes	Response variable, *P*-value, and range of responses					
							Reproduction	Growth, biomass, survival	Leaf *A*	Leaf g_s	Leaf *R*	Leaf N
Sletvold and Agren (2012)	GH	*Arabidopsis lyrata*	Annual forb	Temperate forests	2 treatments (well-watered, altered; 50 ml when 50% of plants were wilting)	6 populations	Flowering propensity (P = 0.70, −85 to −19%)	Total mass (P < 0.0001, −69% to −3%)				
Paccard et al. (2014)	GC	*Arabidopsis lyrata*	Annual forb	Temperate forests	2 treatments (well-watered = > 30%VWC, drought= < 10%VWC)	9 populations	Flowering (P > 0.05; −26 to 118%)	Rosette size (P > 0.05, −18 to −4%)				Mass-basis (P > 0.05; 20 to 43%)
Heschel et al. (2004)	GH	*Polygonum persicaria*	Annual forb	Temperate grassland	2 treatments (wet = 72 ml daily, dry = 24 ml daily)	3 populations	Seed number (P > 0.05; −20 to −5%)		(P > 0.05; −27 to −7)	(P > 0.05; −58 to −31%)		
Total studies: 33		Total species: 30				Range = 2 to 162, mean = 6, Total = 230	n = 11 studies (9 species)	n = 28 studies (26 species)	n = 10 studies (9 species)	n = 13 studies (10 species)	n = 0 studies	n = 6 studies (7 species)

Information on the experimental infrastructure (*GH* greenhouse, *GC* growth chamber, *Field (CG)* one or more common gardens at sites varying in precipitation, field *(RM)* field study with controlled rainfall or irrigation), species, functional type (*BET* broadleaved evergreen tree, *BES* broadleaved evergreen shrub, *BDT* broadleaved deciduous tree, *NET* needle-like evergreen tree, etc.), the biome represented by the species, the treatment conditions (site mean annual precipitation, amount of water applied, volumetric soil water content (VWC), soil water potential), and the number of genotypes, clones, families, population, or provenances tested is provided for each study. For each parameter (reproduction, growth/biomass/survival, net photosynthesis (*A*), stomatal conductance (g_s), dark respiration (*R*), and nitrogen (N)), we provide the *p*-value for the genotype × water interaction (if provided or NA if not available), and the range in responsiveness (plasticity) to water availability among genotypes. Responses to water were always calculated as: 100 × (genotype mean under driest conditions – genotype mean in control or well-watered conditions) ÷ (genotype mean in control or well-watered conditions). We note cases where more than two treatments were included *and* non-linear responses to water were observed († = non-linear response observed). A positive percent change indicates a positive response to decreasing water availability and a negative percent change represents a negative response to decreasing water availability. If data for calculating responses to temperature were not available (NA) in the reference or supplementary material, the range in responsiveness was not calculated. Note: in some cases, experiments tested for three-way interactions or specific traits that do not fit perfectly within one of the categories of response variables. Where appropriate, we provide clarification on the tests conducted and the variables measured

B. *Growth and Biomass Responses to Water Availability*

Twenty-eight studies (n = 26 species) examined intraspecific variation in growth, biomass, or survival responses to water availability. P-values <0.05 for the genotype × water availability interaction term were reported for 20 species. Across genotypes of all species, the average growth/biomass response to decreasing water availability ranged from −45 ± 27% to −21 ± 32% (Fig. 6.2c). There was a consistently larger range of average growth responses to decreased water availability among genotypes of BET species (−43 ± 21% to −20 ± 23%) and C_4 grass species (−47 ± 36% to −21 ± 40%) compared to NET species (−32 ± 24% to −23 ± 24%; Fig. 6.3c). Overall, the available data indicates that there is considerable intraspecific variation in plant growth responses to decreased water availability. As growth is a prerequisite for reproduction, these data suggest that many species show potential for evolutionary responses to changes in water availability.

C. *Photosynthetic, Stomatal, and Respiratory Responses to Water Availability*

Ten studies (n = 9 species) examined intraspecific variation in photosynthetic responses to water availability (Table 6.3). P-values <0.05 for the genotype × water availability interaction term were reported for two species. Across genotypes of all species, the average photosynthetic response to decreasing water availability ranged from −52 ± 20% to −29 ± 17% (Fig. 6.2c). Only a few species of each growth form were available, so we were unable to compare responses between tree, grass, and forb species. However, photosynthetic responses to water were available for multiple temperate and Mediterranean/scrub species. A slightly larger range of photosynthetic responses to reduced water availability were observed among genotypes of Mediterranean/scrub species (−61% to −35%) compared to temperate species (−39% to −20%, Table 6.3).

Thirteen studies (n = 10 species) examined intraspecific variation in g_s responses to water availability. P-values <0.05 for the genotype × water availability interaction term were reported for four species. Across genotypes of all species, the average stomatal response to decreasing water availability ranged from −61 ± 17% to −32 ± 19% (Fig. 6.2c). Most observations came from temperate grass/forb species, where the average stomatal response to decreasing water availability among genotypes ranged from −59 ± 21% to −22 ± 15%.

We were unable to find any studies that examined intraspecific variation in respiratory responses to water availability. However, six studies (n = 7 species) examined intraspecific variation in leaf N responses to water availability. None of the studies reported p-values <0.05 for the genotype × water availability interaction term. Across genotypes of all species, the average response of N_{mass} to decreasing water availability ranged from −3 ± 13% to 13 ± 18% (Fig. 6.2c).

V. Synthesis

Our examination of published data revealed several important trends. First, based solely on whether p-values for G × E interaction terms were less than 0.05, intraspecific variation in plant responses to increasing CO_2 was less extensive than intraspecific variation in plant responses to increasing temperature and decreasing water availability. For example, less than 20% of species reported p-values <0.05 for the genotype × CO_2 interaction term for growth and reproduction. In comparison, more than half of species (>50%) reported p-values <0.05 for the genotype × temperature and genotype × water availability interaction term for growth and reproduction. These results could reflect general differences in the way that plants respond to changes in atmospheric CO_2, temperature, and water. Nearly all C_3 plants respond positively to increasing CO_2, in terms of growth and photosynthesis, such that G × E will be based almost entirely on responses that differ in magnitude. G × E may be more easily detected in temperature studies given that the *direction* and *magnitude* of the response to temperature may differ among genotypes depending upon the genotype's optimal growth temperature. G × E might also be more easily detected in drought studies given that genotypes might show considerable variation in drought strategies (avoidance, tolerance) and traits that modulate responses to reduced water availability (plant size, rooting depth). This does not mean that genotypes within species respond identically to rising atmospheric CO_2; many studies reported a wide range of responses to increasing CO_2 within species (see Fig. 6.2a). There are also limitations to using *p*-values as the sole test of G × E given that significance tests can be influenced by sample sizes (number of genotypes and replication), variance, and the magnitude of the experimental treatment (e.g., the increase in CO_2 relative to ambient). Nonetheless, our comparative approach suggests that warming and reduced water availability may be more likely to reveal differential responses among genotypes than rising CO_2.

Second, intraspecific variation in plant responses to CO_2, temperature, and water generally increased with the scale of the variable considered. Genotypes of many species showed a wider range of reproductive and growth responses to increasing CO_2 and temperature than physiological responses (e.g., A, g_s, N_{mass}, Fig. 6.2). We observed a wide range of reproductive responses to water availability among genotypes of all species, but a similar range of growth and physiological responses to water were observed across genotypes. There appear to be limits to intraspecific variation in leaf-scale responses (many of which are instantaneously measured and may have low heritability), but small differences in physiological responses may be compounded over time resulting in large differences in productivity and reproduction. Thus, moderate intraspecific variation in physiological responses can generate large variation in whole-plant responses to CO_2, temperature, and water.

Third, the extent to which intraspecific variation in plant responses to atmospheric CO_2, temperature, and water availability differ among growth forms, functional groups, and biomes is unclear. Many studies did not provide average values of individual genotypes in different treatments when the *p*-value for the G × E term was greater than 0.05. When data were provided, they were often limited to one or two growth forms (e.g., trees, forbs), functional types, or biomes. Even so, some findings warrant further study. These findings include: (1) genotypes of BDT species showed a broader range of growth responses to increasing CO_2 than genotypes of NET species, (2) most species possessed genotypes that showed decreased or increased g_s in response to increasing CO_2, which will affect water use under warmer, drier conditions, (3) respiratory temperature acclimation is common across species but intraspecific variation in respiratory temperature acclimation is less common, and (4) genotypes of BET, BES, and C_4 grass species showed a consistently larger range of growth responses to water availability than genotypes of NET species. Further studies should be carried out on new species to determine whether these patterns are common.

VI. Genetic Basis of Intraspecific Variation in Response to Environmental Change

Our examination of published data revealed that intraspecific variation in plant responses to global change (i.e., G × E) was dependent upon the global change factor, scale of the trait, and partially on functional group. But what genetic features underlie variable responses among genotypes? Most traits of ecological and evolutionary importance are controlled by many loci (i.e., polygenic) and environmental variation (Lynch and Walsh 1998). As such, quantitative genetic estimates of G × E are "composite effects" generated from independent molecular action at many underlying loci that are additive together. Three types of gene action can underlie G × E. First, the magnitude of the additive effect of alternative alleles at a locus may depend on environmental context. This type of architecture is often termed differential sensitivity. As one extreme case of differences in magnitude, significant additive effects may be restricted to only certain environmental contexts while "neutral" in others (e.g., conditional neutrality, Anderson et al. 2011). Second, the sign of the additive effect of alternative alleles may depend on the environment. For example, a genetic effect may lead to an increase in a trait value in one environment and a decrease in the trait value in an alternative environment. This type of architecture is often termed "antagonistic pleiotropy" and is the primary gene action underlying tradeoffs and constraints across environments. Third, the occurrence of non-additive gene interactions or epistasis may depend on environmental context. The G × E observed in natural populations as composite effects is no doubt a complex mixture of magnitude, sign, and epistatic interactions that are contingent on environment.

Quantitative trait locus (QTL) and genome-wide association studies (GWAS) have begun to tease apart some of these aggregate effects and provide an initial glimpse at the genetic architecture underlying composite G × E (Des Marais et al. 2013; Laitinen and Nikoloski 2018). A review of 700 carefully filtered QTL studies (largely dominated by crop systems) suggests that most traits are polygenic, that most QTL have relatively small effects, and that the majority of QTL (~60%) exhibit G × E (Des Marais et al. 2013). The predominant genetic architecture underlying the reported G × E was differential sensitivity. Similarly, Wadgymer et al. (2017) reviewed studies of local adaptation and G × E for fitness in natural populations and found that G × E is most often caused by conditionally neutral patterns rather than antagonistic pleiotropy. Unfortunately, few studies have searched for environmentally dependent epistasis and most QTL experiments are underpowered for detecting non-additive interactions. Importantly, the evolutionary impacts of G × E will depend in part on the relative abundance of these genetic architectures.

At the molecular level, adaptive G × E likely requires sensing of the environment, signal transduction, and subsequent expression of effector genes that alter plant growth, development, and physiology. In natural populations, both environmental sensitivity and antagonistic patterns at the allelic level must drive differences in these molecular responses that scale to the whole organism. Molecular biologists often seek to understand the molecular mechanisms by which plants sense and respond to their environments by taking advantage of forward and reverse genetic screens, a variety of genomic assays, and in some cases positional cloning of natural alleles. Some important questions are: What type of genes exhibit G × E? How are magnitude or sign-change patterns manifested? Are certain molecular phenomena more important in generating G × E? At a mechanistic level, natural variation in environment-sensing receptors might result in differential responses to the environment. For example, some individuals may carry genetic variants that result in broken or altered receptor functions that drive differential physiological responses in different environments (e.g., El-Din El Assal et al. 2001). Several other mechanisms might also underlie natural allelic G × E including, for example, environmental regulation of gene expression through cis- or trans- factors (Smith and Kruglyak 2008; Lovell et al.

2018), loss or gain of function mutations in coding proteins (Gujas et al. 2012; Zhu et al. 2015), environmentally altered epi-genetic marks (Li et al. 2018), or environment-specific alternative splicing or posttranslational modifications (Kesari et al. 2012; Kalladan et al. 2019). Molecular biologists have made remarkable progress studying plant responses to the environment using extreme alleles. However, we do not know if information from mutagenesis, knockout, or overexpression experiments are accurate reflections of the natural alleles exhibiting G × E and of the loci that would be important in local adaptation to climate change. As such, an important direction in the field is continued effort to clone G × E QTL as this will provide much needed information on molecular mechanisms, as well as enable population genetic studies of alleles of known function and their potential role in climate adaptation.

VII. Outlook

Several areas of research will help improve our understanding of the patterns and consequences of intraspecific variation in plant responses to climate change. First, many studies examine genotype or population responses to a single aspect of climate change. But climate change involves co-occurring increases in CO_2, warming, and altered water availability. Combined, these factors may trigger different responses than are observed or predicted by a single factor. Very few studies we identified examined intraspecific variation in response to multiple climate change factors; two reported genotypic differences in growth responses to multiple factors (CO_2 × drought, Townend 1993; temperature × drought, Arend et al. 2011). To make more realistic predictions of species responses to climate change, future studies should consider how genotypes or populations from different environments vary in their response to multiple aspects of climate change. Given that climate change effects might be most apparent at species range margins, these studies should include genotypes or populations from rear and leading edges of the species range (Mamet et al. 2019).

Second, we advocate for better reporting of results and full access to data from G × E experiments. In our examination of the literature, genotype or population means under each environment were often not reported when *p*-values for G × E were > 0.05. Use of *p*-values as the sole test of G × E has its limitations (as discussed above) and better access to genotype response data and statistics (sample sizes, variance) would aid meta-analyses that quantitatively determine the extent of intraspecific variation in plant responses to climate change, as well as sources of variation among experiments. For instance, the number of genotypes, experimental setup (glasshouse pot study, common garden, open-top chamber, etc.), and the magnitude of the experimental treatment(s) could all influence observed differences in genotype responses to CO_2, temperature, and water availability. Data limitations did not allow us to carefully account for these sources of variation. We also suggest that studies consider using environmental gradients rather than two seemingly arbitrarily selected treatments, which will help more fully reveal genotype or population responses and thresholds to climate change. Very few studies we identified were designed to explore nonlinear responses, making quantitative syntheses of threshold responses to increasing CO_2 and temperature, and decreasing water availability difficult.

Third, future studies should examine plant population responses to short-term stress events, especially heatwaves. Heatwaves are likely to exert stronger selective pressures than gradual warming, yet very few studies have examined how natural populations within species respond to heatwaves (but see Loik et al. 2017; Maher et al. 2019). Genetic variation in leaf cooling and maintenance of photosynthetic integrity during these short-term stress events could have important consequences for plant growth and fitness, and thus evolution and adaptation to climate change (Aspinwall et al. 2019).

Fourth, future studies that combine molecular (i.e., gene expression, proteomics) and leaf-scale responses will improve our understanding of the mechanisms underlying intraspecific variation in plant responses to climate change. We argue that moderate variation in leaf-scale responses can result in large variation in whole-plant responses. Combining molecular responses with leaf- and whole-plant responses would advance our understanding of the causes of intraspecific variation in plant responses to global change. Along these lines, we should also continue to harness the power of GWAS established at different sites, or under different conditions, to track intraspecific variation in growth and functional traits, and identify genomic regions associated with plasticity or acclimation to environmental change (Franks et al. 2014). Ideally, many of these studies should be conducted over longer time periods and include measurements of reproduction, which can provide direct insight into potential evolutionary responses.

Fifth, we advocate for continued analysis of data from long-term common gardens (i.e., provenance trials), and establishment of new common gardens, with populations growing at contrasting sites over long time periods. In long-lived species, it may take decades for genotype or population differences in growth, survival and reproduction to become apparent under different conditions (Germino et al. 2019). Moreover, the longer the experiment, the greater the odds of identifying thresholds in population performance or acclimation capacity.

Lastly, it is widely known that most species exist as many different populations, with different traits, and different sensitivities to changes in the environment (as emphasized throughout this chapter). Yet, models that predict climate change impacts on species distributions often fail to account for intraspecific differences. Accounting for differences in adaptation and plasticity within species may result in widely varying forecasts of species distributions into the future (Valladares et al. 2014). We suggest that species distribution models work to integrate information on adaptive and plastic differences among populations. These approaches, and more, will help us improve our understanding of the causes and consequences of intraspecific variation in plant responses to global change.

Acknowledgements

MJA was financially supported by The United States Department of Agriculture – National Institute of Food and Agriculture (award number: 2019-67013-29161), the University of North Florida, and Auburn University. DTT, PDR and MJA were financially supported by the Australian Science Industry Endowment Fund (award number: RP04–122). TEJ was financially supported by the U.S. Department of Energy, Office of Science, Office of Biological and Environmental Research (award number: DE-SC0014156) and a National Science Foundation Plant Genome Research Program Award (award number: IOS-1444533).

References

Ahrens CW, Byrne M, Rymer PD (2019) Standing genomic variation within coding and regulatory regions contributes to the adaptive capacity to climate in a foundation tree species. Mol Ecol 28:2502–2516

Ainsworth EA, Long SP (2005) What have we learned from 15 years of free-air CO_2 enrichment (FACE)? A meta-analytic review of the responses of photosynthesis, canopy properties and plant production to rising CO_2. New Phytol 165:351–372

Ainsworth EA, Rogers A (2007) The response of photosynthesis and stomatal conductance to rising [CO_2]: mechanisms and environmental interactions. Plant Cell Environ 30:258–270

Aitken SN, Yeaman S, Holliday JA, Wang T, Curtis-McLane S (2008) Adaptation, migration or extirpation: climate change outcomes for tree populations. Evol Appl 1:95–111

Anderson JT, Willis JH, Mitchell-Olds T (2011) Evolutionary genetics of plant adaptation. Trends Genet 7:258–266

Arend M, Kuster T, Günthardt-Goerg MS, Dobbertin M (2011) Provenance-specific growth responses to drought and air warming in three European oak species (*Quercus robur*, *Q. petraea* and *Q. pubescens*). Tree Physiol 31:287–297

Aspinwall MJ, Drake JE, Campany C, Varhammar A, Ghannoum O, Tissue DT, Reich PB, Tjoelker MG (2016) Convergent acclimation of leaf photosynthesis and respiration to prevailing ambient temperatures under current and warmer climates in *Eucalyptus tereticornis*. New Phytol 212:354–367

Aspinwall MJ, Varhammar A, Blackman CJ, Tjoelker MG, Ahrens C, Byrne M, Tissue DT, Rymer PD (2017a) Adaptation and acclimation both influence photosynthetic and respiratory temperature responses in *Corymbia calophylla*. Tree Physiol 37:1095–1112

Aspinwall MJ, Jacob VK, Blackman CJ, Smith RA, Tjoelker MG, Tissue DT (2017b) The temperature response of leaf dark respiration in 15 provenances of *Eucalyptus grandis* grown in ambient and elevated CO_2. Funct Plant Biol 44:1075–1086

Aspinwall MJ, Fay PA, Hawkes CV, Lowry DB, Khasanova A, Whitaker BK, Bonnette J, …, Juenger TE (2017c) Intraspecific variation in precipitation responses of a widespread C_4 grass depend on site water limitation. J Plant Ecol 10:310–321

Aspinwall MJ, Blackman CJ, Resco de Dios V, Busch FA, Rymer PD, Loik ME, Drake JE, …, Tissue DT (2018) Photosynthesis and carbon allocation are both important predictors of genotype productivity responses to elevated CO_2 in *Eucalyptus camaldulensis*. Tree Physiol 38:1286–1301

Aspinwall MJ, Pfautsch S, Tjoelker MG, Varhammar A, Possell M, Drake JE, Reich PB, …, Dennison S (2019) Range size and growth temperature influence *Eucalyptus* species responses to an experimental heatwave. Glob Chang Biol 25:1665–1684

Atkin OK, Tjoelker MG (2003) Thermal acclimation and the dynamic response of plant respiration to temperature. Trends Plant Sci 8:343–351

Atkin OK, Bloomfield KJ, Reich PB, Tjoelker MG, Asner GP, Bonal D, Bonisch G, …, Zaragoza-Castells (2015) Global variability in leaf respiration in relation to climate, plant functional types and leaf traits. New Phytol 206:614–636

Avolio ML, Smith MD (2013) Mechanisms of selection: phenotypic differences among genotypes explain patterns of selection in a dominant species. Ecology 94:953–965

Bansal S, Harrington CA, Gould PJ, St. Clair JB (2015) Climate-related genetic variation in drought-resistance of Douglas-fir (*Pseudotsuga menziesii*). Glob Chang Biol 21:947–958

Baquedano FJ, Valladares F, Castillo FJ (2008) Phenotypic plasticity blurs ecotypic divergence in the response of *Quercus coccifera* and *Pinus halepensis* to water stress. Eur J For Res 127:495–506

Barney JN, Mann JJ, Kyser GB, Blumwald E, Van Deynze A, DiTomaso J (2009) Tolerance of switchgrass to extreme soil moisture stress: ecological implications. Plant Sci 177:724–732

Bazzaz FA, Jasienski M, Thomas SC, Wayne P (1995) Microevolutionary responses to experimental populations of plants to CO_2-enriched environments: parallel results from two model systems. Proc Natl Acad Sci 92:8161–8165

Benomar L, Lamhamedi MS, Rainville A, Beaulieu J, Bousquet J, Margolis HA (2016) Genetic adaptation vs. ecophysiological plasticity of photosynthetic-related traits in young *Picea glauca* trees along a regional climatic gradient. Front Plant Sci 7. https://doi.org/10.3389/fpls.2016.00048

Berry J, Björkman O (1980) Photosynthetic response and adaptation to temperature in higher plants. Annu Rev Plant Physiol 31:491–543

Bigras FJ, Bertrand A (2006) Responses of *Picea mariana* to elevated CO_2 concentration during growth, cold hardening and dehardening: phenology, cold tolerance, photosynthesis and growth. Tree Physiol 26:875–888

Blackman CJ, Aspinwall MJ, Resco de Dios V, Smith R, Tissue DT (2016) Leaf photosynthetic, economics and hydraulic traits are decoupled among genotypes of a widespread species of eucalypt grown under ambient and elevated CO_2. Funct Ecol 30:1491–1500

Bradshaw AD (1965) Evolutionary significance of phenotypic plasticity in plants. Adv Genet 13:115–155

Bresson CC, Vitasse Y, Kremer A, Delzon S (2011) To what extent is altitudinal variation of functional traits driven by genetic adaptation in European oak and beech? Tree Physiol 31:1164–1174

Bruhn D, Wiskich JT, Atkin OK (2007) Contrasting responses by respiration to elevated CO_2 in intact tissue and isolated mitochondria. Funct Plant Biol 34:112–117

Cantin D, Tremblay MF, Lechowicz MJ, Potvin C (1996) Effects of CO_2 enrichment, elevated temperature, and nitrogen availability on the growth and gas exchange of different families of jack pine seedlings. Can J For Res 27:510–520

Collins M, Knutti R, Arblaster J, Dufresne J-L, Fichefet T, Friedlingstein P, Gao X, …, Wehner M (2013) Long-term climate change: projections commitments and irreversibility. In: Stocker TF, Qin D, Plattner M, Tignor SK, Allen J, Boschung A, Nauels Y, Xia Y, Bex V, Midgley PM (eds) Climate Change 2013: The Physical Science Basis. Contribution of Working Group I to the Fifth Assessment Report of the Intergovernmental Panel on Climate Change. Cambridge University Press, New York/Cambridge

Cregg BM (1994) Carbon allocation, gas exchange, and needle morphology of *Pinus ponderosa* genotypes known to differ in growth and survival under imposed drought. Tree Physiol 14:883–898

Crous KY, Drake JE, Aspinwall MJ, Sharwood RE, Tjoelker MG, Ghannoum O (2018) Photosynthetic capacity and leaf nitrogen decline along a controlled climate gradient in provenances of two widely distributed *Eucalyptus* species. Glob Chang Biol 24:4626–4644

Cseke LJ, Tsai C-J, Rogers A, Nelson MP, White HL, Karnosky DF, Podila GP (2009) Transcriptomic comparison in the leaves of two aspen genotypes having similar carbon assimilation rates but different partitioning patterns under elevated [CO_2]. New Phytol 182:891–911

Curtis PS, Snow AA, Miller AS (1994) Genotype-specific effects of elevated CO_2 on fecundity in wild radish (*Raphanus raphinistrum*). Oecologia 97:100–105

Des Marais DL, Hernandez KM, Juenger TE (2013) Genotype-by-environment interaction and plasticity: exploring genomic responses of plants to the abiotic environment. Annu Rev Ecol Evol Syst 44:5–29

Drake BG, González-Meler MA, Long SP (1997) More efficient plants: a consequence of rising atmospheric CO_2? Annu Rev Plant Phsiol Plant Mol Biol 48:609–639

Drake JE, Aspinwall MJ, Pfautsch S, Rymer PD, Reich PB, Smith RA, Crous KY, …, Tjoelker MG (2015) The capacity to cope with climate warming declines from temperate to tropical latitudes in two widely distributed *Eucalyptus* species. Glob Chang Biol 21: 459–472

Drake JE, Power SA, Duursma RA, Medlyn BE, Aspinwall MJ, Choat B, Creek D, …, Tissue DT (2017) Stomatal and non-stomatal limitations of photosynthesis for four tree species under drought: a comparison of model formulations. Agric For Meteorol 247: 454–466

Dusenge ME, Duarte AG, Way DA (2018) Plant carbon metabolism and climate change: elevated CO_2 and temperature impacts on photosynthesis, photorespiration and respiration. New Phytol 221:32–49

El-Din El Assal S, Alonso-Blanco C, Peeters AJ, Raz V, Koornneef M (2001) A QTL for flowering time in *Arabidopsis* reveals a novel allele of CRY2. Nat Genet 29:435–440

Ensslin A, Fischer M (2015) Variation in life-history traits and their plasticities to elevational transplantation among seed families suggests potential for adaptative evolution of 15 tropical plant species to climate change. Am J Bot 102:1371–1379

Feeley KJ, Silman MR (2009) Extinction risk of Amazonian plant species. Proc Natl Acad Sci 106:12382–12387

Franks SJ, Weber JJ, Aitken SN (2014) Evolutionary and plastic responses to climate change in terrestrial plant populations. Evol Appl 7:123–139

Frei ER, Ghazoul J, Pluess AR (2014) Plastic responses to elevated temperature in low and high elevation populations of three grassland species. PLoS One 9:e98677

Germino MJ, Moser AM, Sands AR (2019) Adaptive variation, including local adaptation, requires decades to become evident in common gardens. Ecol Appl 29:e01842

Ghalambor CK, McKay JK, Carroll SP, Reznick DN (2007) Adaptive versus non-adaptive phenotypic plasticity and the potential for contemporary adaptation in new environments. Funct Ecol 21:394–407

Gianoli E (2004) Plasticity of traits and correlations in two populations of *Convolvulus arvensis* (Convuolvulaceae) differing in environmental heterogeneity. Int J Plant Sci 165:825–832

Gianoli E, González-Teuber M (2005) Environmental heterogeneity and population differentiation in plasticity to drought in *Convolvulus chilensis* (Convolvulaceae). Evol Ecol 19:603–613

Gimeno TE, Pías B, Lemos-Filho VF (2009) Plasticity and stress tolerance override local adaptation in the responses of Mediterranean holm oak seedlings to drought and cold. Tree Physiol 29:87–98

Grant OM, Incoll LD, McNeilly T (2005) Variation in growth responses to availability of water in *Cistus albidus* populations from different habitats. Funct Plant Biol 32:817–829

Griffin KL, Anderson OR, Gastrich MD, Lewis JD, Lin G, Schuster W, Seemann JR, …, Whitehead D (2001a) Plant growth in elevated CO_2 alters mitochondrial number and chloroplast fine structure. Proc Natl Acad Sci USA 98:2473–2478

Griffin KL, Tissue DT, Turnbull MH, Schuster W, Whitehead D (2001b) Leaf dark respiration as a function of canopy position in *Nothofagus fusca* trees grown at ambient and elevated CO_2 partial pressures for 5 years. Funct Ecol 15:497–505

Groisman PY, Knight RW, Easterling DR, Karl TR, Hegerl GC, Razuvaev VN (2005) Trends in intense precipitation in the climate record. J Clim 18:1326–1350

Guarnaschelli AB, Lemcoff JH, Prystupa P, Basci SO (2003) Responses to drought preconditioning in *Eucalyptus globulus* Labill. Provenances. Trees 17:501–509

Gujas B, Alonso-Blanco C, Hardtke C (2012) Natural *Arabidopsis* brx loss-of-function alleles confer root adaptation to acidic soil. Curr Biol 22:1962–1968

Gunderson CA, Norby RJ, Wullschleger SD (2000) Acclimation of photosynthesis and respiration to simulated climatic warming in northern and southern populations of *Acer saccharum*: laboratory and field evidence. Tree Physiol 20:87–96

Hansen J, Sato M, Ruedy R (2012) Perception of climate change. Proc Natl Acad Sci U S A 109:14726–14727

Hartman JC, Nippert JB, Springer CJ (2012) Ecotypic responses of switchgrass to altered precipitation. Funct Plant Biol 39:126–136

He M, Dijkstra FA (2014) Drought effect on plant nitrogen and phosphorus: a meta-analysis. New Phytol 204:924–931

Heschel MS, Sultan SE, Glover S, Sloan D (2004) Population differentiation and plastic responses to drought stress in the generalist annual *Polygonum persicaria*. Int J Plant Sci 165:817–824

Heskel MA, O'Sullivan OS, Reich PB, Tjoelker MG, Weerasinghe LK, Penillard A, Egerton JJG, …, Atkin OK (2016) Convergence in the temperature response of leaf respiration across biomes and plant functional types. Proc Nat Acad Sci USA 113:3832–3837

Higgins SI, Clark JS, Nathan R, Hovestadt T, Schurr F, Fragoso JMV, Aguiar MR, …, Lavorel S (2003) Forecasting plant migration rates: managing uncertainty for risk assessment. J Ecol 91:341–347

Hoffmann AA, Srgo CM (2011) Climate change and evolutionary adaptation. Nature 470:479–485

Hsiao TC (1973) Plant responses to water stress. Annu Rev Plant Physiol 24:519–570

Huang G, Rymer PD, Duan H, Smith RA, Tissue DT (2015) Elevated temperature is more effective than elevated [CO_2] in exposing genotypic variation in *Telopea speciosissima* growth plasticity: implications for woody plant populations under climate change. Glob Chang Biol 21:3800–3813

Jenkins Klus D, Kalisz S, Curtis PS, Teeri JA, Tonsor SJ (2001) Family- and population-level responses to atmospheric CO_2 concentration: gas exchange and the allocation of C, N, and biomass in *Plantago lanceolata* (Plantaginaceae). Am J Bot 88:1080–1087

Jump AS, Penuelas J, Rico L, Ramallo E, Estiarte M, Martinez-Izquierdo JA, Lloret F (2008) Simulated climate change provokes rapid genetic change in the Mediterranean shrub *Fumana thymifolia*. Glob Chang Biol 14:637–643

Kahl SM, Lenhard M, Joshi J (2019) Compensatory mechanisms to climate change in the widely distributed species *Silene vulgaris*. J Ecol. https://doi.org/10.1111/1365-2745.13133

Kalladan R, Lasky JR, Sharma S, Kumar MN, Juenger TE, Des Marais DL, Verslues PE (2019) Natural variation in 9-cis-epoxycartenoid dioxygenase 3 and ABA accumulation. Plant Physiol. (In press)

Kattge J, Knorr W (2007) Temperature acclimation in a biochemical model of photosynthesis: a reanalysis of data from 36 species. Plant Cell Environ 30:1176–1190

Kerr KL, Meinzer FC, McCulloh KA, Woodruff DR, Marias DE (2015) Expression of functional traits during seedling establishment in two populations of *Pinus ponderosa* from contrasting climates. Tree Physiol 35:535–548

Kesari R, Lasky J, Villamor JG, Des Marais D, Chen Y-J C, Liu TW, Lin W, …, Verslues P (2012) Intron mediated alternative splicing of *Arabidopsis* P5CS1 and its association with natural variation in proline and climate adaptation. Proc Natl Acad Sci 109:9197–9202

Kubiske ME, Pregitzer KS, Zak DR, Mikan CJ (1998) Growth and C allocation of *Populus tremuloides* genotypes in response to atmospheric CO_2 and soil N availability. New Phytol 140:251–260

Kumarathunge DP, Medlyn BE, Drake JE, Tjoelker MG, Aspinwall MJ, Battaglia M, Cano FJ, …, Way DA (2019) Acclimation and adaptation components of the temperature dependence of plant photosynthesis at the global scale. New Phytol 222:768–784

Laitinen RAE, Nikoloski Z (2018) Genetic basis of plasticity in plants. J Exp Bot 70:739–745

Lamy J-B, Delzon S, Bouche PS, Alia R, Vendramin GG, Cochard H, Plomion C (2013) Limited genetic variability and phenotypic plasticity detected for cavitation resistance in a Mediterranean pine. New Phytol 201:874–886

Lau JA, Shaw RG, Reich PB, Shaw FH, Tiffin P (2007) Strong ecological but weak evolutionary effects of elevated CO_2 on a recombinant inbred population of *Arabidopsis thaliana*. New Phytol 175:351–362

Lázaro-Nogal A, Matesanz S, Godoy A, Pérez-Trautman F, Gianoli E, Valladares F (2015) Environmental heterogeneity leads to higher plasticity in dry-edge populations of a semi-arid Chilean shrub: insights into climate change responses. J Ecol 103:338–350

Lee TD, Reich PB, Bolstad PV (2005) Acclimation of leaf respiration to temperature is rapid and related to specific leaf area, soluble sugars and leaf nitrogen across three temperate deciduous tree species. Funct Ecol 19:640–647

Lei Y, Yin C, Li C (2006) Differences in some morphological, physiological, and biochemical responses to drought stress in two contrasting populations of *Populus przewalskii*. Physiol Plant 127:182–191

Li J, Guan W, Yu F-H, van Kleunen M (2016) Latitudinal and longitudinal clines of phenotypic plasticity in the invasive herb *Solidago canadensis* in China. Oecologia 182:755–764

Li Z, Jiang D, He Y (2018) FRIGIDA establishes a local chromosome environment for FLOWERING LOCUS C mRNA production. Nat Plant 4:836–846

Loik ME, Resco de Dios V, Smith R, Tissue DT (2017) Relationships between climate of origin and photosynthetic responses to an episodic heatwave depend on growth CO_2 concentration for *Eucalyptus camaldulensis* var. *camaldulensis*. Funct Plant Biol 44:1053–1062

López R, Rodríguez-Calcerrada J, Gil L (2009) Physiological and morphological response to water deficit in seedlings of five provenances of *Pinus canariensis*: potential to detect variation in drought-tolerance. Trees 23:509–519

Loreti J, Oesterheld M (1996) Intraspecific variation in the resistance to flooding and drought in populations of *Paspalum dilatatum* from different topographic positions. Oecologia 108:279–284

Lovell JT, Jenkins J, Lowry DB, Mamidi S, Sreedasyam A, Weng X, Barry K, …, Juenger TE (2018) The genomic landscape of molecular responses to drought stress in *Panicum hallii*. Nat Commun 9:5213

Lüscher A, Nösberger J (1997) Interspecific and intraspecific variability in the response of grasses and legumes to free air CO_2 enrichment. Acta Oecol 18:269–275

Lynch M, Walsh B (1998) Genetics and Analysis of Quantitative Traits. Sinauer Associates, Sunderland

Maher T, Mirzaei M, Pascovici D, Wright IJ, Haynes PA, Gallagher RV (2019) Evidence from the proteome for local adaptation to extreme heat in a widespread tree species. Funct Ecol 33:436–446

Maherali H, Caruso CM, Sherrard ME, Latta RG (2010) Adaptive value and costs of physiological plasticity to soil moisture limitation in recombinant inbred lines of *Avena barbata*. Am Nat 175:211–224

Mamet SD, Brown CD, Trant AJ, Laroque CP (2019) Shifting global *Larix* distributions: northern expansion and southern retraction as species respond to changing climate. J Biogeogr 46:30–44

McLean EH, Prober SM, Stock WD, Steane DA, Potts BM, Vaillancourt RE, Byrne M (2014) Plasticity of functional traits varies clinally along a rainfall gradient in *Eucalyptus tricarpa*. Plant Cell Environ 37:1440–1451

McKeirnan AB, Potts BM, Brodribb TJ, Hovenden MJ, Davies NW, McAdam SAM, Ross JJ, …, O'Reilly-Wapstra JM (2015) Responses to mild water deficit and rewatering differ among secondary metabolites but are similar among provenances within *Eucalyptus* species. Tree Physiol 36:133–147

Medlyn BE, Barton CVM, Broadmeadow MSJ, Ceulemans R, De Angelis P, Forstreuter M, Freeman M, …, Jarvis PG (2001) Stomatal conductance of forest species after long-term exposure to elevated CO_2 concentration: a synthesis. New Phytol 149:247–264

Mitchell PJ, O'Grady AP, Tissue DT, White DA, Ottenschlaeger ML, Pinkard EA (2013) Drought response strategies define the relative contributions of hydraulic dysfunction and carbohydrate depletion during tree mortality. New Phytol 197:862–872

Mohan JE, Clark JS, Schlesinger WH (2004) Genetic variation in germination, growth, and survivorship of red maple in response to subambient through elevated atmospheric CO_2. Glob Chang Biol 10:233–247

Molina-Montenegro MA, Palma-Rojas C, Alcayaga-Olivares Y, Oses R, Corcuera LJ, Cavieres LA, Gianoli E (2013) Ecophysiological plasticity and local differentiation help explain the invasion success of *Taraxacum officinale* (dandelion) in South America. Ecography 36:718–730

Mycroft EE, Zhang J, Adams G, Reekie E (2009) Elevated CO_2 will not select for enhanced growth in white spruce despite genotypic variation in response. Basic Appl Ecol 10:349–357

Nicotra AB, Atkin OK, Bonser SP, Davidson AM, Finnegan EJ, Mathesius U, Poot P, …, van Kleunen M (2010) Plant phenotypic plasticity in a changing climate. Trends Plant Sci 15:684–692

Norton LR, Firbank LG, Gray AJ, Watkinson AR (1999) Responses to elevated temperature and CO_2 in the perennial grass *Agrostis curtisii* in relation to population origin. Funct Ecol 13:29–37

O'Keefe K, Tomeo N, Nippert JB, Springer CJ (2013) Population origin and genome size do not impact *Panicum virgatum* (switchgrass) responses to variable precipitation. Ecosphere 4:37. https://doi.org/10.1890/ES12-00339.1

Oleksyn J, Modrzynski J, Tjoelker MG, Zytkowiak R, Reich PB, Karolewski P (1998) Growth and physiology of *Picea abies* populations from elevational transects: common garden evidence for altitudinal ecotypes and cold adaptation. Funct Ecol 12:573–590

Paccard A, Fruleux A, Willi Y (2014) Latitudinal trait variation and responses to drought in *Arabidopsis lyrata*. Oecologia 175:577–587

Polley HW, Tischler CR, Johnson HB, Pennington RE (1999) Growth, water relations, and survival of drought-exposed seedlings from six maternal families of honey mesquite (*Prosopis glandulosa*): responses to CO_2 enrichment. Tree Physiol 19:359–366

Pratt JD, Mooney KA (2013) Clinal adaptation and adaptive plasticity in *Artemisia californica*: implications for the response of a foundation species to predicted climate change. Glob Chang Biol 19:2454–2466

Ramírez-Valiente JA, Sáchez-Gómez D, Aranda I, Valladares F (2010) Phenotypic plasticity and local adaptation in leaf ecophysiological traits of 13 contrasting cork oak populations under different water availabilities. Tree Physiol 30:618–627

Riikonen J, Lindsberg M-M, Holopainen T, Oksanen E, Lappi J, Peltonen P, Vapaavuori E (2004) Silver birch and climate change: variable growth and carbon allocation responses to elevated concentrations of carbon dioxide and ozone. Tree Physiol 24:1227–1237

Riikonen J, Holopainen T, Oksanen E, Vapaavuori E (2005) Leaf photosynthetic characteristics of silver birch during three years of exposure to elevated concentrations of CO_2 and O_3 in the field. Tree Physiol 25:621–632

Rose L, Leuschner C, Köckemann B, Buschmann H (2009) Are marginal beech (*Fagus sylvatica* L.) provenances a source for drought tolerant ecotypes? Eur J For Res 128:335–343

Roumet C, Laurent G, Roy J (1999) Leaf structure and chemical composition as affected by elevated CO_2: genotypic responses of two perennial grasses. New Phytol 143:73–81

Roumet C, Laurent G, Canivenc G, Roy J (2002) Genotypic variation in the response of two perennial grass species to elevated carbon dioxide. Oecologia 133:342–348

Sherrard ME, Maherali H, Latta RG (2009) Water stress alters the genetic architecture of functional traits associated with drought adaptation in *Avena barbata*. Evolution 63:702–715

Silim SN, Ryan N, Kubien DS (2010) Temperature responses of photosynthesis and respiration in *Populus balsamifera* L.: acclimation versus adaptation. Photosynth Res 104:19–30

Sletvold N, Ågren J (2012) Variation in tolerance to drought among Scandinavian populations of *Arabidopsis lyrata*. Evol Ecol 26:559–577

Slot M, Kitajima K (2015) General patterns of acclimation of leaf respiration to elevated temperatures across biomes and plant types. Oecologia 177:885–900

Smith EN, Kruglyak L (2008) Gene-environment interaction in yeast expression. PLoS Biol 6:e83

Spinnler D, Egli P, Körner C (2003) Provenance effects and allometry in beech and spruce under elevated CO_2 and nitrogen on two different forest soils. Basic Appl Ecol 4:467–478

Springate DA, Scarcelli N, Rowntree J, Kover PX (2011) Correlated response in plasticity to selection for early flowering in *Arabidopsis thaliana*. J Evol Biol 24:2280–2288

Springate DA, Kover PX (2014) Plant responses to elevated temperatures: a field study on phenological sensitivity and fitness responses to simulated climate warming. Glob Chang Biol 20:456–465

Thomas CD, Cameron A, Green RE, Bakkenes M, Beaumont LJ, Collingham YC, Erasmus BFN, …, Williams SE (2004) Extinction risk from climate change. Nature 427:145–148

Thuiller W, Albert C, Araujo MB, Berry PM, Cabeza M, Guisan A, Hickler T, …, Zimmermann NE (2008) Predicting global change impacts on plant species' distributions: future challenges. Perspect Plant Ecol Evol Syst 9:137–152

Tissue DT, Griffin KL, Ball JT (1999) Photosynthetic adjustment in fieldgrown ponderosa pine trees after six years of exposure to elevated CO_2. Tree Physiol 19:221–228

Tjoelker MG, Oleksyn J, Lorenc-Plucinska G, Reich PB (2009) Acclimation of respiratory temperature responses in northern and southern populations of *Pinus banksiana*. New Phytol 181:218–229

Tonsor SJ, Scheiner SM (2007) Plastic trait integration across a CO_2 gradient in *Arabidopsis thaliana*. Am Nat 169:119–140

Townend J (1993) Effects of elevated carbon dioxide and drought on the growth and physiology of clonal Sitka spruce plants (*Picea sitchensis* (Bong.) Carr.). Tree Physiol 13:389–399

Valladares F, Gianoli E, Gomez JM (2007) Ecological limits to plant phenotypic plasticity. New Phytol 176:749–763

Valladares F, Matesanz S, Guilhaumon F, Araujo MG, Balaguer L, Benito-Garzon M, Cornwell W, …, Zavala MA (2014) The effects of phenotypic plasticity and local adaptation on forecasts of species range shifts under climate change. Ecol Lett 17:1351–1364

Volaire F (1995) Growth, carbohydrate reserves and drought survival strategies of contrasting *Dactylis glomerata* populations in a Mediterranean environment. J Appl Ecol 32:56–66

Wadgymar SM, Lowry DB, Gould BA, Byron CN, Mactavish RM, Anderson JT (2017) Identifying targets and agents of selection: innovative methods to evaluate the processes that contribute to local adaptation. Methods Ecol Evol 8:738–749

Wang X, Curtis PS, Pregitzer KS, Zak DR (2000) Genotypic variation in physiological and growth responses of *Populus tremuloides* to elevated atmospheric CO_2 concentration. Tree Physiol 20:1019–1028

Ward JK, Strain BR (1997) Effects of low and elevated CO_2 partial pressure on growth and reproduction of *Arabidopsis thaliana* from different elevations. Plant Cell Environ 20:254–260

Way DA, Oren R (2010) Differential responses to changes in growth temperature between trees from different functional groups and biomes: a review and synthesis of data. Tree Physiol 30:669–688

Way DA, Yamori W (2014) Thermal acclimation of photosynthesis: on the importance of adjusting our definitions and accounting for thermal acclimation of respiration. Photosynth Res 119:89–100

Weih M, Bonosi L, Ghelardini L, Ronnberg-Wastljung AC (2011) Optimizing nitrogen economy under drought: increased leaf nitrogen is an acclimation to water stress in willow (*Salix* spp.). Ann Bot 108:1347–1353

Weston DJ, Bauerle WL (2007) Inhibition and acclimation of C_3 photosynthesis to moderate heat: a perspective from thermally contrasting genotypes of *Acer rubrum* (red maple). Tree Physiol 27:1083–1092

Wulff RD, Alexander HM (1985) Intraspecific variation in the response to CO_2 enrichment in seeds and seedlings of *Plantago lanceolata* L. Oecologia 66:458–460

Wullschleger SD, Norby RJ, Gunderson CA (1992) Growth and maintenance respiration in leaves of *Liriodendron tulipifera* L. exposed to long-term carbon dioxide enrichment in the field. New Phytol 121:515–523

Zhu W, Ausin I, Seleznev A, Mendez-Vigo B, Pico FX, Sureshkumar S, Sundaramoorthi V, …, Balasubramanian S (2015) Natural variation identified ICARUS1, a universal gene required for cell proliferation and growth at high temperatures in *Arabidopsis thaliana*. PLoS Genet 11:e1005085

Chapter 7

Tree Physiology and Intraspecific Responses to Extreme Events: Insights from the Most Extreme Heat Year in U.S. History

Jacob M. Carter
Center for Science & Democracy, Union of Concerned Scientists, Washington, DC, USA

Department of Ecology and Evolutionary Biology, University of Kansas, Lawrence, KS, USA

Timothy E. Burnette
Organismal Biology, Ecology, & Evolution, Division of Biological Sciences, University of Montana, Missoula, MT, USA

Department of Ecology and Evolutionary Biology, University of Kansas, Lawrence, KS, USA

and

Joy K. Ward*
College of Arts and Sciences, Department of Biology, Case Western Reserve University, Cleveland, OH, USA

Department of Ecology and Evolutionary Biology, University of Kansas, Lawrence, KS, USA

*Author for correspondence, e-mail: jkw78@case.edu

K. M. Becklin et al. (eds.), *Photosynthesis, Respiration, and Climate Change*, Advances in Photosynthesis and Respiration 48, https://doi.org/10.1007/978-3-030-64926-5_7

Summary

Extreme weather events, including heat waves and droughts, are increasing as a result of climate change, and these events are occurring within the context of gradual increases in global mean temperatures over decadal to century timescales. Trees are especially vulnerable to extreme events and their physiological responses to these conditions may dictate their ability to survive and persist under future climate change. Recent studies indicate that extreme weather events can directly or indirectly weaken trees to the point of mortality, and when compounded over space and time, this has the potential to alter ecosystem functioning and the global carbon cycle more broadly. Furthermore, extreme events can potentially impact the evolutionary trajectory of trees, with rapid selection favoring a subset of survivors with altered physiologies, assuming genetic variation is sufficient to allow for survival in the first place. Unfortunately, the impacts of extreme years on the physiological functioning of trees at the intraspecific level are not well understood, hindering our ability to predict long-term consequences of extreme events on the survival and evolutionary trajectories of tree species.

Here we review our understanding of the physiological responses of trees to extreme events at the intraspecific level, and we provide a new study that measures intraspecific variation in physiology using carbon isotope ratios among 41 populations of white ash (*Fraxinus americana*). Experimental trees were growing in a common garden during the hottest year in U.S. history (2012) and physiological responses during this extreme year are compared with relatively non-extreme years. Such studies with a large number of populations are critical for determining if sufficient genetic variation is present for adaptation to extreme events and to understand if population rank order may shift under these conditions based on leaf-level physiological responses. Overall, we found significant effects of population and year, although there was a narrow range in magnitude of carbon isotopes across populations and years. Furthermore, we did not find a significant population by year interaction, indicating that physiological functioning of white ash trees varied across populations and across extreme and non-extreme years, but the *relative* response patterns and rank order of populations generally remained stable across years. This suggests that trees may not shift their physiological responses to extreme years as a result of changing evolutionary trajectories, although more work is needed to relate physiological responses in extreme years to fitness outcomes.

I. Introduction

A. *Extreme Events and Climate Change*

Over the next century, it has been predicted that climate change will occur 10X faster than it has at any time in at least the last 65 million years (Diffenbaugh and Field 2013). As a result, average global temperatures are predicted to increase at a rate of 0.25 °C per decade by 2020 in the northern hemisphere (Smith et al. 2015). In addition, high seasonal and annual temperature extremes are expected to become more frequent (Hansen et al. 2012; IPCC 2012). For the majority of the planet, global climate models predict that every 1-in-2 years will have mean summer temperatures above the late twentieth century maximum by 2046–2065 (Diffenbaugh and Field 2013). Furthermore, the past 30 years constitute the warmest conditions of the past 1700 years (Donat et al. 2013; Weaver et al. 2014; Chang et al. 2015).

Within the U.S., average temperatures have increased between 0.7 and 1 °C since U.S. record keeping began in 1895 (Walsh et al. 2014). In addition, the number of heat waves has increased from 2/year in the 1960s to well above 5/year in the 2010s within urban areas (GlobalChange.gov: https://www.globalchange.gov/browse/indicators/us-heat-waves). There is also increasing certainty regarding dry spells whereby the annual number of consecutive dry days is predicted to increase across much of the U.S., especially for southern and northwestern regions (Walsh et al. 2014; Wuebbles et al. 2014). In some years, severe droughts will coincide with extreme warm years and this will serve to amplify physiological stress and mortality of long-lived trees (Park Williams et al. 2013; Will et al. 2013). Thus, the presence of intraspecific variation for physiological responses to extreme conditions, both within and between tree populations, will be critical for adaptation that will promote the survival of tree species under future climate change scenarios.

B. *Physiological Responses of Trees to Extreme Events*

It is well known that weather extremes can affect tree physiology, and the effects of more prevalent, intense, and longer heat waves are concerning due to their potential to exceed trees' physiological thresholds for survival (Meehl and Tebaldi 2004; Perkins et al. 2012; Allen et al. 2015). As a result of heat waves, trees exhibit overall reductions in net photosynthesis and growth, with specific responses such as increased photooxidative stress, increased mitochondrial respiration, reduced activity of Rubisco activase and photosystem II, and damage to thylakoid membranes (Teskey et al. 2015). Isoprene levels represent a potentially important trait that helps trees tolerate heat and ozone stress (Sharkey and Singsaas 1995; Sharkey and Yeh 2001) by acting as a signaling molecule that alters gene expression (Zuo et al. 2019) and can act as antioxidants (Loreto and Velikova 2001) during heat waves and have been shown to exhibit substantial intraspecific variation in species such as *Populus tremuloides* (Darbah et al. 2010). Decreased net photosynthetic rates during heat waves may also result from increased photorespiration due to changes in Rubisco specificity and solubilities of CO_2 and O_2 (Teskey et al. 2015). Mitochondrial respiration may also increase due to an increased temperature optimum, reducing net carbon gain; for example, *Populus tremula* optimizes mitochondrial respiration at 55 °C (Hüve et al. 2012; Teskey et al. 2015) which is well above the 20–30 °C optimum for photosynthesis of most temperate trees (Teskey et al. 2015). In addition, Hozain et al. (2010) found that the percentage of Rubisco in the active state decreases by 30% during heat waves (40 °C) compared to individuals grown at lower temperature (27 °C) in both *Populus balsamifera* and *P. deltoides*. Furthermore, reduced photosystem II activity also limits photosynthesis in trees during heat waves and this damage may be irreversible if heat waves exceed 40 °C (Teskey et al. 2015). Some researchers have also found that stomatal closure increases at extreme high temperatures (e.g., Weston and Bauerle 2007), whereas others have shown no change (Cerasoli et al. 2014; von Caemmerer and Evans 2015). Interestingly, Urban et al. (2017a) showed that *P. deltoides* x *nigra* increased stomatal conductance during heat waves, effectively decoupling stomatal conductance from photosynthesis, which was declining or constant as a result of limited water availability. A similar decoupling was observed in *Eucalyptus parramattensis* (Drake et al. 2018). By opening stomata, *P. deltoides* x *nigra* leaves cool nearly 9 °C in well-watered conditions, but during drought, stomatal conductance was reduced, and cooling effects were limited to 1 °C (Urban et al. 2017b). This compensatory cooling mechanism that is limited under drought can drive trees to extreme stress when coupled with high ambient temperatures, compounding the negative impacts of extreme heat events that can lead to tree mortality (Allen et al. 2015).

C. *Intraspecific Variation in Response to Extreme Events*

Intraspecific variation for physiological and growth traits is pronounced across a range of species. Among 112 provenance trial studies involving 59 tree species and 19 traits (physiological, growth, phenological, and frost hardiness), significant genetic variation among populations, as well as clinal variation along environmental gradients, was observed in 90% and 78% of cases, respectively. This suggests that local adaptation may potentially allow for a wide range of responses to future climate extremes within individual species (Alberto et al. 2013). Such variation will be crucial to plant evolution if, as Bréda and Badeau (2008) suggest, climate extremes are an especially strong form of natural selection that rapidly eliminates less tolerant individuals. This may have wide-ranging effects, such as reducing taller white spruce trees that have been found to be more sensitive to heat (Bigras 2000), likely due to a tradeoff between heat tolerance and productivity (Teskey et al. 2015). There are also implications for trees with wide ranges such as *Pseudotsuga menziesii*, whereby Jansen et al. (2012) found that four of six populations at a low-elevation site exhibited large growth reductions, and all populations exhibited increased leaf $\delta^{13}C$ ratios from a major heat wave. Furthermore, other work demonstrated that *P. menziesii* (Junker et al. 2017) and *Fraxinus americana* (Marchin et al. 2008) populations differed in traits such as stomatal conductance, water use efficiency, and stable carbon isotope measurements when grown under common conditions. This was likely due to adaptations to contrasting water availabilities in their original locations, which would be predicted to influence responses to extreme heat events in the future.

Although high levels of variation in physiological traits among populations are needed to increase the potential for adaptation of tree species to more frequent, more intense, and longer extreme heat waves in the future (Aitken et al. 2008), major concerns have been raised about the relatively slow rate of evolution compared with the rapidity of climate change. Furthermore, it is still questionable in many species if there is sufficient genetic variation with respect to physiological traits that will allow adaptation and survival in response to extreme weather (Gutschick and BassiriRad 2003; Allen et al. 2015). In many cases, intraspecific variation in physiological responses may be a result of local adaptation. Fortunately, a plethora of common garden studies have provided information on population differentiation of trees for many physiological traits that are relevant to survival under extreme heat events (Alberto et al. 2013). A number of these studies have found tight correlations between physiological traits and a populations' climate of origin (Marchin et al. 2008; Arend et al. 2011; Carter et al. 2017). This is likely because tree populations are often (although not always) adapted to local environmental conditions, having the highest fitness near their home environments (Kawecki and Ebert 2004; Savolainen et al. 2007; Alberto et al. 2013). For example, Sánchez-Salguero et al. (2018) found that while wet-site populations of *Pinus pinaster* tolerated extreme drought periods best, populations from dry sites exhibited improved recovery likely due to xylem acclimation to stress (but see Hacke et al. 2001; Bréda and Badeau 2008). Further, *Acer rubrum* L. populations originating from hotter sites had higher photosynthetic rates, higher stomatal conductance, more open photosystem II reaction centers, and an increased photosystem II quantum yield (Weston and Bauerle 2007). Still more evidence of intraspecific variation has been found in morphological traits of *Abies balsamifera* (L.), where diameter at breast height, height, and crown width varied in response to temperature across populations from the north and south ends of the species range (Akalusi and Bourque 2018).

D. Carbon Isotope Ratios: A Tool to Assess Intraspecific Variation for Response to Extreme Events

The stable carbon isotope ratios ($\delta^{13}C$) of leaf tissue is an integrative measure of the ratio of intercellular to atmospheric CO_2 concentration (c_i/c_a) (Farquhar et al. 1982; Ehleringer and Cerling 1995). This ratio represents a balance between stomatal conductance and photosynthetic capacity over the development of a leaf, and is therefore more integrative than instantaneous gas exchange measurements (Farquhar et al. 1989) and highly useful for measuring population-level differences in response to the environment. In addition, $\delta^{13}C$ values are often correlated with intrinsic water-use efficiency (WUE) when vapor pressure deficit is held constant, since c_i/c_a reflects the trade-off between photosynthetic carbon gain to water loss at the leaf-level (Ehleringer and Dawson 1992; Ehleringer and Cerling 1995). A higher WUE has been shown to be favored among species growing in hotter, drier areas (Ehleringer 1993a; Ehleringer and Monson 1993; Heschel and Riginos 2005; Knight et al. 2006), and therefore this trait may be predictive of the capacity for populations to adapt to increases in extreme heat events (Nicotra and Davidson 2010). Furthermore, because leaf-level $\delta^{13}C$ is an integrative measure of physiological functioning, it tends to correlate with plant fitness across a range of environments (Casper et al. 2005).

Carbon isotope ratios have indicated in a number of cases that populations may show similar relative rankings for physiological responses under both stressful and nonstressful conditions. For example, Ehleringer (1993a) found that the relative ranking of leaf c_i/c_a from carbon isotope ratios among populations of *Encelia farinosa* shrubs was relatively uniform over time when temperature and precipitation were typical of long-term trends for west central Arizona. Voltas et al. (2008) also observed that the ranking of $\delta^{13}C$ among 25 provenances of *Pinus halepensis* was relatively uniform across two sites with contrasting precipitation and temperature. These observations suggest that leaf-level $\delta^{13}C$ responses may be useful in extrapolating leaf-level responses to extreme events (Ehleringer 1990, 1993a, b; Ehleringer and Cerling 1995). However, as environmental conditions become more extreme with respect to temperature and precipitation, some populations may no longer be physiologically adapted to their home environment and may not maintain their rank order for physiology, likely with fitness consequences (see Fig. 7.1 of Anderson et al. 2012). Furthermore, it is unclear if population (or genotype) rankings will be maintained under extreme years more broadly since controlled intraspecific measurements under these conditions are relatively rare (but see Ehleringer 1990, 1993a, b; Sandquist and Ehleringer 2003).

II. Common Garden Study: Population-Level Variation of White Ash During the Most Extreme Year in U.S. History

In the remainder of this chapter, we report on a common garden study to investigate the physiological responses (via carbon isotopes) of white ash (*Fraxinus Americana*) across 41 populations of mature trees growing in a common garden at the University of Kansas Field Station (Fig. 7.1). This study advances our limited understanding of levels of intraspecific variation of trees in response to extreme heat years. Measurements for this study were conducted across 8 years, including the most extreme heat year in U.S. recorded history (2012), which was also the most extreme year at the common garden (Figs. 7.2a and 7.3a); this year was also drier than average compared with long-term conditions at this field site (Figs. 7.2b and 7.3b). We addressed the following questions: (1) What is the magnitude of change in carbon

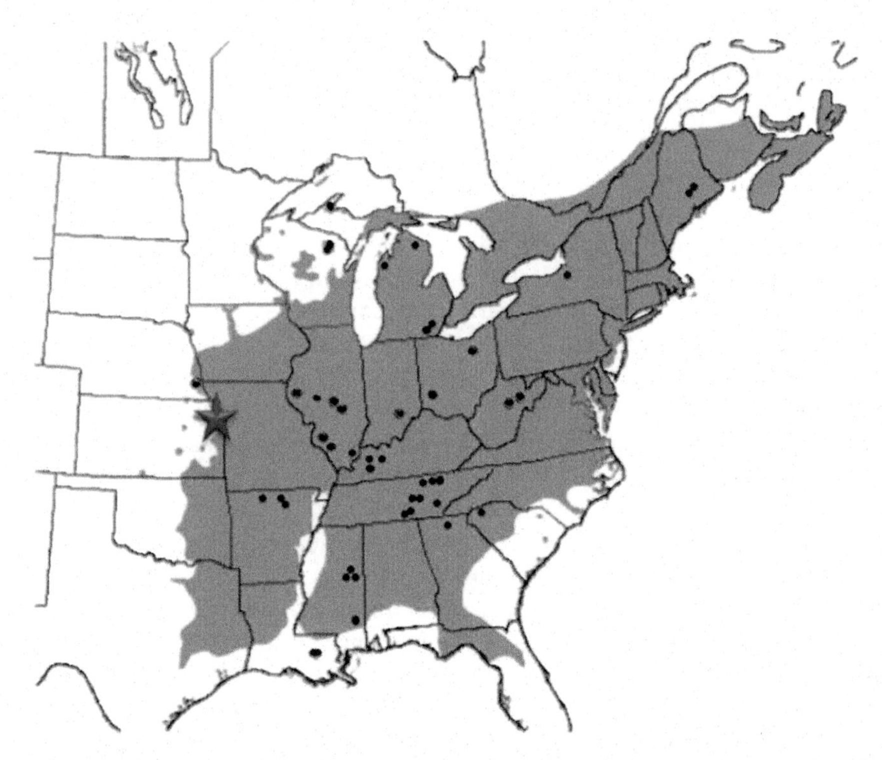

Fig. 7.1. The range of *Fraxinus americana* (white ash) is shown in the shaded grey area. Locations from which white ash populations originated are shown by black circles. Trees from each of these populations were grown in the University of Kansas Field Station common garden, Lawrence, KS (star). Trees from all populations grew side-by-side in this common garden for 42 years to date. (Adapted from Marchin et al. 2008 and USDA Forest Service (www.na.fs.fed.us))

isotope ratios (that are indicative of c_i/c_a) across extreme and non-extreme years in a large number of white ash populations from throughout the species range when all were grown under common garden conditions, and (2) is there evidence for population-level variation in physiological response patterns across extreme heat years and non-extreme years in the common garden?

A. Materials and Methods

1. Study Species: *Fraxinus americana* (White Ash)

White ash (*Fraxinus americana*) is one of the most widely distributed (Fig. 7.1) and commercially important ash species in the U.S. (MacFarlane and Meyer 2005). Across its range, populations experience average annual temperatures ranging from ~4 to 19 °C and average total annual precipitation ranging from ~750 to 1600 mm (see Table 1 of Marchin et al. 2008). Current climate change scenarios indicate that the range of white ash must advance northward in order for this species to maintain its current level of survival and representation within forest ecosystems (Iverson et al. 2008). In addition, mean total annual precipitation is expected to decrease over much of the species range and severe droughts are expected to become more common over this same area during the next century (IPCC 2013). Thus, the ability of white ash to persist under future climate change will not only depend on its potential to respond to mean shifts in climate, but also to interannual variation in temperature and

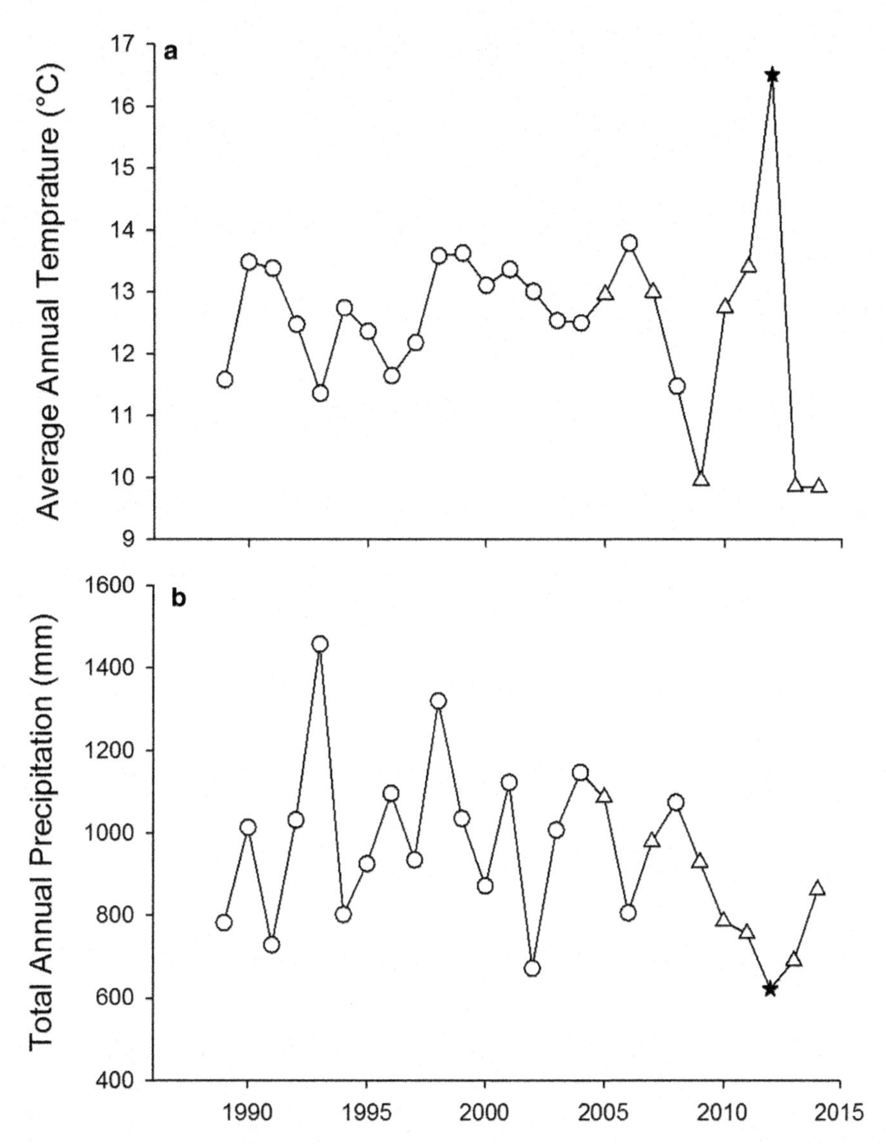

Fig. 7.2. **(a)** Average annual temperature (°C), and (**b**) total annual precipitation (mm) for all recorded years at the white ash common garden site in Lawrence, KS. Triangles represent years in which physiological measurements were made, and the black stars represents the extreme year of 2012. (Data were obtained from the University of Kansas Field Station and Kansas Biological Survey)

precipitation, including extreme events. If this were not concern enough, many populations of ash trees (*Fraxinus spp.)* are rapidly declining or have been eliminated in the U.S. due to the introduced emerald ash borer beetle (EAB) that is most effective at killing trees that are already undergoing physiological stress (MacFarlane and Meyer 2005; Poland and McCullough 2006; McCullough et al. 2009). The loss of ash trees is expected to have substantial economic and ecological impacts within North American eastern forests (Poland and McCullough 2006; Kovacs et al. 2010), and this is a substantial threat to the 7.5 billion ash trees that grow in the U.S. with an estimated value of $282.3 billion (Federal Register 2003; Poland and McCullough 2006).

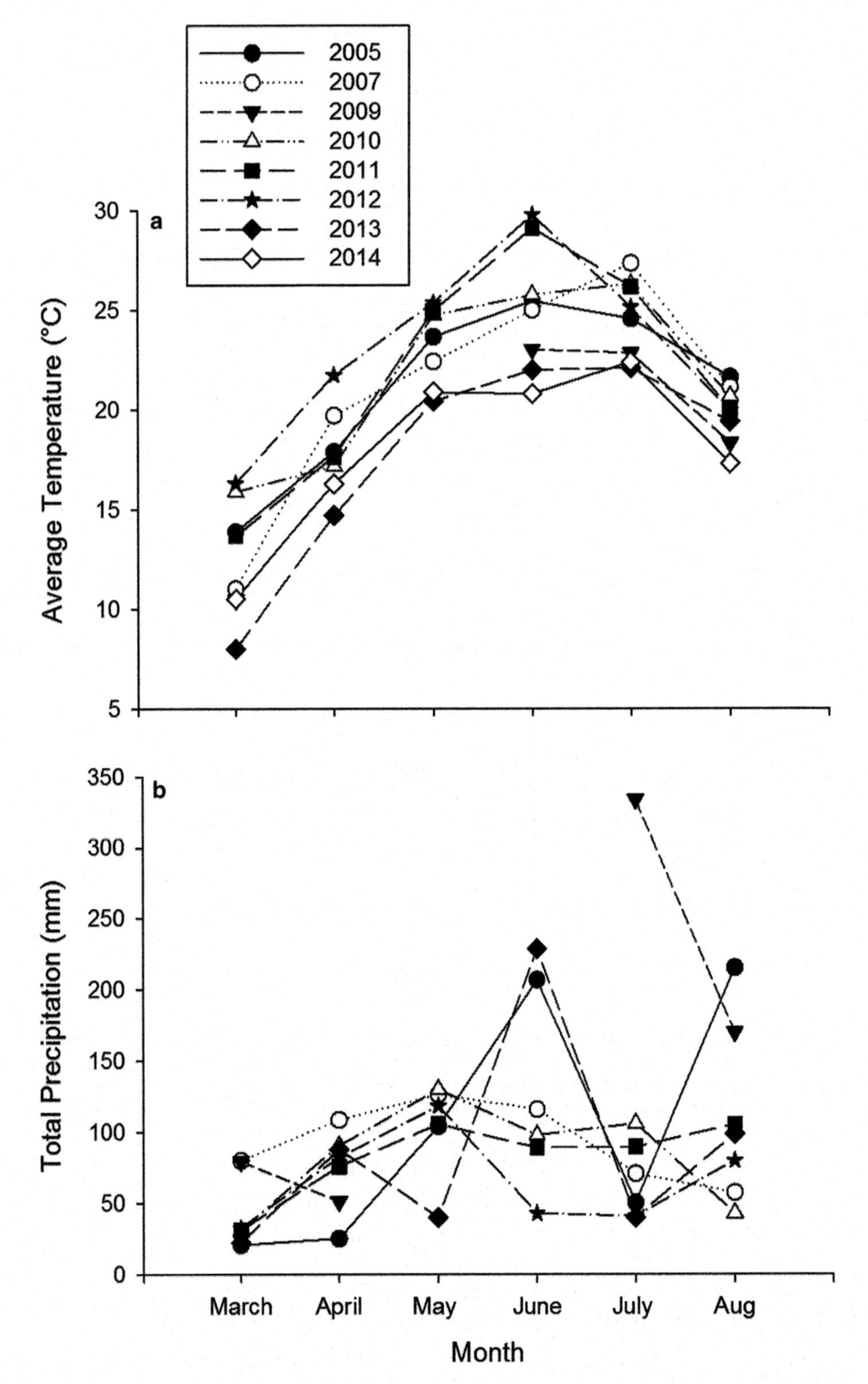

Fig. 7.3. **(a)** Average monthly temperature (°C), and **(b)** total precipitation (mm) across the growing season (March-Aug.) at the white ash common garden site in Lawrence, KS for experimental years (2005, 2007, 2009, 2010, 2011, 2012, 2013, and 2014)

2. *Common Garden and Experimental Site Conditions*

White ash seeds originating from 43 populations from throughout the species range were collected by the U.S. Forest Service from open-pollinated native parent trees. Seeds were planted in 1976 at the University of Kansas Field Station near Lawrence, KS (35.0°N, 95.1°W, 299 m a.s.l.; Fig. 7.1). All trees measured in this study were growing in this common garden, which occurs on the western edge of the species range (Fig. 7.1; Marchin et al. 2008). Trees were highly variable in size, with average diameters at breast height by population ranging from 0.45 cm to 22.86 cm and an overall average diameter of 8.94 cm (see also Marchin et al. 2008).

The field station receives an average annual precipitation of ~900 mm per year with more than 70% of this falling in the growing season from March–August. Temperature fluctuations are relatively large for both diurnal and annual cycles with monthly average temperatures ranging from below −7 °C in January to slightly above 32 °C in July. The extreme year in this study (2012) was the warmest year for the U.S. since record keeping began in 1895. At the common garden, the average annual temperature during the extreme warm year of 2012 was 16.5 °C, which was 4.8 °C warmer than the average of non-extreme years in this study (2005, 2007, 2009, 2010, 2011, 2013, and 2014; Fig. 7.2a). Total annual precipitation was also lower during the extreme year of 2012 (622 mm) relative to the average total annual precipitation during non-extreme years (869 mm; Fig. 7.2b). With respect to growing season conditions (March–August), precipitation was also lower during 2012 (312 mm) relative to the average growing season precipitation during non-extreme years (592 mm; Fig. 7.3b). Growing season vapor pressure deficit (VPD) was calculated using the saturated vapor pressure at the mean daily temperature subtracted from the actual vapor pressure (Murray 1967) and varied from 0.452 to 1.028 kPa across all years of the study.

3. *Leaf Stable Carbon Isotope Measurements*

We measured stable carbon isotope ratios on leaf tissue from 41 living populations of white ash trees that were growing in the common garden in Lawrence, KS (Fig. 7.1, star; two populations were eliminated due to total mortality since the time of planting) within an interior block that lacked edge effects. Isotope measurements were performed at the Keck Paleoenvironmental and Environmental Stable Isotope Laboratory (KPESIL) at the University of Kansas for years 2005, 2007, 2009, 2010, 2011, 2012, and 2013, and measurements for 2014 were performed at the Central Appalachian Stable Isotope Facility (CASIF) at the University of Maryland Center for Environmental Science (UMCES) Appalachian Laboratory located in Frostburg, Maryland (with standards to allow for precision between labs). $\delta^{13}C$ were calculated using the following formula:

$$\delta = (R_{sample} / R_{standard}) - 1 \quad (1)$$

where R is the ratio of $^{13}C/^{12}C$, using ammonium sulfate, caffeine, and ANU sucrose as primary standards, as well as atropine and dogfish muscle as secondary standards. Data were converted to "per mil" (‰) notation by multiplying δ values by 1000. Source air $\delta^{13}C$ for the trees was similar in the outdoor, open-air common garden. Leaves used in the analysis were collected during August (2005, 2007, 2009, 2010, 2011, 2012, 2013, and 2014) from fully mature sunlit leaves on the south side of trees. Two leaves were collected from each tree and dried at 70 °C for 48 h, and then homogenized together in liquid nitrogen to a fine powder. Within-population sample sizes varied from n = 1 to n = 11 trees for each year depending on long-term survivorship (see Fig. 7.4). Tissues were then weighed (3–3.5 mg) and loaded into tin capsules (3.5 × 5.5 mm). A ThermoFinnigan MAT 253 Dual Inlet System was used for $\delta^{13}C$ analysis at the University of Kansas, whereas a ThermoFisher Delta V+ was used for $\delta^{13}C$ analysis at CASIF.

4. *Statistical Analysis*

To assess the effects of both population and year on leaf $\delta^{13}C$, we used a two-way ANOVA with year and population as main effects, along with the interaction term. To test the internal reliability of rank among populations across years, a Cronbach's alpha internal reliability measure was calculated. We used linear regression to test for the relationship between VPD and average leaf $\delta^{13}C$ across population means within a year. All statistical analyses were carried out in SAS 9.2 (Cary, North Carolina, USA).

B. *Results and Discussion*

1. *Overall Responses*

Overall, we found that year ($p < 0.0001$, from ANOVA) and population ($p < 0.0001$) had significant effects on leaf $\delta^{13}C$, although the interaction between the two was not significant ($p = 0.56$). These results suggest that while leaf $\delta^{13}C$ of white ash differs among populations and across years, *relative changes in physiology across extreme and non-extreme years is similar among populations* (Fig. 7.4). To further support this finding, we found that the rank order of populations with respect to leaf $\delta^{13}C$ remained stable across extreme and non-extreme years (Cronbach's alpha >0.8), indicating a similar relative line-up in physiological responses of white ash populations across years (Fig. 7.4). Thus, we can expect some level of predictability of relative physiological functioning of white ash populations across extreme and non-extreme years in the future.

2. *Leaf-Level $\delta^{13}C$ Response During the Extreme Year of 2012 Relative to Non-extreme Years*

Leaf-level $\delta^{13}C$ values of white ash populations significantly varied due to the effect of year ($p < 0.0001$; Figs. 7.4 and 7.5). Across all populations, the 2012 extreme year averaged −28.7‰, whereas non-extreme years averaged −29.6‰. Furthermore, $\delta^{13}C$ values of the extreme year of 2012 differed significantly from many (although not all) of the non-extreme years including 2009, 2010, 2011, and 2014 (Fig. 7.5c, d, e, g). Average conditions across these non-extreme years were wetter (833 mm precipitation) and cooler (11.5 °C) than 2012 (Fig. 7.2). Additionally, VPD was much lower across these years (average of 0.581 kPa) compared with 2012 (1.028 kPa; Fig. 7.6). Relative to these non-extreme years, 2012 was characterized by decreased c_i/c_a as indicated by higher (less negative) $\delta^{13}C$ values, likely due to reduced stomatal conductance that would have reduced water loss under extreme dry, warm conditions.

On the other hand, leaf $\delta^{13}C$ during extreme year 2012 (−28.68 ± 0.074‰) did not vary significantly from non-extreme years 2005 (−28.64 ± 0.12‰) and 2007 (−28.74 ± 0.092‰) (Fig. 7.5a, b), possibly due to relatively higher VPD during these non-extreme years compared to the others (average of 0.754 kPa across 2005 and 2007). Higher VPD in 2005 and 2007 led to lower stomatal conductance, producing similar c_i/c_a as recorded for the extreme year of 2012. Importantly, leaf $\delta^{13}C$ during the extreme year of 2012 may not have reflected the full extent of hot and dry conditions during that growing season, and this may have produced lower (more negative) leaf $\delta^{13}C$ values than expected based on the extreme conditions of that year. More specifically, we measured instantaneous leaf-level gas exchange rates on a subset of 7 populations (n = 2 leaves for each of 21 trees across 7 populations) during

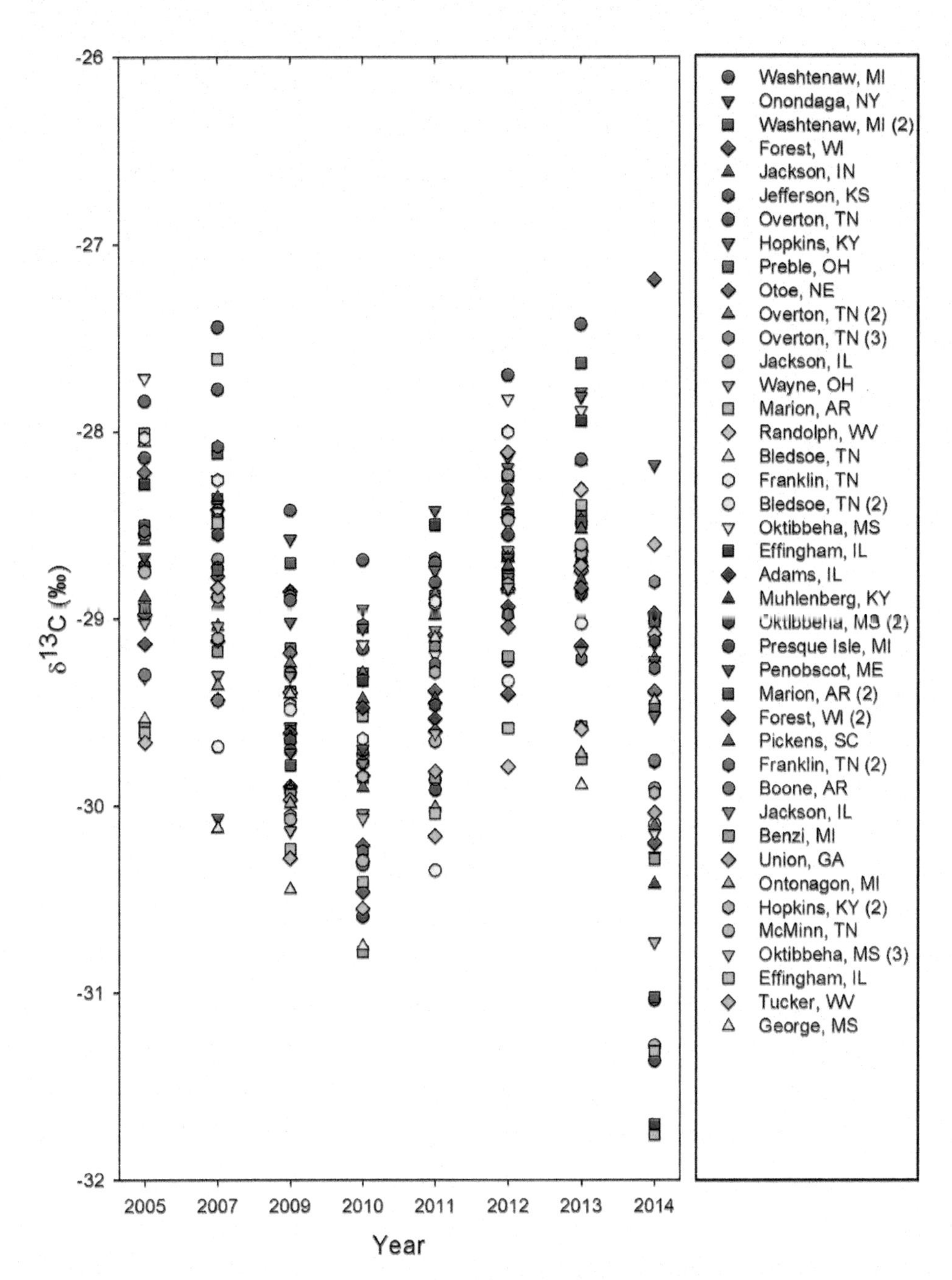

Fig. 7.4. Average leaf-level $\delta^{13}C$ values by year for 41 white ash populations growing in a common garden at the University of Kansas Field Station during non-extreme years (2005, 2007, 2009, 2010, 2011, 2013, 2014) and the extreme year of 2012, which was the hottest recorded year in U.S. history. Populations with more depleted leaf-level $\delta^{13}C$ values in 2009 (randomly chosen) are coded by cooler colors (blues and cyans) relative to populations with more enriched leaf-level $\delta^{13}C$ values during 2009, which are coded in warmer colors (reds and yellows). Numbers in parentheses indicate the number of families that were sourced from a population if greater than 1. Average sample sizes within populations averaged across years ranged from 1–11 depending on long-term survivorship (average sample size by population: $n_{Otoe, NE} = 5$, $n_{Jefferson, KS} = 9$, $n_{Boone, AR} = 5$, $n_{Marion, AR} = 4$, $n_{Marion,AR (2)} = 3$, $n_{Adams, IL} = 4$, $n_{Ontonagon, MI} = 2$, $n_{Jackson, IL} = 5$, $n_{Jackson, IL (2)} = 5$, $n_{Forest, WI} = 3$, $n_{Forest, WI (2)} = 2$, $n_{George, MS} = 1$, $n_{Oktibbeha, MS} = 3$, $n_{Oktibbeha, MS (2)} = 3$, $n_{Oktibbeha, MS (3)} = 4$, $n_{Effingham, IL} = 5$, $n_{Effingham, IL (2)} = 4$, $n_{Hopkins, KY} = 2$, $n_{Hopkins, KY (2)} = 5$, $n_{Muhlberg, KY} = 4$, $n_{Jackson, IN} = 3$, $n_{Benzi, MI} = 4$, $n_{Franklin, TN} = 5$, $n_{Franklin, TN (2)} = 4$, $n_{Overton, TN} = 5$, $n_{Overton, TN (2)} = 4$, $n_{Overton, TN (3)} = 3$, $n_{Bledsoe, TN} = 5$, $n_{Bledsoe, TN (2)} = 3$, $n_{Preble, OH} = 5$, $n_{McMinn, TN} = 5$, $n_{Union, GA} = 4$, $n_{Washtenaw, MI} = 5$, $n_{Washtenaw, MI (2)} = 3$, $n_{Presque Isle, MI} = 3$, $n_{Pickens, SC} = 5$, $n_{Wayne, OH} = 4$, $n_{Randolph, WV} = 4$, $n_{Tucker, WV} = 4$, $n_{Onondaga, NY} = 3$, $n_{Penobscot, ME} = 1$)

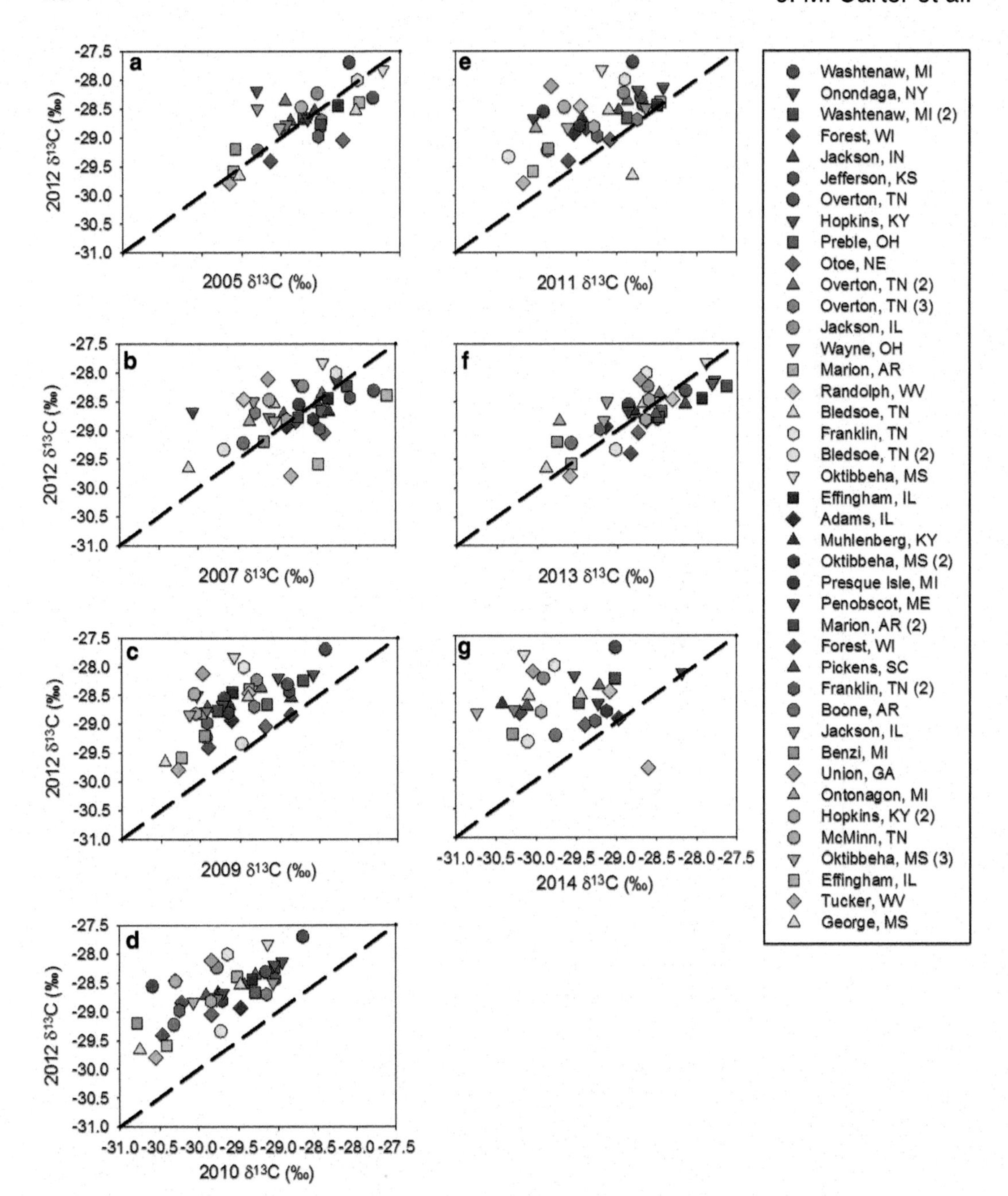

Fig. 7.5. Average leaf-level $\delta^{13}C$ values for each population during the extreme year of 2012 relative to each non-extreme year (a-g). The extreme year 2012 had significantly higher $\delta^{13}C$ values than 2009, 2010, 2011, and 2014 (note deviation off of the one-to-one line), where there were no significant differences in 2005, 2007, and 2013 (see text for details). Numbers in parentheses indicate the number of families that were sourced from a population if greater than 1. Sample sizes and population symbols/colors are the same as in Fig. 7.4

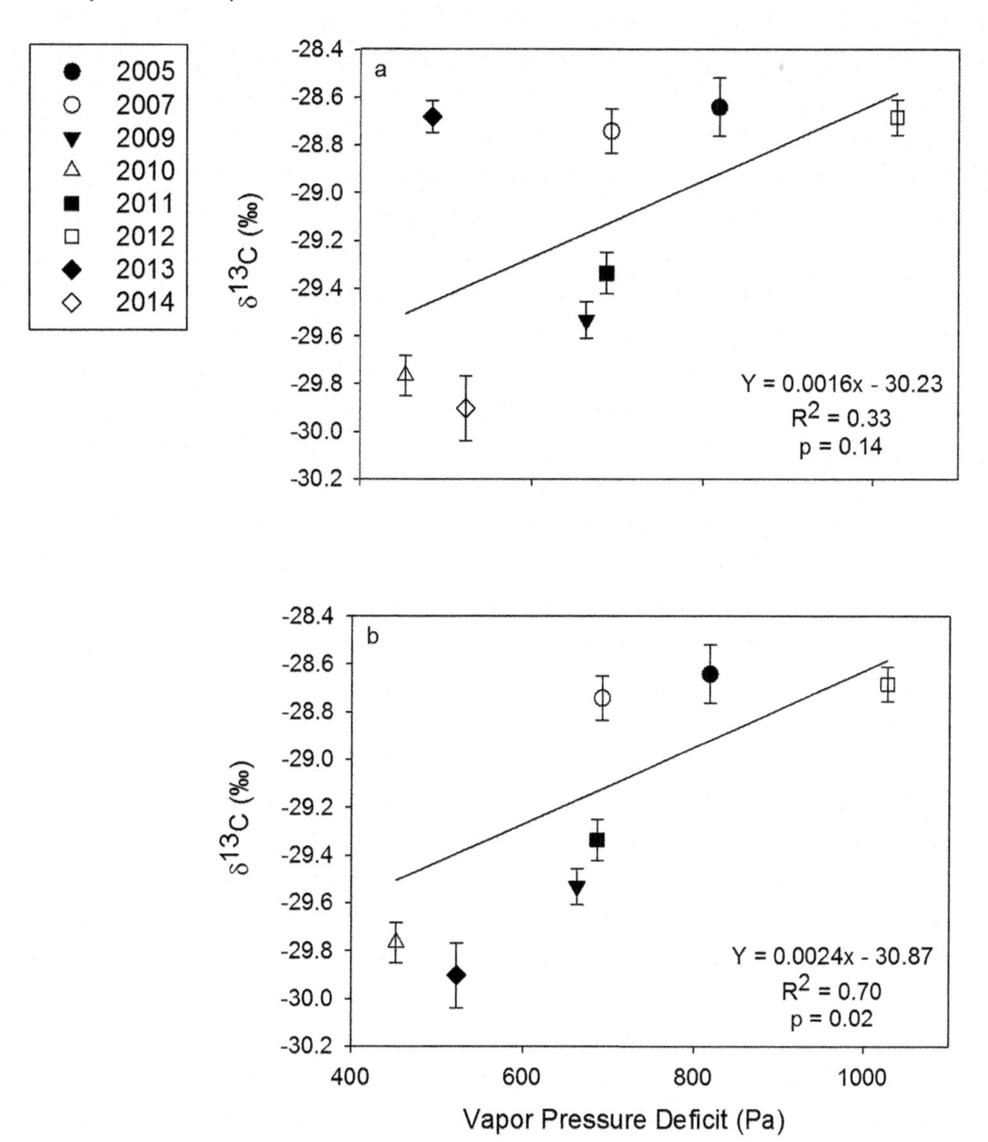

Fig. 7.6. **(a)** The relationship between average leaf-level $\delta^{13}C$ values for 41 populations of white ash trees and growing season (March–August) vapor pressure deficit (VPD) across 8 years of study (2005, 2007, 2009, 2010, 2011, 2012, 2013, and 2014), and **(b)** the same linear relationship between average leaf-level $\delta^{13}C$ values and VPD when the 2013 outlier is excluded. The outlier was excluded in **(b)** to make the point that potential lag effects from the extreme year (2012) may decouple the relationship between $\delta^{13}C$ values and more immediate environmental conditions (see text)

May and June of 2012 (data not shown, Li-cor 6400, Lincoln, NE). Leaf-level gas exchange rates were similar among populations during May and June of 2012 with an average photosynthetic rate of 7.03μmol CO_2 m^{-2} s^{-1} and average stomatal conductance of 0.14 mol H_2O m^{-2} s^{-1}. Interestingly, however, leaf-level gas exchange ceased during July 2012 and this was quickly followed by leaf abscission and leaf fall, which was very premature for that time of year. As a result, the full effects of high temperatures and decreased precipitation during 2012 may not have been reflected by leaf-level $\delta^{13}C$ since photosynthate production greatly declined in response to extreme conditions.

This is reflective of the fact that $\delta^{13}C$ is an integration of physiological response that does not necessarily occur continuously, especially under water and heat stress, and therefore c_i/c_a responses are proportional to the timing of when carbon is actually fixed. Along this line, Zavalloni et al. (2009) found that a lack of larger differences in leaf-level $\delta^{13}C$ between simulated wet and extreme drought treatments were likely due to significant decreases in photosynthesis and growth rates under extreme drought conditions across prairie plant species. The authors suggested that these decreased rates limited the amount of carbon incorporated into leaf tissues under extreme drought such that the isotopic signatures were not fully reflective of the extreme drought conditions. Similarly, reductions in photosynthesis and stomatal conductance were recorded under simulated extreme drought and warm temperature conditions for *Pinus edulis* and *Juniperus monosperma* (Grossiord et al. 2017), although this study did not include $\delta^{13}C$ measurements.

Measurements of $\delta^{13}C$ during the non-extreme year of 2013 (−28.68 ± 0.093‰) were also not significantly different from the extreme year of 2012 (Fig. 7.5f), but likely for a different reason than 2005 and 2007 (described above). This result was initially surprising since precipitation across the growing season was much higher in 2013 (477.5 mm) than in 2012 (312.4 mm), and average monthly temperatures were much cooler in 2013 (17.7 °C) than in 2012 (23 °C; Fig. 7.3). As a result, VPD was also much lower during the non-extreme year of 2013 (0.484 kPa) relative to 2012 (1.028 kPa; Fig. 7.6a). However, it has been proposed that carry-over effects from a previous extreme event may result in hysteresis, whereby lag effects can potentially decouple physiological responses from the immediate environment and reduce carbon assimilation and plant productivity relative to expectations. Although this is generally viewed as an effect that operates at daily to weekly times scales (Gutschick and BassiriRad 2003), the potential is there for hysteresis to be more long-term, especially after a full growing season of extreme conditions as in our study where photosynthesis ceased for several months. The linear relationship between $\delta^{13}C$ and growing season VPD exhibited a marginally significant correlation ($p = 0.1$) when all years were included (Fig. 7.6a); however, when 2013 data are excluded, there is a highly significant relationship between $\delta^{13}C$ and VPD that explains much of the variation in average $\delta^{13}C$ across years ($R^2 = 0.70$; $p < 0.001$; Fig. 7.6b, showing removal of the 2013 outlier). Thus, 2013 is a strong outlier in this relationship, suggesting a carry-over effect from 2012 that may not fully reflect the immediate environmental conditions of 2013. There are several possible mechanisms for such a carry-over effect. One possibility is that water resources were so severely depleted during the extreme year that the increase in growing season precipitation during 2013 did not recharge water resources at the common garden. This would have limited water availability in 2013 and subsequently decreased stomatal conductance, resulting in enriched $\delta^{13}C$ values. It is also possible that there were contributions of older carbon from 2012 that were relatively more positive in $\delta^{13}C$ than in newly developing leaves in 2013 that contributed to this effect. This may have been particularly pronounced in 2013 since the majority of stored photosynthate from 2012 may have been consumed to accommodate the onset of the 2013 growing season, as a result of premature leaf drop in the prior growing season. Generally, a majority of leaf structural tissues in fully grown and mature leaves are composed of carbon assimilated during the current season (Bruggemann et al. 2011), and this would be reflected in $\delta^{13}C$. However, the contribution of carbon to leaves in the following year is variable in deciduous species, but on average is around 30% (Keel et al. 2006) and may have had large effects on 2013 values in our case. Carry-over effects of extreme years on trees have also been commonly shown in other species with respect to xylem functioning. For example, a severe drought that occurred from 2000–2003 across western North America led to widespread die-off of *Populus tremuloides* that commenced in 2004–2005 and continued until 2011 (Anderegg et al. 2013). This carry-over effect of mortality was attributed to the result of damage to xylem vessels leading to increases in xylem fatigue, and thus less resistance to xylem vessel cavitation in subsequent years after the drought (Anderegg et al. 2013). Although not measured in this study, potential effects on xylem from the extreme year may have affected physiological responses into the following year of 2013, particularly since these conditions were so extreme. Furthermore, in a prior study on phenology with the same trees as in the current study, the timing of leaf emergence also showed evidence that the effects of extreme year 2012 persisted into 2013 for determining the timing of leaf bud break at this common garden (Carter et al. 2017). Taken together, potential implications for water depletion, carry-over carbon effects, phenological factors, and effects on xylem functioning may have potentially driven higher $\delta^{13}C$ values in 2013 than would be predicted from immediate growing season conditions.

3. *Population-Level Variation in $\delta^{13}C$*

We found that leaf $\delta^{13}C$ ranged from −27‰ to −32‰ across all years, including the extreme year (2012), and across all 41 populations represented in the common garden. In addition, $\delta^{13}C$ significantly varied by population ($p < 0.0001$ from ANOVA; Fig. 7.4). Within years, leaf $\delta^{13}C$ varied by an average of ~2‰ across populations. Despite significant differences among populations, this physiological range is rather constrained within years, when considering the high number of white ash populations included. Because the populations used in this study originated from the majority of the species range (Fig. 7.1), these $\delta^{13}C$ ranges are likely representative of the majority of physiological variation within white ash as a whole and under the common garden conditions. Thus, the potential for white ash trees to physiologically respond across future extreme and non-extreme heat years may be quite limited based on the range of physiological responses observed in this study.

Population-level variation was likely driven by adaptations to climate conditions at the sites from which the 41 populations originated. In a previous study conducted with a subset of these same trees and in the same common garden, Marchin et al. (2008) found a significant relationship between leaf $\delta^{13}C$ and the amount of precipitation at the site from which trees originated across an east-to-west gradient (wetter to drier, respectively). More specifically, trees from more eastern locations exhibited relatively lower (more negative) leaf $\delta^{13}C$ than trees originating from more western sites when all were assessed in the common garden (Marchin et al. 2008). The higher $\delta^{13}C$ of populations sourced from western locations is likely indicative of a more conservative water use strategy, which would be advantageous at the white ash common garden site since it is warmer and drier than the majority of the species range (Fig. 7.1). However, despite this response across a precipitation gradient based on site of origin, there was still a relatively narrow range in leaf-level $\delta^{13}C$ across populations (approximately 2‰), suggesting evidence for only modest levels of leaf physiological adjustment to extreme conditions (Marchin et al. 2008).

4. *Population Rank Order Across Years*

Importantly, we did not find a significant population by year interaction ($p > 0.05$, based on ANOVA) for leaf $\delta^{13}C$ among the 41 white ash populations across the extreme and non-extreme years. We also found that relative rank order was generally maintained across extreme and non-extreme years for leaf $\delta^{13}C$ (Fig. 7.4; Cronbach's alpha >0.8). Taken together, these results indicate that populations exhibited similar response patterns across years, and the potential for a subset of populations to make major physiological adjustments under extreme conditions seems unlikely. Importantly, this also suggests that the *relative* responses of leaf $\delta^{13}C$ among populations may be highly predictable, even during extreme years. Such information is particularly valuable for maintaining and/or restoring populations of white ash trees via assisted migration and re-plantings, since trees that have more enriched $\delta^{13}C$ (less negative) would be predicted to have better outcomes in drier and warmer environments relative to trees with more depleted $\delta^{13}C$ values (Ehleringer 1990; Comstock and Ehleringer 1992; Ehleringer 1993a, b). Similarly, Flanagan and Johnsen (1995) studied mature *Picea mariana* from four full-sib families growing across three sites with different levels of soil moisture. They found significant differences in carbon discrimination across sites but did not find significant family x site interactions. Also, Voltas et al. (2008) found that the rank order of $\delta^{13}C$ was maintained across 56 populations of *Pinus halepensis* planted at two common garden sites with contrasting temperature and precipitation. Lastly, in a classic study, Ehleringer (1993a) found that the ranking of carbon discrimination was maintained in *Encelia farinose* at the population level over time (measured in early March under typical conditions in the Sonoran Desert and then again on newly produced leaves in late April). While carbon discrimination did change over the sampling period, relative rankings remained uniform over time at the population level (Ehleringer 1993b).

III. Conclusions and Future Directions

To conclude, forests ecosystems are ecologically and economically important, but will be especially vulnerable to decline or extinction under future climate change, especially as a result of extreme heat and drought years (Allen et al. 2010; Anderegg et al. 2012, 2015). The impact of extreme years on the physiological functioning of trees is not well understood, especially at the intraspecific level. Because extreme years are expected to become much more common over the next century, it is critical that we understand how physiological traits will respond to large extremes in climate. This knowledge will be even more impactful at the intraspecific level, as the rate of climate change is likely to outpace the migration capacity of tree species, and this variation may be crucial for determining future tree survival. It will be important, therefore, to quantify both inter- and intra- levels of population variation, and to determine linkages between physiological responses and fitness levels (survival and reproductive output). This can be most effective when using highly integrative physiological measures such as carbon isotope ratios that allow population differences to emerge. It will also be important to determine the impacts of extreme years on physiological responses over the long-term, as subsequent extreme years with shorter durations for recovery may stress trees to their physiological limits. Lastly, intraspecific variation will be key for maintaining the presence of many tree species, and there is a great need to understand cases where intraspecific variation is limited within a species, because such species may be particularly vulnerable to widespread mortality and extinction in response to climate change and accompanying extreme events as was evidenced in the present study with white ash. Future work should also identify the most important physiological traits for survival in response to extreme events with identification of populations and individuals that harbor such traits.

Acknowledgements

This research was supported by grants from the National Science Foundation (NSF) through the Division of Integrative Organismal Systems to JKW (award numbers 1644618 and 1457236) and through funding from the University of Kansas to JMC and JKW from the Department of Ecology and Evolutionary Biology and the College of Liberal Arts and Sciences. JMC was also supported by funding from the NSF C-CHANGE IGERT grant to the University of Kansas through PI Distinguished Professor Joane Nagel.

References

Aitken SN, Yeaman S, Holliday JA, Wang TL, Curtis-McLane S (2008) Adaptation, migration or extirpation: climate change outcomes for tree populations. Evol Appl 1:95–111

Akalusi ME, Bourque CP (2018) Effect of climatic variation on the morphological characteristics of 37-year-old balsam fir provenances planted in a common garden in New Brunswick, Canada. Ecol Evol 8:3208–3218

Alberto FJ, Aitken SN, Alia R, Gonzalez-Martinez SC, Hanninen H, Kremer A, Lefevre F, … O Savolainen (2013) Potential for evolutionary responses to climate change evidence from tree populations. Glob Chang Biol 19: 1645–1661

Allen CD, Macalady AK, Chenchouni H, Bachelet D, McDowell N, Vennetier M, Kitzberger T … Cobb N (2010) A global overview of drought and heat-induced tree mortality reveals emerging climate change risks for forests. For Ecol Manag 259: 660–684

Allen CD, Breshears DD, McDowell NG (2015) On underestimation of global vulnerability to tree mortality and forest die-off from hotter drought in the Anthropocene. Ecosphere 6:1–55

Anderegg WRL, Anderegg LDL, Sherman C, Karp DS (2012) Effects of widespread drought-induced aspen mortality on understory plants. Conserv Biol 26:1082–1090

Anderegg WRL, Plavcova L, Anderegg LDL, Hacke UG, Berry JA, Field CB (2013) Drought's legacy: multiyear hydraulic deterioration underlies widespread aspen forest die-off and portends increased future risk. Glob Chang Biol 19:1188–1196

Anderegg WRL, Schwalm C, Biondi F, Camarero JJ, Koch G, Litvak M, Ogle K … Pacala S (2015) Pervasive drought legacies in forest ecosystems and their implications for carbon cycle models. Science 349: 528–532

Anderson JT, Panetta AM, Mitchell-Olds T (2012) Evolutionary and ecological responses to anthropogenic climate change. Plant Physiol 160:1728–1740

Arend M, Kuster T, Gunthardt-Goerg MS, Dobbertin M (2011) Provenance-specific growth responses to drought and air warming in three European oak species (*Quercus robur*, *Q. petraea* and *Q. pubescens*). Tree Physiol 31:287–297

Bigras FJ (2000) Selection of white spruce families in the context of climate change: heat tolerance. Tree Physiol 20:1227–1234

Bréda N, Badeau V (2008) Forest tree responses to extreme drought and some biotic events: towards a selection according to hazard tolerance? Compt Rendus Geosci 340:651–662

Bruggemann N, Gessler A, Kayler Z, Keel SG, Badeck F, Barthel M, Boeckx P … Bahn M (2011) Carbon allocation and carbon isotope fluxes in the plant-soil-atmosphere continuum: a review. Biogeosciences 8: 3457–3489

Carter JM, Orive ME, Gerhart LM, Stern JH, Marchin RM, Nagel J, Ward JK (2017) Warmest extreme year in U.S. history alters thermal requirements for tree phenology. Oecologia 183:1197–1210

Casper BB, Forseth IN, Wait DA (2005) Variation in carbon isotope discrimination in relation to plant performance in a natural population of *Cryptantha flava*. Oecologia 145:541–548

Cerasoli S, Wertin T, McGuire MA, Rodrigues A, Aubrey DP, Pereira JS, Teskey RO (2014) Poplar saplings exposed to recurring temperature shifts of different amplitude exhibit differences in leaf gas exchange and growth despite equal mean temperature. AoB Plants 6:1–9

Chang H, Castro CL, Carillo CM (2015) The more extreme nature of U.S. warm season climate in the observational record and two "well-performing" dynamically downscaled CMIP3 models. J Geophys Res 120:8244–8263

Comstock JP, Ehleringer JR (1992) Correlating genetic-variation in carbon isotopic discrimination with complex climatic gradients. Proc Natl Acad Sci U S A 89:7747–7751

Diffenbaugh NS, Field CB (2013) Changes in ecologically critical terrestrial climate conditions. Science 341:486–492

Donat MG, Alexander LV, Yang H, Durre I, Vose R, Dunn RJH, Willett KM … Kitching S (2013) Updated analyses of temperature and precipitation extreme indices since the beginning of the twentieth century: the HadEX2 dataset. J Geophys Res Atmos 118: 2098–2118

Drake JE, Tjoelker MG, Varhammar A, Medlyn BE, Reich PB, Leigh A, Pfautsch S … Barton CVM (2018) Trees tolerate an extreme heatwave via sustained transpirational cooling and increased leaf thermal tolerance. Glob Chang Biol 24: 2390–2402

Ehleringer JR (1990) Correlations between carbon isotope discrimination and leaf conductance to water vapor in common beans. Plant Physiol 93:1422–1425

Ehleringer JR, Dawson TE (1992) Water uptake by plants: perspectives from stable isotope composition. Plant Cell Environ 15:1073–1082

Ehleringer JR (1993a) Carbon and water relations in desert plants: an isotopic perspective. In: Ehleringer JR et al (eds) Stable Isotopes and Plant-Carbon Water Relations. Academic, San Diego, pp 155–172

Ehleringer JR (1993b) Variation in leaf carbon-isotope discrimination in *Encelia farinosa* – implications for growth, competition, and drought survival. Oecologia 95:340–346

Ehleringer JR, Monson RK (1993) Evolutionary and ecological aspects of photosynthetic pathway variation. Annu Rev Ecol Syst 24:411–439

Ehleringer JR, Cerling TE (1995) Atmospheric CO_2 and the ratio of intercellular to ambient CO_2 concentrations in plants. Tree Physiol 15:105–111

Farquhar GD, Oleary MH, Berry JA (1982) On the relationship between carbon isotope discrimination and the inter-cellular carbon-dioxide concentration in leaves. Aust J Plant Physiol 9:121–137

Farquhar GD, Ehleringer JR, Hubick KT (1989) Carbon isotope discrimination and photosynthesis. Annu Rev Plant Physiol Plant Mol Biol 40:503–537

Federal Register (2003) Emerald ash borer: Quarantine and regulations. 7 CFR Part 301 [Docket no. 02-125-1]

Flanagan LB, Johnsen KH (1995) Genetic variation in carbon isotope discrimination and its relationship to growth under field conditions in full-sib families of Piceamariana. Can J For Res 25:39–47

Grossiord C, Sevanto S, Dawson TE, Adams HD, Collins AD, Dickman LT, Newman BD … McDowell NG (2017) Warming combined with more extreme precipitation regimes modifies the water sources used by trees. New Phytol 213: 584–596

Gutschick VP, BassiriRad H (2003) Extreme events as shaping physiology, ecology, and evolution of plants: toward a unified definition and evaluation of their consequences. New Phytol 160:21–42

Hacke UG, Stiller V, Sperry JS, Pitterman J, McCulloh KA (2001) Cavitation fatigue. Embolism and refilling cycles can weaken the cavitation resistance of xylem. Plant Physiol 125

Hansen J, Sato M, Ruedy R (2012) Perception of climate change. Proc Natl Acad Sci U S A 109:E2415–E2423

Heschel MS, Riginos C (2005) Mechanisms of selection for drought stress tolerance and avoidance in Impatiens capensis (Balsaminacea). Am J Bot 92:37–44

Hozain MI, Salvucci ME, Fokar M, Holaday AS (2010) The differential response of photosynthesis to high temperature for a boreal and temperate Populus species relates to differences in Rubisco activation and Rubisco activase properties. Tree Physiol 30:32–44

Hüve K, Bichele I, Ivanova H, Keerberg O, Pärnik T, Rasulov B, Mari T … Niinemets Ü (2012) Temperature responses of dark respiration in relation to leaf sugar concentration. Physiol Plant 144: 320–334

IPCC (2012) Managing the risks of extreme events and disaster to advance climate change adaptation. A special report of working groups I and II of the Intergovernmental Panel on Climate Change. In: Field CB, et al. (eds), Cambridge University Press, Cambridge, UK, and New York, NY, USA, p 582

IPCC (2013) Climate change 2013: the physical science basis. Contribution of working group I to the fifth assessment report of the Intergovernmental Panel on Climate Change. In: Stocker TF, et al. (eds), Cambridge, UK, New York, NY, USA, p 1535

Iverson LR, Prasad AM, Matthews SN, Peters M (2008) Estimating potential habitat for 134 eastern US tree species under six climate scenarios. For Ecol Manag 254:390–406

Jansen K, Sohrt J, Kohnle U, Ensminger I, Gessler A (2012) Tree ring isotopic composition, radial increment and height growth reveal provenance-specific reactions of Douglas-fir towards environmental parameters. Trees 27:37–52

Junker LV, Kleiber A, Jansen K, Wildhagen H, Hess M, Kayler Z, Kammerer B … Ensminger I (2017) Variation in short-term and long-term responses of photosynthesis and isoprenoid-mediated photoprotection to soil water availability in four Douglas-fir provenances. Sci Rep 7: 40145

Kawecki TJ, Ebert D (2004) Conceptual issues in local adaptation. Ecol Lett 7:1225–1241

Keel SG, Siegwolf RTW, Korner C (2006) Canopy CO_2 enrichment permits tracing the fate of recently assimilated carbon in a mature deciduous forest. New Phytol 172:319–329

Knight CA, Vogel H, Kroymann J, Shumate A, Witsenboer H, Mitchell-Olds T (2006) Expression profiling and local adaptation of *Boechera holboellii* populations for water use efficiency across a naturally occurring water stress gradient. Mol Ecol 15:1229–1237

Kovacs KF, Haight RG, McCullough DG, Mercader RJ, Siegert NW, Liebhold AM (2010) Cost of potential emerald ash borer damage in US communities, 2009–2019. Ecol Econ 69:569–578

Loreto F, Velikova V (2001) Isoprene produced by leaves protects the photosynthetic apparatus against ozone damage, quenches ozone products, and reduces lipid peroxidation of cellular membranes. Plant Physiol 127:1781–1787

MacFarlane DW, Meyer SP (2005) Characteristics and distribution of potential ash tree hosts for emerald ash borer. For Ecol Manag 213:15–24

Marchin RM, Sage EL, Ward JK (2008) Population-level variation of *Fraxinus americana* (white ash) is influenced by precipitation differences across the native range. Tree Physiol 28:151–159

McCullough DG, Poland TM, Cappaert D (2009) Attraction of the emerald ash borer to ash trees stressed by girdling, herbicide treatment, or wounding. Canan Journal of Forest Research-Revue Canadienne De Recherche Forestiere 39:1331–1345

Meehl GA, Tebaldi C (2004) More intense, more frequent, and longer lasting heat waves in the 21st century. Science 305:994–997

Murray F (1967) On the computation of saturation vapor pressure. J Appl Metereol 6:203–204

Nicotra AB, Davidson A (2010) Adaptive phenotypic plasticity and plant water use. Funct Plant Biol 37:117–127

Park Williams A, Allen CD, Macalady AK, Griffin D, Woodhouse CA, Meko DM, Swetnam TW... McDowell NG (2013) Temperature as a potent driver of regional forest drought stress and tree mortality. Nat Clim Chang 3: 292–297

Perkins SE, Alexander LV, Nairn JR (2012) Increasing frequency, intensity and duration of observed global heatwaves and warm spells. Geophys Res Lett 39:1–5

Poland TM, McCullough DG (2006) Emerald ash borer: invasion of the urban forest and the threat to North America's ash resource. J For 104:118–124

Sánchez-Salguero R, Camarero JJ, Rozas V, Génova M, Olano JM, Arzac A, Gazol A ... Linares JC (2018) Resist, recover or both? Growth plasticity in response to drought is geographically structured and linked to intraspecific variability in Pinus pinaster. J Biogeogr 45: 1126–1139

Sandquist DR, Ehleringer JR (2003) Carbon isotope discrimination differences within and between contrasting populations of *Encelia farinosa* raised under common-environment conditions. Oecologia 134:463–470

Savolainen O, Pyhajarvi T, Knurr T (2007) Gene flow and local adaptation in trees. Annu Rev Ecol Evol Syst 38:595–619

Sharkey TD, Singsaas EL (1995) Why plants emit isoprene. Nature 374:769

Sharkey TD, Yeh S (2001) Isoprene emission from plants. Annu Rev Plant Physiol Plant Mol Biol 52:407–436

Smith SJ, Edmonds J, Harlin CA, Mundra A, Calvin K (2015) Near-term acceleration in the rate of temperature change. Nat Clim Chang 5:333–336

Teskey R, Wertin T, Bauweraerts I, Ameye M, McGuire MA, Steppe K (2015) Responses of tree species to heat waves and extreme heat events. Plant Cell Environ 38:1699–1712

Urban J, Ingwers M, McGuire MA, Teskey RO (2017a) Stomatal conductance increases with rising temperature. Plant Signal Behav 12:e1356534

Urban J, Ingwers MW, McGuire MA, Teskey RO (2017b) Increase in leaf temperature opens stomata and decouples net photosynthesis from stomatal conductance in *Pinus taeda* and *Populus deltoides* x nigra. J Exp Bot 68:1757–1767

Voltas J, Chambel MR, Prada MA, Ferrio JP (2008) Climate-related variability in carbon and oxygen stable isotopes among populations of Aleppo pine grown in common-garden tests. Trees 22:759–769

von Caemmerer S, Evans JR (2015) Temperature responses of mesophyll conductance differ greatly between species. Plant Cell Environ 38:629–637

Walsh J, Wuebbles D, Hayhoe K, Kossin J, Kunkel K, Stephens G, Thorne P ... Somerville R (2014) Our Changing Climate. In: Melillo JM, et al. (eds) Climate Change Impacts in the United States: The Third National Climate Assessment, pp 19–67. United States Global Change Research Program

Weaver SJ, Kumar A, Chen MY (2014) Recent increases in extreme temperature occurrence over land. Geophys Res Lett 41:4669–4675

Weston DJ, Bauerle WL (2007) Inhibition and acclimation of C_3 photosynthesis to moderate heat: a perspective from thermally contrasting genotypes of *Acer ruäum* (red maple). Tree Physiol 27:1083–1092

Will RE, Wilson SM, Zou CB, Hennessey TC (2013) Increased vapor pressure deficit due to higher temperature leads to greater transpiration and faster mortality during drought for tree seedlings common to the forest–grassland ecotone. New Phytol 200:366–374

Wuebbles D, Meehl G, Hayhoe K, Karl TR, Kunkel K, Santer B, Wehner M … Sun L (2014) CMIP5 climate model analyses: climate extremes in the United States. Bull Am Meteorol Soc 95: 571–583

Zavalloni C, Gielen B, De Boeck HJ, Lemmens C, Ceulemans R, Nijs I (2009) Greater impact of extreme drought on photosynthesis of grasslands exposed to a warmer climate in spite of acclimation. Physiol Plant 136:57–72

Zuo Z, Weraduwage SM, Lantz AT, Sanchez LM, Weise SE, Wang J, Childs KL, Sharkey TD (2019) Expression of isoprene synthase in Arabidopsis alters plant growth and expression of key abiotic and biotic stress-related genes under unstressed conditions. Plant Physiol 180:124–152

Part IV

Responses of Plants with Carbon-Concentrating Mechanisms to Climate Change

Chapter 8

Terrestrial CO_2-Concentrating Mechanisms in a High CO_2 World

Rowan F. Sage* and Matt Stata
Department of Ecology and Evolutionary Biology, University of Toronto, Toronto, ON, Canada

*Author for correspondence, e-mail: r.sage@utoronto.ca

K. M. Becklin et al. (eds.), *Photosynthesis, Respiration, and Climate Change*, Advances in Photosynthesis and Respiration 48, https://doi.org/10.1007/978-3-030-64926-5_8

Summary

Anthropogenic global change threatens the Earth's biodiversity, with the future of plants utilizing carbon-concentrating mechanisms (CCM) being of particular concern. Here, we discuss global change effects on plants utilizing CCMs, relative to plants using the C_3 photosynthesis pathway. Terrestrial CCMs include the C_4, CAM and C_2 photosynthetic pathways, which are collectively utilized by 10% of the world's plant flora. They are considered at risk because CCMs are adaptations to low CO_2 atmospheres which become superfluous at elevated CO_2. Rising atmospheric CO_2 represents one form of anthropogenic global change, along with climate change, land transformation, over-exploitation of natural species, terrestrial eutrophication, and exotic species invasions. While rising CO_2 favors the physiology of C_3 over C_4 photosynthesis in warmer temperatures, in natural stands where multiple global change drivers are active, outcomes often do not follow what would be predicted from physiological responses. Based on present trends, which already include CO_2 enrichment effects, the natural diversity of the C_4, CAM and the C_2 functional types is declining. A leading cause is aggressive infilling of grassland habitats by woody C_3 competitors or invasive species. Woody infilling is the result of a combination of drivers including rising CO_2, overgrazing, overhunting of browsers, and land use change. Invasive species can increase the frequency and intensity of fire disturbance, which benefit the invaders, while accelerating habitat conversion where the diversity of CAM succulents and epiphytes is high. In light of these changes, the future of a diverse C_4 and CAM flora is in question. Some C_4 and CAM species should do well, but the bulk of the CCM flora is at risk from many of the same pressures that threaten Earth's biodiversity in general, notably from habitat loss, invasive species, overharvesting and reduced vigor due to climate change. Elevated atmospheric CO_2 will aggravate these effects, by favoring C_3 plants over CCM species.

I. Introduction

The world today contains around 300,000 photosynthetic species, of which perhaps 80,000 operate mechanisms to concentrate CO_2 around Rubisco and thus suppress photorespiration (Raven and Beardall 2014; Sage and Stata 2015). Three major and one minor CO_2 concentrating mechanisms (CCM) are known (Fig. 8.1). In water, tens of thousands of cyanobacterial and eukaryotic algal species concentrate dissolved inorganic carbon (mostly as CO_2 and bicarbonate) into internal proteinaceous or starch-bound compartments using ATP energy (Fig. 8.1a). By doing so, they elevate CO_2 around Rubisco up to 80 times the concentration in surrounding water (Badger and Spalding 2000). On land, the most productive CCM in terms of biomass is C_4 photosynthesis, which utilizes PEP carboxylase (PEPC) and the C_4 biochemical cycle to concentrate CO_2 into an internal compartment where Rubisco is localized, usually the vascular bundle sheath (BS; Fig. 8.1b; Hatch 1987). Approximately 8000 species in 18 families of flowering plants use C_4 photosynthesis (Sage 2016). C_4 plants account for about 25% of the net terrestrial primary productivity on Earth and make up a significant fraction of the biomass that enters the human food chain (Still et al. 2003; Sage and Zhu 2011). Maize is the world's leading C_4 crop, and the C_4 plant sugar cane is the principle source of raw sugar for the world, while C_4 forage grasses support most of the meat production occurring in low latitudes (Brown 1999). The second major terrestrial CCM is Crassulacean Acid Metabolism (CAM), present in about 6% of all land plants, or 15,000–20,000 spe-

a Dissolved Inorganic Carbon (DIC) Pump

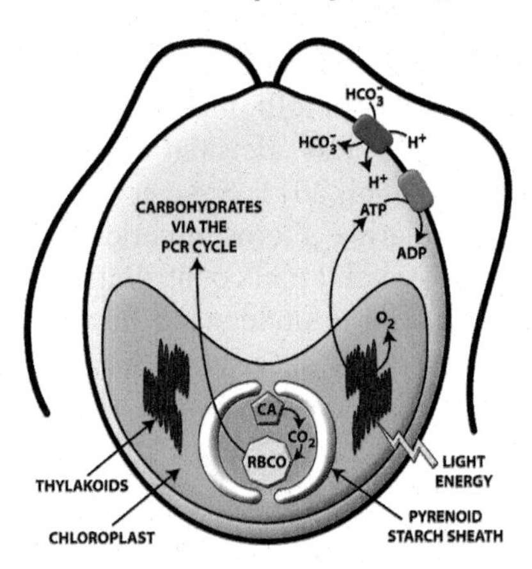

- CO_2 concentration around Rubisco is elevated up to 80x ambient level.
- Photorespiration is greatly suppressed.
- HCO_3^- can be used in alkaline water.
- High radiation use efficiency and quantim yield.
- Single and multicellular green, red, and brown algae species.

b C_4 Photosynthesis

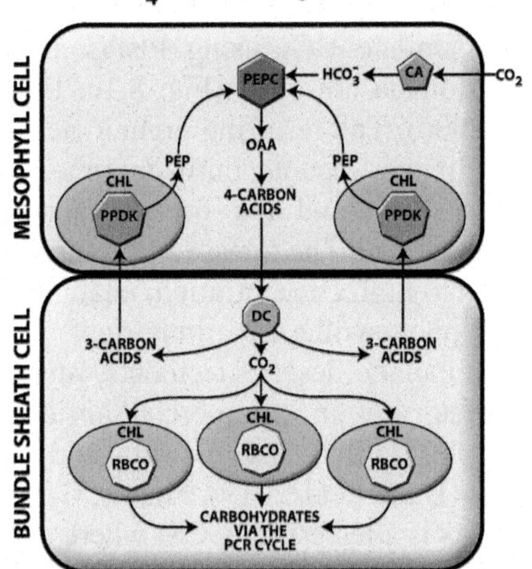

- CO_2 concentration around Rubisco is elevated to 3-10x ambient level.
- Photorespiration is largely suppressed.
- Photosynthesis is enhanced at low CO_2 (< 400 ppm) in warm conditions.
- Radiation, water, and nitrogent use efficiency are enhanced.
- Grasses, sedges, and herbacious to shrubby eudicots.

c CAM Photosynthesis

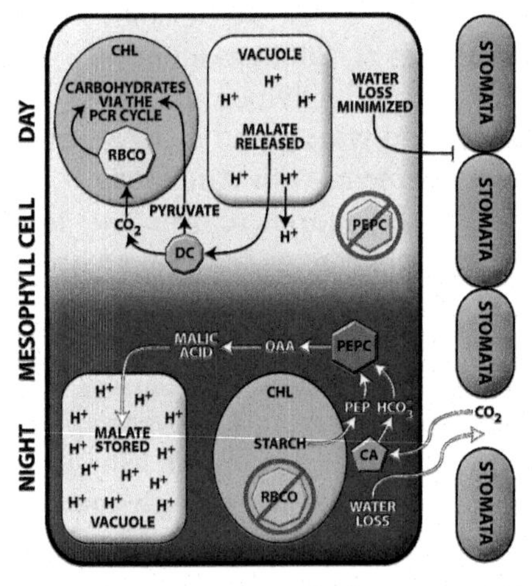

- CO_2 concentration around Rubisco is elevated up to 20x ambient level.
- Photorespiration is largely suppressed.
- Water use efficiency is enhanced 3-10 fold.
- Common in epiphytes and succulents adapted to dry habitats.

d C_2 Photosynthesis

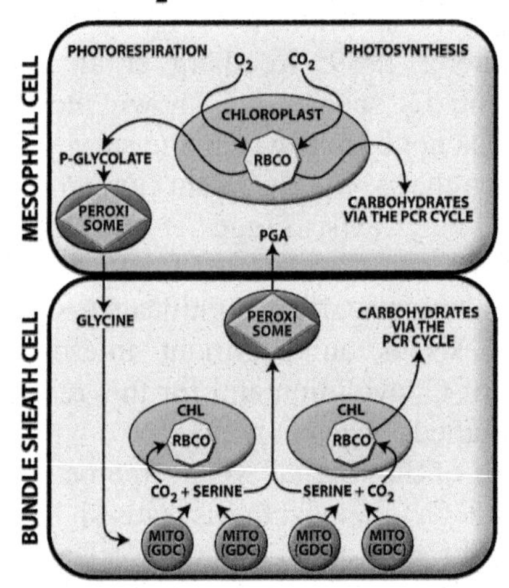

- CO_2 concentration around Rubisco is elevated to 2-3x ambient level.
- Photorespiratory CO_2 loss is reduced.
- Photosynthesis is enhanced at low CO_2 (<350ppm) in hot environments.
- Dozen of species known, mostly herbaceous.

Fig. 8.1. Diagrams outlining the general metabolic and structural features of the four carbon concentrating mechanisms in the Earth's flora. Abbreviations: *CA* Carbonic anhydrase, *CHL* Chloroplast, *DC* Decarboxylating enzyme, *mito* mitochondria, *OAA* Oxaloacetate, *PEPC* PEP Carboxylase, *PPDK* Pyruvate, phosphate dikinase, *RBCO* Rubisco. Reprinted from Sage and Stata (2015)

cies (Lüttge 2011; Winter et al. 2015). CAM plants use a C_4 biochemical mechanism to first assimilate CO_2 using PEPC at night when stomata are open (Fig. 8.1c; Borland et al. 2000). The resulting organic acids are stored in the vacuole until daytime, when stomata close and the organic acids are decarboxylated. The released CO_2 then accumulates to high concentration, enabling efficient photosynthesis. Important CAM species include desert succulents, rainforest epiphytes, and in commercial agriculture, pineapple, vanilla, *Opuntia, Aloe* and *Agave* species (Davis et al. 2019). Finally, C_2 photosynthesis is a terrestrial CCM where expression of the photorespiratory enzyme glycine decarboxylase (GDC) is restricted to the BS cells (Rawsthorne 1992). In C_2 photosynthesis, the photorespiratory metabolite glycine is shuttled from the M to BS cells for decarboxylation, with the resulting accumulation of CO_2 being two to three times the levels in mesophyll chloroplasts (Fig. 8.1d; von Caemmerer 1989; Keerberg et al. 2014). Some 60 C_2 species are known; however, there has not been a systematic survey of C_2 photosynthesis in the plant kingdom, so more likely exist (Sage et al. 2014a ; Lundgren and Christin 2017). While no C_2 species has agricultural significance, C_2 photosynthesis is an important intermediate phase of C_4 evolution and for this reason is well studied (Sage et al. 2014a).

Each of the four CCMs overcome limitations to C_3 photosynthesis caused by low CO_2 in the atmosphere and the costly process of photorespiration (Ehleringer et al. 1991; Raven et al. 2017). However, each incurs costs to operate the CCM and these could be a disadvantage in elevated atmospheric CO_2 (eCO_2), where a CCM is superfluous (Winter and Smith 1996a; Ehleringer et al. 1997; Raven et al. 2017), such that plants with a carbon-concentrating mechanism could be suppressed by enhanced competition from the C_3 vegetation (Henderson et al. 1995; Drennan and Nobel 2000; Raven et al. 2017). These concerns are heightened by paleo-ecological studies that indicate CCMs proliferated in recent geological periods of low atmospheric CO_2, and were uncommon, if not absent, before 30 million years ago when atmospheric CO_2 was elevated relative to today (Arakaki et al. 2011; Sage et al.).

However, relative photosynthetic performance is one of many traits contributing to a plant's fitness, and to understand how eCO_2 may affect a plant's success in its natural environment, the relative response of C_3 versus C_4 photosynthesis to rising CO_2 has to be considered in an ecological context where competitive interactions, disturbance, disease, mutualisms and herbivory also contribute to a species fitness (Kubien and Sage 2003; Griffith et al. 2015). Interactions between species and people must now be considered as well, since human management has become so pervasive as to influence the structure and function of most landscapes where CCM species occur (Ramankutty et al. 2008). Among the most important human activities are those that have become so widespread as to affect the global environment, with dangerous consequences for the biosphere (Steffen et al. 2004). Anthropogenic global change drivers that are relevant to species with CCMs include: (1) increasing atmospheric CO_2; (2) climate change, including global warming and altered precipitation; (3) over-exploitation due to hunting, grazing or fishing; (4) eutrophication of land and water by nitrogen (N) and phosphorous (P) pollution; (5) aggressive bioinvasions by exotic species; and (6) widespread transformation of land cover (Vitousek 1994; Vitousek et al. 1997b; Sala et al. 2000; Steffen et al. 2004). Each driver has unique characteristics that distinguish it from natural processes and each is rapidly changing, often with complex interactions (Vitousek 1994; Sage 2020).

To understand the future prospects of the CCM flora in a high CO_2 world, it is necessary to consider more than just responses of photosynthesis and growth to eCO_2. In this review, we address the prospects for CCM species in a high CO_2 world, considering multiple forms of anthropogenic change and the ecological networks in which the CCM species live. Our focus is terrestrial plants, due to length restrictions for the chapter. For perspectives on the future of algal CCMs, readers are directed to work by John Raven and colleagues (Raven et al. 2012, 2017; Raven and Beardall 2014). Also, we emphasize responses of the non-cultivated flora, since responses of cultivated species are covered by Watson-Lazowski and Ghannoum (this volume).

II. Physiological Context

The adaptive significance of each of the photosynthetic CCMs is their ability to overcome carbon deficiency in warm and often water limited environments where photorespiration inhibits C_3 photosynthesis, and to overcome limitations associated with subsaturating concentrations of CO_2 in the chloroplasts of C_3 plants and algae (Sage 1999; Raven et al. 2012, 2017). Because RuBP carboxylase/oxygenase (Rubisco) determines the CO_2 response of photosynthesis and initiates the photorespiratory cycle, it is useful to first consider its function in the context of C_3 and C_4 photosynthesis.

A. *Carboxylation and Oxygenation by Rubisco*

All forms of Rubisco oxygenate RuBP if CO_2 is low and O_2 high, producing phosphoglycolate (PG), a two-carbon metabolite with no immediate benefit to the plant. Phosphogylycolate can also be toxic if it accumulates in the cell (Sage and Stata 2015). The equation for the oxygenation to carboxylation ratio of Rubisco is:

$$v_o / v_c = (1/S_{rel}) O/C \qquad (8.1)$$

where v_o is the oxygenation rate, v_c is the carboxylation rate, O is the O_2 concentration in the chloroplast stroma, C is the CO_2 concentration in the chloroplast stroma, and S_{rel} is the specificity of Rubisco for CO_2 relative to O_2 (von Caemmerer and Quick 2000). Equation 1 demonstrates there are only three potential ways to reduce v_o/v_c – increase S_{rel}, increase C, or reduce O. Increasing S_{rel} is an unlikely solution because both CO_2 and O_2 are small, neutrally-charged molecules that present an intractable discrimination challenge for Rubisco; changes to Rubisco structure that reduce oxygenation potential also hinder CO_2 entry into the active site. As a result higher S_{rel} is associated with slower catalytic turnover rate (k_{cat}) such that a low oxygenase-form of Rubisco would be slow (Galmés et al. 2014; Sharwood et al. 2016, Tcherkez et al. 2006). In light of this trade-off, Rubisco is considered optimized for distinct terrestrial environments, which in CCM species involves the use of low S_{rel}/high k_{cat} forms that are adapted to operate in a high CO_2 compartment (Yeoh et al. 1981; Sage 2002b; Tcherkez et al. 2006; Whitney et al. 2011). Reducing O is also not a viable strategy, due to the large energy supply required to significantly reduce O below 210,000 ppm (Sage 2013). Instead, increasing C is relatively easy and cost effective, and hence is the repeated evolutionary solution to high oxygenation potential. Over 125 distinct origins of the various CCMs are apparent, occurring in at least 10 biochemical variants. Six variants occur in algae, three are present in C_4 plants (NADP-malic enzyme, NAD-malic enzyme, and PEP carboxykinase subtypes) and two in CAM (NADP-malic enzyme and NAD-malic enzyme subtypes; Winter and Smith 1996a; Sage et al. 2011a; Raven et al. 2017).

At low temperature, Rubisco has a low affinity for O_2, and hence has a high S_{rel}, which leads to low oxygenase activity in cool climates (Jordan and Ogren 1984; von Caemmerer and Quick 2000). As temperatures warm, S_{rel} declines, which, when coupled to a decline in the solubility of CO_2 relative to O_2, allows v_o/v_c to increase (Jordan and Ogren 1984; Sage 2013). To prevent excess build-up of oxygenation metabolites following RuBP oxygenation, the multistep metabolism of photorespiration evolved to recycle the carbon in oxygenation products back to RuBP (Hagemann et al. 2016). A key step in photorespiration is glycine decarboxylation, catalyzed by the mitochondrial enzyme glycine decarboxylase (GDC), which converts glycine to serine, ammonia and CO_2. Serine is metabolized to RuBP using ATP, while CO_2 can escape the cell. The cost of photorespiration is a loss of previously fixed carbon, the use of energy to regenerate RuBP consumed by Rubisco oxygenation, the energy spent to reassimilate ammonia and the competitive inhibition of the Rubisco catalytic sites by O_2 (Sharkey 1988). In recent atmospheric conditions where CO_2 has been 300–400 ppm, and O_2 has been 210,000 ppm, the oxygenation rate relative to the carboxylation rate in C_3 plants and algae can exceed 50% above 35 °C, leading to an inhibition of C_3 photosynthesis by 25% or more (Ehleringer et al. 1991; Sage 2013). This inhibition of C_3 photosynthesis is even more apparent in warm environments and the relatively low atmospheric CO_2 concentrations of the past two million years (between 300 and 180 ppm; Zhang et al. 2013; Higgins et al. 2015); however, at high CO_2 concentrations estimated for Cretaceous and Paleogene climates (>800–900 ppm between 140 and 30 million years ago; Zachos et al. 2008; Berner 2008), photorespiration is <10% of photosynthesis unless stomatal conductance is also low (Sage 2013).

A parameter that varies with v_o/v_c is the maximum quantum yield of photosynthesis, which can be measured as the initial slope of the net CO_2 assimilation rate (A) to absorbed light intensity (Fig. 8.2a; Ehleringer and Pearcy 1983; Sage et al. 2018a, b). Because of its relationship with v_o/v_c, maximum quantum yield is a robust proxy for the cost of photorespiration. In C_4 plants, maximum quantum yield changes little as CO_2 and temperature vary, reflecting the ability of the C_4 CCM to suppress Rubisco oxygenase activ-

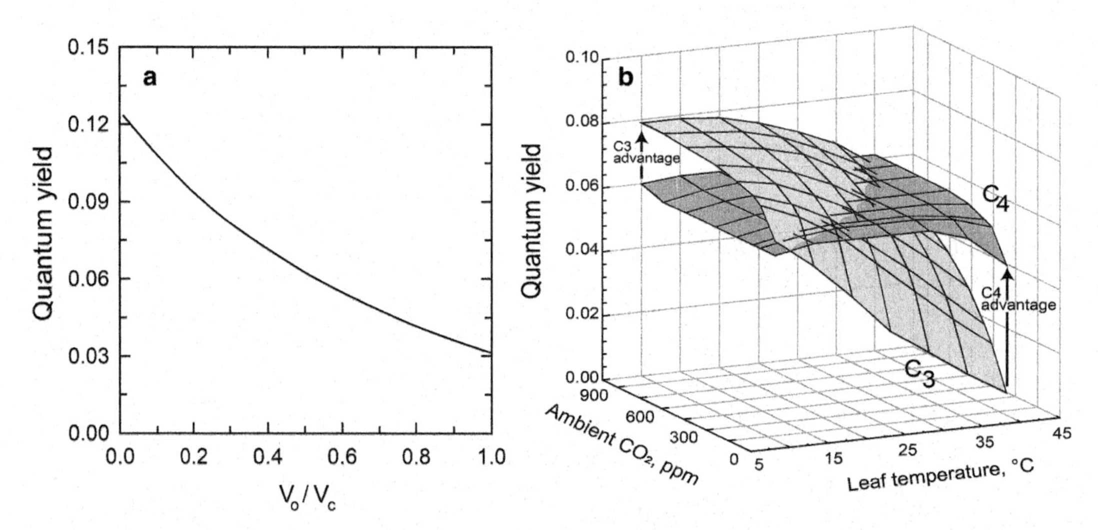

Fig. 8.2. (**a**) The modelled response of the maximum quantum yield of photosynthesis to the ratio of oxygenation (*vo*) to carboxylation (*vc*) rate. The relationship was modelled according to Equation 1 in Sage and Kubien (2003). (**b**) The modelled relationship between the maximum quantum yield of photosynthesis as a function of ambient CO_2 and leaf temperature for typical C_3 and C_4 grass species of the NADP-malic enzyme subtype. Quantum yields in panel b were modelled using the WIMOVAC program (Humphries and Long 1995). Panel a was previously published in Sage et al. 2018; panel b is from Sage (2000)

ity in a wide range of environments (Fig. 8.2b). In C_3 plants at low temperature and elevated CO_2, oxygenase activity is minimal and the plants have a superior quantum yield relative to C_4 plants (Fig. 8.2b). The lower maximum quantum yield of plants using CCMs reflects the energy required to operate a CCM, plus efficiency losses caused by CO_2 leakage from the compartment where Rubisco is localized (Kromdijk et al. 2014; Raven and Beardall 2016). As temperature warms and CO_2 concentration declines, the quantum yield of C_3 photosynthesis declines, falling below C_4 values at moderate temperatures and atmospheric CO_2 levels of the past 100 years (300–415ppm). Quantum yield responses of C_3 and C_4 photosynthesis are easily modelled and readily demonstrate the relative cost of photorespiration, and how temperature and CO_2 availability influence C_3 relative to C_4 photosynthesis (Fig. 8.2b; Ehleringer 1978; Ehleringer et al. 1991, 1997). For this reason, quantum yield-based algorithms have been incorporated into models to predict cross-over temperatures between C_3 and C_4 plants as a function of climate and CO_2 variation, allowing for the estimation of where along a climate gradient C_4 vegetation would be favored, and when past atmospheric CO_2 declines may have first favored C_4 photosynthesis (Ehleringer 1978; Ehleringer et al. 1997; Collatz et al. 1998).

B. Photosynthetic Responses to CO_2 and Temperature in C_3 and C_4 Plants

The performance of C_3, C_4 and C_2 plants in a world of changing CO_2 is also demonstrated by the responses of A to intercellular CO_2 concentration (C_i) and temperature (T), in what are termed A/C_i and A/T curves, respectively (Fig. 8.3). Relative to C_3 plants, A/C_i curves routinely demonstrate that C_4 and C_2 plants have superior A at atmospheric CO_2 levels below 330—380 ppm, the late-twentieth century values, when much pioneering gas exchange research was conducted (Fig. 8.3a). This is because C_4 and C_2 plants exhibit lower CO_2 compensation points of A, and in the case of C_4 photosynthesis, have higher initial slopes of the A/C_i curve (Fig. 8.3a). As CO_2 levels increase above ~350 ppm, C_4 plants exhibit CO_2 saturation, because Rubisco capacity, RuBP regeneration capacity, or in some cases, the regeneration of PEP by pyruvate-P_i-dikinase becomes limiting for A (Sage and Pearcy 2000; von Caemmerer 2000). In C_3 plants by contrast, A continues to increase with rising atmospheric CO_2 concentration above 350 ppm, with the degree of enhancement increasing at higher temperature. In both C_3 and C_4 plants, rising temperatures often increase the concentration of CO_2 required to saturate A, particularly in C_3 plants (see Fig. 8.3b for C_3 plants; Sage and Kubien 2007).

It is often presumed that C_3 plants have higher A than C_4 plants at elevated CO_2 (Kirkham 2011), but this requires the species have similar A at current levels of CO_2, as shown in Fig. 8.3a for closely related *Flaveria* species. C_4 plants often have very high rates of A due to their ruderal or graminoid life-form, and hence often exhibit higher A than many C_3 plants at elevated CO_2 and temperature, as shown in Fig. 8.3b. Ruderals and fast-growing grasses generally have high A, regardless of photosynthetic pathway (Field and Mooney 1986). This ability of many C_4 species to maintain higher A at elevated CO_2 could enable them to maintain their competitive edge in warmer climates. Where A in C_4 plants is superior at both low and high CO_2, the photosynthetic differences between the C_3 and C_4 plants decline with rising CO_2 (Fig. 8.3b), unless acclimation to high CO_2 reduces A in the C_3 species, as will be discussed below.

Temperature variation differentially affects C_3 and C_4 photosynthesis, because the strong thermal dependency of the oxygenase reaction by Rubisco increases photorespiratory inhibition in C_3 species as temperatures rise (Sage and Kubien 2007). In both C_3 and C_4 plants, the initial slope of the A/C_i response shows only a slight change with temperature,

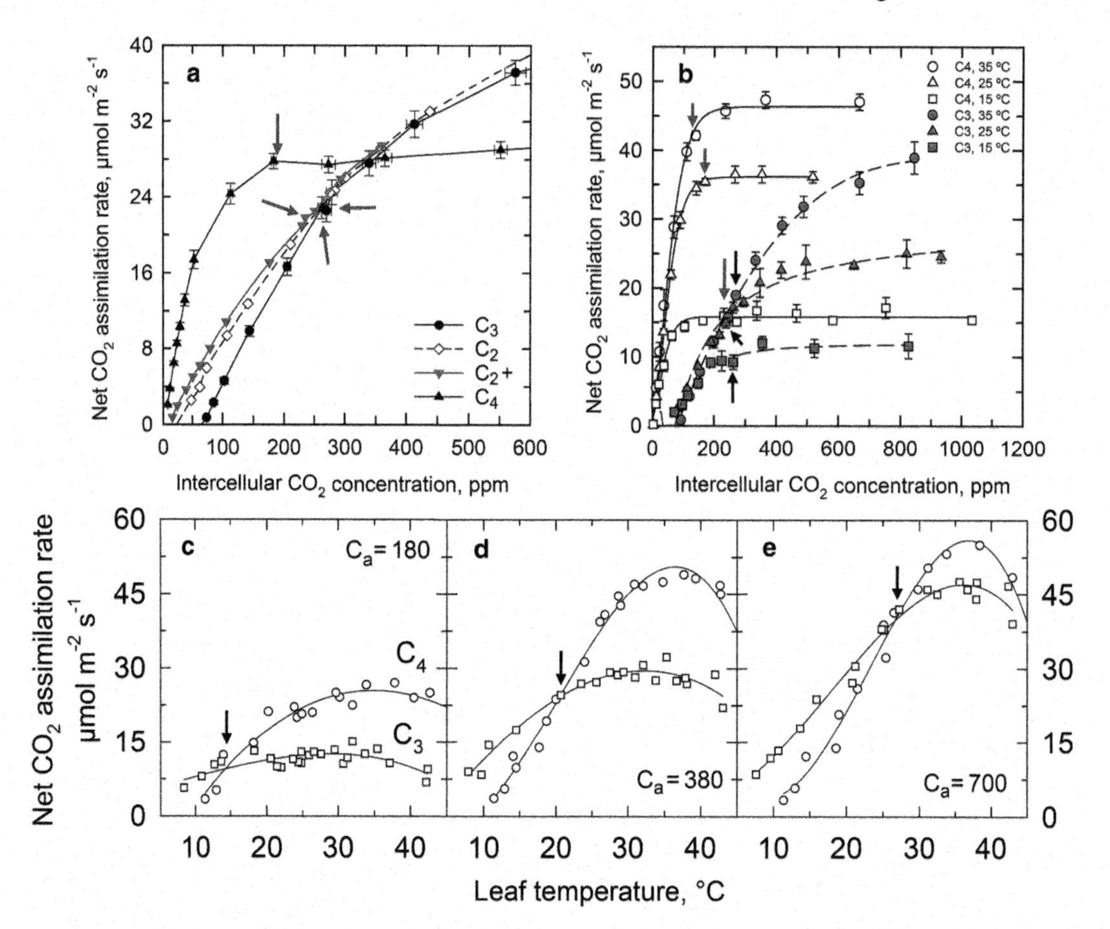

Fig. 8.3 (**a**) The response of net CO_2 assimilation rate to intercellular CO_2 concentration at 33 °C and light saturation in a C_4 species of *Flaveria* (*Flaveria trinervia*), a C_2 species lacking a C_4 metabolic cycle (*Flaveria angustifolia*), a C_2+ species with a weak C_4 metabolic cycle (*Flaveria floridana*), and a C_3 species (*Flaveria cronquistii*). Arrows indicate the intercellular CO_2 corresponding to 400 ppm, the prevailing ambient CO_2 at the time of measurement (redrawn from Sage et al. 2018). Note the greater CO_2 assimilation rate below 250 ppm CO_2 in the C_2 relative to C_3 species. (**b**) The response of net CO_2 assimilation rate to intercellular CO_2 concentration in sugarcane cv. RB835486 (solid lines and open symbols) and cassava cv. BGM1721 (dashed lines and filled symbols) at 15 °C (squares), 25 °C (triangles), and 35 °C (circles). Measurements were conducted at saturating light intensity using greenhouse grown plants at 390 ppm and 30 °C. Arrows indicate the intercellular CO_2 corresponding to 390 ppm, the prevailing ambient CO_2 at the time of measurement. Grey arrows correspond to cassava, black arrows to sugarcane. Note how net CO_2 assimilation rate is CO_2 saturated at 15 °C in both the C_3 and C_4 species. Redrawn from Sage et al. (2014b). (**c**–**e**) Net CO_2 assimilation rate as a function of leaf temperature in the C_4 plant *Amaranthus retroflexus* (circles) and the C_3 plant *Chenopodum album* (squares) measured at light saturation and an ambient CO_2 concentration of 180 ppm (panel c), 380 ppm (panel d) and 700 ppm (panel e). The arrows indicate the cross-over temperature which increases with measurement CO_2 content. Redrawn from Sage and Pearcy (2000)

while the CO_2 saturated rate increases dramatically with temperature, such that the initial slope region of the A/C_i response extends to higher C_i (Fig. 8.3b). Because the initial slope region has a modest thermal response, A will also have a modest thermal response at low CO_2 levels corresponding to the initial slope (Sage and Kubien 2007). Hence, both C_3 and C_4 species have relatively shallow A/T responses above 15 °C at low CO_2 (Fig. 8.3c), with an important difference being that increasing photorespiration causes the CO_2 compensation point to increase with rising temperature in C_3 plants. As a result, A begins to decline at lower temperature in C_3 than C_4 plants. At CO_2 concentrations above the initial slope region of the *A/Ci* response, *A* become highly responsive to rising temperature, and the *A/T* response of both C_3 and C_4 plants assumes a bell-shape that becomes steeper above and below the thermal optimum, and with a narrower thermal optimum (Fig. 8.3d, e). Because they have lower CO_2-saturation points of photosynthesis, C_4 plants shift from the low thermal response mode to a high thermal response mode at lower CO_2 concentrations than C_3 plants, which allows them to have a more pronounced *A/T* response at intermediate values of CO_2 (Fig. 8.3d). Rising CO_2 narrows the thermal range where C_4 photosynthesis is favored, but at the thermal optimum, C_4 plants at high CO_2 can still exhibit superior *A* values than C_3 species of equivalent life form (Fig. 8.3e). While the range of temperatures favoring C_4 photosynthesis may narrow at high CO_2, in a globally warmed-world, thermal conditions will correspond to the C_4 thermal optimum more often, which may offset the greater CO_2 stimulation of C_3 photosynthesis.

C. Water and Nitrogen Use Efficiencies of C_3 and C_4 Plants

C_4 plants have higher water use efficiencies (WUE) than C_3 plants, due to the ability of the CCM to maintain high rates of *A* while allowing for lower stomatal conductance (g_s) (Huxman and Monson 2003; Kocacinar et al. 2008; Ghannoum et al. 2011). Typically, the C_4 WUE is 1.5 to 4 times greater than C_3 WUE at moderate temperature, but will be much higher in hot conditions above 35 °C (Larcher 2003). In addition to boosting yield on limited quantities of water, increases in WUE attenuate drought and lengthen growing seasons (Polley 1997; Leakey 2009; Morgan et al. 2011). Relative to C_3 species, C_4 species often exhibit similar declines in stomatal conductance (g_s), and increases in WUE as atmospheric CO_2 concentration increases (Fig. 8.4; Sage 1994; Polley et al. 1994; Wand et al. 1999; Anderson et al. 2001; Maherali et al. 2003; Kirkham 2011; Taylor et al. 2018). The boost in WUE can be more beneficial for C_4 than co-occurring C_3 species because C_4 plants are typically active in warmer, frequently drier seasons where water deficiency is common and the benefits of higher WUE more significant (Polley et al. 2002; Sage and Kubien 2003). WUE responses to rising CO_2 are most important in episodically dry environments where exhaustion of soil water intensifies drought stress and initiates a dry season that shuts down photosynthesis and growth (Polley et al. 1994, 2002; Morgan et al. 2004, 2011). Paradoxically however, enhanced WUE of C_4 plants could benefit their C_3 competitors, particularly in semi-arid landscapes where competition for water is high. With higher WUE, water consumption by the C_4 flora could decline, increasing water availability for the C_3 competition, and increasing the amount of water percolating to deeper soil horizons more likely to be populated by roots of eudicot C_3 species (Blumenthal et al. 2013). For example, in the upper soil region of semi-arid savannas, C_4 grasses are commonly the superior competitors for water relative to seedlings of C_3 woody species at recent CO_2 levels (Scholes and Archer 1997). The higher WUE of C_4 plants at elevated CO_2 benefits the woody seedlings by reducing the intensity of belowground competition, enabling them to survive and extend their tap roots below the grass root zone and thus escape critically dry situations where they might perish before establishment (Polley et al. 1997; Morgan et al. 2011).

C_4 plants also have a higher N use efficiency of photosynthesis (PNUE, expressed as *A/N*) than C_3 plants, typically by 25–50% (Brown 1978; Sage and Pearcy 1987a, b; Sage et al. 1987). This is largely because of reduced investment in photorespiratory enzymes and Rubisco relative to *A* (Sage et al. 1987; Ghannoum et al. 2011). Higher Rubisco investment in C_3 plants compensates for limitations due to low CO_2 and pho-

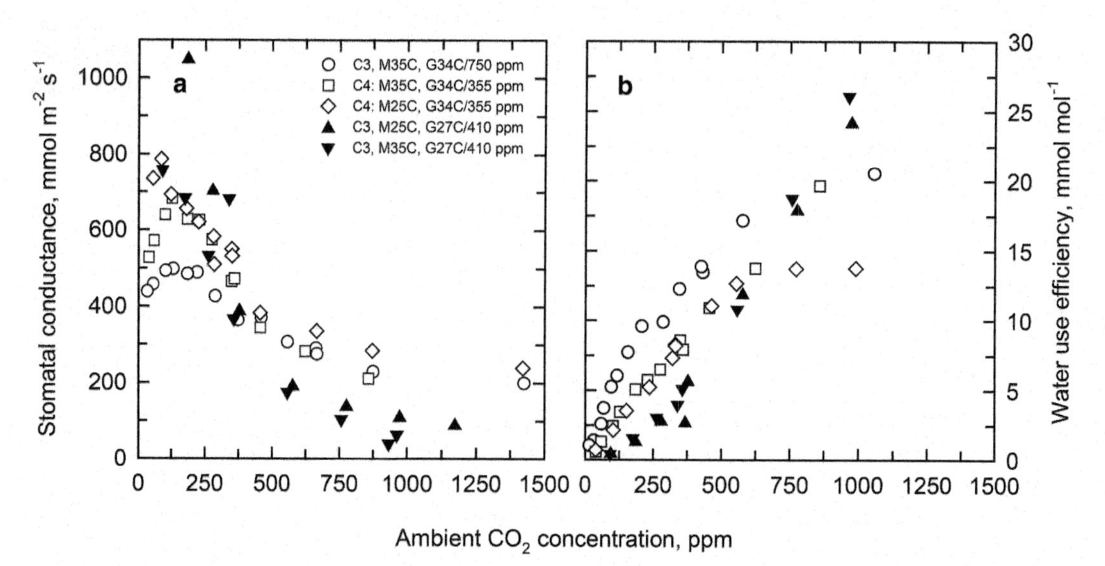

Fig. 8.4. The response of (**a**) stomatal conductance and (**b**) water use efficiency (WUE) to ambient CO_2 concentration at measurement temperatures of 25 °C or 35 °C. The C_4 species is *Amaranthus retroflexus* grown at either 34 °C or 27 °C day temperature and a CO_2 concentration of 750 or 355 ppm (in 1992) in plant growth chambers. The C_3 species is the similar weed *Chenopodium album* grown at a daytime temperature of 27 °C and a CO_2 concentration of 410 ppm in 2019. The photosynthetic type, measurement (M) temperature, and growth conditions (G) of CO_2 concentration and temperatures for each curve are shown in panel a. See Sage et al. (1995) for methods used for *A. retroflexus*. *Chenopodium album* was measured using a Li-Cor 6400 following a similar procedure. The results show a reduction in stomatal conductance with rising CO_2 around the plants, with stomatal acclimation to CO_2 and temperature having a small effect on the response. Water use efficiency is higher in the C_4 plants at low to intermediate ambient CO_2, but becomes higher in the C_3 plants at the highest measurement CO_2 concentration

torespiration. By investing more of their N reserves into Rubisco, C_3 plants can match the photosynthetic capacity of C_4 plants (Sage and Pearcy 1987a; Ghannoum et al. 2011). This strategy, however, limits allocation of N to other functions, such as leaf area or root production. C_4 plants typically maintain less leaf N than C_3 plants of similar life form and ecological habit, enabling them to redistribute their N reserves to more critical fitness functions such as leaf area production in resource-rich environments, or root growth in N limited soils (Sage and Pearcy 1987b; Kochanicar and Sage 2005; Atkinson et al. 2016). In grasslands, the greater NUE of C_4 grasses enables significant productivity on relatively low leaf N content, but also limits N availability for C_3 competitors (Wedin 2004). C_4 leaves with reduced leaf N content decompose less rapidly, which sequesters N in detritus and soil organic matter, which in turn reduces establishment success of C_3 grasses and woody seedlings in a C_4 grass sward (Long 1999; Wedin and Tilman 1993; Silva et al. 2013). Slower decomposition also promotes litter accumulation, which supports hotter and more frequent fires that torch woody plants in C_4 grasslands (Knapp and Medina 1999). The NUE differences between C_3 and C_4 species thus highlight how photosynthetic physiology can cascade into ecological interactions to influence species success.

D. CAM Photosynthesis

The ecophysiology of CAM plants with respect to eCO_2 is influenced by the ability of most CAM plants to periodically function in a C_3 mode by directly assimilating CO_2 diffusing into the leaf through open stomata during the day. During a diel cycle, C_3 modes alternate with the CAM biochemical mode (Borland et al. 2000). The CAM mode is defined by nighttime CO_2 assimilation by PEPC with organic acid storage, followed by daytime decarboxylation and refixation by Rubisco when stomata are closed. In a subset of CAM photosynthesis, termed facultative CAM, the CAM mode is absent when plants are unstressed, with all CO_2 assimilated via C_3 photosynthesis. CAM is then activated when drought or salinity stress intensify (Osmond 1978; Winter and Smith 1996a, b; Dodd et al. 2002; Winter et al. 2015). Obligate CAM species always utilize a CAM mode, although in extreme stress stomata are continuously closed and the plant scavenges night-time respiratory CO_2 using PEPC in a mode termed CAM idling (Borland et al. 2000). C_3 plants can also recycle respiratory CO_2 using a weak CAM-like metabolism termed CAM cycling, but they should not be classified as CAM plants because they don't open their stomata at night nor significantly concentrate CO_2 (Winter et al. 2015).

When CAM is fully engaged, plants exhibit a pronounced diel cycle of net CO_2 assimilation rate in which four distinct phases are recognized, termed I through IV (Fig. 8.5; Osmond 1978). Phase I is night-time CO_2 uptake when stomata are open and PEPC assimilates carbon into organic acids that are then stored in the vacuole. Phase III occurs during daytime when stomata are closed, decarboxylation occurs, and the released CO_2 is refixed by Rubisco. CO_2 concentrations in chloroplasts can rise to 4000–10,000 ppm during phase III (Osmond et al. 1982). In phases II and IV, typical C_3 photosynthesis occurs. Phase II occurs in the morning as plants transit from phase I to phase III. Phase IV begins late in the afternoon when the stored pool of C_4 acids in the vacuole is exhausted; stomata then reopen and the plant conducts C_3 photosynthesis until dark. Phase II is usually brief (<2 h), while phase IV can last a few hours prior to sundown (Borland et al. 2011). All CAM plants exhibit phase I and III, while the length and rate of phases II and IV vary with taxa and stress intensity (Dodd et al. 2002; Borland et al. 2011). During severe water stress, phases II and IV typically weaken or

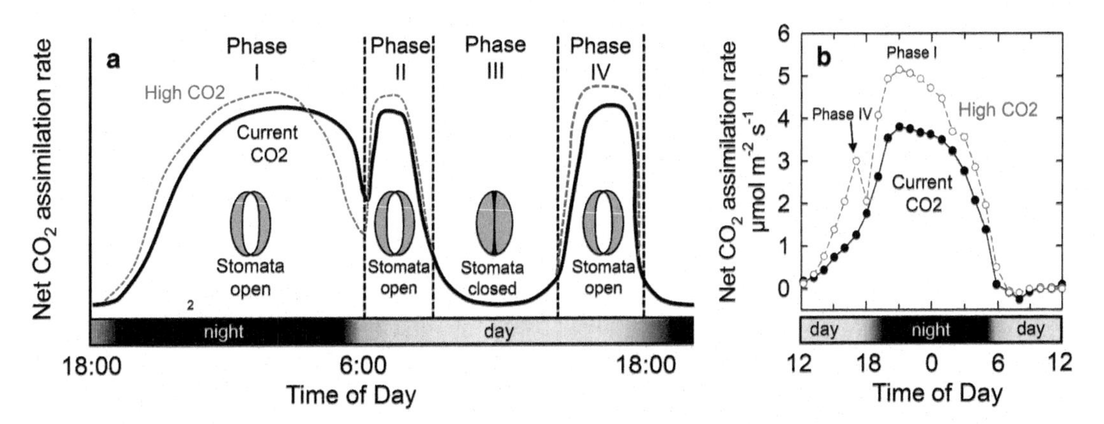

Fig. 8.5. (**a**) A diagram illustrating the CO_2 assimilation rate during a diel cycle in CAM. Four distinct phases are often apparent in CAM species. Phase I represents night-time CO_2 uptake by PEP carboxylase when stomata are closed. In Phase 2, the stomata open in the morning and a C_3-mode of photosynthesis briefly occurs. Stomata close in phase III, and the stored organic acids are decarboxylated to release CO_2 for Rubisco carboxylation. In phase IV, the organic acid pool is used up and stomata will re-open and CO_2 will be assimilated in a C_3-mode if plants are not severely stressed. Not all CAM plants exhibit phases II and IV. Adapted from Borland et al. (2018). (**b**) Diurnal profiles of net CO_2 assimilation rate in the CAM plant *Agave desertii* grown and measured at either 370 ppm (solid line) or 750 ppm CO_2 (dashed line) Redrawn from Graham and Nobel (1996). Note the appearance of phase IV in the high CO_2-grown plants. Daily water use efficiency at 370 ppm was 0.02 mol CO_2 mol^{-1} H_20 while at 750 ppm it was 0.042 mol mol^{-1}, a 110% increase

disappear (Borland and Griffiths 1996; Dodd et al. 2002). In desert succulents such as cacti and agave with obligate CAM, most carbon enters the plants at night and phases II and IV are small, while in many other CAM species, phases II and IV can be pronounced, with phase IV contributing much of the carbon used for growth (Borland and Griffiths 1996; Winter and Smith 1996a).

On average, daily net CO_2 uptake is 35% greater in CAM plants grown and measured at eCO_2 compared to values in CAM plants grown and measured in atmospheric CO_2 concentrations of the late-twentieth century, which is similar to the mean stimulation of *A* in C_3 crops grown at elevated relative to recent CO_2 concentrations (Drennan and Nobel 2000; Long et al. 2004; Ceusters and Borland 2011). Much of this enhancement occurs during phases II and IV when the plants function in a C_3 mode. During phases II and IV, Rubisco in CAM plants is more CO_2 limited than in C_3 plants, because M cells are more tightly packed in CAM tissues, which reduces diffusive influx of CO_2 (Maxwell et al. 1997; Nelson and Sage 2008). Because of the stimulation of *A* during phases II and IV by eCO_2, the proportion of daytime CO_2 uptake in high CO_2 air increases from 2% to 150% depending upon species and temperature, with an average being near 40% (Drennan and Nobel 2000; Ceusters and Borland 2011). Much of this increase occurs during phase IV, when carbon assimilation rate is greater in eCO_2 (Graham and Nobel 1996; Zhu et al. 1999). During phase I, CAM species can also exhibit enhanced night-time CO_2 fixation under eCO_2, particularly those with weak phase II and IV activity. In the prickly pear cactus *Opuntia ficus-indica*, phase I CO_2 fixation doubles at 700 ppm atmospheric CO_2 relative to 350 ppm (Cui et al. 1993; Drennan and Nobel 2000). Many CAM species show limited stimulation of night-time CO_2 uptake, however, which is probably due to a finite storage capacity for organic acids or exhaustion of PEP reserves (Winter and Smith 1996a; Drennan and Nobel 2000). In these cases, relative to plants at current CO_2 levels, more rapid CO_2 uptake occurs at eCO_2 early in the night due to enhanced PEPC activity, but this is offset by less uptake in the second half of the night, resulting in little net increase in phase I uptake.

The inverted rhythm of stomatal opening in CAM photosynthesis leads to high WUE of *A* relative to C_3 photosynthesis, because CAM plants mostly transpire at night when temperature is cool and relative humidity high. Where nights are warm and transpiration potential is high, CAM plants are uncommon, as water loss rates are too high (Borland et al. 2018). Water use efficiencies of CAM plants are 2–10 times higher than in C_3 plants, which confers evolutionary fitness for CAM species in drought-prone landscapes and as epiphytes on dry microsites such as tree branches in tropical rainforests (Kluge and Ting 1978; Larcher 2003). At eCO_2, the WUE of CAM plants is further increased 1.5 to two-fold relative to current conditions. This reflects reductions in stomatal conductance in phases I, II and IV, and the enhanced CO_2 uptake in each phase. Improved WUE delays drought-induced suppression of overall physiological activity, and phase IV in particular (Drennan and Nobel 2000).

E. C_2 Photosynthesis

C_2 plants exhibit superior photosynthesis below current CO_2 levels, largely because their CCM reduces by half the CO_2 compensation point of photosynthesis (Fig. 8.3a; Dai et al. 1996; Sage et al. 2013). A key feature of the C_2 CCM is that Rubisco oxygenase activity and photorespiration produce the glycine metabolite that shuttles CO_2 into the BS compartment. As CO_2 increases, the production of glycine declines and with it, the strength of the CCM (von Caemmerer 1989). Consequently, the photosynthetic stimulation by the C_2-CCM is lost at CO_2 concentrations above the current ambient, as is evident from similar A/C_i responses of related C_3 and C_2 species above 300 ppm C_i (Fig. 8.3a; Monson 1989; Vogan and Sage 2011, 2012; Pinto et al. 2011; Sage et al. 2011b). In essence, C_2 species at elevated CO_2 essentially function as C_3 species, but with the added cost of the C_2 pathway, which would be manifested as underutilized investment in the bundle sheath tissue. WUE and photosynthetic NUE (pNUE) differences between C_2 and C_3 species are not pronounced, and both parameters increase with rising CO_2 in a similar manner in C_2 as in C_3 plants (Vogan et al. 2007; Pinto et al. 2011; Vogan and Sage 2012).

III. Acclimation and Adaptation to CO_2 Enrichment

A. Acclimation to Elevated CO_2

Above 25°C, the enhancement of A in the minutes following an increase in ambient CO_2 is substantial in C_3 plants and modest in C_4 plants as shown by the A/Ci responses (Fig. 8.3). Following extended periods of growth at eCO_2 (days to months), the enhancement of A is often less than the short-term enhancement, reflecting regulatory adjustments at the whole-plant level which are commonly termed CO_2 acclimation, down-regulation or CO_2 homeostasis (Fig. 8.6). The attenuation of A by acclimation processes alters the degree to which rising CO_2 will favor C_3 over C_4 photosynthesis, and hence is a central issue in predicting how C_3 and C_4 plants will perform relative to each other in eCO_2 (Sage et al. 1989; Moore et al. 1999; Leakey et al. 2012). In C_3 plants, acclimation to eCO_2 can reduce the photosynthetic stimulation following a doubling of atmospheric CO_2 from over 35% to below 25% (Sage 1994; Gunderson and Wullschleger 1994; Moore et al. 1999). Acclimation-induced reductions in A often occur within days of CO_2 enhancement in herbaceous C_3 plants, and months to years in shrubs and trees (Moore et al. 1999; Leakey et al. 2012; Way et al. 2015). Roots and other sinks sense the enriched carbohydrate status, and alter long-distance signaling patterns that induce CO_2 acclimation by reducing photosynthetic protein synthesis (Sims et al. 1998; Moore et al. 1999; Sage 2002a). C_4 plants also acclimate to eCO_2, but by a small magnitude if at all ($<10\%$ reduction in A in high compared to low CO_2-grown plants; Ainsworth and Long 2005; Leakey 2009). CAM plants exhibit little in the way of down-regulation of A at eCO_2, and some even show long-term stimulation, apparently due to anatomical changes that increase leaf thickness and the potential to store CO_2 at night (Drennan and Nobel 2000; Ceusters and Borland 2011).

Photosynthetic down-regulation in eCO_2 is determined by genotype, sink strength, phloem transport mode, mycorrhizal colonization, carbohydrate storage capacity, and availability of limiting resources such as N and P (Moore et al. 1999; Norby et al. 2010; Bishop et al. 2018; Terrer et al. 2019). Species with low storage capacity, nutrient supply, and sink strength show strong down-regulation, which in the extreme results in no CO_2 enhancement of A in plants grown in eCO_2 compared to lower CO_2 (Fig. 8.6; Sage 1994; Sims et al. 1998; Ainsworth and Long 2005). Where nutrients are limiting, their supply rate as determined by decomposition rate, N deposition, N fixation, mycorrhizae and disturbances such as fire control the degree of CO_2 acclimation (Ainsworth and Long 2005; Reich et al. 2006; Norby et al. 2010; Terrer et al. 2016, 2018, 2019). Should nutrient deficiency induce strong CO_2 acclimation in C_3 but not C_4 plants, C_4 species can exhibit greater CO_2 responsiveness, which when combined with the higher WUE and NUE of C_4 vegetation could lead to superior performance in eCO_2. Also, CO_2 responsiveness reflects plant-rhizosphere interactions in a manner that can determine relative C_3 and C_4 performance. For example, enhancement of N cycling in the C_4 root zone at eCO_2

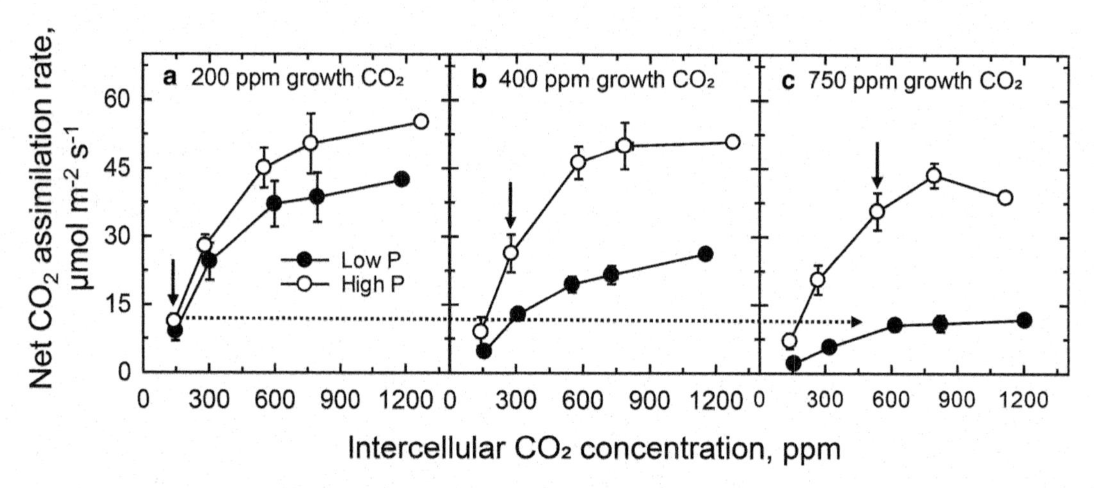

Fig. 8.6. The response of net CO_2 assimilation rate to variation in intercellular CO_2 concentration at 30 °C in white lupine (*Lupinus alba*) grown at abundant or deficient supply of soil phosphorous (P). Arrow indicates the net CO_2 assimilation rate at the intercellular CO_2 concentration corresponding to the growth CO_2 concentrations of 200 ppm (panel a); 400 ppm (panel b) or 750 ppm (panel c). The dotted horizontal arrow indicates the net CO_2 assimilation rate at the growth CO_2 did not vary between 200. 400 and 750 ppm growth CO_2 in the low P plants, demonstrating complete CO_2 homeostasis at low P. Nutrient limitations such as those shown here would negate the CO_2 fertilization of carbon gain that tends to benefit C_3 over C_4 plants. From Campbell and Sage (2006)

can favor growth of C_4 over C_3 grasses (Reich et al. 2018). C_4 grasses can also sequester ecosystem N into fibrous leaves that degrade slowly, and in doing so deprive N to neighboring C_3 plants (Wedin 2004). Alternatively, if eCO_2 accelerates nutrient acquisition in C_3 plants, they can exhibit an enhanced response to rising CO_2. This is best observed in legumes and other N-fixing species, where eCO_2 can stimulate N_2-fixation (Tissue et al. 1996; Millett et al. 2012; Pastore et al. 2019). This response explains in part the widespread encroachment of woody C_3 legumes such as mesquite and *Acacia* into C_4 grasslands in recent decades (Polley et al. 1994; Van Auken 2000; Archer et al. 2017).

Of note, acclimation by turning off C_4 photosynthesis at elevated CO_2 has never been documented, and in all but a few sedges and grasses of wetlands, the C_4 CCM is constitutive (Sage et al. 2011a). C_2 species also do not appear to acclimate to variation in growth CO_2 concentration by altering investment in the C_2 CCM (Vogan and Sage 2012). This contrasts with algae, which often lose their bicarbonate pumps at high CO_2, and CAM plants where the time spent in the C_3 mode can increase at eCO_2 (Raven et al. 2017; Drennan and Nobel, 2000).

B. Adaptation to Elevated CO_2

Over many generations, natural selection in eCO_2 will create novel genotypes with different CO_2 responses than present in existing genotypes. In micro-algae, for example, eCO_2 favors genotypes lacking a CCM (Bell and Collins 2008; Raven et al. 2017). What evolutionary patterns might be expected in terrestrial plants in elevated CO_2, especially those with CCMs?

One evolutionary scenario is CCM plants might revert to C_3 photosynthesis. In the case of C_4 photosynthesis, reversion is unlikely due to the difficulty of reassembling the C_3 cycle as the C_4 cycle is dismantled in a series of mutational steps (Christin et al. 2010; Oakley et al. 2014). In CAM plants, reversion is more likely, as a C_3 system already functions in phase II and IV, and during wet periods in facultative CAM species. Phylogenetic evidence indicates reversions from CAM to C_3 happened in the bromeliads and orchids (Crayn et al. 2004; Silvera et al.

2009). C_2 plants might also revert to C_3 in eCO_2, since they maintain a C_3 cycle in the mesophyll tissue.

In eCO_2, the enriched carbon supply reduces the fitness benefits of high photosynthetic investment, such that selection favors other functions within the plant that are more limiting in high CO_2 atmospheres (Woodrow 1994; Sage and Coleman 2001). In situations where non-carbon limitations such as nutrient or water supply are significant, C_4 plants may initially be favored in eCO_2, since they maintain superior WUE and NUE, and appear to have already adapted their whole-plant performance for a high CO_2-like situation. Indeed, C_4 plants may serve as useful proxies for what a high-CO_2 adapted phenotype may look like in the C_3 flora (Kocacinar and Sage 2005). C_4 plants have low Rubisco investment relative to C_3 plants, low photosynthetic investment per leaf, and exhibit hydraulic transport properties that indicate they have optimized their xylem to exploit high WUE (Sage et al. 1989; Kocacinar and Sage 2005; Taylor et al. 2018). A big difference is that C_4 plants have additional costs to operate the CCM, which C_3 species won't have in a high CO_2 world. Through natural selection, an optimizing C_3 flora should overcome excess carbohydrate supply at eCO_2 and reduce feedbacks that attenuate responses to elevated CO_2. By doing so, the C_3 ability to suppress the C_4 flora could gradually increase, such that any initial advantages the C_4 flora may have as CO_2 increases could eventually disappear. How long this might take is uncertain, but could rapidly occur in species with short generation times, as has been documented in invasive annual grasses grown at elevated CO_2 (Peñuelas et al. 2013; Grossman and Rice 2014). Adaptation based on both existing genetic variation and new mutations is typically faster and more effective in larger populations (Kimura et al. 1963; Gravel 2016). Because many CCM species are niche specialists, and given the substantial habitat and population declines occurring in many CCM communities (Parr et al. 2014; Hultine et al. 2016), their capacity to adapt to global change could be low.

IV. The Terrestrial CCM Flora

Because C_4, CAM and C_2 species do not have the same distribution of life-forms as the C_3 flora, life form becomes an important co-factor determining the outcome of competition between CCM and C_3 plants. To realistically consider the future of CCM species in a high CO_2 world, it is necessary to review taxa and life forms that make up the various CCM floras, and the biomes where they commonly occur (Tables 8.1 and 8.2).

A. *C_4 Life-Forms*

C_3 photosynthesis is present in all life forms, and is particularly dominant in shrubs and forest trees. C_4 photosynthesis by contrast, is not common in the woody flora, but is well represented in grasses and sedges that together comprise the herbaceous, graminoid life form (Tables 8.1 and 8.2). Approximately 6400 C_4 species are graminoids, of which around 5000 are grasses and 1400 species are sedges (Sage 2016). C_4 sedges are common in marshlands, where flooding suppresses forest dominance. Approximately 1800 eudicot species are known, with the Amaranthaceae (=Amaranthaceae sensu stricto + Chenopodiacae) accounting for 800 species, including herbaceous ruderals and most C_4 shrubs. With the exception of desert shrubs in *Caligonum* (80 species of the Polygonaceae), C_4 eudicots outside the Amaranthaceae are largely herbs of ruderal habitats or locations of extreme heat, drought, and salinity (Sage 2016).

In contrast to these numbers, there are about 6000 C_3 grasses and sedges, most of which are either shade-adapted species of forests interiors or cool-adapted species of higher latitude, altitude, or cold-season habi-

Table 8.1. Principal life-forms of C_4 plants, with a note on C_3 life forms

Life-form	Species number	Examples	References
Trees	4	Two rare *Euphorbia* spp. (Euphorbiaceae) of Hawaii, < 15 m Two *Haloxylon* spp. (Amaranthaceae) of Asia/Africa <12 m	Sage and Sultmanis (2016)
Sub-trees (<5 m)	~10	Five *Euphorbia* species of Hawaii and about 5 *Haloxylon* species of Asia and Africa	Sage and Sultmanis (2016)
Woody vines	None	Some C_4 herbs are creeping vines along the ground especially in *Boerhavia* and *Allionia* (Nyctaginaceae) and *Tribulus* (Zygophyllaceae). None are woody climbing vines	Sage and Sage (2013)
Shrubs	400	Amaranthaceae: numerous xerophytic or halophytic *Aerva, Anabasis, Atriplex, Gomphrena*; *Haloxylon, Salsola, Suada* spp. Euphorbiaceae: numerous *Euphorbia* spp. Polygonaceae: 80 *Calligonum* spp.	Sage and Sultmanis (2016)
Herbaceous eudicots, including subshrubs with woody bases	1400	All eudicot C_4 families except Polygonaceae have herbaceous species, and many form woody tissues in rootstocks and stems, including some that are low stature, subshrubs. Life forms include summer annuals, desert ephemerals, annual and perennial herbs of low latitudes, tropical and summer-active weeds, psammophytic herbs of low latitudes, tropical beach herbs and many agricultural weeds	Sage et al. (1999), Sage and Sage (2013), Sage and Sultmanis (2016)
Graminoids (none arborescent)	~6360	5040 grasses (Poaceae, all in PACMAD clade) 1320 sedges (Cyperaceae, mostly in wetlands)	Sage (2016)
C_3 angiosperms	~225,000	All life forms – trees, shrubs, woody vines, herbaceous subshrubs and herbaceous annuals	Sage et al. (1999), Sage and Stata (2015). The species number assumes there are 250,000 angiosperms

Updated from Sage and Sage (2013)

tats (Table 8.1; Edwards and Smith 2010; Osborne et al. 2014; Kellogg 2015). These are habitats where C_4 species are uncommon, so the likelihood of direct competition between C_3 and C_4 graminoids is relatively low outside of the C_3 and C_4 ecotones in the temperate zone and higher elevations in the tropics (Sage et al. 1999; Griffith et al. 2015). On a global basis, the main interactions between C_3 and C_4 vegetation occur between species of different life form, notably, between C_4 graminoids and C_3 shrubs and trees (Bond 2008). Tens of thousands of C_3 woody species are present, of which most occur in the tropics and sub-tropics, where they are physiologically active under the same environmental conditions of high light, heat and drought as the C_4 flora, even in the 180 ppm CO_2 levels of the last ice age (Prentice et al. 2011; Sage 2013). As such, they experience high potentials for photorespiration, which means potentially large

Table 8.2. Major biomes of the world with high C_4 and CAM species richness, and the principle global change threats as discussed in the text

Biome	Major location	Principle global change threat	Current status of change
Biomes with high C4 representation			
Tropical, subtropical grassland and savanna	Central and South America, Africa, India, Southeast Asia, Australia, Oceanic islands of low latitude	Land transformation, overgrazing, CO_2 enrichment, woodland encroachment, invasive grasses	>70% lost to conversion to crop land, timber plantation and pasture. Woodland encroachment widespread, with invasive species infesting many remaining natural stands in the Americas.
Warm temperate grassland and savanna	Central and southeastern North America, northern Argentina, South Africa, China, southern Australia	Land transformation, overgrazing, CO_2 enrichment, woodland encroachment, invasive grasses	Heavily converted to row crop agriculture and pasture. >75% of original grasslands and savanna now converted. Remainder in degraded state from overgrazing and woodland encroachment, often assisted by fire control to protect rural homesteads.
Beach dunes, bluffs and coastal grassland	Global at mid-to-low latitude	CO_2 enrichment, climate change and sea level rise, coastal erosion, land transformation with vacation home and resort development	Uncertain, although coastal development is widespread in developed nations, particularly for vacation homes and resorts.
Subtropical wetlands (freshwater, non-arborescent)	Global, especially South America, Central Africa, Southeast Asia	Climate warming, CO_2 enrichment, land transformation, water diversion, groundwater pumping, and damming of watercourses; eutrophication	Lakes of semi-arid regions disappearing due to water diversion, while dams flood marsh and riverine habitat; large areas lost to rice cultivation.
Salt marsh (warm temperate to tropical)	Global, but less in tropics due to mangrove dominance	Mangrove encroachment, climate change with sea level rise, elevate CO_2, eutrophication, conversion to aquaculture farms, herbivore outbreaks (for example, by snails as in Silliman et al. 2005)	Uncertain, but poleward mangrove expansion documented. Temperate marshes also declining due to snail outbreak, erosion and eutrophication. Widespread conversion of estuaries to shrimp and fish farms ongoing in tropics.
Salt desert (<45° latitude)	Western North America, Central Asia, arid Africa, Central Australia	Climate change, CO_2 enrichment, overgrazing, wood cutting, water withdrawal for irrigation	Central Asian landscapes degraded by overgrazing and wood cutting, although expansion has occurred worldwide due to ground water pumping, water diversion and salinization following irrigation.

(continued)

Table 8.2. (continued)

Biome	Major location	Principle global change threat	Current status of change
Hot deserts and semi-deserts, arid steppe	Southwestern North America, Africa, southwest and Central Asia, India, Australia	Climate change, overgrazing, CO_2 enrichment, invasive species, woodland encroachment	Nearly all dry grasslands are in a degraded state due to overgrazing and/or shrub and tree encroachment. Invasive species encroachment widespread in semiarid regions.
Disturbed ground (low latitudes, low altitude)	Global	CO_2 enrichment (competition from C_3 species)	Expanding with human land use change and increasing exploitation. In particular, invasive, fire-responsive C_4 grasses have created large areas of low-diversity grassland formerly dominated by forest or natural grassland.
Biomes with high CAM representation			
Tropical Forest epiphytes on tree branches and trunks	Wet tropics and subtropics, worldwide	Deforestation, land use change, tree plantations, and invasive grasses accelerating destruction of forests. Often, runaway fire disturbance accelerates loss of the epiphytic habitat	40% of high-diversity tropical rainforest is converted and another 30% exists in a degraded state. Pace of forest conversion remains high, such that some regions will lose all undisturbed rainforest by 2050.
Semi-arid to arid desert, steppe, and woodland, to include south American Catiinga, Madagascar thorn-scrub, and succulent forests and seasonally dry forests	Temperature to tropical landscapes across the globe	Invasive grasses with fire cycle enhancement, land transformation for ag development, cutting for wood and charcoal, woodland encroachment, CO_2 enrichment, climate warming	Suitable dry habitat for CAM desert succulents may expand with climate change, but habitat loss due to land conversion, invasive grasses (both C_3 and C_4), woodland encroachment, fire, and grazing disturbance is widespread.
C_3 plants dominate all canopy forming forests worldwide; all polar, boreal and alpine landscapes; all cool temperate grasslands, savannas and saltmarshes; all mangrove swamps; all Mediterranean-type shrublands (chaparral and mattoral) and grasslands in California, southern Europe, southwestern and northern-most Africa, the western Middle East, southern Chile, Southwest Australia; and temperate deserts and semi-deserts			

Modified from Sage and Sage (2013).

responses to rising CO_2 (Sage and Kubien 2003; Bond and Midgley 2012).

Given the importance of C_4 grasslands, there has been a tendency to ignore landscapes where C_4 shrubs become important floristic elements. Approximately 400 C_4 eudicot species grow as robust shrubs (Table 8.1; Sage 2016) (> 0.2 m high; Sage et al. 1999; Sage and Sage 2013; Sage 2016). All are from rocky, sandy or clay soils in arid, semi-arid, and/or saline zones of temperate to tropical latitudes, where they can form dominant stands, particularly in arid Australia, western North America, Central and western Asia, and North Africa (Sage et al. 1999; Sage and Sage, 2013). About 10 of these shrub species, mostly in the Amaranthaceae genus *Haloxylon*, become tree-like (arborescent) with age, reaching statures above 5 m, and potentially forming tall thickets that resemble woodlands along watercourses in the dry landscapes of central Asia (Walter and Box 1986; Sage and Sultmanis 2016). With the exception of these arborescent thickets, woody C_4 plants do not form forests. Only four true C_4 trees have been identified (Table 8.1). Two are rare *Euphorbia* species restricted to the Hawaiian Islands and two are taller-stature species in *Haloxylon* (Sage and Sultmanis 2016). Thus, forested landscapes will be dominated by C_3 photosynthesis regardless of any other parameter that might otherwise favor C_4 plants, with the exception of the *Haloxylon* thickets. With some exceptions, where C_4 grasses can grow, C_3 woody species can also proliferate (Fig. 8.7). The exceptions are alpine tundra, deep-water swamps, temperate and boreal salt-marshes, and arid deserts. As a result, the probability that a C_3 woody canopy will establish and suppress a C_4 flora typically depends upon a gap in the occurrence of episodic factors that suppress woody vegetation, such as fire, floods, storms, waves, browsing, shallow soils, lethal drought, and human action (Sage et al. 1999; Bond 2008).

B. *CAM Life-Forms*

An estimated 6% of the world's flora use CAM, which is some 15,000–18,000 species (Borland et al. 2018). About a third are terrestrial succulents of deserts and dry microsites, particularly in arid-to semi-arid regions of the Americas and South Africa, while two-thirds are rainforest epiphytes in the tropics to subtropics, with the exception of Australia (Table 8.3; Smith and Winter 1996; Winter and Smith 1996b; Borland et al. 2018). Some

Fig. 8.7. Woody vegetation in the arid landscape of western Namibia, along road D3716 near Grosse Spitzkuppe, 30 km NW of Usakos, Namibia. The photo demonstrates the ability of woody species to establish in all but the most extreme deserts, including most areas where C_4 plants can proliferate. As a result, dominance of a C_3 woody versus C_4 grass floras will usually depend upon ecological controlling factors, not just the ability of a functional type to establish and grow. (Photo by R.F. Sage)

Table 8.3. Six major global change parameters of significance for future C_4 and CAM floras

Global change parameter	Time until significant impact	General effects
A. Atmospheric CO_2 rise	Now	Direct stimulation of photosynthesis (C_3 and CAM more than C_4) Direct stimulation of growth (C_3 and CAM more than C_4) Reduction in stomatal conductance and water use (all CCM spp.) Main effect to date is to promote woody species establishment in grasslands
B. Climate warming	<30 years	Stimulates C_4 photosynthesis, especially at intermediate levels of CO_2. Stimulates growth of C_4 plants and all CCM plants at eCO_2 if H_20 adequate Increases water consumption and drought probability. Effects on C_3 versus C_4 will depend upon timing of warming and moisture
C. Land transformation	Ongoing	Ag development has transformed large areas of grasslands and savanna. C_3-dominated forests in tropics and subtropics converted to low diversity C_4-dominated pasture. Forest loss is depleting CAM epiphyte habitat and diversity
D. Species exploitation	Ongoing	Overhunting reducing browsing, removing a check on woody encroachment Overgrazing selectively removes C_4 graminoid vegetation, disrupting fire cycles and enabling woody encroachment. Logging reduces habitat for CAM epiphytes
E. Terrestrial eutrophication	Ongoing	Enables C_3 plants to overcome C_4 advantage in N use efficiency. Promotes rapid Promotes growth and competition of weedy C_3 and C_4 species and reduces the magnitude of CO_2 acclimation in C_3 plants. Reduces establishment time of woody species, weakening fire and drought traps
F. Bioinvasions	Ongoing	Dozens of C_4 and CAM plants are aggressive weeds and exotic invaders Tropical C_4 grasses increase fire frequency and intensity, displacing native vegetation and promoting conversion to exotic-dominated C_4 grasslands Exotic C_3 and C_4 grasses are a major threat to CAM diversity through fire mortality in semi-arid rangelands and tropical forests

Developed from Vitousek (1994), Sala et al. (2000) Sage and Kubien (2003), and Sage (2020)
Abbreviations – *Ag* Agriculture, *eCO_2* Elevated atmospheric CO_2, *N* nitrogen

CAM plants, notably *Agave*, ice-plants, and *Opuntia* species, have naturalized beyond their range and in some cases are now invasive pests (Fig. 8.10c; Osmond et al. 2008). CAM species are uncommon where winter cold is severe, nights are warm, and moisture is both infrequent and unpredictable (Lüttge 2004; Holtum et al. 2016). Extreme deserts such as the inland Sahara and Arabian deserts, and the Australian deserts are therefore low in CAM diversity, possibly because of extreme heat, warm nights, and unpredictable precipitation with long gaps between rain (Holtum et al. 2016). These regions could become more suited to CAM if monsoons shift to bring more moisture. Life forms include the large columnar cacti of the Americas and similarly shaped *Euphobia* succulents in Africa. Other notable life forms include barrel-shaped stem succulents, the prickly-pear cacti, massive leaf rosettes such as Agave and Aloes, low-growing ice plants of the Aizoaceae, and small, surface-level succulents such as stone plants in the genus *Lithops* (Kluge and Ting 1978). CAM species in semi-arid regions attain tree-like stature in the eudicot genus *Clusia*, the monocot genus *Yucca*, and a number of the large stem succulents such as the Saguaro cactus. Epiphytes include succulent rosettes such as bromeliads, and a diverse array of orchids, *Peperomia* spp., and numerous eudicots including 124 cacti species (Lüttge 2007; Zotz 2016).

C. C_2 Life-Forms

Of the 60 or so C_2 species characterized, the large majority are herbaceous, growing in ruderal habitats, or on harsh desert sand-dunes, rocky hardpans, limestone caliche, or salinized soils (Sage et al Lundgren and Christin 2017). A few species such as *Mollugo verticillata* are widespread ruderals of warm habitats, but most occur in specialized habitats, for example, *Flaveria linearis* is commonly found along the salinized edges of coastal strands, mangrove swamps, or saltmarshes, and C_2 *Heliotropium convolvulaceum* is a specialist of deep sand in arid regions of the Americas. New C_2 species are being discovered each year, including a high number in the genus *Blepharis* (Acanthaceae), where Fisher et al. (2015) identified over a dozen candidate C_2 species. In *Blepharis,* putative C_2 taxa include shrubs, which occur in arid regions of southern Africa, in similarly dry conditions where C_4 shrubs occur (Vollesen 2000). C_2 species are poor competitors in general, and appear to be readily suppressed by C_3 and C_4 floras (Sage et al. 2014a). C_2 plants may be particularly vulnerable in a high CO_2 world if C_3 and C_4 ruderals more aggressively colonize areas where C_2 species are common.

V. History of Carbon-Concentrating Mechanisms

Paleoecology perspectives are relevant to global change discussions because past eras experienced high atmospheric CO_2 and warmer climates. Understanding how past global change affected CCM vegetation can thus inform how future change may influence CCM versus non-CCM plants.

Rubisco oxygenase activity was not widely significant in the early history of the earth when C_3 photosynthesis evolved, because atmospheric CO_2 was high and O_2 negligible (Sage 1999). Subsequent low CO_2 episodes in Earth's history would have increased photorespiration, particularly in warm climates, facilitating CCM evolution (Raven et al. 2017). In the 450 million years of terrestrial plant evolution, two low CO_2 episodes are inferred, the first occurring in the Carboniferous period 310 ± 40 mya, and the second in the past 30 mya, following global climate deterioration in the late-Oligocene epoch (Fig. 8.8a; Franks et al. 2013). No definitive evidence has been presented for any CCM prior to 30 million years ago (Ma), including the low CO_2 episode of the Carboniferous, although it is reasonable to assume that algal CCMs are ancient, and CAM in ferns and the aquatic lycophyte *Isoetes* suggests early land plants were capable of evolving CAM (Lüttge 2002; Raven et al. 2008).

Molecular clock analysis of phylogenetic trees indicate the earliest C_4 clades arose 25–30 Ma, in synchrony with the late-Oligocene CO_2 decline and global climate deterioration that included drier conditions at low-to-mid latitudes and cooling at higher latitudes (Fig. 8.8b; Vicentini et al. 2008; Christin et al. 2008, 2011). Similarly, molecular clock analyses indicate repeated origins of CAM in higher plants after the late-Oligocene (Arakaki et al. 2011; Horn et al. 2014). While there is some controversy on the precise date of the first C_4 origin due to calibration assumptions of the molecular clock methods (Vicentini et al. 2008), it is clear that throughout the Miocene epoch and into the Pliocene, C_4 and CAM repeatedly arose in a wide range of plant families (Strömberg 2011; Christin et al. 2011; Arakaki et al. 2011). The first clear isotopic evidence for C_4 vegetation dates to the early Miocene, by 20–23 Ma, and isotopic evidence for C_4 presence becomes more widespread and abundant towards the later Miocene (Fox and Koch 2003; Strömberg 2011). Atmospheric CO_2 levels do not appear to have changed appreciably throughout the Miocene epoch, remaining around 300–400 ppm (5–25 Ma; Kürschner et al. 2008; Zhang et al. 2013). Dry habitats expanded as the Miocene pro-

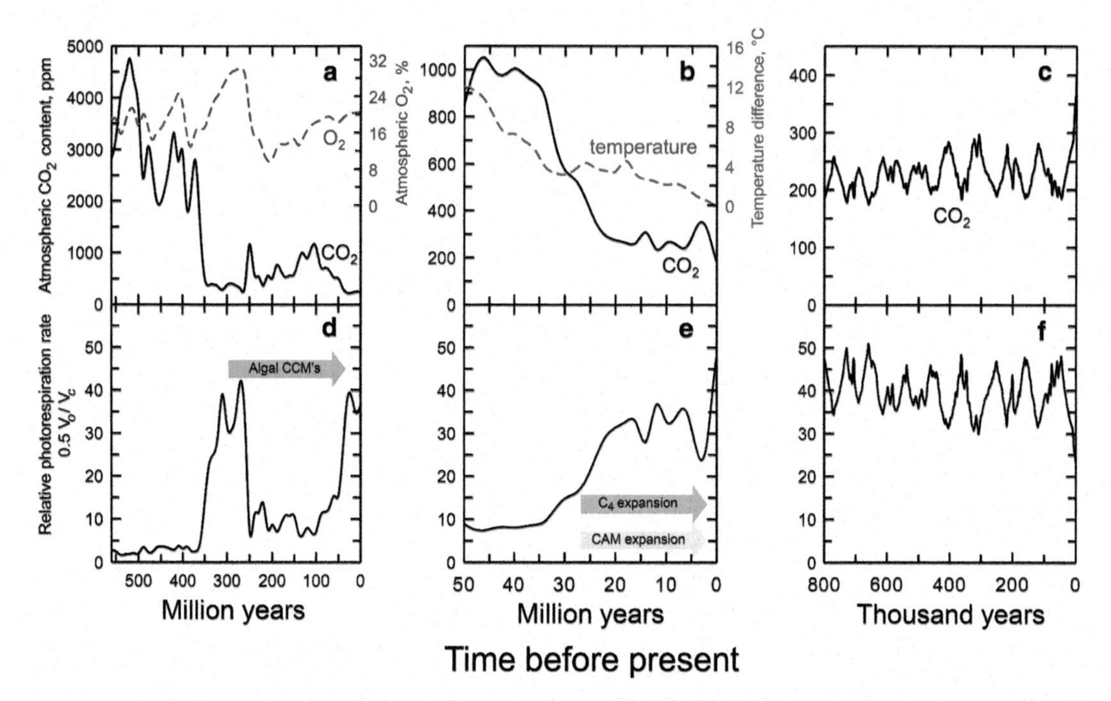

Fig. 8.8. Atmospheric CO_2 and O_2 concentrations for: (**a**) the past 560 million years; (**b**) the past 50 million years, and (**c**) the past 800,000 years. Panels d, e, and f show the corresponding modelled value of relative photorespiration (0.5*Vo/Vc*) at 32 °C, assuming a model C_3 plant Rubisco (van Cammerer and Quick 2000), and a 40% reduction in CO_2 concentration from the modelled atmospheric value to the chloroplast value. Atmospheric CO_2 and O_2 values in panel a are midpoint estimates from the model of Berner (2008) (simulation basalt/granite weathering = 2, NV = 0 and fB(0) = 5). Atmospheric CO_2 estimates in panel b are based on a estimates from Royer et al. (2012), and in panel C are from the Vostok ice core in Antarctica as shown by Gerhart and Ward (2010). Biological radiations are delineated for algal carbon concentrating mechanisms (Badger and Price 2003; Raven et al. 2012), arid-adapted taxa (Sage 2001), C_4 radiations (Edwards et al. 2010; Christin et al., 2011), and CAM expansion (Edwards and Ogburn 2012)

gressed, particularly in the sub-tropics, which promoted radiation of xeromorphic taxa, including clades where C_4 and CAM eventually evolved (Zachos et al. 2008; Senut et al. 2009; Graham 2010; Pound et al. 2012). By the late Miocene, arid landscapes such as the Sahara, Mojave and South African deserts appeared, and grassland/savanna systems expanded across the Earth (van Devender 2000; Schuster et al. 2006; Senut et al. 2009; Strömberg 2011; Pound et al. 2012). In southern Asia, Africa, and the Americas, the carbon isotope ratio ($\delta^{13}C$) of fossilized soil carbonates, plant phytoliths, and animal remains demonstrate a pronounced expansion of C_4-dominated grasslands between 5 and 10 Ma (Cerling et al. 1997; Strömberg 2011; Bouchenak-Khelladi et al. 2014). Phylogenetic evidence also indicates a third of all C_4 origins correspond to the late-Miocene at 5–10 Ma (Sage 2016). Because evidence for a late-Miocene CO_2 decline is uncertain, the predominant view is that C_4 grasslands expanded at 5–10 Ma in response to further climate deterioration that increased seasonality and fire frequency, and promoted the radiation of mammalian megafauna that favored grass over woodland cover (Beerling and Osborne 2006; Tipple and Pagani 2007; Edwards et al. 2010; Osborne 2011; Karp et al. 2018).

In the past five million years, atmospheric CO_2 declined below late-Miocene values to under 300 ppm, which persisted over the past two million years (Fig. 8.8b; Bartoli et al. 2011; Zhang et al. 2013; Higgins et al. 2015).

By the late-Pleistocene epoch (800 to 12 thousand years ago), atmospheric CO_2 levels were oscillating between 180 ppm and 300 ppm in phase with glacial to interglacial cycles (Fig. 8.8c; Higgins et al. 2015). Climates also became colder as Earth slipped into the Pleistocene ice age, with additional increases in aridity at low latitudes (Martin 2006; Feakins and Demenocal 2010; Liddy et al. 2016). Further expansion of C_4 grasslands occurred 2 to 4 Ma in eastern and southern Africa (Cerling et al. 2015; Uno et al. 2016). By the Pleistocene epoch (2 to 0.12 Ma), atmospheric CO_2 levels periodically fell below 200 ppm, representing the lowest value in the history of land plants (Higgins et al. 2015). These low CO_2 episodes are predicted to have favored further expansion of C_4 grasslands and savannas in low-to-mid latitudes (Collatz et al. 1998; Prentice et al. 2011; Hoogakker et al. 2016). Across low latitudes during the last glacial maximum (LGM) 20,000 years ago, Prentice et al. (2011) model a greater extent of desert, dry grass and shrubland, and less woodland compared to recent, pre-industrial time, and summarize many paleo-ecological studies that support the modelled output (see also Harrison and Prentice 2003; Hoogakker et al. 2016). Prentice et al. (2011) note that consistency between model and paleo-data only occurs when the low CO_2 of the LGM is included in the model. While C_4-dominated landscapes appear more widespread at the LGM, it is important to also note C_3-dominated forests were still widespread in the tropics and subtropics, and even expanded at higher elevation where the climate at the time was cooler than now (Bush et al. 2007; Dupont 2011; Miller et al. 2016; Shanahan et al. 2016). Thus, although atmospheric conditions markedly increased photorespiratory potential and inhibited C_3 photosynthesis relative to C_4 photosynthesis, the far-richer C_3 flora was able to persist and still dominate much of the low-latitude landscape of the late-Pleistocene (Bush et al. 2007; Sage 2013).

The termination of the Pleistocene ice age and the entry into the Holocene epoch (last 11,500 years) provides a good case study on the impacts of climate warming and CO_2 increase on the relative success of C_3 and C_4 species. At the end of the Pleistocene and during the early Holocene, the Earth's annual mean temperature on land warmed by 5–7 °C, with most of the warming coming at high latitudes (Collatz et al. 1998; Masson-Delmotte et al. 2013). Atmospheric CO_2 increased from 180 ppm 18,000 years ago to 260 ppm 10,000 years ago, and stabilized near 275 ppm in the mid-to-late Holocene (<5000 years ago, Gerhart and Ward 2010). This increase in CO_2 has been experimentally shown to improve C_3 relative to C_4 performance (Sage 1995; Dippery et al. 1995; Polley et al. 2003a; Gerhart and Ward 2010). In the transition from Pleistocene to Holocene conditions, C_4 grasslands expanded across central North America, as periglacial boreal- and cool-temperate ecosystems retreated northward with climate warming (Kelly et al. 1991; Nordt et al. 2008; Prentice et al. 2011; Cotton et al. 2016). The warming appears to have removed a cold-filter that kept all but the most cold-tolerant of C_4 species out of the mid-latitude boreal and tundra zones of the Ice Age. In contrast to temperature, the change in CO_2 had no clear influence over the movement of C_4 grasslands in the temperate zone (Cotton et al. 2016; Still et al. 2019). Elsewhere during the mid-Holocene warm episodes, C_4-dominated grasslands expanded into the Sahara during a moist interlude between ice age and late Holocene hyper-aridity, while forests expanded across sub-Saharan West Africa (Dupont 2011; Hoag and Svenning 2017). Across the Holocene, human-caused fires become more important, and increased the area of derived grasslands and savanna in tropical and temperate latitudes (Bond 2008; Bird et al. 2013; Garcin et al. 2018). The advent of anthropogenic fires, which first burned on all continents in just the past 15 millennia, coupled with low atmospheric

CO_2, persistent dry climates, and interglacial warmth, suggest that globally, C_4 and CAM vegetation may have reached their greatest diversity and extent in just the past few thousand years.

Interestingly, in the grassland regions of today, there has not been a detectable expansion of C_3 or C_4 grasses in the past 300 years, when CO_2 levels have increased some 50% (Griffith et al. 2017). However, woodland encroachment into grasslands has been widespread, for a number of reasons including rising atmospheric CO_2 (Bond and Midgley 2012; Archer et al. 2017).

VI. Global Change Drivers

To address how plants with CCMs will fare in a high CO_2 world, we must consider effects of each global change driver on CCM performance (Sage and Kubien 2003). Six anthropogenic global change drivers are relevant to CCM success: (1) atmospheric CO_2 enrichment; (2) climate warming and associated precipitation change; (3) land transformation; (4) overexploitation of natural species and associated disruptions to ecological networks; (5) eutrophication of land and water due to N emissions from agriculture and industry; and (6) ecosystem meltdown from exotic species invasion (Table 8.3; Figs. 8.9 and 8.10).

A. *Atmospheric CO_2 Enrichment*

Due largely to fossil fuel consumption and deforestation, atmospheric CO_2 concentration has increased from a pre-industrial value near 280 ppm to 415 ppm in 2019, with a current rate of increase between 2 and 3 ppm per year (www.CO2.earth, 2019). With the exception of the past 120 years, direct measurements of CO_2 trapped in polar ice show that atmospheric CO_2 over the past 800,000 years ranged between 180 and 300 ppm with an average near 240 ppm (Sage and Coleman 2001; Higgins et al. 2015). These low CO_2 values occurred over a sufficient timespan that they represent the evolutionary baseline for the current flora (Sage and Coleman 2001). The atmosphere is now approaching twice the evolutionary baseline. Assuming a mean 2.5 ppm per year increase for the next few decades, atmospheric CO_2 will double the 240 ppm mean of the past 800,000 years by 2046. Because CO_2 is a substrate for Rubisco carboxylation and an inhibitor of Rubisco oxygenation, its proportional effect is greater at lower CO_2 concentration than high, following Michaelis-Menton kinetics (von Caemmerer 2000). As a result, the first doubling of CO_2 above 240 ppm should have greater impacts on photosynthesis and growth than any subsequent doubling, as enzyme saturation attenuates responses at higher values. Because of this, the world's flora already exists in a CO_2-enriched world and enrichment effects should already be evident.

Predicting future CO_2 change is highly uncertain given the underlying economic, social and technological trends that determine CO_2 emissions. Business-as-usual forecasts such as the IPCC RCP 8.5 indicate the atmosphere could be experiencing Cretaceous-like CO_2 levels above 800 ppm after 2100 (Collins et al. 2013; Nazarenko et al. 2015). Realistic low growth scenarios (RCP 4.5) predict CO_2 levels approaching 550 ppm by 2100 (Nazarenko et al. 2015). While the advent of high CO_2 concentrations in the atmosphere is noteworthy, the duration of elevated CO_2 also has to be considered, as the fate of the CCM flora will depend upon

Fig. 8.9 (continued) that once dominated the landscape. (f) **Exotic invasions:** In the photo, weedy C_4 grasses (*Saccharum* spp) have displaced native woodlands on the north shore of Oahu, Hawaii through aggressive growth and intensification of destructive fire cycles. Remnants of the woodland are evident in the background

Fig. 8.9. Global Change and C_4 Plants. The six photos illustrate impacts of individual global change drivers as discussed in the text. All photos by RF Sage. (**a**) **Atmospheric CO_2 enrichment**: The photo shows tree and shrub encroachment into formerly C_4-dominated desert grasslands of southwest Texas. Remnant C_4 grasses are present in the herbaceous layer in the foreground. (**b**) **Climate change**: Moderation of temperatures in the cool-temperate zones is creating opportunities for northward expansion of C_4 vegetation. The photo shows a *Spartina*-dominated saltmarsh forming along a bay near Gambo, Newfoundland, with a dense sward along the upper tideline (white arrows), and patches of grass colonizing the tidal mudflats (black arrows). Local residents told R.F. Sage that *Spartina* was absent in the bay a generation ago. (**c**) **Land transformation:** Champaign-Urbana, central Illinois USA. Central Illinois was the location of the Tall-grass Prairie peninsula, an eastward extension of the vast tall-grass prairie biome that has been transformed into maize and soybean farms, replacing an ocean of prairie with a sea of maize, as shown in the photo. The conversion of the tall-grass prairie biome demonstrates the ability of humanity to completely eradicate entire biomes. (**d**) **Overexploitation of C_4 grasses**: The photo shows an over-grazed landscape in southern Africa, where cattle and goat populations remain poorly regulated. The removal of grasses opens up sites for colonization by woody species, erosion and desertification. (**e**) **Terrestrial eutrophication**: The photo shows the border of a maize field dominated by C_4 weeds, notably *Amaranthus retroflexus*, the broad leaf weed in the foreground. The weed responds aggressively to fertilizers blowing off the cultivated field, crowding out native species to create low diversity, weed-dominated landscapes. In farming regions, fertilizer pollution enables weedy, often exotic species to suppress remnant populations of former grassland natives

Fig. 8.10. Global Change and CAM Plants. The four photos illustrate impacts of global change of relevance to terrestrial CAM plants. (**a**) A prickly pear cactus (*Opuntia* spp.) that expired following a record heat wave in the Mojave desert in 2014, 30 km north of Las Vegas, Nevada. Cacti and other CAM succulents of hot, semi-arid regions live near the upper thermal limit of plant life. With record heat waves now occurring in places such as the Mojave desert, succulents are beginning to die from climate stress, particularly smaller individuals in the boundary layer where heat is trapped and lethal thresholds are more likely to be breached. (**b**) Fires driven by invasive grasses and increasing vegetative density in general are major threat for CAM succulents in semi-arid landscapes, as shown here for the western Mojave desert region of California, where Joshua tree (*Yucca brevifolia*) is abundant. Dense stands of Joshua trees occur in the foreground and middle of the photo, but are absent on the light-colored burn scars in the upper middle region of the photo (arrow). Red brome infestations have filled in the fuel layer in the foreground, enabling frequent hot fires that destroy the Joshua trees and other succulents such as cacti, as shown in panel (d). (**c**) A prickly pear cactus infestation in central South Africa, near Marydale. Prickly Pears remain one of the most invasive of CAM plants due in part to the ability to colonize degraded land via asexual reproduction from rooted cladodes. (d) Mojave desert, California – an infestation by the exotic invader *Bromus rubrens* (red brome, a C_3 grass) forms dense stands that compete for water, increasing stress levels in the cacti and other CAM species (as indicated by red pigmentation in the cactus stems and spines in the foreground). The extent of the brome infestation is evident along the hillside behind the cacti.

the time required for photosynthetic and stomatal responses to alter ecological and evolutionary outcomes. If C_3 dominance requires evolutionary optimization of the C_3 flora to elevated CO_2, t than the period of enriched atmospheric CO_2 may be less than the time required for complete suppression of the C_4 flora.

Because Earth system controls over atmospheric CO_2 have not changed, CO_2 levels will eventually return to pre-industrial levels once humanity exhausts or replaces fossil fuels, perhaps after a few millennia (Archer and Brovkin 2008; Eby et al. 2009). A few thousand years may be too short for evolutionary optimization, but more than enough time for altered ecological outcomes. One critical variable is the impact of other global change drivers, which could on the one hand offset benefits of eCO_2 for the C_3 flora (for example, from climate warming and drying), but could also aggravate relative CO_2 impacts through habitat loss, disruption of essential ecological interactions, and increased stress. In any case, atmospheric CO_2 concentrations will eventually return to the low values of recent geological time. If in the meantime global change substantially reduces CCM diversity, there may be few C_4, CAM and C_2 species left to exploit low CO_2 atmospheres once they return.

B. *Climate Warming*

With the rise in greenhouse gas content in the atmosphere, global mean temperature has already warmed by almost 1 °C and is expected to increase by another 1 1-4.8 °C by the end of the century, depending upon emission scenario (Collins et al. 2013). Warming alone generally favors species with CCMs over C_3 species because increased photorespiration hinders the performance of C_3 plants. This advantage will attenuate in a high CO_2 world, which theoretically should reduce the thermal range where C_4 species will be favored. $_2$. However, the relative impact of warming on C_4 versus C_3 vegetation will vary with latitude and season, and associated changes in water availability.

An important aspect of global warming is the variation in space and time, with winters predicted to warm more than summer, nights more than days, and high latitudes more than low latitudes (Collins et al. 2013). The change in the frequency and severity of extreme temperature events will be critical, with fewer cold extremes and more frequent heat extremes predicted and already underway (Cubasch et al. 2013). This becomes important for C_4 and CAM species because cold filters that currently exclude them from higher latitude and elevation could be relaxed (Fig. 8.9b), while summer warming could lead to more severe heat waves and drought that could weaken summer-active species, which include much of the temperate zone C_4 and CAM flora (Fig. 8.10a). If warming leads to milder winters, then a cool growing season favoring C_3 plants could expand. Alternatively, heat-waves and drought events in the cool growing season could stress the C_3 flora and open up a longer growing season for the C_4 and CAM floras, thereby offsetting any C_3 advantage conferred by eCO_2. Nighttime warming could be particularly bad for CAM species, because increased transpiration at night could offset the CAM advantage, unless CO_2-induced stomatal closure compensates for increased evapotranspiration (Hultine et al. 2016). An additional aspect of extreme heat is greater heat-induced sterility, which could be particularly damaging for summer flowering species in hot environments, such as C_4 and CAM plants. Exposure to >35 °C during pollination is sufficient to sterilize a wide range of C_3 and C_4 species (Sage et al. 2015).

With climate change comes altered precipitation, of which the direction and timing of change will be paramount. If additional moisture falls as winter and spring rain, a C_3 flora would be favored. If it falls in summer, a C_4 flora should be favored as long as woody species do not dominate. Likewise, if winters become drier, the C_3 flora could suffer, while drier summers could disproportionately harm a C_4 flora unless fire regimes become more severe and increase mortality of C_3 woody plants and CAM species. Increased precipitation extremes are predicted for warmer climates, with more intense drought events, and a greater frequency of heavy precipitation, particularly in monsoon climates (Hartmann et al. 2013). Monsoons may strengthen, which could bring more moisture to arid landscapes in summer, although in heavier, less frequent rain events (Lin et al.

1996; Kitoh et al. 2013; Knapp et al. 2017). Heavy precipitation events are noted to facilitate shrub establishment in semi-arid steppes with a dominant C_4 grass flora, because water can percolate to deeper soil zones where the tap roots of C_3 shrubs and trees proliferate (Polley et al. 1997; Archer et al. 2017). Increased drought intensity due to less precipitation and/or greater evapotranspiration could aggravate drought, shifting already hot, dry landscapes where CAM and C_4 plants are prevalent into barren deserts, where C_4 and CAM diversity is low (Lüttge 2004). Alternatively, new habitat for C_4 and CAM should open up where moist landscapes become drier, as predicted for the central USA, or where extremely arid landscapes gain moisture, for example from strengthened monsoons, as may occur in the Saharan desert region (Collins et al. 2013).

C. *Land Transformation*

Land-use and land-cover change is a leading threat to biodiversity across the planet because it entails severe habitat degradation if not outright transformation to a human-dominated landscape (Sala et al. 2000; Newbold et al. 2017). So pervasive is land appropriation by humans that natural landscapes are no longer predominant over many of the regions where CCM plants occur (Ramankutty et al. 2008). Instead, varying degrees of human-affected landscapes cover most of the planet outside of arid deserts and polar tundra, with the predominant land use being row-crop agriculture, forestry, and grazing (Foley et al. 2005; Ellis and Ramankutty 2008). C_4-dominated grasslands and savannas are particularly impacted, since they have features desired by farmers, pastoralists and ranchers, such as open landscapes with deep loamy soils. By the year 2000, over 70% of the world's grasslands were converted to cropland, pasture, grazed rangeland, or existed in a degraded condition (Fig. 8.9c,d; Ramankutty et al. 2008). In North America, half of the temperate grassland and savanna that once covered the continent is now converted, and specific biomes dominated by native C_4 grasses are now largely eliminated. Over 95% of the native tall-grass prairie biome has been lost to cropland, cities, and pasture, with isolated remnants found in small preserves, roadsides and graveyards (Hoekstra et al. 2005). In the southeastern USA, the longleaf pine-wiregrass savanna has also been largely converted to crop, pasture or timber plantations (Noss 2013). In South America, over half of the Brazilian savanna (termed cerrado) has been converted to croplands and sown pasture (Lehmann and Parr 2016; Stevens et al. 2017). In Minas Gerais state, Brazil, 17% of remaining cerrado was lost between 2001 and 2015, with conversion to pasture and forestry plantations being major causes (Espirito-Santos et al. 2016). Secondary cerrado can form once pasture is abandoned, but it is considered degraded, with low diversity (Espirito-Santos et al. 2016; Veldman 2016). C_4 grasses often are present on former grasslands and savannas, but as cultivated monocultures of maize, sugar cane, millets, and sorghum, or as low-diversity stands of C_4 crop weeds, pasture grasses, and invasive species (Lehmann and Parr 2016). In drier regions, rangeland grazing often causes landscape degradation (Ramankutty et al. 2008). Fire cycles are typically disrupted by heavy grazing, due either to active fire suppression, fragmentation, or a loss of the continuous fuel layer (Stevens et al. 2016; Venter et al. 2017). This, in combination with reduction of the C_4 grass competition, facilitates woody plant encroachment, and eventual conversion to dense woody thickets, as is widespread in the American Southwest, southern Africa, and dryland Australia (Van Auken 2000; Briggs et al. 2005; Stevens et al. 2016; Skowno et al. 2017; Venter et al. 2017; Andela et al. 2017). Restoration of degraded or converted grassland is of limited value from a biodiversity standpoint, as the species planted are typically a poor reflection of the original diversity, especially when grazing is the restoration objective (Veldman 2016).

Deforestation potentially creates new habitat for C_4 and CAM species, however, deforestation is often followed by agricultural or forestry plantations, or invasion by exotic C_4 grasses that arrest forest recovery by promoting destructive fire cycles (Cochrane et al. 1999; Sage and Kubien 2003; D'Antonio et al. 2017). In the New World tropics and subtropics, almost 60% of new grassland was created at the expense of forest between 2001 and 2013 (Graesser et al. 2015). This derived, secondary grassland is typically of low diversity and dominated by commercially significant C_4 grasses or invasive C_4 exotics (Bond and Parr 2010; Veldman 2016).

Tropical deforestation is perhaps the single greatest threat to the global diversity of CAM photosynthesis, because it results in habitat loss and destruction of the epiphytic flora, which represents the majority of CAM species (Lüttge 2004). With tropical deforestation exceeding 75% in many regions of the world and with more than 60% of low latitude forests in a degraded condition (Sanderson et al. 2002; Ellis et al. 2010; Hansen et al. 2013; Zotz 2016), it can be assumed that large reductions in CAM epiphyte diversity have already occurred, with many species likely lost before they could be identified. CAM epiphytes will recolonize secondary forest and tree plantations, but the rate of recovery can be slow and represents a fraction of the original diversity (Zotz 2016). Epiphyte diversity is particularly hindered when rainforest is reduced to isolated patches, due to increased drying of the patch edges and reduced availability of epiphyte propagules to reseed recovering forest patches.

D. *Exploitation of Natural Species*

Humans now appropriate close to a third of the primary productivity on Earth, a value that increases year-by-year with population and economic growth (Vitousek et al. 1997b; Krausmann et al. 2013). For plants with CCMs, this exploitation has a number of consequences, the first being habitat loss as open landscapes are converted to managed grasslands, farms, and forest (Ramankutty et al. 2008; Campbell et al. 2017). A second is widespread grazing by cattle and sheep (Fig. 8.9d), and the third is reduction of wild populations via hunting and poaching (Archer et al. 2017; Hempson et al. 2017). Overgrazing selectively removes C_4 graminoids, potentially shifting the balance to favor C_3 forbs and woody plants. Nearly all remaining C_4-dominated grasslands and savannas not set aside as preserves are currently impacted by grazing, and as a result, a large majority of this vegetation exists in a degraded condition (Ramankutty et al. 2008; Parr et al. 2014; Veldman 2016). Overhunting of the natural fauna is a serious threat for terrestrial CCM species because trophic interactions are disrupted, resulting in ecosystem cascades that contribute to switches from C_4-grass to woodland-dominated ecosystems (Stevens et al. 2016; Hempson et al. 2017; Osborne et al. 2018; Davies and Asner 2019). In the case of grasslands and savanna, large browsing animals such as elephants can physically destroy woodlands by knocking down trees and shredding branches. Browsers in general thin canopy foliage and selectively eat seedlings of woody species establishing in grass swards (Dublin et al. 1990; Dublin 1995; Davies and Asner 2019). By doing so, they reduce established woodlands and arrest woodland encroachment into existing grasslands. In sub-Saharan Africa, elephant populations have declined 85% since European colonization, removing their impact outside of preserves (Ripple et al. 2015; Chase et al. 2016). Since the middle of the last century, woodlands have increased 11–17% in savanna biomes of sub-Saharan Africa, in part due to disrupted trophic dynamics caused by elephant loss (Stevens et al. 2016).

E. Terrestrial Eutrophication

A particularly limiting nutrient for terrestrial plants is N, which is available in reactive forms such as nitrate and ammonia. Bioavailable N is naturally replenished in ecosystems by biological fixation of atmospheric N_2 and abiotic N_2 fixation by lightning. However, humans are now producing more available N through industrial fixation of N_2 for fertilizers, cultivation of legume crops, and fossil fuel combustion than all natural processes combined (Vitousek et al. 1997a). By the mid-1990s, anthropogenic production of reactive N equaled all natural sources, and by 2050 may exceed natural N production over three-fold (Vitousek et al. 1997a; Galloway and Cowling 2002; Fowler et al. 2013). Once released into the environment, anthropogenic N disperses into land and water, increasing ecosystem productivity in a process termed eutrophication. While all plants, including those with CCMs, are potentially stimulated by N addition, species adapted to exploit high nutrients through rapid growth suppress more conservative species that comprise most of the floristic diversity on a landscape (Fig. 8.9e; Vitousek et al. 1997a; Simkin et al. 2016). In the case of C_4-dominated ecosystems, N addition can be particularly threatening because the higher NUE of C_4 plants and CAM plants is offset by the ability of C_3 competitors to acquire enough N to overwhelm the C_4 advantage (Wedin and Tilman 1996; Sage and Pearcy 2000; Isbell et al. 2013). In the ranglelands of North America, N deposition enables C_3 grasses to grow prolifically in the spring, before C_4 grasses break dormancy, and in doing so, form a dense sward that suppresses the C_4 grasses once they become active (Sage et al. 1999). Nitrogen deposition also favors C_3 grasses by sustaining a high CO_2 stimulation of photosynthesis, and in doing so can favor rapidly growing annual grasses that are highly N responsive (Vitousek et al. 1997a). In drier climates, increased biomass following N deposition intensifies fire cycles, which threaten native CAM and C_4 species. In savannas, N addition stimulates woody species growth and establishment, facilitating woody plant establishment in C_4 grasslands (Skowno et al. 2017).

F. Exotic Species Invasions

Exotic species invasions are also critical threats to C_4 biodiversity through competitive displacement, altered nutrient cycling, changes in disturbance regimes, and altered patterns of herbivory (Mack et al. 2000; Ehrenfeld 2010; Walsh et al. 2016). The worst invaders are the so-called meltdown invaders that can completely displace native ecosystems. Meltdown invasions caused by exotic C_4 grasses are common in the Americas, southeast Asia and Pacific Islands (Fig. 8.9f). Typically, these exotic grasses facilitate greater fire frequency and intensity at the expense of native C_3, C_4 and CAM species (Figs. 8.9f and 8.10b; D'Antonio and Vitousek 1992; Baruch 1996; D'Antonio et al. 2017). African grasses are commonly the worst invaders, due to high growth rates and prior adaptation to human activity (Mack and D'Antonio 1998; Morales-Romero et al. 2019). As an example of their impact on native grassland diversity, invasion of South American savannas by the C_4 grass *Hyparrhenia rufa* reduced herbaceous species richness by 80%; superior leaf production and greater tolerance of defoliation by grazing explained the success of *H. rufa* (Baruch 1996). In the North American prairies, the African *Anrodpogon bladhii* has become problematic (Reed et al. 2005). This species responds well to fire, creating a dilemma for preserve managers in that burning controls required to suppress woody encroachment also promote *A. bladhii* infestation, leading to an overall decline in native diversity.

For CAM species, invasive grasses are one of the greatest threats, as they establish intense fire activity that destroys epiphyte-rich forests and desert succulents. In Mexico and the American Southwest, exotic annual C_4 grasses (*Pennisetum ciliare, Eragrostis lehmanniana*), and an annual C_3 grass (*Bromus rubens*), are major contributors to reductions in CAM succulent diversity (Figs. 8.10b, d; Humphrey 1974; D'Antonio and Vitousek 1992; Archer and Predick 2008; Defalco et al. 2010; Abatzoglou and Kolden 2011). In tropical forests, loss of epiphytic habitat is often driven by invasive C_4 grasses that can intensify destructive fire cycles that promote deforestation, often in concert with selective logging and road construction (D'Antonio and Vitousek 1992; Cochrane et al. 1999; Laurance and Arrea 2017).

VII. The Future of Terrestrial Carbon Concentrating Mechanisms

Because large fractions of natural habitat have already been transformed into low diversity, human-dominated landscapes, consideration of future CCM diversity has to focus on habitat not yet transformed into farms, pasture, cities, timber plantations, and weed fields (Parr et al. 2014; Veldman et al. 2015; Hultine et al. 2016). Some of this landscape resides in nature reserves where species and natural processes are protected from many forms of global change, though not from climate warming and atmospheric CO_2 enrichment. On the broader landscape beyond reserves, discussions of CCM futures have to consider the full spectrum of global change drivers, a daunting task given the number of interacting agents. To address how CCMs might actually fare in future ecosystems, we take the following approach: we first address results of CO_2 enrichment studies examining C_3 and C_4 biomass responses in order to consider how this one global change agent might influence relatively undisturbed ecosystems. To consider cases where all global change drivers are present, we consider two biomes – C_4 dominated saltmarsh and C_4-dominated grasslands and savannas – where ongoing trends, coupled with modelling and experimental results, allow us to predict realistic futures despite the complexity. A broader summary of global change effects on biomes rich in C_4 diversity is presented in Table 8.3. We then address how CAM and C_2 species may fare in a globally-changed world.

A. CO_2 Enrichment of Natural Communities with C_3 and C_4 Plants

Table 8.4 highlights the results of 14 studies examining eCO_2 effects on C_3 and C_4 vegetation. Three are meta-analyses that summarize results of most CO_2 enrichment studies that evaluated C_3 and C_4 responses before 2010. Ten present results of C_3 versus C_4 responses to CO_2 enrichment in more realistic settings, namely multi-species communities using outdoor FACE (Free-air CO_2 Enrichment), semi-open chambers, or tent systems to elevate atmospheric CO_2 concentration. Eight of these outdoor studies were conducted over multiple growing seasons.

Results from two-thirds of the 14 studies in Table 8.4 support the hypothesis that eCO_2 will favor C_3 biomass production over C_4 production, particularly over longer periods. One clearly shows C_4 plants respond more to eCO_2 (Reich et al. 2018), and four have no clear eCO_2 effect on C_3 versus C_4 biomass performance. The comprehensive meta-analysis of Wand et al. (1999) summarizing the first wave of eCO_2 studies (many of which involve potted plants in growth chamber or greenhouse environments) indicate eCO_2 enhances both C_3 and C_4 growth, with C_3 growth being greater (44% for C_3 to 33% mean enhancement for C_4 plants). The Ainsworth and Long (2004) meta-analysis of FACE studies prior to 2004 observed less dramatic responses, with no clear promotion of growth by eCO_2 in C_4 grasses but a 10% stimulation of C_3 productivity. Notably, they observed a strong (24% to 28%) stimulation of biomass production in eCO_2 by C_3 legumes

Table 8.4. A summary of results from studies directly assessing effects of elevated CO_2 on C_3 and C_4 vegetation

Location	CO_2 enrichment method	Years in operation	Relevant results	References
Meta-analysis of >30 studies (exact number depends upon co-variables considered)	Largely growth chamber and greenhouse, mostly potted plants. Some open-top chambers and FACE	Covers studies form 1981–1997	Mean C_3 stimulation of growth by eCO_2 was 44%, while mean C_4 stimulation was 33%. Stomatal conductance in eCO_2 was reduced in C_3 by 24% and in C_4 by 29%, C_4 WUE in eCO_2 was 72% higher than in ambient CO_2.	Wand et al. (1999)
Meta-analysis of 12 studies, but only 5 C_4 species from two studies	Only FACE	Single season to multi-year	No average stimulation of C_4 crop or wild grass biomass production by eCO_3, while C_3 grasses responded to eCO_2 with a 10% enhancement. C_3 trees and legumes had a substantial biomass response to eCO_2 (24% to 28% growth stimulation).	Ainsworth and Long (2004)
Meta-analysis of over a dozen studies with both elevated CO_2 and temperature treatments.	Potted plants in chamber and greenhouse, naturally rooted in semi-open chambers	Studies before 2010, usually short duration	C_3 biomass stimulated more by eCO_2 at ambient T than C_4, while at elevated T, the C_4 and C_3 biomass responses to eCO_2 were similar; however, root biomass responded more to eCO_2 in C_4 than C_3 at elevated T. Photosynthesis in woody C_3 plants exhibited the strongest response to eCO_2 of any functional group.	Wang et al. (2012)
Chesapeake Bay Salt Marsh, Maryland USA	Open-top chamber	28	Increase in C_3 dominant sedge and reduction in C_4 dominant grass in 28 years of elevated CO_2; sea level rise also favored C_3 sedge over C_4 grass dominants, particularly in combination with elevated CO_2.	Drake (2014)
Eucalyptus Savanna, New South Wales, Australia	FACE	3	C_4 grass cover in the *Eucalptus* understory was 59% lower than C_3 graminoid cover in eCO_2. The C_4:C_3 ratio in eCO_2 was inversely related to soil N.	Hasegawa et al. (2017)

(continued)

Table 8.4. (continued)

Location	CO_2 enrichment method	Years in operation	Relevant results	References
Tallgrass Prairie, Kansas USA	Open-top chamber	8	eCO_2 boosted productivity due in part to delayed drought stress. No eCO_2 effect on patterns of dominance by warm-season C_4 grasses, and no clear eCO_2 enhancement of the C_3 functional type, due to improved soil water in the warm season.	Owensby et al. (1999)
Short-grass steppe, Colorado USA	Open-top chamber	5	Co-dominant C_3 grass stimulated by eCO_2, while C_4 grass co-dominant showed no biomass response. C_3 and C_4 grasses experienced elevate WUE at eCO_2 and improved soil water in this dry setting. Gradual shift to C_3 dominance predicted at this site in eCO_2.	Morgan et al. (2004), Nelson et al. (2004)
Mixed-grass Prairie, Wyoming USA	Prairie Heating and CO_2 Enrichment (PHACE) rings, 2006–2014	7	C_4 grasses by year three showed superior growth at eCO_2 and elevated T relative to C_3 grasses, due to improved soil water, but by year seven C_3 grasses were surging in eCO_2 at the expense of C_4 grasses, particularly in the enhanced T treatment.	Morgan et al. (2011), Mueller et al. (2016)
Temperate Savanna, Minnesota, USA	Free-air CO_2 enrichment (FACE) rings	20	C_3 grasses responded more to eCO_2 than C_4 in first 12 years, but in last 8 years C_4 grasses showed superior growth than C_3 at eCO_2 due to greater nitrogen mineralization in C_4 root zone.	Reich et al. (2018)

(continued)

Table 8.4. (continued)

Location	CO_2 enrichment method	Years in operation	Relevant results	References
Central Texas Prairie, USA	CO_2 tents in a tunnel configuration	(a) 10	a) C_4 dominant *Sorgastrum nutens* increased its relative abundance in eCO_2 relative to other C_4 and C_3 grassland species. The grassland became more mesic in structure as the taller C_4 grass displaced shorter, more drought – adapted C_4 grasses.	(a) Polley et al. (2012a, b, 2019);
		(b) 4	b) C_4 grass dominants declined at eCO_2 while C_3 perennial forbs increased. The forb increase may reflect accelerated succession at eCO_2 following removal of grazing.	(b) Polley et al. (2002, 2003a);
		(c) <1	(c) Survival of C_3 mesquite shrubs in C_4 grassland increased in eCO_2 due to improved soil water.	(c) Polley et al. (2003b)
South African Savanna species	Greenhouse chambers with potted plants	1	C_3 shrub *Acacia karoo* responded strongly to eCO_2 while C_4 grass *Themada triandra* had little response. C_4 had greater yield at low CO_2 but less yield than *Acacia* at eCO_2. *Acacia* root systems were particularly responsive to eCO_2.	Kgope et al. (2009), Buitenwerf et al. (2012)
Brazilian Cerrado	Open-top chambers	330 days	Competitive ability of C_3 Cerrado tree *Hymenaea stignocarpa* against alien C_4 grass *Melinus minutiflora* enhanced by eCO_2.	Melo et al. (2018)

Abbreviations: *eCO_2* Elevated atmospheric CO_2 treatment (value varies with study); *T* Temperature treatment

and trees. The meta-analysis of Wang et al. (2012) supports the long-standing hypothesis that eCO_2 at moderate temperatures favor C_3 over C_4 plant production, while rising temperature will favor C_4 over C_3 plants, such that the two offset and there is no advantage of one photosynthetic type over the other (Ehleringer et al. 1997). In a study of growth responses to both eCO_2 and temperature in similar C_3 and C_4 weeds (C_3 *Chenopodium album* and C_4 *Amaranthus retroflexus*), Grise (1996) also observed that the C_3 advantage under eCO_2 (700 and 1000 ppm) at moderate temperature (23 °C daytime) was not present at elevated temperature (34 °C daytime temperature). While these results might indicate no clear C_3 advantage in a warmer, CO_2-enriched world, the offsetting responses have to be considered in the context of realistic field environments. Today, C_4 species often require warm episodes to gain an advantage over the C_3

competition, since C_3 vegetation is favored during cooler periods of the day and growing season (Pearcy et al. 1981; Sage et al. 1999). If the C_3 advantage during these cool periods is enhanced by eCO_2, while the relative C_4 advantage in warm episodes is reduced, then overall, the net result of warming and CO_2 enrichment could be a net advantage for C_3 competitors, not simply an offsetting of trends.

Outdoor CO_2 enrichment using open-top chambers, tent-like tunnels, and FACE studies provides the most realistic assessment of responses to future change by enabling CO_2 enrichment of multi-species communities where plants are rooted in the ground and experience realistic environmental variation and biotic interactions (Leadley and Drake 1993; Polley et al. 1994; McLeod and Long 1999). These outdoor studies show that C_3 species usually outperform C_4 species, allowing communities to shift away from C_4 dominance to either C_3 and C_4 co-dominance or outright C_3 dominance (Table 8.4). Where C_3 woody species are included, they do particularly well in terms of increased seedling survival, resprouting, early growth and establishment (Polley et al. 2002; Davis et al. 2007; Melo et al. 2018; Mueller et al. 2016). CO_2 enrichment is not directly harmful to C_4 vegetation, and is generally beneficial, particularly in terms of enhanced water use efficiency and water status. Reductions in C_4 performance that were observed in eCO_2 were usually due to competitive interactions with the C_3 vegetation and other C_4 species, or confounding factors such as increased sea level, summer drought, or reduced N availability. Notable observations from the outdoor studies include the following. In the semi-arid grassland of central Texas, a long-term winner in eCO_2 was the C_4 prairie grass *Sorghastrum nutans*, which is commonly found as a co-dominant in the more mesic tall-grass prairie of the American Midwest (Polley et al. 2019). Its improved growth came at the expense of more xeric, shorter-stature C_4 grasses from dry grasslands, leading Polley et al. (2012b) to hypothesize that improved soil water in eCO_2 may shift the semi-arid Texas grassland towards a more mesic grassland similar to that found in the tall-grass prairie region of the central USA. In the mixed grass prairie of Wyoming and Kansas, observations of superior C_4 performance relative to C_3 plants in eCO_2 were associated with reduced stomatal conductance and improved soil water status in the summer growing season, when the C_4 flora is most active and drought impacts most severe (Morgan et al. 2004, 2011; Owensby et al. 1999). In Wyoming, the superior C_4 plant growth occurred in the first few years of the eCO_2 treatment but did not persist, as the C_3 forbs surged at eCO_2 in the later years of the seven-year study in both warming and non-warming treatments (Morgan et al. 2011; Mueller et al. 2016). The C_3 gain may have been due in part to improved N supply. In the salt marshes of the Chesapeake Bay, Maryland, rising sea level favored a C_3 sedge (*Scirpus olneyi*) over a common C_4 marsh grass (*Spartina patens*), with eCO_2 and rising sea level acting in concert to favor a consistently stronger response of the C_3 sedge in the later years of the 28-year study (Drake 2014). In Australian savannas, higher soil N promoted richness of C_3 over C_4 grasses at eCO_2, possibly by compensating for the C_4 NUE advantage (Hasegawa et al. 2017). On the prairie savanna of Minnesota, biomass production by C_3 grasses was initially greater at eCO_2 than productivity of the C_4 grasses. However, N cycling was faster in soils around the C_4 grasses, which allowed them to capture more N and produce more biomass at eCO_2 than the C_3 grasses in the second decade of the 20-year study (Reich et al. 2018). This shift from C_3 to C_4 success highlights the importance of long-term experiments that are able to encompass ecological processes than can reverse initial, short-term responses to eCO_2. However, no eCO_2 study has been long or extensive enough to account for the many factors that might impact species success in natural ecosystems, let alone account for multiple global change drivers. Owensby et al. (1999), for example, noted the improved C_3 performance at eCO_2 on Kansas grasslands would likely be offset by springtime fires that disproportionally harm the earlier-active C_3 grasses.

B. *Case Studies of Natural C_3 and C_4 Vegetation in a High-CO_2 World*

To evaluate realistic futures, we can turn to natural systems where global change is already occurring, and trends now point towards likely outcomes in a high CO_2 world. In this section, we discuss two systems where eco-physiological understanding, predictive models and field observations of recent change are robust enough to allow clear projections of future outcomes. The first is the coastal saltmarsh biome now dominated by C_4 *Spartina* grasses in the warm temperate zone. The second is the widespread C_4 grassland/savanna biome of the tropics to warm-temperate zone, where most of the Earth's C_4 biomass and a large fraction of the C_4 floristic diversity are located.

1. *The Future of C_4-Dominated Saltmarshes*

At subtropical and tropical latitudes, coastal estuaries can be dominated by C_4 grasses, but these marshlands tend to be transitory and establish after major disturbances. In time, most of the estuaries succeed to C_3 mangrove forests, such that the global pattern is for mangroves to dominate estuaries at low latitudes, C_4 grasses to dominate estuaries at mid-latitudes, and C_3 grasses and sedges to dominate higher latitude estuaries (Archibold 1995; Sage et al. 1999). Mangroves would also dominate the mid-latitude estuaries but are excluded by winter cold in the temperate zone (McKee et al. 2012; Osland et al. 2013; Chen et al. 2017). At the mangrove x saltmarsh ecotone in the temperate zone, mangrove seedlings regularly establish in the C_4 marshlands, but are killed by a hard winter frost, arresting succession to mangrove forests (Cavanugh et al. 2014). A simple prediction, therefore, is that as lethal cold retreats with climate warming, mangroves forests will move poleward, compressing temperate-zone salt-marshes at lower latitude margins (Osland et al. 2013; 2017; Cavanaugh et al. 2015; 2019; Lovelock et al. 2016). This prediction is now supported by recent observations of poleward migration of mangroves into the C_4-dominated saltmarshes where lethal frost is no longer present (Cavanaugh et al. 2014, 2015, 2019; Saintilan et al. 2014, 2019; Osland et al. 2017). As well, mangroves are stimulated by CO_2 enrichment and eutrophication, both of which are increasing as winters become less severe (Silliman et al. 2005; McKee et al. 2012; Lovelock et al. 2016; Jacotot et al. 2018). In northern Florida, the current rate of poleward migration by mangroves averages 2–3 km per year (Cavanaugh et al. 2015), while in the Gulf of Mexico, most salt marshes are predicted to be invaded by mangroves by 2100 under higher CO_2 emission scenarios that lead to greater climate warming (Osland et al. 2013).

Moderating cold could compensate for habitat loss at lower latitude by enabling poleward migration of C_4 saltmarshes, which appears to be occurring based on observations of C_4 *Spartina* invasion in estuaries of Newfoundland (Fig. 8.9b) and Wales (Sage, personal observation). The fate of the C_4 saltmarsh outside of the mangrove zone will likely depend upon the relative impact of rising CO_2, sea level, increased grazing pressure, and eutrophication on C_3 and C_4 saltmarsh species, which the eCO_2 research from Chesapeake Bay indicates will favor the C_3 species (Silliman et al. 2005; Drake 2014). A hypothesis is therefore that C_4 saltmarsh distribution will be compressed by ongoing trends in climate, sea level rise, eutrophication and eCO_2. In the extreme, C_4-dominated saltmarshes may become a fugitive ecosystem in eCO_2, establishing in the aftermath of storms and other disturbances that create opportunities for colonization, but disappearing as mangroves and C_3 marsh plants assert their competitive dominance.

2. *The Future of C_4-Dominated Grasslands and Savannas*

The situation at the C_3 woodland x C_4 grassland interface is more complicated than mangrove and C_4 saltmarsh because there is not a single environmental parameter such as lethal cold that prevents the dominance of woody species. Instead, multiple global change drivers are impacting grassland/savanna systems, where the combined effect tends to favor C_3 woody species over C_4 grasses, with the exception of where invasive, pyrophytic C_4 grasses have been able to establish destructive fire cycles (D'Antonio and Vitousek 1992; Cochrane and Schulz 1999; Sage and Kubien 2003). In the remaining C_4-dominated grasslands and savannas, the most pervasive threat from the combination of global change drivers is promotion of woody C_3 plants over C_4 grasses, such that woody species are now encroaching into C_4-dominated grasslands (Fig. 8.9a), where they promote succession to shrub or forest ecosystems in wetter areas, and mixed woodland and grassland savannas in semi-arid regions (Stevens et al. 2016; Scheiter et al. 2018). In Africa, for example, forests are predicted to increase from 31% land cover in 1850 to 47% cover by 2100, while savanna declines from 23% to 14% of land cover by 2100 (Higgins and Schleiter 2012).

In recent decades, woody encroachment is estimated to have occurred on half of the non-woody biomes of sub-Saharan Africa (Venter et al. 2018). In South America, the rate of encroachment is three times greater than in African savannas and seven times the encroachment in Australia (Stevens et al. 2016). In North America, shrub encroachment has converted much of the southern plains grasslands to woody thorn-scrub (Van Auken 2000). Savanna systems are particularly vulnerable to woodland conversion as they are mosaics of grass- and woody-dominated patches, which allows woody propagules to readily disperse into the adjacent grass swards, creating a pool of seedlings that promote woodland encroachment. Factors that historically determined woody versus grass dominance include disturbances such as fire, storms and flooding, herbivory by large grazers and browsers, and severe drought. Reduction in the C_4 grass cover by overgrazing creates colonization sites for woody propagules while reducing competition from the grasses for light, water and nutrients, while browsers, notably elephants, consume woody plants and thus deter woody succession (Dublin 1995; Stevens et al. 2016; Venter et al. 2017; Hempson et al. 2017; Davies and Asner 2019). Soil depth also plays a role, with shallow soils favoring grasslands by preventing woody tap roots from escaping belowground competition with grasses (Archibold 1995; Noss 2013; Bond 2015). More frequent and intense fires suppress the woody element in favor of grass swards by killing the top growth and propagules of woody species (Bond 2008; Venter et al. 2017). This control is more significant in wetter than dry landscapes, because woody succession is promoted by moisture (Stevens et al. 2016). In drier grasslands, drought mortality of woody seedlings is common, and fire is more haphazard due to limiting fuel loads (Sankaran et al. 2005; Staver et al. 2011; Devine et al. 2015). Grazers also influence fire cycles by reducing grass fuels that carry a fire (Venter et al. 2017). Seedlings and saplings of woody species are most vulnerable to episodic events from fire or drought, as they have not attained sufficient size to sink tap roots below the grass root zone, nor are they tall enough to survive the flames of a grass fire (Bond and Midgley 2000; Bond et al. 2003; Saintilan and Rogers 2015). The situations where woody seedlings are vulnerable to drought, fire, or browsers are termed mortality traps (Bond 2008). If a mortality trap is strong, then there is high probability the lethal event will return before the seedlings can grow sufficiently to escape fire, grass competition, and browsing, and the grassland can persist. If mortality traps are weak and trees and shrubs can establish, then the C_4 grasses are shaded, preventing accumulation of fine fuels while retaining surface moisture in the cool understory. At this point, the woodland can lock in as an alternative stable state (Bond 2008; Staver et al. 2011; Osborne et al. 2018; Sankaran 2019). Environmental factors that accelerate growth (for example, elevated CO_2, N deposition, and additional moisture) enhance woody establishment and in doing so, weaken fire, drought and herbivory traps by reducing the length of time when woody seedlings are vulnerable (Bond et al. 2003). In light of this, atmospheric CO_2 decline in the Neogene coupled with increasing drought and seasonality could have increased the time a tree seedling existed within a mortality trap, and in doing so, could have contributed to the worldwide expansion of grasslands five to ten million years ago (Beerling and Osborne 2006; Edwards et al. 2010; Bond 2016).

While accelerated growth can shorten the duration of a mortality trap, equally important are impacts of global change drivers that directly harm the grasses and remove the sources of mortality for woody seedlings. Overhunting of wild animal populations has largely removed the native megafauna, with replacement by cattle, sheep and goats (Owen-Smith 1988; Stevens et al. 2016; Hempson et al. 2017; Osborne et al. 2018). This favors woody establishment, except near human settlements where heavy wood gathering reduces the woody vegetation (Venter et al. 2018). Intensification of grazing in recent decades has been particularly destructive for C_4 grass swards, particularly in dry years when grazing remains high despite lower primary productivity (Fig. 8.9d). Economic development also favors woody encroachment by promoting road networks, which create fire breaks, facilitates active fire suppression and promotes further development (Estes et al. 2016; Osborne et al. 2018; Laurance and Arrea 2017).

Projecting into the future, woodland encroachment is modelled to accelerate as atmospheric CO_2 levels continue to increase, such that in Africa, natural C_4 grasslands will largely disappear in regions where there is more than 800 mm of precipitation, and their extent will decline where there is less than 600 mm of rain (Higgens and Scheiter 2012). Woody encroachment can be prevented, but this requires significant financial investment, and in more remote places may require creative solutions such as strategic use of browsers in combination with intense fire to kill the woody vegetation, followed by restrictions on grazing (Osborne et al. 2018).

Offsetting the loss of natural C_4 grasslands could be the expansion of C_4-dominated ecosystems into areas that are currently too dry to support significant grass production, due to improved WUE and altered precipitation patterns (Higgins and Scheiter 2012; Pausetta et al. 2019). The Sahara region, for example, is predicted to green-up due to a stronger monsoon in tandem with improved WUE of C_4 species (Pausata et al. 2017, 2020). These predictions are contingent upon human land use decisions (Higgins and Scheiter 2012), which will be pervasive in a world of nine to ten billion or more people, many of whom could be refugees from sea level rise. In the Sahara, a greener landscape could become a magnet for intensive grazing as currently occurs in the Sahel region south of the Sahara. Alternatively, the Sahara and other arid regions represent potential opportunities to develop vast carbon and energy farms in a CO_2-enriched world (Keller et al. 2014; Li et al. 2018). Such developments could preclude expansion of natural C_4-dominated grasslands into these regions, unless the economic development can also support the establishment of novel, high-diversity grasslands designed to protect regional floras.

3. The Future of the Earth's CAM Diversity

Elevated CO_2 is directly beneficial for the CAM flora, which exhibits photosynthetic and growth responses that rival or exceed those of C_3 species (Drennan and Nobel 2000). CAM water use efficiency is markedly improved in eCO_2, which could lead to increased establishment and density of CAM plants in the dry habitats where they currently occur, and enable expansion of CAM succulents into hyper-arid deserts currently too dry for CAM, or colonization by CAM epiphytes of dry branches and forests where they are currently excluded by insufficient moisture (Drennan and Nobel 2000; Ceusters and Borland 2011). Ironically, the improved fortunes of CAM plants in eCO_2 may largely arise from becoming more C_3 like, due to increased carbon gain in phase II and IV, or, for facultative CAM species, a delay in CAM induction due to slower water depletion (Drennan and Nobel 2000).

When scaling to the ecosystem level, however, the future of the CAM flora is not promising, as evidenced by ongoing trends (Table 8.3). The biggest threat to CAM diversity overall is a depletion of epiphytic floras by tropical deforestation, which is already widespread with losses of virgin forests approaching 90% in certain regions such as Indonesia and Southeast Asia, and by the recent acceleration of deforestation in Amazonia (Zotz 2016; Hughes 2018). Secondary forests do host CAM epiphytes, but these tend to be depleted in species richness unless propagules can readily disperse from old growth remnants (Zotz 2016). In coming decades, deforestation driven by agricultural development, palm oil plantations, timber harvesting, and exotic-caused wildfires will continue to degrade the epiphytic inventory as well as increase fragmentation, which reduces propagule availability (Barlow et al. 2018). A realistic yet poorly studied possibility is C_3 epiphytes may respond more to eCO_2 than CAM epiphytes, and as a result may become more competitive in eCO_2. CAM epiphytes tend to dominate the drier, more exposed portions of the epiphytic niche, usually higher in the canopy, while C_3 epiphytes dominate lower and more interior branches and the trunks of trees (Winter et al. 1983; Zotz and Ziegler 1997; Silvera and Lasso 2016). With higher WUE in eCO_2, C_3 epiphytes could encroach into the CAM niche, lowering CAM diversity. This would be a particular concern in re-establishing secondary forests where initial growth rates influence establishment success.

For terrestrial CAM succulents of arid and semi-arid lands, diverging trends lead to an uncertain future. On the one hand, warmer and drier conditions could expand CAM opportunities for dominating existing habitat, while also opening up new habitats where conditions currently don't favor CAM. In the American Southwest, for example, existing CAM landscapes are predicted to warm and further dry, which could favor CAM succulents at the expense of C_3 shrubs (Seager et al. 2007). Consistent with this prediction, cactus recruitment has increased in the Sonoran Desert of Arizona during the late-twentieth century, in a pattern that is correlated with climate warming and drying, and possible eCO_2 (Munson et al. 2012). Drying and warming in the mesic grasslands of the American interior should also enable CAM expansion, particularly of *Opuntia* species that can rapidly disperse via vegetative propagules (Munson et al. 2012; Owen et al. 2016; Tietjen et al. 2017). Most CAM species are intolerant of freezing, so loss of sub-zero cold will enable poleward expansion of CAM assuming human land use, invasive species and a more a robust C_3 flora does not prevent CAM establishment (Weiss and Overpeck 2005; Springer et al. 2015).

Trends that threaten the desert CAM flora include habitat degradation from land use change, invasive species competition, and infilling of CAM habitats by C_3 shrubs (Turner et al. 2003; Munson et al. 2012).

Diverse CAM floras found in Caatinga , a seasonally-dry and succulent thorn-scrub forest of low latitude, are particularly impacted by conversion to agriculture, wood cutting, and fire. In western Mexico, over 60% of the dry forest vegetation is considered lost, while in Brazil, Caatinga losses may exceed 50% (Trejo and Dirzo 2000; Leal et al. 2005). In Madagascar, dryland biomes housing uniquely rich CAM floras are considered critically endangered due to habitat loss and species harvesting (Burgess et al. 2004). Night-time warming also poses a threat, as it increases transpiration during phase I, offsetting the CAM advantage (Drennan and Nobel 2000), while daytime warming could be a particularly significant threat for CAM species living in hot deserts close to their thermal tolerance limits. For these species, increased frequency and intensity of extreme heat and drought could be particularly lethal (Fig. 8.10a; Weiss and Overpeck 2005; Hultine et al. 2016). A significant threat is competition from pyrophytic C_3 and C_4 grasses (Humphrey 1974; Archer and Predick 2008; Abatzoglou and Kolden 2011). CAM succulents are prone to fire injury as they lack fire-resistant adaptations such as thick bark (Humphrey 1974), while the pyrophytic grasses can fill in the space between plants on the desert floor, enabling a hot fire to carry across the landscape once the grasses senesce. Invasive C_3 and C_4 grasses have already caused extensive damage to CAM succulents in southwestern North America through competition for water and increased wildfire (Fig. 8.10b, d; Humphrey 1974; Defalco et al. 2010; Springer et al. 2015). As a result, habitats rich in CAM succulents are becoming low diversity landscapes dominated by invasive grasses (Brooks and Matchett 2006; Franklin and Molina-Freaner 2010; Abatzoglou and Kolden 2011). In the Sonoran Desert, infestation of rangelands by buffelgrass (*Pennisetum ciliare*) have eliminated regeneration of the columnar cactus *Pachycereus pectin-aboriginum*, and will thus eventually extirpate these cacti in infested areas (Morales-Romero and Molina-Freaner 2008). Greater productivity in eCO_2 will boost fuel loads, causing hotter, more lethal fires (Sage 1995; Smith et al. 2000).

Future CAM winners should include horticultural succulents that represent a broad sampling of the world's CAM flora and are often planted in gardens and for landscaping; some of these have escaped to become established exotics (Osmond et al. 2008; Hultine et al. 2016; Pangburn et al. 2016). To conserve CAM species at risk, gardeners and other enthusiasts are a recognized resource that can be co-opted to work with botanical gardens and conservation agencies (Hultine et al. 2016). The main winners, however, will be cultivated CAM species such as pineapple, agaves, *Aloe* spp. and potentially, CAM biofuel feedstocks (Borland et al. 2009; Davis et al. 2014; Owen et al. 2016). CAM biofuel candidates include *Opuntia, Agave,* and *Euphorbia* species which couple the high WUE of CAM with good production on marginal sites (Borland et al. 2009). However, their cultivation also could open up natural areas with diverse CAM and C_4 floras for agricultural development, and could drive loss of CCM species in semi-arid grasslands should CAM biofuel plantationsbecome common (Owen et al. 2016). Weedy CAM species may also infest CAM-rich sites, causing loss of native diversity. The most notorious CAM weed is prickly pear cactus (*Opuntia stricta*), which severely infested eastern Australia a century ago and is now a problem in Africa and subtropical oceanic islands (Fig. 8.10c; Osmond et al. 2008). The Australian outbreak was controlled by the moth *Cactoblastus cactorum* introduced from North America. *Cactoblastus* uses slight CO_2 variations caused by phase I activity to identify favorable egg-laying sites on *Opuntia*, but this sensory capacity is saturated at eCO_2, suggesting biocontrol by the moth could weaken in a high-CO_2 world (Osmond et al. 2008).

4. *The Future of the Earth's C_2 Flora*

Elevated CO_2 will reduce photorespiratory glycine production in C_2 plants and thus attenuate the capacity to concentrate CO_2, such that the photosynthetic enzymes in the BS cells will be underutilized (von Caemmerer 1989; Vogan and Sage 2012). Since they are largely specialized for hot, disturbed or stressed climates where competition is low, C_2 species should continue to persist, but their options for success may narrow. For C_2 ruderals, the threat should come from more rapid and aggressive establishment by C_3 and C_4 plants, which will reduce the window of opportunity for a C_2 population to establish and repopulate a seedbank. For C_2 specialists on harsh sites, such as salt flats and arid sand-dunes, three threats are likely. One is biotic, arising from increased competition on the harsh sites from C_3 and C_4 species that benefit from high-CO_2 enhancements of A and WUE. Second, the greater density of vegetation could lead to more fires for which C_2 species are poorly adapted. In Africa, woodland infilling could threaten the C_2 ecotype of the grass *Alloteropsis semialata*, which grows in dry woodlands and savannas (Lundgren et al. 2015, 2016). TThird, C_2 plants could experience more extreme heat and drought stress. Since C_2 species are often specialists on particularly hot sites and live close to their thermal limits (Schuster and Monson 1990; Sage et al. 2018), they may be vulnerable to heat waves that push them over a survival threshold. Following a record heat spell in the Mojave Desert, for example, we witnessed extreme die-back in the C_2 sand dune specialist *Heliotropium convolvulaceum*, an event which collapsed the local population for the next few years.

VIII. Conclusion

A number of observations are particularly relevant for the future of the CCM floras. First, global change is accelerating due to the intensification of multiple, interacting drivers that are already of unprecedented magnitude, such that further declines in CCM species and habitat can be expected as the drivers intensify (Sage 2020). Second, while species with CCMs might be able to adjust to one or two drivers, the rapid pace of change and the combination of multiple drivers threaten to overwhelm their response capacity. Third, while the accelerating change is a major threat to Earth's biodiversity in general (Barnosky et al. 2012), for the terrestrial CCM flora, the threat is proportionally greater because CCM species represent a relatively small fraction of the Earth's flora (Sage 2016). A loss of 6000 C_4 species would represent 75% of the world's C_4 diversity, while a loss of 6000 C_3 species would just be a 3% loss of the C_3 flora. Fourth, a large fraction of existing C_4- and CAM-rich habitat is already lost and remaining populations often exist in degraded, fragmented places where ecological processes are compromised. With fewer individuals to resist the accelerating onslaught of global change, small population problems arise such as inbreeding depression, reduced reproductive opportunities, vulnerability to extreme events, and gradual erosion of species and ecosystem integrity (Brook et al. 2008). Extinction debts will multiply, particularly in the considerable fraction of the CCM floras that are niche specialists of limited population and geographical extent (Sage 2020). Extinction debts in the CCM flora may also arise if critical mutualists are impacted. Epiphytic CAM orchids are particularly vulnerable to loss of pollinator services since they often have a specialized relationship with just one to a few pollinator species (Nilsson 1992; Gaskett 2011).

In the future, preservation of the Earth's CCM diversity will require that people specifically recognize the unique threats to CCM species and develop effective conservation strategies. Not all CCM species will become extinct of course, as demonstrated by the exotics that respond well to global change. How much the CCM flora may decline is impossible to say, but given ongoing loss of rainforest and grasslands habitat, and increasing effects of eCO_2, the fraction at risk could exceed the 75% threshold required for a mass extinction event (Barnosky et al. 2011). To promote conservation of the CCM flora, an important task is to fully catalogue CCM species and ensure the species lists are widely available to conservation biologists who manage the landscapes where CCM diversity resides (Sage 2016 and Smith and Winter 1996 present lists in progress of the World's C_4 and CAM genera, respectively). Such lists will facilitate the identification and preservation of hotspots of CCM diversity, thus optimizing conservation efficiency (Myers et al. 2000).

In closing, our species has become so pervasive across the Earth that the vegetation on future landscapes will in some way depend on human action. If people want to preserve the C_4, C_2 and CAM floras, they can make it so, but this will require recognizing the members of the flora and prioritizing their conservation. , Alternatively, people can do nothing, but this is also management decision that will have powerful and often negative consequences.

Quo vadis C_4 et CAM meis amicis? Et in manibus meis hominibus amicis.

References

Abatzoglou JT, Kolden CA (2011) Climate change in western us deserts: potential for increased wildfire and invasive annual grasses. Rangel For Ecol Manag 64:471–478. https://doi.org/10.2111/REM-D-09-00151.1

Ainsworth EA, Long SP (2005) What have we learned from 15 years of free-air CO_2 enrichment (FACE)? A meta-analytic review of the responses of photosynthesis, canopy properties and plant production to rising CO_2. New Phytol 165:351–371. https://doi.org/10.1111/j.1469-8137.2004.01224.x

Andela N, Morton DC, Giglio L, Chen Y, van der Werf GR, Kasibhatla PS, DeFries RS et al (2017) A human-driven decline in global burned area. Science 356:1356–1361. https://doi.org/10.1126/science.aal4108

Anderson LJ, Maherali H, Johnson HB, Polley HW, Jackson RB (2001) Gas exchange and photosynthetic acclimation over subambient to elevated CO_2 in a C_3-C_4 grassland. Glob Chang Biol 7:693–707. https://doi.org/10.1046/j.1354-1013.2001.00438.x

Arakaki M, Christin PA, Nyffeler R, Lendel A, Eggli U, Ogburn RM, Spriggs E et al (2011) Contemporaneous and recent radiations of the World's major succulent plant lineages. Proc Natl Acad Sci U S A 108:8379–8384. https://doi.org/10.1073/pnas.1100628108

Archer D, Brovkin V (2008) The millennial atmospheric lifetime of anthropogenic CO_2. Clim Chang 90:283–297. https://doi.org/10.1007/s10584-008-9413-1

Archer SR, Predick KI (2008) Climate change and ecosystems of the southwestern United States. Rangelands 30:23–28. https://doi.org/10.2111/1551-501X(2008)30[23:CCAEOT]2.0.CO;2

Archer SR, Andersen EM, Predick KI, Schwinning S, Steidl RJ, Woods SR (2017) Woody plant encroachment: causes and consequences. In: Briske DD (ed) Rangeland Systems: Processes, Management and Challenges. Springer, Berlin, pp 25–84

Archibold OW (1995) Ecology of World Vegetation. Springer, Dordrecht

Atkinson RRL, Mockford EJ, Bennett C, Christin PA, Spriggs EL, Freckleton RP, Thompson K et al (2016) C_4 photosynthesis boosts growth by altering physiology, allocation and size. Nat Plants 2:16038. https://doi.org/10.1038/nplants.2016.38

Badger MR, Price GD (2003) CO_2 concentrating mechanisms in cyanobacteria: molecular components. their diversity and evolution. J Exp Bot 54:609–622. https://doi.org/10.1093/jxb/erg076

Badger MR, Spalding MH (2000) CO_2 acquisition, concentration and fixation in cyanobacteria and algae. In: Leegood RC, Sharkey TD, von Caemmerer S (eds) Photosynthesis: Physiology and Metabolism. Springer, Dordrecht, pp 369–397

Badger MR, Price GD (2003b) CO concentrating mechanisms in cyanobacteria: molecular components. their diversity and evolution. J Exp Bot 54:609–622. https://doi.org/10.1093/jxb/erg076

Barlow J, França F, Gardner TA, Hicks CC, Lennox GD, Berenguer E, Castello L et al (2018) The future of hyperdiverse tropical ecosystems. Nature 559:517. https://doi.org/10.1038/s41586-018-0301-1

Barnosky AD, Matzke N, Tomiya S, Wogan GOU, Swartz B, Quental TB, Marshall C, McGuire JL, Lindsey EL, Maguire KC, Mersey B, Ferrer EA

(2011) Has the Earth's sixth mass extinction already arrived? Nature 471:51–57. https://doi.org/10.1038/nature09678

Bartoli G, Hönisch B, Zeebe RE (2011) Atmospheric CO_2 decline during the Pliocene intensification of Northern Hemisphere glaciations. Paleoceanography 26:PA4213. https://doi.org/10.1029/2010PA002055

Baruch Z (1996) Ecophysiological Aspects of the invasion by African grasses and their impact on biodiversity and function of Neotropical savannas. In: Solbrig OT, Medina E, Silva JF (eds) Biodiversity and Savanna Ecosystem Processes: A Global Perspective. Springer, Berlin, pp 79–93

Beerling DJ, Osborne CP (2006) The origin of the savanna biome. Glob Chang Biol 12:2023–2031. https://doi.org/10.1111/j.1365-2486.2006.01239.x

Bell G, Collins S (2008) Adaptation, extinction and global change. Evol Appl 1:3–16. https://doi.org/10.1111/j.1752-4571.2007.00011.x

Berner RA (2008) Addendum to "inclusion of the weathering of volcanic rocks in the GEOCARBSULF model". Am J Sci 308:100–103. https://doi.org/10.2475/01.2008.04

Bird MI, Hutley LB, Lawes MJ, Michael J, Lloyd J, Luly JG, Ridd PV et al (2013) Humans, megafauna and environmental change in tropical Australia. J Quat Sci 28:439–452. https://doi.org/10.1002/jqs.2639

Bishop KA, Lemonnier P, Quebedeaux JC, Montes CM, Leakey ADB, Ainsworth EA (2018) Similar photosynthetic response to elevated carbon dioxide concentration in species with different phloem loading strategies. Photosynth Res 137:453–464. https://doi.org/10.1007/s11120-018-0524-x

Blumenthal DM, Resco V, Morgan JA, Williams DG, LeCain DR, Hardy EM, Pendall E, Bladyka E (2013) Invasive forb benefits from water savings by native plants and carbon fertilization under elevated CO_2 and warming. New Phytol 200:1156–1165. https://doi.org/10.1111/nph.12459

Bond WJ (2008) What limits trees in C_4 grasslands and savannas? Annu Rev Ecol Evol Syst 39:641–659. https://doi.org/10.1146/annurev.ecolsys.39.110707.173411

Bond WJ (2015) Fires in the Cenozoic: a late flowering of flammable ecosystems. Front Plant Sci 5:749. https://doi.org/10.3389/fpls.2014.00749

Bond WJ (2016) Ancient grasslands at risk. Science 351:120–122. https://doi.org/10.1126/science.aad5132

Bond WJ, Midgley GF (2000) A proposed CO_2-controlled mechanism of woody plant invasion in grasslands and savannas. Glob Chang Biol 6:865–869. https://doi.org/10.1046/j.1365-2486.2000.00365.x

Bond WJ, Midgley GF (2012) Carbon dioxide and the uneasy interactions of trees and savannah grasses. Philos Trans R Soc Lond Ser B Biol Sci 367:601–612. https://doi.org/10.1098/rstb.2011.0182

Bond WJ, Parr CL (2010) Beyond the forest edge: ecology, diversity and conservation of the grassy biomes. Biol Conserv 143:2395–2404. https://doi.org/10.1016/j.biocon.2009.12.012

Bond WJ, Midgley GF, Woodward FI (2003) The importance of low atmospheric CO_2 and fire in promoting the spread of grasslands and savannas. Glob Chang Biol 9:973–982. https://doi.org/10.1046/j.1365-2486.2003.00577.x

Borland AM, Griffiths H (1996) Variations in the phases of crassulacean acid metabolism and regulation of carboxylation patterns determined by carbon-isotope-discrimination techniques. In: Winter K, Smith JAC (eds) Crassulacean Acid Metabolism: Biochemistry, Ecophysiology and Evolution. Springer, Berlin, pp 230–249

Borland AM, Maxwell K, Griffiths H (2000) Ecophysiology of plants with crassulacean acid metabolism. In: Leegood RC, Sharkey TD, von Caemmerer S (eds) Photosynthesis: Physiology and Metabolism. Springer, Dordrecht, pp 583–605

Borland AM, Griffiths H, Hartwell J, Smith JAC (2009) Exploiting the potential of plants with crassulacean acid metabolism for bioenergy production on marginal lands. J Exp Bot 60:2879–2896. https://doi.org/10.1093/jxb/erp118

Borland AM, Barrera Zambrano VA, Ceusters J, Shorrock K (2011) The photosynthetic plasticity of crassulacean acid metabolism: an evolutionary innovation for sustainable productivity in a changing world. New Phytol 191:619–633. https://doi.org/10.1111/j.1469-8137.2011.03781.x

Borland AM, Leverett A, Hurtado-Castano N, Hu R, Yang X (2018) Functional anatomical traits of the photosynthetic organs of plants with crassulacean acid metabolism. In: Adams WW, Terashima I (eds) The Leaf: A Platform for Performing Photosynthesis. Springer, Berlin, pp 281–305

Bouchenak-Khelladi Y, Slingsby JA, Verboom GA, Bond WJ (2014) Diversification of C_4 grasses (Poaceae) does not coincide with their ecological dominance. Am J Bot 101:300–307. https://doi.org/10.3732/ajb.1300439

Briggs JM, Knapp AK, Blair JM, Heisler JL, Hoch GA, Lett MS, McCarron JK (2005) An ecosystem in transition: causes and consequences of the conversion of mesic grassland to shrubland. Bioscience 55:243–254. https://doi.org/10.1641/0006-3568(2005)055[0243:AEITCA]2.0.CO;2

Brook BW, Sodhi NS, Bradshaw CJA (2008) Synergies among extinction drivers under global change. Trends Ecol Evol 23:453–460. https://doi.org/10.1016/j.tree.2008.03.011

Brooks ML, Matchett JR (2006) Spatial and temporal patterns of wildfires in the Mojave Desert, 1980–2004. J Arid Environ 67:148–164. https://doi.org/10.1016/j.jaridenv.2006.09.027

Brown RH (1978) A difference in N use efficiency in C_3 and C_4 plants and its implications in adaptation and evolution. Crop Sci 18:93–98. https://doi.org/10.2135/cropsci1978.0011183X001800010025x

Brown RH (1999) Agronomic implications of C_4 photosynthesis. In: Sage RF, Monson RK (eds) C_4 Plant Biology. Academic, San Diego, pp 473–507

Buitenwerf R, Bond WJ, Stevens N, Trollope WSW (2012) Increased tree densities in South African savannas: >50 years of data suggests CO_2 as a driver. Glob Change Biol 18:675–684. https://doi.org/10.1111/j.1365-2486.2011.02561.x

Burgess N, D'Amico Hales J, Underwood E, Dinerstein E, Olson D, Itoua I, Schipper J et al (2004) Terrestrial Ecoregions of Africa and Madagascar: A Conservation Assessment. Island Press, Washington

Bush MB, Gosling WD, Colinvaux PA (2007) Climate change in the lowlands of the Amazon basin. In: Bush MB, Flenley JR (eds) Tropical Rainforest Responses to Climatic Change. Springer, Berlin, pp 55–76

Campbell CD, Sage RF (2006) Interactions between the effects of atmospheric CO_2 content and P nutrition on photosynthesis in white lupin (*Lupinus albus* L.). Plant Cell Environ 29:844–853

Campbell B, Beare D, Bennett E, Hall-Spencer JM, Ingram JSI, Jaramillo F, Ortiz R et al (2017) Agriculture production as a major driver of the Earth system exceeding planetary boundaries. Ecol Soc 22:8. https://doi.org/10.5751/ES-09595-220408

Cavanaugh KC, Dangremond EM, Doughty CL, William AP, Parker JD, Hayes MA, Rodrigues W, Feller IC (2019) Climate-driven regime shifts in a mangrove–salt marsh ecotone over the past 250 years. Proc Natl Acad Sci USA 116:21602–21608. https://doi.org/10.1073/pnas.1902181116

Cavanaugh KC, Parker JD, Cook-Patton SC, Feller IC, Williams AP, Kellner JR (2015) Integrating physiological threshold experiments with climate modeling to project mangrove species' range expansion. Glob Change Biol 21:1928–1938. https://doi.org/10.1111/gcb.12843

Cavanaugh KC, Kellner JR, Forde AJ, Gruner DS, Parker JD, Rodriguez W, Feller IC (2014) Poleward expansion of mangroves is a threshold response to decreased frequency of extreme cold events. Proc Nat Acad Sci USA 111:723–727. https://doi.org/10.1073/pnas.1315800111

Cerling TE, Harris JM, MacFadden BJ, Leakey MG, Quade J, Eisenmann V, Ehleringer JR (1997) Global vegetation change through the Miocene/Pliocene boundary. Nature 389:153. https://doi.org/10.1038/38229

Cerling TE, Andanje SA, Blumenthal SA, Brown FH, Chritz KL, Harris JM, Hart JA et al (2015) Dietary changes of large herbivores in the Turkana Basin, Kenya from 4 to 1 Ma. Proc Natl Acad Sci U S A 112:11467–11472. https://doi.org/10.1073/pnas.1513075112

Ceusters J, Borland AM (2011) Impacts of elevated CO_2 on the growth and physiology of plants with crassulacean acid metabolism. In: Lüttge UE, Beyschlag W, Büdel B, Francis D (eds) Progress in Botany, vol 72. Springer, Berlin, pp 163–181

Chase MJ, Schlossberg S, Griffin CR, Bouché PJC, Djene SW, Elkan PW, Ferreira S et al (2016) Continent-wide survey reveals massive decline in African savannah elephants. PeerJ 4:e2354. https://doi.org/10.7717/peerj.2354

Chen L, Wang W, Li QQ, Zhang Y, Yang S, Osland MJ, Huang J, Peng C (2017) Mangrove species' responses to winter air temperature extremes in China. Ecosphere 8:e01865. https://doi.org/10.1002/ecs2.1865

Christin P-A, Besnard G, Samaritani E, Duvall MR, Hodkinson TR, Savolainen V, Salamin N (2008) Oligocene CO_2 decline promoted C_4 photosynthesis in grasses. Curr Biol 18:37–43. https://doi.org/10.1016/j.cub.2007.11.058

Christin P-A, Freckleton RP, Osborne CP (2010) Can phylogenetics identify C_4 origins and reversals? Trends Ecol Evol 25:403–409. https://doi.org/10.1016/j.tree.2010.04.007

Christin P-A, Osborne CP, Sage RF, Arakaki M, Edwards EJ (2011) C_4 eudicots are not younger than C_4 monocots. J Exp Bot 62:3171–3181. https://doi.org/10.1093/jxb/err041

Cochrane MA, Alencar A, Schulze M, Souza CM Jr, Nepstad DC, Lefebvre P, Davidson EA (1999) Positive feedbacks in the fire dynamic of closed canopy tropical forests. Science 284:1832–1835. https://doi.org/10.1126/science.284.5421.1832

Cochrane MA, Schulze MD (1999) Fire as a recurrent event in tropical forests of the eastern Amazon: effects on forest structure, biomass. and species composition. Biotropica 31:2–16. https://doi.org/10.1111/j.1744-7429.1999.tb00112.x

Collatz GJ, Berry JA, Clark JS (1998) Effects of climate and atmospheric CO_2 partial pressure on the global distribution of C_4 grasses: present, past, and future. Oecologia 114:441–454. https://doi.org/10.1007/s004420050468

Collins M, Knutti R, Arblaster J, Dufresne J-L, Fichefet T, Friedlingstein P et al (2013) Long-term climate change: projections, commitments and irreversibility. In: Stocker TF, Qin D, Plattner G-K, Tignor M, Allen SK, Boschung J et al (eds) Climate Change 2013: The Physical Science Basis. Contribution of Working Group I to the Fifth Assessment Report of the Intergovernmental Panel on Climate Change. Cambridge University Press, Cambridge, pp 1029–1136

Cotton JM, Cerling TE, Cotton JM, Still CJ, Cotton JM, Hoppe KA, Hoppe KA, Mosier TM (2016) Climate, CO_2, and the history of North American grasses since the Last Glacial Maximum. Sci Adv 2:e1501346. https://doi.org/10.1126/sciadv.1501346

Coupland RT (1992) Mixed prairie. In: Coupland RT (ed) Natural Grasslands: Introduction and Western Hemisphere. Ecosystems of the World, vol 8A. Elsevier, Amsterdam, pp 151–182

Crayn DM, Winter K, Smith JAC (2004) Multiple origins of crassulacean acid metabolism and the epiphytic habit in the neotropical family Bromeliaceae. Proc Natl Acad Sci U S A 101:3703–3708. https://doi.org/10.1073/pnas.0400366101

Cubasch U, Wuebbles D, Chen D, Facchini MC, Frame D, Mahowald N, Winther J-G (2013) Introduction. In: Stocker TF, Qin D, Plattner G-K, Platner M, Tignor M, Allen SK, Boschung J, Nauels A, Xia Y, Bex V, Midgley PB (eds) Climate Change 2013: The Physical Science Basis. Contribution of Working Group I to the Fifth Assessment Report of the Intergovernmental Panel on Climate Change. Cambridge University Press, Cambridge, pp 119–158

Cui M, Miller PM, Nobel PS (1993) CO_2 exchange and growth of the crassulacean acid metabolism plant *Opuntia ficus-indica* under elevated CO_2 in opent-Top chambers. Plant Physiol 103:519–524. https://doi.org/10.1104/pp.103.2.519

D'Antonio CM, Vitousek PM (1992) Biological invasions by exotic grasses, the grass/fire cycle, and global change. Annu Rev Ecol Syst 23:63–87. https://doi.org/10.1146/annurev.es.23.110192.000431

D'Antonio CM, Yelenik SG, Mack MC (2017) Ecosystem vs. community recovery 25 years after grass invasions and fire in a subtropical woodland. J Ecol 105:14621474. https://doi.org/10.1111/1365-2745.12855

Dai Z, Ku MSB, Edwards GE (1996) Oxygen sensitivity of photosynthesis and photorespiration in different photosynthetic types in the genus *Flaveria*. Planta 198:563–571. https://doi.org/10.1007/BF00262643

Davies AB, Asner GP (2019) Elephants limit aboveground carbon gains in African savannas. Glob Chang Biol 25:1368–1382. https://doi.org/10.1111/gcb.14585

Davis MA, Reich PB, Knoll MJB, Dooley L, Hundtoft M, Attleson I (2007) Elevated atmospheric CO_2: a nurse plant substitute for oak seedlings establishing in old fields. Glob Chang Biol 13:2308–2316. https://doi.org/10.1111/j.1365-2486.2007.01444.x

Davis SC, LeBauer DS, Long SP (2014) Light to liquid fuel: theoretical and realized energy conversion efficiency of plants using crassulacean acid metabolism (CAM) in arid conditions. J Exp Bot 65:3471–3478. https://doi.org/10.1093/jxb/eru163

Davis SC, Simpson J, Gil-Vega KC, Niechayev NA, van Tongerlo E, Castano NH, Dever LV, Búrquez A (2019) Undervalued potential of crassulacean acid metabolism for current and future agricultural production. J Exp Bot 70:6521–6537. https://doi.org/10.1093/jxb/erz223

Defalco LA, Esque TC, Scoles-Sciulla SJ, Rodgers J (2010) Desert wildfire and severe drought diminish survivorship of the long-lived Joshua tree (*Yucca brevifolia*; Agavaceae). Am J Bot 97:243–250. https://doi.org/10.3732/ajb.0900032

Derner JD, Tischler CR, Polley HW, Johnson HB (2005) Seedling growth of two honey mesquite varieties under CO_2 enrichment. Rangel Ecol Manag 58:292–298

Devine AP, Stott I, McDonald RA, Maclean IMD (2015) Woody cover in wet and dry African savannas after six decades of experimental fires. J Ecol 103:473–478. https://doi.org/10.1111/1365-2745.12367

Dippery JK, Tissue DT, Thomas RB, Strain BR (1995) Effects of low and elevated CO_2 on C_3 and C_4 annuals. Oecologia 101:13–20. https://doi.org/10.1007/BF00328894

Dirzo R, Young HS, Galetti M, Ceballos G, Isaac NJB, Collen B (2014) Defaunation in the Anthropocene. Science 345:401–406. https://doi.org/10.1126/science.1251817

Dodd AN, Borland AM, Haslam RP, Griffiths H, Maxwell K (2002) Crassulacean acid metabolism: plastic, fantastic. J Exp Bot 53:569–580. https://doi.org/10.1093/jexbot/53.369.569

Drake BG (2014) Rising sea level, temperature, and precipitation impact plant and ecosystem responses to elevated CO_2 on a Chesapeake Bay wetland: review of a 28-year study. Glob Chang Biol 20:3329–3343. https://doi.org/10.1111/gcb.12631

Drennan PM, Nobel PS (2000) Responses of CAM species to increasing atmospheric CO_2 concentrations. Plant Cell Environ 23:767–781. https://doi.org/10.1046/j.1365-3040.2000.00588.x

Dublin HT (1995) Vegetation dynamics in the Serengeti-Mara ecosystem: the role of elephants, fire, and other factors. In: Sinclair ARE, Arcese P (eds) Serengeti II: Dynamics, Management, and Conservation of an Ecosystem. University of Chicago Press, Chicago, pp 71–90

Dublin HT, Sinclair ARE, Mcglade J (1990) Elephants and fire as causes of multiple stable states in the Serengeti-Mara woodlands. J Anim Ecol 59:1147–1164. https://doi.org/10.2307/5037

Dupont L (2011) Orbital scale vegetation change in Africa. Quat Sci Rev 30:3589–3602. https://doi.org/10.1016/j.quascirev.2011.09.019

Eby M, Zickfeld K, Montenegro A, Archer D, Meissner KJ, Weaver AJ (2009) Lifetime of anthropogenic climate change: millennial time scales of potential CO_2 and surface temperature perturbations. J Clim 22:2501–2511. https://doi.org/10.1175/2008JCLI2554.1

Edwards EJ, Ogburn RM (2012) Angiosperm responses to a low-CO_2 world: CAM and C_4 photosynthesis as parallel evolutionary trajectories. Int J Plant Sci 173:724–733

Edwards EJ, Smith SA (2010) Phylogenetic analyses reveal the shady history of C_4 grasses. Proc Natl Acad Sci U S A 107:2532–2537. https://doi.org/10.1073/pnas.0909672107

Edwards EJ, Osborne CP, Strömberg CAE, Smith SA, Bond WJ, Christin PA, Cousins AB et al (2010) The origins of C_4 grasslands: integrating evolutionary and ecosystem science. Science 328:587–591. https://doi.org/10.1126/science.1177216

Ehleringer JR (1978) Implications of quantum yield differences on the distributions of C_3 and C_4 grasses. Oecologia 31:255–267. https://doi.org/10.1007/BF00346246

Ehleringer J, Pearcy RW (1983) Variation in quantum yield for CO_2 uptake among C_3 and C_4 plants. Plant Physiol 73:555–559. https://doi.org/10.1104/pp.73.3.555

Ehleringer JR, Sage RF, Flanagan LB, Pearcy RW (1991) Climate change and the evolution of C_4 photosynthesis. Trends Ecol Evol 6:95–99. https://doi.org/10.1016/0169-5347(91)90183-X

Ehleringer JR, Cerling TE, Helliker BR (1997) C_4 photosynthesis, atmospheric CO_2, and climate. Oecologia 112:285–299. https://doi.org/10.1007/s004420050311

Ehrenfeld JG (2010) Ecosystem consequences of biological invasions. Annu Rev Ecol Evol Syst 41:59–80. https://doi.org/10.1146/annurev-ecolsys-102209-144650

Ellis EC, Ramankutty N (2008) Putting people in the map: anthropogenic biomes of the world. Front Ecol Environ 6:439–447. https://doi.org/10.1890/070062

Ellis EC, Goldewijk KK, Siebert S, Lightman D, Ramankutty N (2010) Anthropogenic transformation of the biomes, 1700 to 2000. Glob Ecol Biogeogr 19:589–606. https://doi.org/10.1111/j.1466-8238.2010.00540.x

Espirito-Santos MM, Leite ME, Silva JO, Barbosa RS, Rocha AM, Anaya FC, Dupin MG (2016) Understanding patterns of land-cover change in the Brazilian Cerrado from 2000 to 2015. Philos Trans R Soc Lond Ser B Biol Sci 371:20150435. https://doi.org/10.1098/rstb.2015.0435

Estes LD, Searchinger T, Spiegel M, Tian D, Sichinga S, Mwale M, Kehoe L et al (2016) Reconciling agriculture, carbon and biodiversity in a savannah transformation frontier. Philos Trans R Soc Lond Ser B Biol Sci 371:20150316. https://doi.org/10.1098/rstb.2015.0316

Feakins SJ, Demenocal PB (2010) Global and African regional climate during the Cenozoic. In: Wedelin L, William JS (eds) Cenezoic Mammals of Africa. University of California Press, Berkeley

Field C, Mooney H (1986) The photosynthesis-nitrogen relationship in wild plants. In: Givinish TD (ed) On the Economy of Plant Form and Function. Cambridge University Press, Cambridge, pp 25–55

Fisher AE, McDade LA, Kiel CA, Khoshravesh R, Johnson MA, Stata M, Sage TL, Sage RF (2015) Evolutionary history of *Blepharis* (Acanthaceae) and the origin of C_4 photosynthesis in section *Acanthodium*. Int J Plant Sci 176:770–790. https://doi.org/10.1086/683011

Foley JA, DeFries R, Asner GP, Barford C, Bonan G, Carpenter SR, Chapin FS et al (2005) Global consequences of land use. Science 309:570–574. https://doi.org/10.1126/science.1111772

Fowler D, Coyle M, Skiba U, Sutton MA, Cape JN, Reis S, Sheppard LJ et al (2013) The global nitrogen cycle in the twenty-first century. Philos Trans R Soc Lond Ser B Biol Sci 368:20130164. https://doi.org/10.1098/rstb.2013.0164

Fox DL, Koch PL (2003) Tertiary history of C_4 biomass in the Great Plains, USA. Geology 31:809–812. https://doi.org/10.1130/G19580.1

Franklin K, Molina-Freaner F (2010) Consequences of buffelgrass pasture development for primary productivity, perennial plant richness, and vegetation structure in the drylands of Sonora, Mexico. Conserv Biol 24:1664–1673. https://doi.org/10.1111/j.1523-1739.2010.01540.x

Franks PJ, Adams MA, Amthor JS, Barbour MM, Berry JA, Ellsworth DS, Farquhar GD et al (2013) Sensitivity of plants to changing atmospheric CO_2 concentration: from the geological past to the next century. New Phytol 197:1077–1094. https://doi.org/10.1111/nph.12104

Galloway JN, Cowling EB (2002) Reactive nitrogen and the world: 200 years of change. Ambio 31:64–71. https://doi.org/10.1579/0044-7447-31.2.64

Galmés J, Andralojc PJ, Kapralov MV, Maxim V, Flexas J, Keys AJ, Molins A et al (2014) Environmentally driven evolution of Rubisco and improved photosynthesis and growth within the C_3 genus *Limonium* (Plumbaginaceae). New Phytol 203:989–999. https://doi.org/10.1111/nph.12858

Garcin Y, Deschamps P, Ménot G, De Saulieu G, Oslisly R, Schefuss E, Dupont LM et al (2018) Early anthropogenic impact on western Central African rainforests 2,600 y ago. Proc Natl Acad Sci U S A 115:3261–3266. https://doi.org/10.1073/pnas.1715336115

Gaskett AC (2011) Orchid pollination by sexual deception: pollinator perspectives. Biol Rev 86:33–75. https://doi.org/10.1111/j.1469-185X.2010.00134.x

Gerhart LM, Ward JK (2010) Plant responses to low [CO_2] of the past. New Phytol 188:674–695. https://doi.org/10.1111/j.1469-8137.2010.03441.x

Ghannoum O, Evans JR, Von Caemmerer S (2011) Nitrogen and water use efficiency of C_4 plants. In: Raghavendra AS, Sage RF (eds) C_4 Photosynthesis and Related CO_2 Concentrating Mechanisms. Springer, Dordrecht, pp 129–146

Graesser J, Aide TM, Grau HR, Ramankutty N (2015) Cropland/pastureland dynamics and the slowdown of deforestation in Latin America. Environ Res Lett 10:034017. https://doi.org/10.1088/1748-9326/10/3/034017

Graham A (2010) Late Cretaceous and Cenozoic history of Latin American Vegetation and Terrestrial Environments. Missouri Bot Garden Press, St. Louis

Graham EA, Nobel PS (1996) Long-term effects of a doubled atmospheric CO_2 concentration on the CAM species *Agave deserti*. J Exp Bot 47:61–69. https://doi.org/10.1093/jxb/47.1.61

Gravel S (2016) When Is Selection Effective? Genetics 203:451–462. https://doi.org/10.1534/genetics.115.184630

Griffith DM, Anderson TM, Osborne CP, Stromberg CAE, Forrestel EJ, Still CJ (2015) Biogeographically distinct controls on C_3 and C_4 grass distributions: merging community and physiological ecology. Glob Ecol Biogeogr 24:304–313. https://doi.org/10.1111/geb.12265

Griffith DM, Cotton JM, Powell RL, Sheldon ND, Still CJ (2017) Multi-century stasis in C_3 and C_4 grass distributions across the contiguous United States since the industrial revolution. J Biogeogr 44:2564–2574. https://doi.org/10.1111/jbi.13061

Grossman JD, Rice KJ (2014) Contemporary evolution of an invasive grass in response to elevated atmospheric CO_2 at a Mojave Desert FACE site. Ecol Lett 17:710–716. https://doi.org/10.1111/ele.12274

Grise DJ (1996) Effects of elevated CO_2 and high temperature on the relative growth rates and competitive interactions between a C_3 (*Chenopodium album*) and a C_4. Amaranthus hybridus) annual. PhD thesis, University of Georgia, Athens, Georgia

Gunderson CA, Wullschleger SD (1994) Photosynthetic acclimation in trees to rising atmospheric CO_2: a broader perspective. Photosynth Res 39:369–388. https://doi.org/10.1007/BF00014592

Haddad NM, Brudvig LA, Clobert J, Davies KF, Gonzalez A, Holt RD, Lovejoy TE et al (2015) Habitat fragmentation and its lasting impact on Earth's ecosystems. Sci Adv 1:e1500052. https://doi.org/10.1126/sciadv.1500052

Hagemann M, Kern R, Maurino VG, Veronica G, Hanson DT, Weber APM, Sage RF, Bauwe H (2016) Evolution of photorespiration from cyanobacteria to land plants, considering protein phylogenies and acquisition of carbon concentrating mechanisms. J Exp Bot 67:2963–2976. https://doi.org/10.1093/jxb/erw063

Hansen MC, Potapov PV, Moore R, Hancher M, Turubanova SA, Tyukavina A, Thau D et al (2013) High-resolution global maps of 21st-century forest cover change. Science 342:850–853. https://doi.org/10.1126/science.1244693

Harrison SP, Prentice CI (2003) Climate and CO_2 controls on global vegetation distribution at the last glacial maximum: analysis based on palaeovegetation data, biome modelling and palaeoclimate simulations. Glob Chang Biol 9:983–1004. https://doi.org/10.1046/j.1365-2486.2003.00640.x

Hartmann DL, Tank AMGK, Rusticucci M, Alexander LV, Broenniman B, Charabi Y, Dentener FJ et al (2013) Observations: atmosphere and surface. In: Stocker TF, Qin D, Plattner G-K, Tignor M, Allen SK, Boschung J et al (eds) Climate Change 2013: The Physical Science Basis. Contribution of Working Group I to the Fifth Assessment Report of the Intergovernmental Panel on Climate Change. Cambridge University Press, Cambridge, pp 159–254. https://doi.org/10.1017/CBO9781107415324.008

Hasegawa S, Piñeiro J, Ochoa-Hueso R, Am H, Rymer PD, Barnettt KL, Power SA (2017) Elevated CO_2 concentrations reduce C $_4$ cover and decrease diversity of understorey plant community in a *Eucalyptus* woodland. J Ecol 106:1483–1494. https://doi.org/10.1111/1365-2745.12943

Hatch MD (1987) C_4 photosynthesis: a unique blend of modified biochemistry, anatomy and ultrastructure. Biochim Biophys Acta 895:81–106. https://doi.org/10.1016/S0304-4173(87)80009-5

Hempson GP, Archibald S, Bond WJ (2017) The consequences of replacing wildlife with livestock in Africa. Sci Rep 7:17196. https://doi.org/10.1038/s41598-017-17348-4

Henderson S, Hattersley P, von Caemmerer S, Osmond CB (1995) Are C_4 pathway plants threatened by global climatic change? In: Schulze E-D, Caldwell MM (eds) Ecophysiology of Photosynthesis. Springer, Berlin, pp 529–549

Higgins JA, Kurbatov AV, Spaulding NE, Brook E, Introne DS, Chimiak LM, Yan Y et al (2015) Atmospheric composition 1 million years ago from blue ice in the Allan Hills, Antarctica. Proc Natl Acad Sci U S A 112:6887–6891. https://doi.org/10.1073/pnas.1420232112

Higgins SI, Scheiter S (2012) Atmospheric CO_2 forces abrupt vegetation shifts locally, but not globally. Nature 488:209–212. https://doi.org/10.1038/nature11238

Hoag C, Svenning J-C (2017) African environmental change from the Pleistocene to the Anthropocene. Annu Rev Environ Resour 42:27–54. https://doi.org/10.1146/annurev-environ-102016-060653

Hoekstra JM, Boucher TM, Ricketts TH, Roberts C (2005) Confronting a biome crisis: global disparities of habitat loss and protection. Ecol Lett 8:23–29. https://doi.org/10.1111/j.1461-0248.2004.00686.x

Holtum JA, Hancock LP, Edwards EJ, Crisp MD, Crayn DM, Sage R, Winter K (2016) Australia lacks stem succulents but is it depauperate in plants with crassulacean acid metabolism (CAM)? Curr Opin Plant Biol 31:109–117. https://doi.org/10.1016/j.pbi.2016.03.018

Hoogakker BAA, Smith RS, Singarayer JS, Marchant R, Prentice IC, Allen JRM, Anderson RS et al (2016) Terrestrial biosphere changes over the last 120 kyr. Clim Past 12:51–73. https://doi.org/10.5194/cp-12-51-2016

Horn JW, Xi Z, Riina R, Riina R, Peirson JA, Yang Y, Dorsey BL et al (2014) Evolutionary bursts in *Euphorbia* (Euphorbiaceae) are linked with photosynthetic pathway. Evolution 68:3485–3504. https://doi.org/10.1111/evo.12534

Hughes AC (2018) Have Indo-Malaysian forests reached the end of the road? Biol Cons 223:129–137. https://doi.org/10.1016/j.biocon.2018.04.029

Hultine KR, Majure LC, Nixon VS, Puente-Martinez R, Arias S, Burquez A, Goettsch B, Zavala-Hurtado JA (2016) The role of botanical gardens in the conservation of Cactaceae. Bioscience 66:1057–1065. https://doi.org/10.1093/biosci/biw128

Humphrey RR (1974) Fire in the deserts and desert grassland of North America. In: Kozlowksi T, Ahlgren CE (eds) Fire and Ecosystems. Academic, New York, pp 365–400

Humphries SW, Long SP (1995) WIMOVAC: a software package for modelling the dynamics of plant leaf an canopy photosynthesis. Bioinformatics 11:361–371. https://doi.org/10.1093/bioinformatics/11.4.361

Huxman TE, Monson RK (2003) Stomatal responses of C_3, C_3- C_4 and C_4 *Flaveria* species to light and intercellular CO_2 concentration: implications for the evolution of stomatal behaviour. Plant Cell Environ 26:313–322. https://doi.org/10.1046/j.1365-3040.2003.00964.x

Isbell F, Reich PB, Tilman D, Hobbie SE, Polasky S, Binder S (2013) Nutrient enrichment, biodiversity loss, and consequent declines in ecosystem productivity. Proc Natl Acad Sci U S A 110:11911–11916. https://doi.org/10.1073/pnas.1310880110

Jacotot A, Marchand C, Gensous S, Allenbach M (2018) Effects of elevated atmospheric CO_2 and increased tidal flooding on leaf gas-exchange parameters of two common mangrove species: *Avicennia marina* and *Rhizophora stylosa*. Photosynth Res 138:249–260. https://doi.org/10.1007/s11120-018-0570-4

Jordan DB, Ogren WL (1984) The CO_2/O_2 specificity of ribulose 1,5-bisphosphate carboxylase/oxygenase. Planta 161:308–313. https://doi.org/10.1007/BF00398720

Karp AT, Behrensmeyer AK, Freeman KH (2018) Grassland fire ecology has roots in the late Miocene. Proc Natl Acad Sci U S A 115:12130–12135. https://doi.org/10.1073/pnas.1809758115

Keller DP, Feng EY, Oschlies A (2014) Potential climate engineering effectiveness and side effects during a high carbon dioxide-emission scenario. Nat Commun 5:3304. https://doi.org/10.1038/ncomms4304

Keerberg O, Pärnik T, Ivanova H, Bassüner B, Bauwe H (2014) C_2 photosynthesis generates about 3-fold elevated leaf CO_2 levels in the C_3- C_4 intermediate species *Flaveria pubescens*. J Exp Bot 65:3649–3656. https://doi.org/10.1093/jxb/eru239

Kellogg EA (2015) Flowering Plants. Monocots: Poaceae. Springer International Publishing, Cham

Kelly EF, Amundson RG, Marino BD, DeNiro MJ (1991) Stable carbon isotopic composition of carbonate in Holocene grassland soils. Soil Sci Soc Am J 55:1651–1658. https://doi.org/10.2136/sssaj1991.03615995005500060025x

Kgope BS, Bond WJ, Midgley GF (2009) Growth responses of African savanna trees implicate atmospheric CO_2 as a driver of past and current changes in savanna tree cover: response of African savanna trees to CO_2. Austral Ecology 35:451–463. https://doi.org/10.1111/j.1442-9993.2009.02046.x

Kimura M, Maruyama T, Crow JF (1963) The mutation load in small populations. Genetics 48:1303–1312

Kirkham MB (2011) Elevated Carbon Dioxide: Impacts on Soil and Plant Water Relations. CRC Press, Boca Raton

Kitoh A, Endo H, Kumar KK, Cavalcanti IFA, Goswami P, Zhou T (2013) Monsoons in a changing world: a regional perspective in a global context. J Geophys Res 118:3053–3065. https://doi.org/10.1002/jgrd.50258

Kluge M, Ting IP (1978) Crassulacean Acid Metabolism: Analysis of an Ecological Adaptation. Springer, Berlin

Knapp AK, Medina E (1999) Success of C_4 photosynthesis in the field: lessons from communities dominated by C_4 plants. In: Sage RF, Monson RK (eds) C_4 Plant Biology. Academic, San Diego, pp 251–283

Knapp AK, Avolio ML, Beier C, Carroll CJ, Collins SL, Dukes JS, Fraser LH et al (2017) Pushing precipitation to the extremes in distributed experiments: recommendations for simulating wet and dry years. Glob Chang Biol 23:1774–1782. https://doi.org/10.1111/gcb.13504

Kocacinar F, Sage RF (2005) 25 – hydraulic properties of the xylem in plants of different photosynthetic pathways. In: Holbrook NM, Zwieniecki MA (eds) Vascular Transport in Plants. Academic, Burlington, pp 517–533

Kocacinar F, McKown AD, Sage TL, Sage RF (2008) Photosynthetic pathway influences xylem structure and function in *Flaveria* (Asteraceae). Plant Cell Environ 31:1363–1376. https://doi.org/10.1111/j.1365-3040.2008.01847.x

Krausmann F, Erb K-H, Gingrich S, Haberl H, Bondeau A, Gaube V, Lauk C et al (2013) Global human appropriation of net primary production doubled in the 20th century. Proc Natl Acad Sci U S A 110:10324–10329. https://doi.org/10.1073/pnas.1211349110

Kromdijk J, Ubierna N, Cousins AB, Griffiths H (2014) Bundle-sheath leakiness in C_4 photosynthesis: a careful balancing act between CO_2 concentration and assimilation. J Exp Bot 65:3443–3457. https://doi.org/10.1093/jxb/eru157

Kubien DS, Sage RF (2003) C_4 grasses in boreal fens: their occurrence in relation to microsite characteristics. Oecologia 137:330–337. https://doi.org/10.1007/s00442-003-1369-2

Kürschner WM, Kvacek Z, Dilcher DL (2008) The impact of Miocene atmospheric carbon dioxide fluctuations on climate and the evolution of terrestrial ecosystems. Proc Natl Acad Sci U S A 105:449–453. https://doi.org/10.1073/pnas.0708588105

Larcher W (2003) Physiological Plant Ecology: Ecophysiology and Stress Physiology of Functional Groups, 4th edn. Springer, Berlin

Laurance WF, Arrea IB (2017) Roads to riches or ruin? Science 358:442–444. https://doi.org/10.1126/science.aao0312

Leadley PW, Drake BG (1993) Open top chambers for exposing plant canopies to elevated atmospheric CO2 and for measuring net gas exchange. Vegetatio 104:3–15

Leakey ADB (2009) Rising atmospheric carbon dioxide concentration and the future of C_4 crops for food and fuel. Proc Biol Sci 276:2333–2343. https://doi.org/10.1098/rspb.2008.1517

Leakey ADB, Ainsworth EA, Bernacchi CJ, Zhu X, Long SP, Ort DR (2012) Photosynthesis in a CO_2-rich atmosphere. In: Eaton-Rye JJ, Tripathy BC, Sharkey TD (eds) Photosynthesis: Plastid Biology, Energy Conversion and Carbon Assimilation. Springer, Dordrecht, pp 733–768

Leal IR, Da Silva JMC, Tanarelli M, Lacher TE Jr (2005) Changing course of biodiversity conservation in the Caatinga of northeastern Brazil. Conserv Biol 19:701–706

Lehmann CE, Parr CL (2016) Tropical grassy biomes: linking ecology, human use and conservation. Philos Trans R Soc B 371:20160329

Liddy HM, Feakins SJ, Tierney JE (2016) Cooling and drying in Northeast Africa across the Pliocene. Earth Planet Sci Lett 449:430–438. https://doi.org/10.1016/j.epsl.2016.05.005

Li Y, Kalnay E, Motesharrei S, Rivas J, Kucharski F, Kirk-Davidoff D, Bach E, Zeng N (2018) Climate model shows large-scale wind and solar farms in the Sahara increase rain and vegetation. Science 361:1019–1022. https://doi.org/10.1126/science.aar5629

Lin G, Phillips SL, Ehleringer JR (1996) Monosoonal precipitation responses of shrubs in a cold desert community on the Colorado Plateau. Oecologia 106:8–17. https://doi.org/10.1007/BF00334402

Long SP (1999) Environmental responses. In: Sage RF, Monson RK (eds) C_4 Plant Biology. Academic, San Diego, pp 215–249

Long SP, Ainsworth EA, Rogers A, Ort DR (2004) Rising atmospheric carbond dioxide: plants FACE the future. Annu Rev Plant Biol 55:591–628. https://doi.org/10.1146/annurev.arplant.55.031903.141610

Lovelock CE, Krauss KW, Osland MJ, Reef R, Ball MC (2016) The physiology of mangrove trees with changing climate. In: Goldstein GH, Santiago LS (eds) Tropical Tree Physiology: Adaptations and Responses in a Changing Environment. Springer, Berlin, pp 149–179

Lundgren MR, Christin P-A, Escobar EG, Ripley BS, Besnard G, Long CM, Hattersley PW, Ellis RP, Leegood RC, Osborne CP (2016) Evolutionary implications of C $_3$ -C $_4$ intermediates in the grass *Alloteropsis semialata*: C_3-C_4 *Alloteropsis semialata*. Plant, Cell & Environment 39:1874–1885. https://doi.org/10.1111/pce.12665

Lundgren MR, Christin P-A (2017) Despite phylogenetic effects, C_3- C_4 lineages bridge the ecological gap to C_4 photosynthesis. J Exp Bot 68:241–254. https://doi.org/10.1093/jxb/erw451

Lundgren MR, Besnard G, Ripley BS, Brad S, Lehmann CER, Chatelet DS, Kynast RG et al (2015) Photosynthetic innovation broadens the niche within a single species. Ecol Lett 18:1021–1029. https://doi.org/10.1111/ele.12484

Lüttge U (2002) CO_2-concentrating: consequences in crassulacean acid metabolism. J Exp Bot 53:2131–2142. https://doi.org/10.1093/jxb/erf081

Lüttge U (2004) Ecophysiology of crassulacean acid metabolism (CAM). Ann Bot 93:629–652. https://doi.org/10.1093/aob/mch087

Lüttge U (ed) (2007) *Clusia*: A Woody Neotropical Genus of Remarkable Plasticity and Diversity. Springer, Berlin

Lüttge U (2011) Photorespiration in phase III of crassulacean acid metabolism: evolutionary and ecophysiological implications. In: Lüttge UE, Beyschlag W, Büdel B, Francis D (eds) Progress in Botany, vol 72. Springer, Berlin, pp 371–384

Mack MC, D'Antonio CM (1998) Impacts of biological invasions on disturbance regimes. Trends Ecol Evol 13:195–198. https://doi.org/10.1016/S0169-5347(97)01286-X

Mack RN, Simberloff D, Lonsdale WM, Evans H, Clout M, Bazzaz FA (2000) Biotic invasions: causes, epidemiology, global consequences, and control. Ecol Appl 10:689–710. https://doi.org/10.1890/1051-0761(2000)010[0689:BICEGC]2.0.CO;2

Maherali H, Johnson HB, Jackson RB (2003) Stomatal sensitivity to vapour pressure difference over a subambient to elevated CO_2 gradient in a C_3/ C_4 grassland. Plant Cell Environ 26:1297–1306. https://doi.org/10.1046/j.1365-3040.2003.01054.x

Martin HA (2006) Cenozoic climatic change and the development of the arid vegetation in Australia. J Arid Environ 66:533–563. https://doi.org/10.1016/j.jaridenv.2006.01.009

Masson-Delmotte V, Schulz M, Abe-Ouchi J, Beer J, Ganopolski A, González Rucco JF, Jansen E, Lambeck K, Luterbacher J, Naish T, Osborn T, Otto-Bleisner B, Quinn T, Ramesh R, Rojas M, Shao X, Timmermann A (2013) Information from paleoclimate archives. In: Stocker TF, Qin D, Plattner GK, Tignor M, Allen SK, Boshung J, Nauels A, Xia Y, Bex V, Midgley PM (eds) Climate Change 2013: The Physical Science Basis. Contribution of Working Group I to the Fifth Assessment Report of the Intergovernmental Panel on Climate Change. Cambridge Univ Press, Cambridge, UK, pp 383–464

Maxwell K, von Caemmerer S, Evans JR (1997) Is a low internal conductance to CO_2 diffusion a consequence of succulence in plants with crassulacean acid metabolism? Aust J Plant Physiol 24:777–786

McKee K, Rogers K, Saintilan N (2012) Response of salt marsh and mangrove wetlands to changes in atmospheric CO_2, climate and sea level. In: Middleton BA (ed) Global Change and the function of and distribution of wetlands. Global Change Ecology and Wetlands, vol 1. Springer, Dordrecht, pp 63–96. https://doi.org/10.1007/978-94-007-4494-3_2

McLeod AR, Long SP (1999) Free-air carbon dioxide enrichment (FACE) in global change research: a review. Adv Ecol Res 28:1–56. https://doi.org/10.1016/S0065-2504(08)60028-8

Melo NMJ, Rosa RS-EG, Pereira EG, Souza JP (2018) Rising CO_2 changes competition relationships between native woody and alien herbaceous Cerrado species. Functional Plant Biol 45:854. https://doi.org/10.1071/FP17333

Miller CS, Gosling WD, Kemp DB, Coe AL, Gilmour I (2016) Drivers of ecosystem and climate change in tropical West Africa over the past ~540 000 years. J Quat Sci 31:671–677. https://doi.org/10.1002/jqs.2893

Millett J, Godbold D, Smith AR, Grant H (2012) N_2 fixation and cycling in *Alnus glutinosa*, *Betula pendula* and *Fagus sylvatica* woodland exposed to free air CO_2 enrichment. Oecologia 169:541–552. https://doi.org/10.1007/s00442-011-2197-4

Monson RK (1989) On the evolutionary pathways resulting in C_4 photosynthesis and crassulacean acid metabolism (CAM). In: Begon M, Fitter AH, Ford ED, MacFadyen A (eds) Advances in Ecological Research. Academic, San Diego, pp 57–110

Moore BD, Cheng S-H, Sims D, Seemann JR (1999) The biochemical and molecular basis for photosynthetic acclimation to elevated atmospheric

CO_2. Plant Cell Environ 22:567–582. https://doi.org/10.1046/j.1365-3040.1999.00432.x

Morgan JA, Mosier AR, Milchunas DG, LeCain DR, Nelson JA, Parton WJ (2004a) CO_2 enhances productivity, alters species composition, and reduces digestibility of shortgrass steppe vegetation. Ecological Applications 14:208–219. https://doi.org/10.1890/02-5213

Morales-Romero D, Molina-Freaner F (2008) Influence of buffelgrass pasture conversion on the regeneration and reproduction of the columnar cactus, *Pachycereus pecten-aboriginum*, in northwestern Mexico. J Arid Environ 72:228–237. https://doi.org/10.1016/j.jaridenv.2007.05.012

Morales-Romero D, Lopez-Garcia H, Martinez-Rodriguez J, Molina-Freaner F (2019) Documenting a plant invasion: the influence of land use on buffelgrass invasion along roadsides in Sonora, Mexico. J Arid Environ 164:53–59. https://doi.org/10.1016/j.jaridenv.2019.01.012

Morgan JA, Pataki DE, Korner C, Clark H, Del Grosso SJ, Grunzweig JM, Knapp AK et al (2004) Water relations in grassland and desert ecosystems exposed to elevated atmospheric CO_2. Oecologia 140:11–25. https://doi.org/10.1007/s00442-004-1550-2

Morgan JA, LeCain DR, Pendall E, Blumenthal DM, Kimball BA, Carrillo Y, Williams DG et al (2011) C_4 grasses prosper as carbon dioxide eliminates desiccation in warmed semi-arid grassland. Nature 476:202–U101. https://doi.org/10.1038/nature10274

Munson SM, Webb RH, Belnap J, Andrew Hubbard J, Swann DE, Rutman S (2012) Forecasting climate change impacts to plant community composition in the Sonoran Desert region. Glob Chang Biol 18:10831095. https://doi.org/10.1111/j.1365-2486.2011.02598.x

Mueller KE, Blumenthal DM, Pendall E, Carrillo Y, Dijkstra FA, Williams DG, Follett RF, Morgan JA (2016) Impacts of warming and elevated CO_2 on a semi-arid grassland are non-additive, shift with precipitation, and reverse over time. Ecol Lett 19:956–966. https://doi.org/10.1111/ele.12634

Myers N, Mittermeier RA, Mittermeier CG, da Fonseca GAB, Kent J (2000) Biodiversity hotspots for conservation priorities. Nature 403:853–858. https://doi.org/10.1038/35002501

Nazarenko L, Schmidt GA, Miller RL, Bleck R, Canuto V, Cheng Y, Faluvegi G et al (2015) Future climate change under RCP emission scenarios with GISS ModelE2. J Adv Model Earth Syst 7:244–267. https://doi.org/10.1002/2014MS000403

Nelson EA, Sage RF (2008) Functional constraints of CAM leaf anatomy: tight cell packing is associated with increased CAM function across a gradient of CAM expression. J Exp Bot 59:1841–1850. https://doi.org/10.1093/jxb/erm346

Nelson JA, Morgan JA, LeCain DR, Mosier AR, Milchunas DG, Parton BA (2004) Elevated CO_2 increases soil moisture and enhances plant water relations in a long-term field study in semi-arid shortgrass steppe of Colorado. Plant Soil 259:169–179. https://doi.org/10.1023/B:PLSO.0000020957.83641.62

Newbold T, Boakes EH, Hill SLL, Harfoot MBJ, Collen B (2017) The present and future effects of land use on ecological assemblages in tropical grasslands and savannas in Africa. Oikos 126:1760–1769. https://doi.org/10.1111/oik.04338

Nilsson LA (1992) Orchid pollination biology. Trends Ecol Evol 7:255–259. https://doi.org/10.1016/0169-5347(92)90170-G

Norby RJ, Warren JM, Iversen CM, Medlyn BE, McMurtrie RE (2010) CO_2 enhancement of forest productivity constrained by limited nitrogen availability. Proc Natl Acad Sci U S A 107:19368–19373. https://doi.org/10.1073/pnas.1006463107

Nordt L, Von Fischer J, Tieszen L, Tubbs J (2008) Coherent changes in relative C_4 plant productivity and climate during the late quaternary in the north American Great Plains. Quat Sci Rev 27:16001611. https://doi.org/10.1016/j.quascirev.2008.05.008

Noss RF (2013) Forgotten Grasslands of the South: Natural History and Conservation. Island Press/Center for Resource Economics, Washington, DC

Oakley JC, Sultmanis S, Stinson CR, Sage TL, Sage RF (2014) Comparative studies of C_3 and C_4 Atriplex hybrids in the genomics era: physiological assessments. J Exp Bot 65:3637–3647. https://doi.org/10.1093/jxb/eru106

Osborne CP (2011) Chapter 17: The geologic history of C_4 plants. In: Raghavendra AS, Sage RF (eds) C_4 Photosynthesis and Related CO_2 Concentrating Mechanisms. Springer, Dordrecht, pp 339–357

Osborne CP, Salomaa A, Kluyver TA, Visser V, Kellogg EA, Morrone O, Vorontsova MS et al (2014) A global database of C_4 photosynthesis in grasses. New Phytol 204:441–446. https://doi.org/10.1111/nph.12942

Osborne CP, Charles-Dominique T, Stevens N, Bond WJ, Midgley G, Lehmann CER (2018) Human impacts in African savannas are mediated by plant functional traits. New Phytol 220:10–24. https://doi.org/10.1111/nph.15236

Osland MJ, Enwright N, Day RH, Doyle TW (2013) Winter climate change and coastal wetland foundation species: salt marshes vs. mangrove forests in the southeastern United States. Glob Change Biol 19:1482–1494. https://doi.org/10.1111/gcb.12126

Osland MJ, Day RH, Hall CT, Brumfield MD, Dugas JL, Jones WR (2017) Mangrove expansion and contraction at a poleward range limit: climate extremes and land-ocean temperature gradients. Ecology 98:125–137. https://doi.org/10.1002/ecy.1625

Osmond CB (1978) Crassulacean acid metabolism: a curiosity in context. Annu Rev Plant Physiol 29:379–414. https://doi.org/10.1146/annurev.pp.29.060178.002115

Osmond CB, Winter K, Ziegler H (1982) Functional significance of different pathways of CO_2 fixation in photosynthesis. In: Lange OL, Nobel PS, Osmond CB, Ziegler H (eds) Physiological Plant Ecology II: Water Relations and Carbon Assimilation. Springer, Berlin, pp 479–547

Osmond B, Neales T, Stange G (2008) Curiosity and context revisited: crassulacean acid metabolism in the Anthropocene. J Exp Bot 59:1489–1502. https://doi.org/10.1093/jxb/ern052

Owen-Smith RN (1988) Megaherbivores: The Influence of Very Large Body Size on Ecology. Cambridge University Press, Cambridge

Owensby CE, Ham JM, Knapp AK, Auen LM (1999) Biomass production and species composition change in a tallgrass prairie ecosystem after long-term exposure to elevated atmospheric CO_2. Global Change Biology 5:497–506. https://doi.org/10.1046/j.1365-2486.1999.00245.x

Owen NA, Fahy KF, Griffiths H (2016) Crassulacean acid metabolism (CAM) offers sustainable bioenergy production and resilience to climate change. GCB Bioenergy 8:737–749. https://doi.org/10.1111/gcbb.12272

Pangburn LE, Eberts E, Carmona-Galindo V (2016) Photosynthetic characterization of invasive plant diversity in Los Angeles county from 1830–2010. Loyola-Marymount University Honors thesis, 130. https://digitalcommons.lmu.edu/honors-thesis/130

Parr CL, Lehmann CER, Bond WJ, Hoffmann WA, Andersen AN (2014) Tropical grassy biomes: misunderstood, neglected, and under threat. Trends Ecol Evol 29:205–213. https://doi.org/10.1016/j.tree.2014.02.004

Pastore MA, Lee TD, Hobbie SE, Reich PB (2019) Strong photosynthetic acclimation and enhanced water-use efficiency in grassland functional groups persist over 21 years of CO_2 enrichment, independent of nitrogen supply. Glob Chang Biol:1–14. https://doi.org/10.1111/gcb.14714

Pausata FSR, Emanuel KA, Chiacchio M, Diro GT, Zhang Q, Sushama L, Stager JC, Donnelly JP (2017) Tropical cyclone activity enhanced by Sahara greening and reduced dust emissions during the African Humid Period. Proc Natl Acad Sci USA 114:6221–6226. https://doi.org/10.1073/pnas.1619111114

Pausata FSR, Gaetani M, Messori G, Berg A, de Souza DM, Sage RF, deMenocal PB (2020) The greening of the Sahara: past changes and future implications. One Earth 2:235–250. https://doi.org/10.1016/j.oneear.2020.03.002

Pearcy RW, Tumosa N, William K (1981) Relationships between growth, photosynthesis and competitive interactions for a C_3 and a C_4 plant. Oecologia 48:371–376. https://doi.org/10.1007/BF00346497

Peñuelas J, Sardans J, Estiarte M, Ogaya R, Carnicer J, Coll M, Barbeta A et al (2013) Evidence of current impact of climate change on life: a walk from genes to the biosphere. Glob Chang Biol 19:2303–2338. https://doi.org/10.1111/gcb.12143

Pinto H, Tissue DT, Ghannoum O (2011) *Panicum milioides* (C_3-C_4) does not have improved water or nitrogen economies relative to C_3 and C_4 congeners exposed to industrial-age climate change. J Exp Bot 62:3223–3234. https://doi.org/10.1093/jxb/err005

Polley HW (1997) Implications of rising atmospheric carbon dioxide concentration for rangelands. Rangel Ecol Manag 50:562–577

Polley H, Johnson H, Mayeux H (1994) Increasing CO_2 – comparative responses of the C_4 grass *Schizachyrium* and grassland invader *Prosopis*. Ecology 75:976–988. https://doi.org/10.2307/1939421

Polley HW, Mayeux HS, Johnson HB, Tischler CR (1997) Viewpoint: atmospheric CO_2, soil water, and shrub/grass ratios on rangelands. Rangel Ecol Manag 50:278–284

Polley HW, Johnson HB, Derner JD (2002) Soil- and plant-water dynamics in a C_3/C_4 grassland exposed to a subambient to superambient CO_2 gradient. Glob Chang Biol 8:1118–1129. https://doi.org/10.1046/j.1365-2486.2002.00537.x

Polley HW, Johnson HB, Derner JD (2003a) Increasing CO_2 from subambient to superambient concentrations alters species composition and increases above-ground biomass in a C_3/C_4 grassland. New Phytol 160:319–327. https://doi.org/10.1046/j.1469-8137.2003.00897.x

Polley HW, Johnson HB, Tischler CR (2003b) Woody invasion of grasslands: evidence that CO_2 enrichment indirectly promotes establishment of *Prosopis glandulosa*. Plant Ecol 164:85–94. https://doi.org/10.1023/A:1021271226866

Polley HW, Jin VL, Fay PA (2012a) CO_2-caused change in plant species composition rivals the shift in vegetation between mid-grass and tallgrass prairies. Glob Change Bio 18:700–710

Polley HW, Jin VL, Fay PA (2012b) Feedback from plant species change amplifies CO_2 enhancement of grassland productivity. Glob Change Biol 18:2813–2823. https://doi.org/10.1111/j.1365-2486.2012.02735.x

Polley HW, Aspinwall MJ, Collins HP, Gibson AE, Gill RA, Jackson RB, Jin VL, Khasanova AR, Reichmann LG, Fay PA (2019) CO_2 enrichment and soil type additively regulate grassland productivity. New Phytol 222:183–192. https://doi.org/10.1111/nph.15562

Pound MJ, Haywood AM, Salzmann U, Riding JB (2012) Global vegetation dynamics and latitudinal temperature gradients during the mid to late Miocene (15.97-5.33Ma). Earth-Sci Rev 112:1–22. https://doi.org/10.1016/j.earscirev.2012.02.005

Prentice IC, Harrison SP, Bartlein PJ (2011) Global vegetation and terrestrial carbon cycle changes after the last ice age. New Phytol 189:988–998. https://doi.org/10.1111/j.1469-8137.2010.03620.x

Ramankutty N, Evan AT, Monfreda C, Foley JA (2008) Farming the planet: 1. Geographic distribution of global agricultural lands in the year 2000. Glob Biogeochem Cycles 22:GB1003. https://doi.org/10.1029/2007GB002952

Raven JA, Beardall J (2014) CO_2 concentrating mechanisms and environmental change. Aquat Bot 118:24–37. https://doi.org/10.1016/j.aquabot.2014.05.008

Raven JA, Beardall J (2016) The ins and outs of CO_2. J Exp Bot 67:1–13. https://doi.org/10.1093/jxb/erv451

Raven JA, Cockell CS, De La Rocha CL (2008) The evolution of inorganic carbon concentrating mechanisms in photosynthesis. Philos Trans R Soc Lond Ser B Biol Sci 363:2641–2650. https://doi.org/10.1098/rstb.2008.0020

Raven JA, Giordano M, Beardall J, Maberly SC (2012) Algal evolution in relation to atmospheric CO_2: carboxylases, carbon-concentrating mechanisms and carbon oxidation cycles. Philos Trans R Soc Lond Ser B Biol Sci 367:493–507. https://doi.org/10.1098/rstb.2011.0212

Raven JA, Beardall J, Sanchez-Baracaldo P (2017) The possible evolution and future of CO_2-concentrating mechanisms. J Exp Bot 68:3701–3716. https://doi.org/10.1093/jxb/erx110

Rawsthorne S (1992) C_3-C_4 intermediate photosynthesis: linking physiology to gene expression. Plant J 2:267–274. https://doi.org/10.1111/j.1365-313X.1992.00267.x

Reed HE, Seastedt TR, Blair JM (2005) Ecological consequences of C_4 grass invasion of a C_4 grassland: a dilemma for management. Ecol Appl 15:1560–1569. https://doi.org/10.1890/04-0407

Reich PB, Hungate BA, Luo Y (2006) Carbon-nitrogen interactions in terrestrial ecosystems in response to rising atmospheric carbon dioxide. Annu Rev Ecol Evol Syst 37:611–636. https://doi.org/10.1146/annurev.ecolsys.37.091305.110039

Reich PB, Hobbie SE, Lee TD, Pastore MA (2018) Unexpected reversal of C_3 versus C_4 grass response to elevated CO_2 during a 20-year field experiment. Science 360:317. https://doi.org/10.1126/science.aas9313

Ripple WJ, Newsome TM, Wolf C, Dirzo R, Everatt KT, Galetti M, Hayward MW et al (2015) Collapse of the world's largest herbivores. Sci Adv 1:e1400103. https://doi.org/10.1126/sciadv.1400103

Royer DL, Pagani M, Beerling DJ (2012) Geobiological constraints on earth system sensitivity to CO_2 during the cretaceous and Cenozoic. Geobiology 10:298–310. https://doi.org/10.1111/j.1472-4669.2012.00320.x

Sage RF (1994) Acclimation of photosynthesis to increasing atmospheric CO_2: the gas exchange perspective. Photosynth Res 39:351–368. https://doi.org/10.1007/BF00014591

Sage RF (1995) Was low atmospheric CO_2 during the Pleistocene a limiting factor for the origin of agriculture? Glob Chang Biol 1:93–106. https://doi.org/10.1111/j.1365-2486.1995.tb00009.x

Sage RF (1996) Modification of fire disturbance by elevated CO_2. In: Korner C, Bazzaz FA (eds) Carbon Dioxide, Populations, and Communities. Academic, San Diego, pp 231–249

Sage RF (1999) Why C_4 plants? In: Sage RF, Monson RK (eds) C_4 Plant Biology. Academic, San Diego, pp 3–16

Sage RF (2000) C_3 versus C_4 photosynthesis in rice: ecophysiological perspectives. In: Sheehy JE, Mitchell PL, Hardy B (eds) Redesigning Rice Photosynthesis to Increase Yield. Elsevier Science, Amsterdam, pp 13–35

Sage RF (2001) Environmental and evolutionary preconditions for the origin and diversification of the C_4 photosynthetic syndrome. Plant Biol 3:202–213

Sage RF (2002a) How terrestrial organisms sense, signal, and respond to carbon dioxide. Integr Comp Biol 42:469–480. https://doi.org/10.1093/icb/42.3.469

Sage RF (2002b) Variation in the k_{cat} of Rubisco in C_3 and C_4 plants and some implications for photosynthetic performance at high and low temperature. J Exp Bot 53:609–620. https://doi.org/10.1093/jexbot/53.369.609

Sage RF (2013) Photorespiratory compensation: a driver for biological diversity. Plant Biol 15:624–638. https://doi.org/10.1111/plb.12024

Sage RF (2016) A portrait of the C_4 photosynthetic family on the 50th anniversary of its discovery: species number, evolutionary lineages, and Hall of Fame. J Exp Bot 67:4039–4056. https://doi.org/10.1093/jxb/erw156

Sage RF (2020) Global change biology: a primer. Glob Chang Biol 26:3–30. https://doi.org/10.1111/gcb.14893

Sage RF, Coleman JR (2001) Effects of low atmospheric CO_2 on plants: more than a thing of the past. Trends Plant Sci 6:18–24. https://doi.org/10.1016/S1360-1385(00)01813-6

Sage RF, Kubien DS (2003) Quo vadis C_4? An ecophysiological perspective on global change and the future of C_4 plants. Photosynth Res 77:209–225. https://doi.org/10.1023/A:1025882003661

Sage RF, Kubien DS (2007) The temperature response of C_3 and C_4 photosynthesis. Plant Cell Environ 30:1086–1106. https://doi.org/10.1111/j.1365-3040.2007.01682.x

Sage RF, Pearcy RW (1987a) The nitrogen use efficiency of C_3 and C_4 plants: II. Leaf nitrogen effects on the gas exchange characteristics of *Chenopodium album* (L.) and *Amaranthus retroflexus* (L.). Plant Physiol 84:959–963. https://doi.org/10.1104/pp.84.3.959

Sage RF, Pearcy RW (1987b) The nitrogen use efficiency of C_3 and C_4 plants: I. leaf nitrogen, growth, and biomass partitioning in *Chenopodium album* (L.) and *Amaranthus retroflexus* (L.). Plant Physiol 84:954–958. https://doi.org/10.1104/pp.84.3.954

Sage RF, Pearcy RW (2000) The physiological ecology of C_4 photosynthesis. In: Leegood RC, Sharkey TD, von Caemmerer S (eds) Photosynthesis: Physiology and Metabolism. Springer, Dordrecht, pp 497–532

Sage RF, Sage TL (2013) C_4 plants. In: Levin SA (ed) Encyclopedia of Biodiversity, vol 2, 2nd edn. Academic, Waltham, pp 361–381

Sage RF, Stata M (2015) Photosynthetic diversity meets biodiversity: the C_4 plant example. J Plant Physiol 172:104–119. https://doi.org/10.1016/j.jplph.2014.07.024

Sage RF, Sultmanis S (2016) Why are there no C_4 forests? J Plant Physiol 203:55–68. https://doi.org/10.1016/j.jplph.2016.06.009

Sage RF, Zhu X-G (2011) Exploiting the engine of C_4 photosynthesis. J Exp Bot 62:2989–3000. https://doi.org/10.1093/jxb/err179

Sage RF, Pearcy RW, Seemann JR (1987) The nitrogen use efficiency of C_3 and C_4 plants : III. Leaf nitrogen effects on the activity of carboxylating enzymes in *Chenopodium album* (L.) and *Amaranthus retroflexus* (L.). Plant Physiol 85:355–359. https://doi.org/10.1104/pp.85.2.355

Sage RF, Sharkey TD, Seemann JR (1989) Acclimation of photosynthesis to elevated CO_2 in five C_3 species. Plant Physiol 89:590–596. https://doi.org/10.1104/pp.89.2.590

Sage RF, Santrucek J, Grise DJ (1995) Temperature effects on the photosynthetic response of C_3 plants to long-term CO_2 enrichment. Vegetatio 121:67–77

Sage RF, Wedin DA, Li M (1999) The biogeography of C_4 photosynthesis. In: Sage RF, Monson RK (eds) C_4 Plant Biology. Academic, San Diego, pp 313–373

Sage RF, Christin P-A, Edwards EJ (2011a) The C_4 plant lineages of planet Earth. J Exp Bot 62:3155–3169. https://doi.org/10.1093/jxb/err048

Sage TL, Sage RF, Vogan PJ, Rahman B, Johnson DC, Oakley JC, Heckel MA (2011b) The occurrence of C_2 photosynthesis in *Euphorbia* subgenus *Chamaesyce* (Euphorbiaceae). J Exp Bot 62:3183–3195. https://doi.org/10.1093/jxb/err059

Sage TL, Busch FA, Johnson DC, Friesen PC, Stinson CR, Stata M, Sultmanis S et al (2013) Initial events during the evolution of C_4 photosynthesis in C_3 species of *Flaveria*. Plant Physiol 163:1266–1276. https://doi.org/10.1104/pp.113.221119

Sage RF, Khoshravesh R, Sage TL (2014a) From proto-Kranz to C_4 Kranz: building the bridge to C_4 photosynthesis. J Exp Bot 65:3341–3356. https://doi.org/10.1093/jxb/eru180

Sage RF, Peixoto M, Sage TL (2014b) Photosynthesis in sugarcane. In: Moore PH, Botha F (eds) Physiology of Sugarcane. Wiley-Blackwell, Inc., Oxford, pp 121–154

Sage TL, Bagha S, Lundsgaard-Nielsen V, Branch HA, Sultmanis S, Sage RF (2015) The effect of high temperature stress on male and female reproduction in plants. Field Crop Res 182:30–42. https://doi.org/10.1016/j.fcr.2015.06.011

Sage RF, Monson RK, Ehleringer JR, Adachi S, Pearcy RW (2018) Some like it hot: the physiological ecology of C_4 plant evolution. Oecologia 187:941–966. https://doi.org/10.1007/s00442-018-4191-6

Saintilan N, Wilson NC, Rogers K, Rajkaran A, Krauss KW (2014) Mangrove expansion and salt marsh decline at mangrove poleward limits. Glob Change Biol 20:147–157. https://doi.org/10.1111/gcb.12341

Saintilan N, Rogers K (2015) Woody plant encroachment of grasslands: a comparison of terrestrial and wetland settings. New Phytol 205:1062–1070. https://doi.org/10.1111/nph.13147

Saintilan N, Rogers K, McKee KL (2019) The shifting saltmarsh-Mangrove ecotone in Australasia and the Americas. In: Perillo ME, Wolanski E, Cahoon

DR, Hopkinson CS (eds) Coastal Wetlands, 2nd edn. Elsevier, Amsterdam, pp 915–945. https://doi.org/10.1016/B978-0-444-63893-9.00026-5

Sala OE, Chapin FS III, Armesto JJ, Berlow E, Bloomfield J, Dirzo R, Huber-Sanwald E et al (2000) Global biodiversity scenarios for the year 2100. Science 287:1770–1774. https://doi.org/10.1126/science.287.5459.1770

Sanderson EW, Jaiteh M, Levy MA, Redford KH, Wannebo AV, Woolmer G (2002) The human footprint and the last of the wild. Bioscience 52:891–904. https://doi.org/10.1641/0006--3568(2002)052[0891:THFATL]2.0.CO;2

Sankaran M (2019) Droughts and the ecological future of tropical savanna vegetation. J Ecol 107:1531–1549. https://doi.org/10.1111/1365-2745.13195

Sankaran M, Hanan NP, Scholes RJ, Ratnam J, Augustine DJ, Cade BS, Gignoux J et al (2005) Determinants of woody cover in African savannas. Nature 438:846. https://doi.org/10.1038/nature04070

Scholes RJ, Archer SR (1997) Tree-grass interactions in savannas. Annu Rev Ecol Syst 28:517–544. https://doi.org/10.1146/annurev.ecolsys.28.1.517

Schortemeyer M, Atkin OK, McFarlane N, Evans JR (2002) N_2 fixation by *Acacia* species increases under elevated atmospheric CO_2. Plant Cell Environ 25:567–579. https://doi.org/10.1046/j.1365-3040.2002.00831.x

Schuster WS, Monson R (1990) An examination of the advantages of C_3-C_4 intermediate photosynthesis in warm environments. Plant Cell Environ 13:903–912

Schuster M, Duringer P, Ghienne J-F, Vignaud P, Mackaye HT, Likius A, Brunet M (2006) The age of the Sahara Desert. Science 311:821–821. https://doi.org/10.1126/science.1120161

Seager R, Ting M, Held I, Kushnir Y, Lu J, Vecchi G, Huang H-P et al (2007) Model projections of an imminent transition to a more arid climate in Southwestern North America. Science 316:1181–1184. https://doi.org/10.1126/science.1139601

Senut B, Pickford M, Ségalen L (2009) Neogene desertification of Africa. Compt Rendus Geosci 341:591–602. https://doi.org/10.1016/j.crte.2009.03.008

Shanahan TM, Hughen KA, McKay NP, Overpeck JT, Scholz CA, Gosling WD, Miller CS et al (2016) CO_2 and fire influence tropical ecosystem stability in response to climate change. Sci Rep 6:29587. https://doi.org/10.1038/srep29587

Scheiter S, Gaillard C, Martens C, Erasmus BFN, Pfeiffer M (2018) How vulnerable are ecosystems in the Limpopo province to climate change? S Afr J Bot 116:86–95. https://doi.org/10.1016/j.sajb.2018.02.394

Sharkey TD (1988) Estimating the rate of photorespiration in leaves. Physiol Plant 73:147–152. https://doi.org/10.1111/j.1399-3054.1988.tb09205.x

Sharwood RE, Ghannoum O, Kapralov MV, Gunn LH, Whitney SM (2016) Temperature responses of Rubisco from Paniceae grasses provide opportunities for improving C_3 photosynthesis. Nat Plants 2:16186. https://doi.org/10.1038/nplants.2016.186

Silliman BR, van de Koppel J, Bertness MD, Stanton LE, Mendelssohn IZ (2005) Drought, snails, and large-scale die-off of Southern U.S. Salt Marshes. Science 310:1803–1806. https://doi.org/10.1126/science.1118229

Silva LCR, Hoffmann WA, Rossatto DR, Haridasan M, Franco AC (2013) Can savannas become forests? A coupled analysis of nutrient stocks and fire thresholds in Central Brazil. Plant Soil 373:829–842. https://doi.org/10.1007/s11104-013-1822-x

Silvera K, Lasso E (2016) Ecophysiology and crassulacean acid metabolism of tropical epiphytes. In: Goldstein G, Santiago LS (eds) Tropical Tree Physiology: Adaptations and Responses in a Changing Environment. Springer International Publishing, Cham, pp 25–43

Silvera K, Santiago LS, Cushman JC, Winter K (2009) Crassulacean acid metabolism and epiphytism linked to adaptive radiations in the Orchidaceae. Plant Physiol 149:1838–1847. https://doi.org/10.1104/pp.108.132555

Simkin SM, Allen EB, Bowman WD, Clark C, Belnap J, Brooks M, Cade B, Waller D (2016) Conditional vulnerability of plant diversity to atmospheric nitrogen deposition across the United States. Proc Natl Acad Sci U S A 113:4086–4091. https://doi.org/10.1073/pnas.1515241113

Sims DA, Luo Y, Seemann JR (1998) Comparison of photosynthetic acclimation to elevated CO_2 and limited nitrogen supply in soybean. Plant Cell Environ 21:945–952. https://doi.org/10.1046/j.1365-3040.1998.00334.x

Skowno AL, Thompson MW, Hiestermann J, West AG, Bond WJ, Bond WJ (2017) Woodland expansion in South African grassy biomes based on satellite observations (1990-2013): general patterns and potential drivers. Glob Chang Biol 23:2358–2369. https://doi.org/10.1111/gcb.13529

Smith JAC, Winter K (1996) Taxonomic distribution of crassulacean acid metabolism. In: Winter K, Smith JAC (eds) Crassulacean Acid Metabolism: Biochemistry, Ecophysiology and Evolution. Springer, Berlin, pp 427–436

Smith SD, Huxman TE, Zitzer SF, Charlet TN, Housman DC, Coleman JS, James S et al (2000) Elevated CO_2 increases productivity and invasive species success in an arid ecosystem. Nature 408:79–82. https://doi.org/10.1038/35040544

Springer AC, Swann DE, Crimmins MA (2015) Climate change impacts on high elevation saguaro range expansion. J Arid Environ 116:57–62. https://doi.org/10.1016/j.jaridenv.2015.02.004

Staver AC, Archibald S, Levin SA (2011) The global extent and determinants of savanna and forest as alternative biome states. Science 334:230–232. https://doi.org/10.1126/science.1210465

Steffen W, Sanderson RA, Tyson PD, Jäger J, Matson PA, Moore B III, Oldfield F et al (2004) Global Change and the Earth System: A Planet Under Pressure. Springer, Berlin/Heidelberg. https://doi.org/10.1007/b137870

Stevens N, Erasmus BFN, Archibald S, Bond WJ (2016) Woody encroachment over 70 years in South African savannahs: overgrazing, global change or extinction aftershock? Philos Trans R Soc Lond Ser B Biol Sci 371:1–9. https://doi.org/10.1098/rstb.2015.0437

Stevens N, Lehmann CER, Murphy BP, Durigan G (2017) Savanna woody encroachment is widespread across three continents. Glob Chang Biol 23:235–244. https://doi.org/10.1111/gcb.13409

Still CJ, Berry JA, Collatz GJ, DeFries RS (2003) Global distribution of C_3 and C_4 vegetation: carbon cycle implications. Glob Biogeochem Cycles 17:1006. https://doi.org/10.1029/2001GB001807

Still CJ, Cotton JM, Griffith DM (2019) Assessing earth system model predictions of C_4 grass cover in North America: from the glacial era to the end of this century. Glob Ecol Biogeogr 28:145–157. https://doi.org/10.1111/geb.12830

Strömberg CAE (2011) Evolution of grasses and grassland ecosystems. Annu Rev Earth Planet Sci 39:517–544. https://doi.org/10.1146/annurev-earth-040809-152402

Taylor SH, Aspinwall MJ, Blackman CJ, Choat B, Tissue DT, Ghannoum O (2018) CO_2 availability influences hydraulic function of C_3 and C_4 grass leaves. J Exp Bot 69:2731–2741. https://doi.org/10.1093/jxb/ery095

Tcherkez GGB, Farquhar GD, Andrews TJ (2006) Despite slow catalysis and confused substrate specificity, all ribulose bisphosphate carboxylases may be nearly perfectly optimized. Proc Natl Acad Sci U S A 103:7246–7251. https://doi.org/10.1073/pnas.0600605103

Terrer C, Vicca S, Hungate BA, Phillips RP, Prentice IC (2016) Mycorrhizal association as a primary control of the CO_2 fertilization effect. Science 353:72–74. https://doi.org/10.1126/science.aaf4610

Terrer C, Vicca S, Stocker BD, Hungate BA, Phillips RP, Reich PB, Finzi AC, Prentice IC (2018) Ecosystem responses to elevated CO_2 governed by plant-soil interactions and the cost of nitrogen acquisition. New Phytol 217:507–522. https://doi.org/10.1111/nph.14872

Terrer C, Jackson RB, Prentice IC, Keenan TF, Kaiser C, Vicca S et al (2019) Nitrogen and phosphorus constrain the CO_2 fertilization of global plant biomass. Nat Climate Change 9:684–689. https://doi.org/10.1038/s41558-019-0545-2

Tietjen B, Schlaepfer DR, Bradford JB, Lauenroth WK, Hall SA, Duniway MC, Hochstrasser T et al (2017) Climate change-induced vegetation shifts lead to more ecological droughts despite projected rainfall increases in many global temperate drylands. Glob Chang Biol 23:2743–2754. https://doi.org/10.1111/gcb.13598

Tipple BJ, Pagani M (2007) The early origins of terrestrial C_4 photosynthesis. Annu Rev Earth Planet Sci 35:435–461. https://doi.org/10.1146/annurev.earth.35.031306.140150

Tissue DT, Megonigal JP, Thomas RB (1996) Nitrogenase activity and N_2 fixation are stimulated by elevated CO_2 in a tropical N_2-fixing tree. Oecologia 109:28–33. https://doi.org/10.1007/s004420050054

Trejo I, Dirzo R (2000) Deforestation of seasonally dry tropical forest: a national and local analysis in Mexico. Biol Conserv 94:133–142

Turner RM, Hastings JR, Webb RH, Bowers JE (2003) The Changing Mile Revisited: An Ecological Study of Vegetation Change with Time in the Lower Mile of an Arid and Semiarid Region. University of Arizona Press, Tucson

Uno KT, Polissar PJ, Jackson KE, deMenocal PB (2016) Neogene biomarker record of vegetation change in eastern Africa. Proc Natl Acad Sci U S A 113:6355–6363. https://doi.org/10.1073/pnas.1521267113

Van Auken OW (2000) Shrub invasions of North American semiarid grasslands. Annu Rev Ecol Syst 31:197–215. https://doi.org/10.1146/annurev.ecolsys.31.1.197

van Devender T (2000) The deep history of the Sonoran desert. In: Phillips SJ, Comus PW (eds) A Natural History of the Sonoran Desert. University of California Press, Berkeley, pp 61–69

Veldman JW (2016) Clarifying the confusion: old-growth savannahs and tropical ecosystem degradation. Philos Trans R Soc Lond Ser B Biol

Sci 371:20150306. https://doi.org/10.1098/rstb.2015.0306
Veldman JW, Buisson E, Durigan G, Fernandes GW, Le Stradic S, Mahy G, Negreiros D et al (2015) Toward an old-growth concept for grasslands, savannas, and woodlands. Front Ecol Environ 13:154–162. https://doi.org/10.1890/140270
Venter ZS, Hawkins H-J, Cramer MD (2017) Implications of historical interactions between herbivory and fire for rangeland management in African savannas. Ecosphere 8:e01946. https://doi.org/10.1002/ecs2.1946
Venter ZS, Cramer MD, Hawkins H-J (2018) Drivers of woody plant encroachment over Africa. Nat Commun 9:2272. https://doi.org/10.1038/s41467-018-04616-8
Vicentini A, Barber JC, Aliscioni SS, Giussani LM, Kellogg EA (2008) The age of the grasses and clusters of origins of C_4 photosynthesis. Glob Chang Biol 14:2963–2977. https://doi.org/10.1111/j.1365-2486.2008.01688.x
Vitousek PM (1994) Beyond global warming: ecology and global change. Ecology 75:1861–1876. https://doi.org/10.2307/1941591
Vitousek PM, Aber JD, Howarth RW, Likens GE, Matson PA, Schindler DW, Schlesinger WH, Tilman DG (1997a) Human alteration of the global nitrogen cycle: sources and consequences. Ecol Appl 7:737–750. https://doi.org/10.1890/1051-0761(1997)007[0737:HAOTGN]2.0.CO;2
Vitousek PM, Mooney HA, Lubchenco J, Melillo JM (1997b) Human domination of Earth's ecosystems. Science 277:494–499. https://doi.org/10.1126/science.277.5325.494
Vogan PJ, Sage RF (2011) Water-use efficiency and nitrogen-use efficiency of C_3-C_4 intermediate species of *Flaveria* Juss. (Asteraceae). Plant Cell Environ 34:1415–1430. https://doi.org/10.1111/j.1365-3040.2011.02340.x
Vogan PJ, Sage RF (2012) Effects of low atmospheric CO_2 and elevated temperature during growth on the gas exchange responses of C_3, C_3-C_4 intermediate, and C_4 species from three evolutionary lineages of C_4 photosynthesis. Oecologia 169:341–352. https://doi.org/10.1007/s00442-011-2201-z
Vogan PJ, Frohlich MW, Sage RF (2007) The functional significance of C_3- C_4 intermediate traits in *Heliotropium* L. (Boraginaceae): gas exchange perspectives. Plant Cell Environ 30:1337–1345. https://doi.org/10.1111/j.1365-3040.2007.01706.x
Vollesen K (2000) *Blepharis*: A Taxonomic Revision. Royal Botanic Gardens, Kew
von Caemmerer S (1989) A model of photosynthetic CO_2 assimilation and carbon-isotope discrimination in leaves of certain C_3-C_4 intermediates. Planta 178:463–474. https://doi.org/10.1007/BF00963816
von Caemmerer S (2000) Biochemical Models of Leaf Photosynthesis. Csiro Publishing, Collingwood
von Caemmerer S, Quick WP (2000) Rubisco: physiology *in vivo*. In: Leegood RC, Sharkey TD, von Caemmerer S (eds) Photosynthesis: Physiology and Metabolism. Springer, Dordrecht, pp 85–113
Walsh JR, Carpenter SR, Zanden MJV (2016) Invasive species triggers a massive loss of ecosystem services through a trophic cascade. Proc Natl Acad Sci U S A 113:4081–4085. https://doi.org/10.1073/pnas.1600366113
Walter H, Box EO (1986) Middle Asian deserts. In: West NE (ed) Ecosystems of the World 5: Temperate Deserts and Semi-deserts. Elsevier, Amsterdam, pp 79–104
Wand SJE, Midgley GF, Jones MH, Curtis PS (1999) Responses of wild C_4 and C_3 grass (Poaceae) species to elevated atmospheric CO_2 concentration: a meta-analytic test of current theories and perceptions. Glob Chang Biol 5:723–741. https://doi.org/10.1046/j.1365-2486.1999.00265.x
Wang D, Heckathorn SA, Wang X, Philpott SM (2012) A meta-analysis of plant physiological and growth responses to temperature and elevated CO_2. Oecologia 169:1–13. https://doi.org/10.1007/s00442-011-2172-0
Way DA, Oren R, Kroner Y (2015) The space-time continuum: the effects of elevated CO_2 and temperature on trees and the importance of scaling. Plant Cell Environ 38:991–1007. https://doi.org/10.1111/pce.12527
Wedin DA (2004) C_4 grasses: resource use, ecology, and global change. In: Warm Season C_4 Grasses, Agronomy Monograph, pp 15–50. https://doi.org/10.2134/agronmonogr45.c2
Wedin DA, Tilman D (1993) Competition among grasses along a nitrogen gradient: initial conditions and mechanisms of competition. Ecol Monogr 63:199–229. https://doi.org/10.2307/2937180
Wedin DA, Tilman D (1996) Influence of nitrogen loading and species composition on the carbon balance of grasslands. Science 274:1720–1723. https://doi.org/10.1126/science.274.5293.1720
Weiss JL, Overpeck J (2005) Is the Sonoran Desert losing its cool? Glob Chang Biol 11:2065–2077. https://doi.org/10.1111/j.1365-2486.2005.01020.x
Whitney SM, Houtz RL, Alonso H (2011) Advancing our understanding and capacity to engineer nature's

CO_2-sequestering enzyme, Rubisco. Plant Physiol 155:27–35. https://doi.org/10.1104/pp.110.164814
Winter K, Smith JAC (1996a) Crassulacean ccid metabolism: current status and perspectives. In: Winter K, Smith JAC (eds) Crassulacean Acid Metabolism: Biochemistry, Ecophysiology and Evolution. Springer, Berlin, pp 389–426
Winter K, Smith JAC (1996b) An introduction to crassulacean acid metabolism: biochemical principles and ecological diversity. In: Winter K, Smith JAC (eds) Crassulacean Acid Metabolism: Biochemistry, Ecophysiology and Evolution. Springer, Berlin, pp 1–13
Winter K, Wallace BJ, Stocker GC, Roksandic Z (1983) Crassulacean acid metabolism in Australian vascular epiphytes and some related species. Oecologia 57:129–141. https://doi.org/10.1007/BF00379570
Winter K, Holtum JAM, Smith JAC (2015) Crassulacean acid metabolism: a continuous or discrete trait? New Phytol 208:73–78. https://doi.org/10.1111/nph.13446
Woodrow IE (1994) Optimal acclimation of the C_3 photosynthetic system under enhanced CO_2. Photosynth Res 39:401–412. https://doi.org/10.1007/BF00014594
Yeoh HH, Badger MR, Watson L (1981) Variations in kinetic properties of Ribulose-1,5-bisphosphate carboxylases among plants. Plant Physiol 67:1151–1155. https://doi.org/10.1104/pp.67.6.1151
Zachos JC, Dickens GR, Zeebe RE (2008) An early Cenozoic perspective on greenhouse warming and carbon-cycle dynamics. Nature 451:279–283. https://doi.org/10.1038/nature06588
Zhang YG, Pagani M, Liu Z, Bohaty SM, DeConto R (2013) A 40-million-year history of atmospheric CO_2. Philos Trans R Soc A 371:20130096. https://doi.org/10.1098/rsta.2013.0096
Zhu J, Goldstein G, Bartholomew DP (1999) Gas exchange and carbon isotope composition of *Ananas comosus* in response to elevated CO_2 and temperature. Plant Cell Environ 22:999–1007. https://doi.org/10.1046/j.1365-3040.1999.00451.x
Zotz G (2016) Plants on Plants – The Biology of Vascular Epiphytes. Springer International Publishing, Cham
Zotz G, Ziegler H (1997) The occurrence of crassulacean acid metabolism among vascular epiphytes from Central Panama. New Phytol 137:223–229. https://doi.org/10.1046/j.1469-8137.1997.00800.x

Chapter 9

The Outlook for C_4 Crops in Future Climate Scenarios

Alexander Watson-Lazowski and Oula Ghannoum*
ARC Centre of Excellence for Translational Photosynthesis, Hawkesbury Institute for the Environment, Western Sydney University, Penrith, NSW, Australia

*Author for correspondence, e-mail: o.ghannoum@westernsydney.edu.au

K. M. Becklin et al. (eds.), *Photosynthesis, Respiration, and Climate Change*, Advances in Photosynthesis and Respiration 48, https://doi.org/10.1007/978-3-030-64926-5_9

Summary

Several C_4 crops dominate a large percentage of current agriculture due to the array of physiological benefits provided by C_4 photosynthesis over the ancestral C_3 photosynthesis, such as high productivity, resource use efficiency and stress tolerance. It is therefore imperative that we understand both how these crops will fare under future climate scenarios, and how we can further utilise C_4 crops to help achieve future food security thresholds. The heightened water use efficiency of C_4 plants suggests they may take a more prominent role in agriculture across the world to alleviate some of the future stress on fresh water supplies. However, as temperatures increase, they will begin to exceed the temperature optima of C_4 crops such as *Zea mays* (maize) in key growing areas, which will shape the agricultural landscape of *Z. mays* and alternative C_4 crops. Elevated CO_2 has little direct effect on C_4 plants due to the carbon concentrating mechanism, however, under water stressed conditions elevated CO_2 does provide some drought tolerance to C_4 plants due to decreased transpiration. However, the limited positive effects of elevated CO_2 will likely not outweigh the stress of increased temperature and reduced water availability. Nevertheless, C_4 crops are expected to play a major role in maintaining future food security due to their superior physiological attributes. Looking forward, two approaches to maintain or advance production of C_4 crops seem feasible: (1) continued improvement of current crop lines via marker-assisted breeding, cutting-edge gene editing techniques and beneficial farming practices. This could allow current cropping practices to continue into the future without reduced yields. (2) Changes in the geographic distribution of C_4 crops. As environments become more suitable, C_4 crops will likely be introduced into those areas, while areas currently dominated by C_4 crops may require alternative, more resilient species to maintain economic value.

I. Introduction

The evolution of C_4 photosynthesis has occurred independently over 60 times within the angiosperms (Sage et al. 2011), beginning around 30 million years ago during the Oligocene epoch due to a period of sub-ambient CO_2 concentrations (Sage 2004). C_4 photosynthesis is most prominent among the grasses, which include some of the most productive and prolific weed and crop species (Sage 2004). The high productivity of C_4 plants is due to the evolution of a carbon concentrating mechanism (CCM) which elevates the concentration of carbon dioxide (CO_2) around Rubisco (ribulose-1,5-bisphosphate carboxylase/oxygenase), the ultimate CO_2-fixing enzyme in plants, algae and cyanobacteria. The C_4 CCM functions across two cell types, the mesophyll cell (MC, where CO_2 is transiently fixed) and the bundle sheath cell (BSC, where CO_2 is fixed) (Hatch 1987). Some of the world's most productive crop and pasture species

Abbreviations: C_i – Intercellular CO_2 concentration (μl L$-^1$); E – Leaf transpiration rate (mmol m$-^2$ s$-^1$); NUE – Nitrogen use efficiency; PNUE – Photosynthetic nitrogen use efficiency; Rubisco – Ribulose-1,5-bisphosphate carboxylase/oxygenase; RuBP – Ribulose-1,5-bisphosphate; CCM – Carbon concentrating mechanism; BSC – Bundle sheath cell; MC – Mesophyll cell; NADP-ME – NADP-malic enzyme; NAD-ME – NAD-malic enzyme; PEP-CK – Phosphoenolpyruvate carboxykinase; PEP – Phosphoenolpyruvate; PEPC – PEP carboxylase; PS II – Photosystem II; VPD – Leaf-to-air vapor pressure difference (kPa); WUE – Water use efficiency; k_{cat} – Catalytic turnover rates; T6P – Trehalose-6-phosphate; QTL – Quantitative trait loci; PP – Primary productivity; S – Annual integral of incident solar radiation; ε_i – Efficiency of radiation interception; ε_c – The conversion efficiency of absorbed radiation into biomass; k – Energy content of plant biomass; CWP – Crop water productivity; T_{opt} – Optimum temperature; $S_{c/o}$ – CO_2/O_2 specificity; OTC – Open top chambers; FACE – Free air CO_2 enrichment; CRISPR – Clustered regularly interspaced short palindromic utilise C_4 photosynthesis, including *Zea mays* (maize), *Saccharum officinarum* (sugarcane) and *Sorghum bicolor* (sorghum). Traits which are associated with C_4 photosynthesis include high photosynthetic rates, high light, water and nitrogen use efficiency (NUE), and high growth rates (Osmond et al. 1982; Long 1999; Sage and Pearcy 2000). These characteristics are extremely beneficial from an agronomic point of view, making C_4 crops key to meeting the future food goals required for our growing population (Steffen et al. 2015). In particular, the introduction of C_4 photosynthetic traits into rice is being considered as an avenue for boosting rice yield and food security (von Caemmerer et al. 2012; Wang et al. 2016). Not only do C_4 species have superior physiological characteristics when compared to C_3 species, but within C_4 species there is further diversity. Three subtypes of C_4 photosynthesis have been broadly characterised as NADP-malic enzyme (NADP-ME), NAD-malic enzyme (NAD-ME) and phosphoenolpyruvate carboxykinase (PEP-CK), with each known to be optimised to specific environments (Hattersley 1992; Kanai and Edwards 1999; Ghannoum et al. 2011).

Future climate models predict changes in a wide range of factors directly influencing plant productivity (Mitchell et al. 2004), some of which may have a positive effect, but most likely to contribute to an overall reduction in productivity (Ciais et al. 2005). Three of the key factors are CO_2, temperature and water availability. Knowing how these factors will influence C_4 crops is essential to understanding how we can efficiently utilise C_4 crops in the future. This chapter will summarise our state-of-understanding of how the key factors of climate change will likely impact C_4 crops, particularly maize and sorghum (and to a lesser extent sugarcane and millets), with a focus on field studies. The chapter will also outline the crop improvement strategies available in our breeding and research toolbox for developing climate-ready crops. For the benefit of early career researchers and students, the chapter starts with an updated and detailed introduction on C_4 photosynthesis with a focus on grasses since these constitute the C_4 lineages most closely related to our major C_4 crops.

II. C_4 Grasses Are Ecologically and Economically Important

A. *Rubisco*

Photosynthesis is constrained by the catalytic properties of Rubisco. Rubisco first evolved ~3 billion years ago when atmospheric CO_2 concentrations were high and oxygen (O_2) concentrations were minimal (Whitney et al. 2011). The advent and phenomenal success of oxygen-producing photosynthesis increased atmospheric O_2. However, as O_2 concentrations rose in tandem with decreasing CO_2 concentrations (Foster et al. 2017), a key deficiency of this enzyme was uncovered. Rubisco catalyses the incorporation of both CO_2 and O_2 into RuBP (ribulose-1,5-bisphosphate). Rubisco oxygenation produces one molecule of 3-phosphoglycerate and one molecule of 2-phosphoglycolate, a toxic two-carbon compound which is recycled through photorespiration at a cost. CO_2 and O_2 directly compete for Rubisco catalytic sites, and in ambient air (20% O_2), approximately one molecule of O_2 is fixed for every three molecules of CO_2 by C_3 plants (von Caemmerer 2000; von Caemmerer and Quick 2000).

Photorespiration results in carbon and energy losses, which increase with temperature since O_2 competes more effectively with CO_2 at higher temperatures (Jordan and Ogren 1984; von Caemmerer and Quick 2000). Further to this, the solubility of CO_2 declines relative to that of O_2 as temperature increases, influencing the availability of CO_2 within the leaf (Jordan and Ogren 1984; von Caemmerer and Quick 2000). Consequently, elevated CO_2 exerts a relatively larger effect on net photosynthesis at higher temperatures by reducing photorespiration. To achieve maximum productivity, C_3 plants compensate for Rubisco's catalytic inefficiency by investing large amounts of nitrogen (N) in Rubisco, and by operating with open stomata (to increase intercellular CO_2 concentration, C_i), lowering their water and NUE relative to C_4 plants (Long 1999).

Although Rubisco's efficiency, including its specificity towards CO_2 relative to O_2, varies among photosynthetic organisms (Sharwood et al. 2016a), Rubisco remains a conserved and maladapted catalyst relative to its cardinal role in global carbon uptake, due to its inherent thermodynamic and mechanistic traits (Bathellier et al. 2018). Active Rubisco requires carbamylation of the conserved residue K201 followed by binding of Mg^{2+} (Lorimer et al. 1976; Roy and Andrews 2000). The fixation of CO_2 to RuBP produces two molecules of 3-phosphoglycerate in a five-step catalytic process, making Rubisco prone to the generation of unwanted products and inhibitors, which are removed from the active site of Rubisco by Rubisco activase (von Caemmerer and Quick 2000).

B. The Evolution C_4 Photosynthesis

Atmospheric CO_2 declined 20–30 million years ago during the Oligocene and has oscillated between 180 and 300 ppm for the last one–three million years (Zhang et al. 2013). The Oligocene was also a time when the Earth was largely dry (Willis and McElwain 2014). In response to selection pressure caused by low atmospheric CO_2, possibly in combination with aridity, vascular plants have evolved CCMs, such as C_4 photosynthesis, to alleviate the catalytic inefficiencies of Rubisco (Sage 2004; Christin et al. 2008). The earliest origins of C_4 photosynthesis date back to the Oligocene period (Sage 2004; Sage et al. 2011). C_4 plants remained in low abundance until the late Miocene and Pliocene (3–eight million years ago), when a worldwide expansion of C_4 grasslands and savannas occurred (Edwards et al. 2010).

In the current era, a small proportion of the Earth's angiosperm species (~3%) fix atmospheric CO_2 via the C_4 photosynthetic pathway, yet account for approximately 25% of terrestrial photosynthesis (Lloyd and Farquhar 1994). Other than maize (*Zea mays*), sugarcane (*Saccharum officinarum*) and sorghum (*Sorghum bicolor*), numerous other C_4 crops are utilised as food, fodder or biofuel crops around the world. These include species such as miscanthus (*Miscanthus* × *giganteus*), pearl millet (*Pennisetum glaucum*), proso millet (*Panicum milliaceum*) and switchgrass (*Panicum virgatum*). Some of the world's noxious weeds are also C_4 species such as *Amaranthus* and *Setaria* spp. The high productivity of C_4 plants is a virtue of their CCM.

During C_4 photosynthesis, atmospheric CO_2 diffuses through the stomata into mesophyll cells where it is hydrated into bicarbonate, which reacts with phosphoenolpyruvate (PEP) with the aid of PEP carboxylase (PEPC) to produce oxaloacetate, a C_4 acid (Fig. 9.1a). Oxaloacetate is converted into either malate or aspartate, which then diffuses into bundle sheath cells where it is decarboxylated by one of three C_4 acid decarboxylases, releasing CO_2 for fixation by the C_3 cycle. The C_3 product of the decarboxylation reaction, pyruvate or alanine, returns to the mesophyll, completing the C_3 cycle (Hatch 1987). Sucrose is preferentially (65–100%) synthesised in mesophyll cells, while transitory starch accumulates in bundle sheath cells during the day and is mobilised the following night (Fig. 9.1a) (Lunn and Furbank 1999). The localisation of sucrose synthesis does not seem to correlate with the C_4 decarboxylation type, but maize exhibits a more asymmetric distribution (100% in mesophyll) relative to sorghum (65%) (Lunn and Furbank 1997).

The C_4 cycle acts as a CCM for two main reasons. Firstly, PEPC is a faster carboxylase than Rubisco and is insensitive to O_2

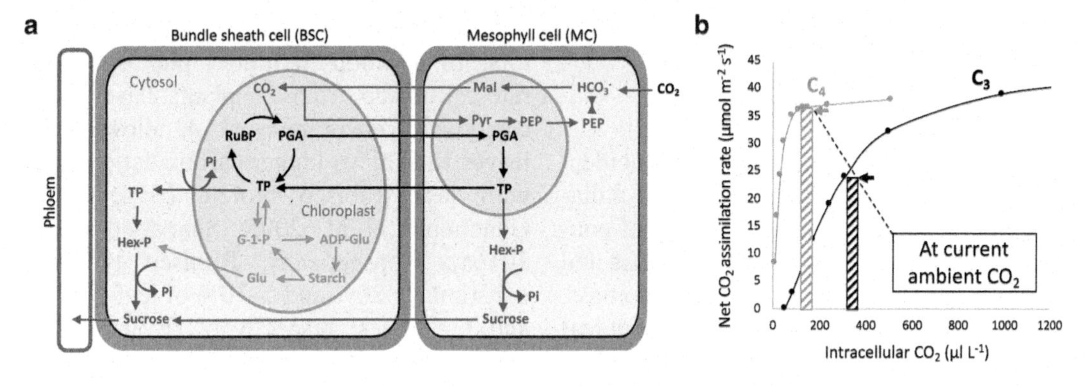

Fig. 9.1. **(a)** Overview of C_4 photosynthetic metabolism in leaves of maize and sorghum. Abbreviations are as follows: *CO_2* carbon dioxide, *HCO_3^-*, bicarbonate, *PEP* phosphoenolpyruvate, *PYR* pyruvate, *Mal* malate, *PGA* phosphoglyceric acid, *RuBP* ribulose-1,5-bisphosphate, *TP* triose phosphate, *G-1-P* glucose-1-phosphate, *ADP-Glu* ADP-glucose, *Glu* glucose, *Hex-P* hexose-phosphate and *Pi* inorganic phosphate. Text and arrow colour relate to the different process during C_4 photosynthesis: blue – C_4 pathway, black – Calvin cycle, green – starch metabolism and red - sucrose metabolism. **(b)** Typical wheat (C_3; black) and maize (C_4; grey) photosynthetic assimilation response curves to increasing intercellular CO_2 concentrations taken at a leaf temperature of 30 °C and 1800μmol m^{-2} s^{-1} of light. The bars of each colour represent a typical current operating assimilation rate for both C_3 (black) and C_4 (grey) plants at the current ambient CO_2 concentration. Created with BioRender.com

(Edwards and Walker 1983; Pengelly et al. 2012; Sharwood et al. 2016b; Sonawane et al. 2018). Secondly, the bundle sheath cell ultra-structure presents a significant gaseous diffusion barrier (von Caemmerer 2000). Consequently, the high CO_2 concentration in the bundle sheath reduces photorespiration and saturates C_4 photosynthesis at a lower ambient CO_2 concentration than for C_3 plants (Fig. 9.1b). Photorespired CO_2 is released within the bundle sheath, contributing to increased bundle sheath CO_2 concentration, where it may be refixed. Under high bundle sheath CO_2 concentration, PEPC kinetics give rise to the characteristic steep initial slope of the photosynthesis-CO_2 response curve of C_4 leaves, while Rubisco kinetics and RuBP and PEP regeneration mostly govern the maximal photosynthetic rates (Fig. 9.1b).

C. C_4 Ecophysiology

The CCM confers direct physiological advantages to C_4 photosynthesis (Björkman 1971; Osmond et al. 1982; Ehleringer and Pearcy 1983; Pearcy and Ehleringer 1984; Long 1999; Sage and Pearcy 2000; Ghannoum et al. 2011). In particular, C_4 plants have higher CO_2 assimilation rates, especially at high temperatures, and higher photosynthetic temperature optima than their C_3 counterparts (Berry and Bjorkman 1980; Sage and Kubien 2007; Sonawane et al. 2017). Although the operation of the CCM incurs additional energy used for the regeneration of PEP (the C_3 precursor in the C_4 cycle), at temperatures higher than ~25 °C, the cost of photorespiration in C_3 leaves exceeds that of operating the CCM in C_4 leaves, leading to higher photosynthetic quantum yield for C_4 relative to C_3 photosynthesis at warmer temperatures under current atmospheric CO_2 (Long 1999). However, this advantage is reduced at elevated CO_2 (Ehleringer and Pearcy 1983; Ehleringer et al. 1997). High photosynthetic rates and quantum yield lead to C_4 plants generally having higher primary productivity and light use efficiency relative to C_3 plants. Primary productivity (PP) can be expressed as:

$$PP = \frac{S\varepsilon_i\varepsilon_c}{k}, \quad (1)$$

where S is the annual integral of incident solar radiation, ε_i is the efficiency of radiation interception, ε_c is the efficiency of converting absorbed radiation into biomass and k is the energy content of plant biomass. When basic assumptions are made about leaf absorptance, the costs of dark respiration, nitrate reduction and the conversion of triose phosphate into sucrose, and the fact that ~50% of sunlight cannot be used in photosynthesis, then the maximal theoretical ε_c for a C_3 and a C_4 (NADP-ME) crop is 4.6% and 6%, respectively under current atmospheric CO_2, leaf temperature of 30 °C and 1000 kJ of incident solar radiation (Long et al. 2006; Zhu et al. 2008).

In addition, the CCM confers a number of indirect physiological and ecological advantages onto C_4 plants. Firstly, the saturation of C_4 photosynthesis at low CO_2 (Fig. 9.1b) allows C_4 leaves to operate with lower stomatal conductance (Taylor et al. 2010, 2012; Pinto et al. 2014). This leads to higher leaf- and plant-level water use efficiency (WUE) in C_4 relative to C_3 plants (Long 1999; Sage and Pearcy 2000; Ghannoum et al. 2011). Lower stomatal conductance in C_4 plants is possibly related to the efficient carbonic anhydrase-PEPC enzyme system of CO_2 uptake, which allows C_4 photosynthesis to operate at a lower C_i relative to C_3 photosynthesis (Taylor et al. 2010). In theory, low CO_2 concentration in the sub-stomatal cavity should lead to higher stomatal conductance in C_4 leaves (Engineer et al. 2016) but can be counter-balanced by other leaf traits. Lower stomatal conductance in C_4 grasses is associated with smaller diurnal deviations in leaf water potential, which implies a greater capacity for hydraulic supply relative to transpiration in C_4 than in C_3 grasses and consequently, greater hydraulic safety (Kocacinar and Sage 2003; Taylor et al. 2010, 2018). Secondly, C_4 plants are generally more NUE than C_3 counterparts due to a number of factors. In addition to higher photosynthetic rates, Rubisco from C_4 plants have faster catalytic turnover rates (k_{cat}), allowing C_4 leaves to achieve higher carboxylation rates with less Rubisco protein (Sage 2002; Ghannoum et al. 2005; Sharwood et al. 2016a). Consequently, Rubisco typically constitutes ~20% and 5–10% of leaf N in C_3 and C_4 leaves, respectively (Evans 1989; Makino et al. 2003; Ghannoum et al. 2005). Hence, C_4 leaves tend to achieve higher photosynthetic rates with less leaf N content (Pinto et al. 2011, 2014). All these physiological properties are reflected in the geographic distribution of the C_4 pathway, which is positively correlated with growing season temperature. Accordingly, C_4 grasses dominate many warm and high light environments such as tropical and sub-tropical rangelands and grasslands (Hattersley 1983; Ehleringer et al. 1997; Pau et al. 2013). The high productivity achieved by C_4 plants under warm conditions leads to an ecological and agricultural importance that is disproportionately high relative to their small taxonomic representation (Lloyd and Farquhar 1994; Brown 1999; Edwards et al. 2010).

D. C_4 Subtypes

C_4 photosynthesis has evolved numerous times, such that C_4 plants are biochemically and phylogenetically diverse (Sage et al. 2011). In grasses, three C_4 biochemical subtypes have been identified based on the primary C_4 acid decarboxylase enzyme operating in the bundle sheath cells. These enzymes are nicotinamide adenine dinucleotide phosphate malic enzyme (NADP–ME), nicotinamide adenine dinucleotide malic enzyme (NAD–ME) and phosphoenolpyruvate carboxykinase (PEP-CK) (Kanai and Edwards 1999; Bräutigam et al. 2014). In many C_4 species, especially those using NADP-ME as a primary decarboxylase, PEP-CK operates as a secondary decarboxylase (Chapman and Hatch 1981; Wingler et al. 1999; Sharwood et al. 2014). Although some PEP-CK activity is assumed to be present within NAD-ME species (Wang et al. 2014), evidence at the genetic and biochemical level regularly finds this not to be the case for monocots (Pinto et al. 2014; Koteyeva et al. 2015; Sonawane et al. 2018; Watson-Lazowski et al. 2018). However, C_4 grasses with high PEP-CK activity seem to frequently operate significant NADP-ME or NAD-ME activity (Burnell and Hatch 1988; Pinto et al. 2014; Sonawane et al. 2018; Watson-Lazowski et al. 2018). Further to this, there is some debate whether it is feasible that the PEP-CK decarboxylase can operate solely, without either of the other two malic acid decarboxylases (Wang et al. 2014). Despite these variable classifications, the primary decarboxylase utilised is still generally associated with a suite of anatomical, biochemical and physiological features (Hattersley 1992; Ghannoum et al. 2011).

Among the grasses, around 24 distinct lineages have independently evolved the C_4 pathway from C_3 ancestors (Grass Phylogeny Working 2012). There is a strong association between the C_4 subtypes and certain grass subfamilies. For example, C_4 species with classical NADP-ME type anatomy in the Panicoideae subfamily occur in the Andropogoneae, Arundinelleae and Paniceae tribes. The classical NAD-ME and PEP-CK type anatomy predominantly occur in the Chloridoideae subfamily, and evolved only once in the *Panicum/Urochloa/Setaria* clade within the Panicoideae subfamily (Hattersley and Watson 1992; Sage et al. 1999; Vicentini et al. 2008; Christin et al. 2009). The grass family includes species from all biochemical subtypes. Among the key C_4 crops, sorghum is strictly NADP-ME, while maize operates NADP-ME and PEP-CK as primary and secondary decarboxylases, respectively. Millets are mostly NADP-ME and NAD-ME with some PEP-CK species (Pinto et al. 2014; Sonawane et al. 2018). In terms of productivity or physiological performance, there seems to be no obvious consequences for the extent and type of the secondary decarboxylase (Pinto et al. 2016; Sonawane et al. 2017, 2018; Watson-Lazowski et al. 2020), despite hypothesised energetic differences (Wang et al. 2014; Ver Sagun et al. 2019). In NADP-ME species such as maize and sorghum, the bundle sheath cell wall is lined with a suberin lamella and the centrifugally-located bundle sheath chloroplasts lack photosystem II (PS II) activity, preventing the evolution of O_2 in the vicinity of Rubisco (Hattersley 1992; Ghannoum et al. 2005; Christin et al. 2013; Hernández-Prieto et al. 2019). This has led to differential adaptation in Rubisco kinetics among the C_4 subtypes (Sharwood et al. 2016a).

In addition to well-documented anatomical and biochemical variation (Dengler et al. 1994; Christin et al. 2013), C_4 grasses are associated with lineage specific traits (Hattersley and Watson 1992; Ghannoum et al. 2011). At the genetic level, unique optimisations have been noted between dicots and monocots, such as the use of a single gene for decarboxylation in NAD-ME monocots (NAD-ME2) compared to two genes in NAD-ME dicots (NAD-ME1 and NAD-ME2) (Bräutigam et al. 2014; Watson-Lazowski et al. 2018). At the physiological level, NADP-ME species tend to have higher photosynthetic nitrogen use efficiency (PNUE) relative to other C_4 grasses (Taub and Lerdau 2000; Ghannoum et al. 2005), except for one NADP-ME lineage (Aristidoideae) which has PNUE similar to C_3 counterparts (Taylor et al. 2010). Higher PNUE in NADP-ME grasses is driven by a faster Rubisco enzyme (Sharwood et al. 2016a). Strong correlations have been observed between average annual precipitation and the percentage of the C_4 grass flora with a particular C_4 subtype. Species with the NAD-ME subtype are predominantly found in the driest habitats, and the percentage of the C_4 grass flora using the NADP-ME subtype increases as aridity declines (Hattersley 1992). In line with these observations, NAD-ME grasses have higher whole-plant WUE relative to their NADP-ME counterparts under water stress (Ghannoum et al. 2002), which is possibly related to stomatal traits (Taylor et al. 2012). Stomatal patterning shows an easily identifiable phylogenetic signal in grasses, especially in terms of their specific dumbbell shape, which likely reflects the ecological adaptation of the grass lineages (Liu et al. 2012; Liu and Osborne 2015). The contrast between C_4 subtypes in their WUE and PNUE is yet to be explored in breeding programs or in an agricultural context.

III. Overview of the Main C_4 Crops

About 60% of C_4 species are grasses and about 40% of grasses utilise the C_4 photosynthetic pathway (Sage et al. 1999; Osborne et al. 2014). C_4 grasses mostly belong to warm-origin taxa (the PACMAD clade) dominating warm-climate grasslands and savannas (Hattersley 1983; Hattersley 1992; Edwards et al. 2010; Sage et al. 2011). C_4 grasses include ecologically and economically important species such as the world's major staple food, fodder and biofuel crops, as well as numerous major weeds (Brown 1999). C_4 crops predominate in warm and drought-prone climates and are becoming increasingly important for food and bioenergy security, with the global production of C_4 maize currently surpassing that of key C_3 cereals such as wheat and rice (Fig. 9.2). Sorghum is the fifth most important grain crop worldwide, and the second most important in semi-arid tropics (Fig. 9.2). Maize, sorghum and millet constitute 37.8%, 3.1% and 1.6% of total global cereal production (FAO 2019), which is a large increase, especially for maize, from what was grown in 1987 (25.6%, 3.7% and 1.4%), shedding light on the changing preference of farmers towards C_4 crops.

A. *Maize*

Increased demand for maize as livestock feed and continued demand for food maize due to population growth are changing global cereal demand. Currently, maize is the most produced cereal crop worldwide (Fig. 9.2b), a position that was long held by wheat. Its global distribution lies heavily within North and South America, as well as Eastern and Southern Africa (Fig. 9.3). Over 60% of maize produced is now under temperate maize production, while the rest is accounted for by tropical maize production (Ranum et al. 2014). Temperate maize production accounts for about 90% and 25% of total crop production in the developed and developing parts of the world, respectively (Pingali 2001). Maize is an important source of nutrition, and its use as a source of food accounts for 25% and 15% of the total maize demand in the developing world and globally, respectively (Pingali 2001). Unlike the developed world, maize production in numerous developing countries is largely destined for human consumption. For example, maize accounts for 73% (East Saharan Africa), 64% (West Saharan Africa), 46% (South Asia), 44% (Mesoamerica), 39% (North Africa), 36% (the Andean region) and 29% (South East Asia) of the total food crop demand. This makes maize particularly important to the poor in many developing regions of Africa, Latin America and Asia. Its high yields (relative to other cereals) make maize particularly attractive to famers in areas with land scarcity and high population pressure (Shiferaw et al. 2011). Therefore, continuous improvements of maize yield under climate change is central to food security of tropical and developing regions.

Similar to other grain crops, maize yield depends on dry matter accumulation (source activity) and allocation to the grain (sink activity). In contrast to other crops such as wheat, all the dry matter allocated to the

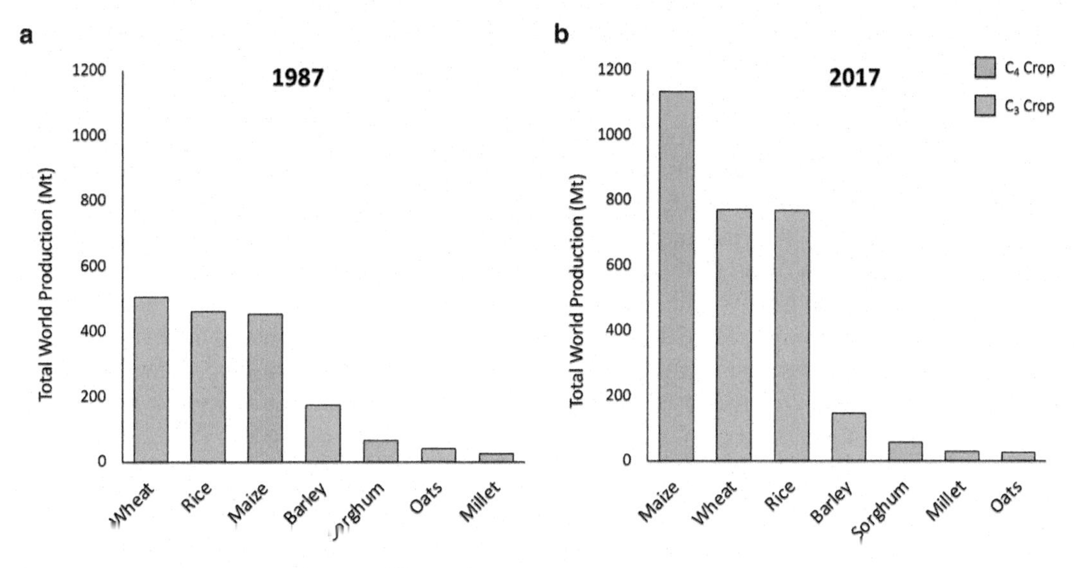

Fig. 9.2. Total world production in millions of tons (Mt) for seven of the world's most productive C_3 and C_4 cereal crops in (**a**) 1987 and (**b**) 2017. Data sourced from FAO (2019)

Fig. 9.3. The production percentage of the top three crops grown in regions across the world. Each pie chart corresponds to the region where it is placed on the map and the percentages are relative to the total production of all crops grown within that region. Data sourced from Leff et al. (2004)

grain of maize is fixed during the grain filling period, i.e., little dry matter is remobilised during grain filling (Tollenaar and Lee 2011). Temperate maize yields have increased more than fivefold during the 1940–2010 period (Roser and Ritchie 2013). In the US, average commercial maize yield has increased from about 1 Mg/ha in the 1930s to about 7 Mg/ha in the 1990s (Tollenaar and Lee 2002). The incremental yield increases for the newly released hybrids were not related to yield potential per plant (determined under optimal conditions), maximum potential photosynthesis or the harvest index (proportion of dry matter allocated to the grain). Instead, increases in maize yield were brought about by two main factors: increased leaf area index and leaf erectness of the newer hybrids (Pepper et al. 1977) and the slower decline of photosynthetic rates during the grain filling period, a phenotype known as stay-green (Tollenaar and Lee 2011). In particular, the North American hybrid breeding program has improved maize yield by increasing stress tolerance, particularly to intensification. In modern maize crops, harvest indices now average 50%, which is close to the theoretical maximum (Sharma-Natu and Ghildiyal 2005). However, temperate maize is less productive than a biomass crop like miscanthus, and the main reason is related to earlier leaf area development in miscanthus. Hence, there is scope for improving maize yields in general by breeding for strong early vigour (Dohleman and Long 2009).

Unlike their temperate counterparts, tropical maize lines have greater genetic diversity and have been bred under wide ranging environments (Anami et al. 2009). The key environmental challenge for tropical maize is water stress. Therefore, attempts to improve drought tolerance in maize have been similar to those used with other crops (Anami et al. 2009). There have been attempts at improving maize drought tolerance by manipulating genes and transcription factors related to glycine betaine biosynthesis, a known metabolic osmoticum (Anami et al. 2009). However, water stress involves a multitude of responses which are not necessarily co-ordinated. For example, the sensitivity of tissue expansion is the determining factor in the response of plant growth to drought. However, the responses of tissue expansion appear loosely coordinated with the responses of photosynthesis or cell division, which occur on different time scale and with different sensitivities under water deficit (Tardieu et al. 2011).

Another promising avenue to improve grain yield under drought in maize, and possibly other cereals including sorghum, is to manipulate sink-source feedbacks and sugar signalling in plants. It has long been known that photosynthesis depends on carbohydrate utilisation by sink tissues. Recently, trehalose-6-phosphate (T6P) has been identified as a sensing sugar which promotes resource mobilisation and sucrose allocation to the grain, which can enhance crop performance under abiotic stresses (Paul et al. 2018). By producing maize plants with genetically reduced levels of T6P in the phloem, more sucrose is moved to developing kernels where it's converted into starch (Oszvald et al. 2018). The mutation also led to increased rates of photosynthesis, thereby increasing sugar production, providing additional sucrose that can be used for grain filling (Oszvald et al. 2018). An exciting new discovery is the development of a chemical spray containing T6P precursors, which has been shown to enhance plant productivity (Griffiths et al. 2016). This research line holds promise for improving grain yield of most cereal crops over the coming years using genetically modified free approaches.

B. *Sorghum*

Sorghum is a summer crop which is generally more drought tolerant than maize (Schittenhelm and Schroetter 2014) and hence grown in drier climates (Leff et al. 2004). It is often a recommended option for farmers operating in harsh environments with limited rainfall and resources to apply fertilizers. It is native to Central and East Africa and is the most produced crop in Sahelian Africa (Fig. 9.3), constituting the principal source of energy and nutrition for millions of the poorest people in these regions. The stalks and foliage are used as fodder, fuel and construction materials and are often as important as the grain in rural areas. Sorghum is also used as fodder for poultry and cattle in developed countries, including Australia, and is considered as an important economic resource. There is also considerable promise for sorghum as a bioenergy crop, as sweet sorghum can produce the most ethanol per unit area of land of all current crops (Regassa and Wortmann 2014). This elevated output of ethanol is especially true when resources are limited (Regassa and Wortmann 2014), hence, sweet sorghum may become increasingly important as water and fertilisers become more scarce.

Sorghum is widely grown in many arid and semi-arid areas of the world due to its high yield under rainfed or water-limited conditions. The drought resistance of sorghum is attributed to a dense and prolific root system, leaf rolling, ability to maintain stomatal opening at low levels of leaf water potential through osmotic adjustment and its ability to delay reproductive development (Borrell et al. 2014). The shorter growth duration of sorghum compared with maize also contributes to its ability to escape drought (Krupa et al. 2017). Although sorghum is drought tolerant, the crop's susceptibility to water stress depends on the developmental stage. A drought period during the early seedling or vegetative stages of sorghum can inhibit establishment of the crop or reduce yield more than 36%, while water stress during the reproductive stage impacts grain development and can reduce yield more than 55% (Assefa et al. 2010). Therefore, the ability to withstand water deficit at these stages is critical to productivity. A comprehensive comparative study comparing maize and sorghum productivity was undertaken for the USA (Assefa et al. 2014). The authors found that the mean and maximum possible yields of maize were greater than that of grain sorghum, while the variation in dryland yield was less for sorghum than for maize. This is because maize was more responsive and sensitive to environmental variation than sorghum. Importantly, approximately 432 mm of total seasonal rainfall (or 533 mm of evapotranspiration) was the threshold value above which maize, and below which sorghum, exhibited better average WUE (Assefa et al. 2014).

C. *Sugarcane and Millets*

Several additional C_4 crops play a key role in current agriculture, including sugarcane and millet species. Sugarcane is known for its high accumulation of sugar, and hence, currently accounts for over 75% of global sugar production (De Souza et al. 2008). Further, the sugar produced by sugarcane can also be used to produce ethanol, allowing sugarcane to also be utilised as a bioenergy crop. Sugarcane leads all other crops in total production, with over 1800 Mt. being produced worldwide in 2017 (FAO 2019). Currently, Brazil and India lead the production of sugarcane, although there is also considerable production across the world (FAO 2019). In line with demands for sugar-based products and alternative fuels, sugarcane production has almost doubled in the last 30 years (FAO 2019). With these demands expected to continue to rise, it is likely production of sugarcane will continue to increase, with additional production likely occurring within Sub Saharan Africa, where sugarcane may provide a viable economic option for growers (Hess et al. 2016).

Millets are another set of C_4 crops that contribute towards both food and fodder production. Several C_4 crop species are classed as millets, including *Pennisetum glaucum* (pearl millet) and *Setaria italica* (foxtail millet). Over 95% of millets are currently produced in developing countries (FAO 2019). They hold a particular importance in parts of Sahelian Africa, where millets contribute a large proportion of total crop production (Fig. 9.3). Their combined drought and heat tolerance make them suitable alternatives to maize and sorghum in areas experiencing harsh conditions (Serba and Yadav 2016; Varshney et al. 2017).

IV. Climate Change Interacts with Global Food Security

By the end of the twenty-first century, if minimal policy change occurs, atmospheric CO_2 concentrations are likely to exceed 900 ppm, bringing with it an increase of around 4 °C in global mean surface temperature (Nordhaus 2018). Steps are currently been taken to reduce temperature increases (Estrada et al. 2017), and it is hoped temperature increases are limited to 2 °C and atmospheric CO_2 concentrations to 550 ppm by end of the twenty-first century (Gao et al. 2017). However, even if these targets are met, they still represent marked increases when compared to today's climate and will likely lead to increases in the frequency and intensity of climate extremes, including heat and drought events (Lhotka et al. 2018; Weber et al. 2018). These climatic events will have profound and complex impacts on the world's agriculture (Reddy and Hodges 2000; Tubiello et al. 2007; Wheeler and Von Braun 2013). Agriculture is the main form of human land use, with crops accounting for ~12% of the Earth's land area (Leff et al. 2004), consumes about 70% of total human water usage, and is an important source (CO_2, H_2O, CH_4, NO_3) and sink (CO_2, H_2O) of greenhouse gases (FAO 2001). Therefore, it is crucial during this century to both have a clear understanding of how global climate change will impact agriculture and what crop improvement and management strategies are available to cope with unfolding climate change.

V. How Tolerant Are C_4 Plants to Water Stress?

A. *Effect of Water Stress on C_4 Photosynthesis*

Water stress is one of the most limiting environmental factors to plant productivity worldwide (Chaves et al. 2011). The response of C_3 photosynthesis to water stress depends on the severity of the stress. In the early, mild phase of water stress, the decline in CO_2 assimilation rate is largely the result of reduced C_i due to decreased stomatal conductance. Under these conditions, photosynthetic inhibition usually recovers following re-hydration. In the moderate to severe phases of water stress, the loss of photosynthetic activity becomes less responsive to increased CO_2 supply and photosynthesis fails to partially or fully recover following the removal of water stress. The two phases are generally labelled as stomatal and non-stomatal (or metabolic) inhibition, with the latter involving photoinhibition, damage to photosynthetic enzymes and electron transport capacity (Tezara et al. 1999; Chaves et al. 2009; Lawlor and Tezara 2009). At the whole plant level, water stress inhibits plant growth via multiple parallel factors including carbon starvation due to reduced photosynthesis (McDowell and Sevanto 2010), reduced cellular and tissue growth (Tardieu et al. 2011) and hydraulic failure (McDowell 2011; Choat et al. 2012).

Due to the CCM, low stomatal conductance and high WUE, C_4 plants are expected to show higher drought tolerance relative to C_3 plants. This expectation is strengthened by the geographic dominance of C_4 grasses and crops in arid and semi-arid regions. However, the geographic dominance of C_4 grasses in tropical and sub-tropical regions is generally correlated with summer rainfall, rather than aridity *per se*, as opposed to regions with winter rainfall which are more suitable for C_3 activity (Hasegawa et al. 2018). High WUE can reduce crop water consumption, hence reduce the rate at which water stress develops. Alternatively, crops with high WUE often accumulate more biomass rather than save water, especially in areas with high potential evapotranspiration (Ghannoum 2016). In addition, modern high yielding maize lines tend to drain the soil water supplies earlier (Lobell et al. 2014).

Hence, and contrary to expectations, accumulating evidence suggests that C_4 photosynthetic metabolism is highly sensitive to water stress (Ghannoum et al. 2003; Carmo-Silva et al. 2007; Ripley et al. 2007, 2010). With declining leaf water status, CO_2 assimilation rates and stomatal conductance decrease rapidly (Farquhar and Sharkey 1982). The initial, mainly stomatal phase, may not be apparent because the CCM will offset small stomatal declines, ensuring near saturation of C_4 photosynthesis under relatively low C_i. However, with greater water stress, C_4 photosynthesis is reduced due to stomatal and non-stomatal factors. Overall, C_4 photosynthesis is equally sensitive to moderate-to-severe water stress as its C_3 counterpart, in spite of the greater capacity and WUE of the C_4 photosynthetic pathway (Ghannoum 2009).

B. *Can C_4 Crops Help Sustain Fresh Water Supplies?*

If climate change and farming practices continue in their current fashion, it is essential to plan for future water shortages and look to how we can take steps to limit water deficit. Crop Water Productivity (CWP) is a measure of how much food is produced per water input and is a common measure to identify crops which require less water to produce sufficient yields. The CWP of several major crops was reviewed (Zwart and Bastiaanssen 2004), and although ranges were high, maize exhibited the highest CWP amongst the crops measured (including wheat and rice). The lower water requirements of C_4 species when compared to rice suggest C_4 species may become more sought after in rice dominated countries, and suggestions are now being made in countries such as India to switch to maize rather than the comparably water hungry rice (Davis et al. 2018). Davis et al. (2018) showed that growing maize instead of rice in regions of India could save up to 33% of the irrigated water currently used for agriculture, whilst increasing the nutrient supply to the community. Changes such as this can help alleviate part of the current drain on fresh water supplies in these parts of the world.

Although the effects of severe drought on C_4 crops are substantial, within C_4 crops, notable differences in drought tolerance have been found, with sorghum exhibiting higher drought tolerance than maize (Schittenhelm and Schroetter 2014; Amaducci et al. 2016) and millets providing yields in areas too arid to grow either maize or sorghum (Burton 1983). These species can be used as alternative C_4 crops as water supplies become limited. When droughts become severe, the economic benefit of sorghum or millets becomes greater than that of maize (Staggenborg et al. 2008), therefore if water restrictions occur or are put in place these crops may become more economically viable across the world.

Watering regimes and crop traits need to be considered together in order to use the lowest amounts of water while maintaining yields. Various watering regimes and methods have been investigated using maize, some of which can be used to reduce irrigation requirements (Liu et al. 2010; Zhang et al. 2017). One of the most beneficial methods was film mulching, whereby a plastic film is placed over the soil to both supress weeds and reduce moisture loss. When compared to rain fed controls, results showed film mulching increased yields across Chinese maize fields (Liu et al. 2010; Zhang et al. 2017). Methods such as these are viable practices to help sustain fresh water supplies, and so should be considered even before water shortages become apparent. Looking forward, understanding how these methods will be effective under future climates is key to determining future best practices for farming.

VI. Role of Elevated Temperatures on Shifting Future Geographic Distributions of C_4 Crops

A. *Effect of High Temperature on C_4 Photosynthesis*

Understanding the effects of temperature on C_4 plants is essential for predicting how C_4 crops will fare in future climate scenarios. Rubisco, the first and rate-limiting enzyme of the Calvin (C_3) cycle, catalyses two competing reactions with CO_2 (carboxylation) and O_2 (oxygenation). The oxygenation reaction of Rubisco (photorespiration) takes catalytic sites away from carboxylation and results in net energy and carbon loss (von Caemmerer 2000). The ratio of oxygenation to carboxylation increases with temperature as the CO_2/O_2 specificity ($S_{c/o}$) of Rubisco decreases while the solubility of CO_2 decreases with temperature faster than that of O_2, offsetting carbon gains from increased catalytic activity (Jordan and Ogren 1984; Brooks and Farquhar 1985; Sharwood et al. 2016a). The CCM of C_4 photosynthesis overcomes the problems of photorespiration at high temperature by concentrating CO_2 around Rubisco in bundle sheath cells, allowing the enzyme to fix CO_2 close to its saturated carboxylation rate, which increases exponentially with temperature (Kubien et al. 2003; Sharwood et al. 2016a; Sonawane et al. 2017).

Similar to their C_3 counterparts, C_4 plants exhibit a wide range of short- and long-term responses to warmer temperatures depending on genotype and habitat (Berry and Bjorkman 1980; Sage and Kubien 2007; Yamori et al. 2014). C_4 photosynthesis is highly sensitive to short-term increases in temperature due to higher optimum temperatures (T_{opt}) relative to their C_3 counterparts (Long and Woolhouse 1978; Henning and Brown 1986; Pittermann and Sage 2000; Kubien et al. 2003; Kubien and Sage 2004; Sonawane et al. 2017). This plasticity applies to the two C_4 crops, maize and sorghum (Fig. 9.4). Out of the two C_4 crops, sorghum appears to have the highest temperature optima for photosynthesis (Fig. 9.4e), as well as a less steep rise in dark respiration with temperature (Fig. 9.4b); both features are in line with the more heat and drought tolerant characteristics of this crop (Peacock 1982; Schittenhelm and Schroetter 2014). Nevertheless, both crops maintain high photosynthetic efficiency and coordination between the C_3 and C_4 cycles at warm temperatures as demonstrated by roughly constant leakiness (Sonawane et al. 2017), which is a measure of CCM efficiency and reflects the high efficiency of the NADP-ME subtype (Siebke et al. 2003; von Caemmerer et al. 2014). Sharwood et al. (2016a) demonstrated that NADP-ME species, which include maize and sorghum, are characterised by superior Rubisco catalytic properties at warm temperatures relative to C_3 and other C_4 counterparts. With increasing temperatures, NADP-ME species showed a steeper k_{cat} increase accompanied by a less pronounced decline of Rubisco $S_{c/o}$, which would be advantageous in warmer climates.

Exposure to long term warming modulates short-term photosynthesis and growth responses to temperature. In various C_4 species, CO_2 assimilation rates often undergo significant thermal acclimation, such that when compared at growth temperatures, photosynthesis increases less than what would be expected given the strong photosynthesis response to short-term changes in leaf temperature. Thermal photosynthetic acclimation is generally accompanied by higher temperature optimum and lower leaf N and Rubisco content (Berry and Bjorkman 1980; Dwyer et al. 2007; Yamori et al. 2014). Both maize and sorghum show significant acclimation in response to growth at high temperature, however, sorghum appears to increase biomass and leaf area more than maize at high temperature (Fig. 9.4f). In addition, sorghum has a more favourable pho tosynthesis/respiration ratio at high temperature relative to maize (Fig. 9.4f). Again, these results highlight the greater heat tolerance that sorghum is generally known for when compared to maize.

Heat stress is a major factor influencing plant productivity in most parts of the world. Heat tolerance varies between crops depending on the species, genotype, seed origin as well as within a crop depending on the physiological process in question (Wahid et al. 2007). Crop species of tropical and subtropical origin are sensitive to high temperatures in the range 30–55 °C. In the semi-arid tropics where sorghum is mostly grown, air tem-

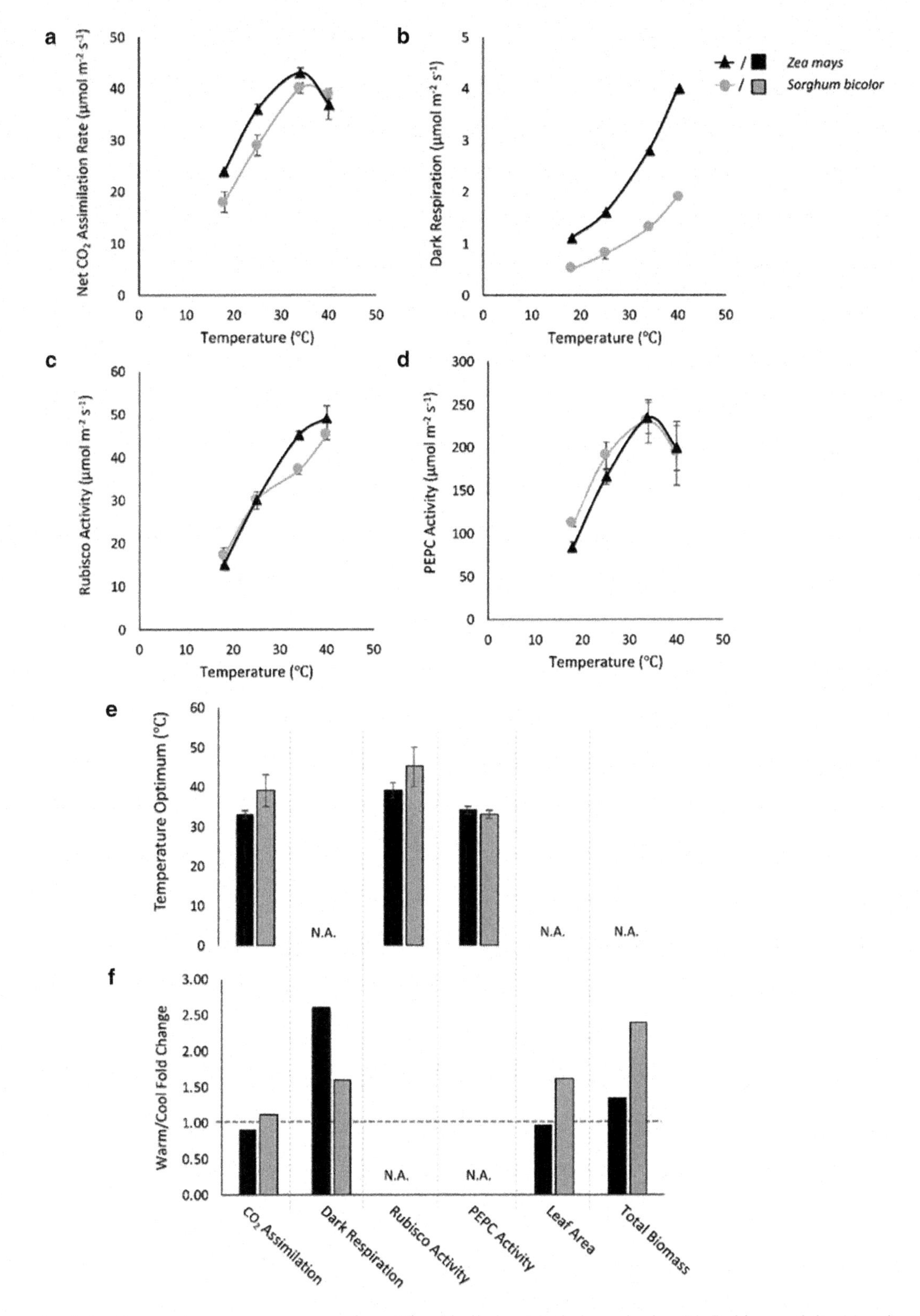

Fig. 9.4. Temperature response curves of net CO_2 assimilation (**a**), dark respiration (**b**), Rubisco activity (**c**) and PEPC activity (**d**) for leaves of *Sorghum bicolor* and *Zea mays*. Temperature optima of select parameters are also shown (**e**). Data for **a–e** were extracted from Sonawane et al. (2017). A comparison of warm (34 °C) and cool (25 °C) grown *Z. mays* and *S. bicolor* represented by the warm/cool fold change (**f**) (Sonawane BV, Ghannoum O; unpublished). Plants in **f** were grown under similar conditions as in a-e except for the growth temperature. N.A. highlights where measurements were not collected

peratures often exceed 40 °C and leaf temperatures of 55 °C have been measured (Peacock 1982). Temperatures above 40–48 °C can be lethal for seed germination in sorghum, while the T_{opt} for leaf growth for sorghum is around 32 °C. Photosynthesis declines when leaf temperatures exceeds ~33 °C (Fig. 9.4a), and is 75% and 95% inhibited above 44 °C and 48 °C, respectively (Peacock 1982; Yan et al. 2013). Above 38 °C, the inhibition of maize photosynthesis is partly due to Rubisco inactivation. Above 45 °C, the maximum quantum yield of PS II is inhibited as a sign of long-term photosynthetic damage (Crafts-Brandner and Salvucci 2002).

B. *Warming Experiments with C_4 Crops*

The influence of elevated temperature on C_4 plants has been widely studied in controlled environments, but field-based studies using artificial warming are much less common. Although fewer, these field-based studies give the most 'realistic' insight into how warming will affect yields across the world. The C_4 crop maize has been studied using field based artificial warming in the two countries which have the largest total maize production (North America and China). The results show the antagonistic effects elevated temperatures can have on photosynthesis and yield (Table 9.1). In North America, where ambient temperatures are already high, continual increased temperature has a negative effect on both photosynthesis and yield (Ruiz-Vera et al. 2015). This is also true for short term heatwaves, although the timing of these is critical to whether yield is affected (Siebers et al. 2017). When a heatwave was applied at the vegetative stage, photosynthesis decreased but yield was unaffected. When applied at the reproductive stage, both photosynthesis and yield significantly decreased. However, when similar experiments have been carried out in China, where ambient temperatures sit around 13 °C, elevating temperatures by 2 °C had a positive effect on photosynthesis (Zheng et al. 2013, 2018b). However, this increased assimilation was not realised in crop yield (Zheng et al. 2018a), inhibiting the potential to offset yield losses. In fact, models have predicted a 7.4% reduction in global yields of maize per one degree Celsius increase (Zhao et al. 2017), highlighting the extent of the impact increased temperatures may have. Field studies help identify how temperature will affect some of the key growing regions across the world, and as an increase in temperature between 2

Table 9.1. Results of field studies which have conducted warming and heat stress treatments using *Zea mays* in North America and China (Zheng et al. 2013; Ruiz-Vera et al., 2015; Siebers et al. 2017; Zheng et al. 2018a, b)

Study	Location	Ambient Temperature (°C)	Continual warming (°C)	Heatwave (°C / hours / growth stage)	Photosynthesis	Yield
Zheng et al. (2013)	Yucheng, China	~13.1	~2	N.A.	Increase	N.A.
Zheng et al. (2018a)	Yucheng, China	~13.1	~2	N.A.	Increase	n.s.
Zheng et al. (2018b)	Yucheng, China	~13.1	~2	N.A.	Increase	N.A.
Siebers et al. (2017)	Illinois, USA	~22.7	~2.64	N.A.	Decrease	Decrease
Siebers et al. (2017)	Illinois, USA	~22.7	N.A.	6 / 72 / vegetative	Decrease	n.s.
Ruiz-Vera et al. (2015)	Illinois, USA	~22.7	N.A.	6 / 72 / reproductive	Decrease	Decrease

Increase/decrease represents a significant change in direction ($P \leq 0.05$) while n.s. means data was collected but there was no significant change. N.A. highlights where measurements were not collected within the study and 3 °C is expected by 2100 within these regions (Mcdowell et al. 2016; Hui et al. 2018), gives a realistic insight into the changes we may expect. Looking forward, in cooler climates increased temperatures will likely be beneficial and so efforts here should be directed at enhancing the positive effects so they are translated into yield. In countries with warmer environments the increase in temperature will likely be detrimental. Here, there needs to be a focus on improving heat tolerance within C_4 crops in an attempt to limit negative impacts, ensuring yields can remain cost effective for the growers, sustaining production.

Another aspect of climate warming is rising vapour pressure deficit (VPD) which is the major driver of crop water loss or evapotranspiration. In the US Corn Belt, VPD is expected to increase from ~2.2 kPa to ~2.7 kPa by mid-century primarily due to warmer temperatures (DeLucia et al. 2019). This projected increase in VPD will create a ceiling to future increases in maize yields. Maintaining current maize yields (2013–2016) in the Midwest USA will require more than threefold expansion of areas currently under irrigation. In order to continue making yield increases in maize and other economically-valuable grasses, developing cultivars with greater WUE will be required (DeLucia et al. 2019).

It is worth considering geographical flexibility to alleviate heat stress, such as shifting growing regions, so crops are grown in climates that are more suitable for high productivity. A recent study used future climate predictions to model land which will likely have climate suitability for maize growth in 2100 across the world (Ramirez-Cabral et al. 2017). One impactful finding of this study was out of the top five maize growing countries, two (Brazil and Mexico) will lose over 48% of the land which is currently suitable for maize growth by 2100 (Table 9.2). This will limit the agricultural potential of these countries in relation to maize and may lead to these countries being unable to continue to grow the crop. In North America and China, land with a suitable climate is predicted to increase, which may provide an avenue for North America to maintain yields and for China to further their production capacity. However, one caveat of this study is that the land which may become suitable may not be appropriate for agricultural conversion due to current anthropogenic land use or soil type. This aspect needs further investigation, and it would be beneficial to assess the land which will become climatically suitable to quantify the true availability of land suitable for maize growth.

These results highlight two possible impacts in terms of future temperatures which need to be considered: (1) as maize is currently mostly grown in temperate climates, if no advances are made, we will see an overall decrease in yield due to increased temperature (and hence, VPD) which may lead to an increased economic benefit of other C_4 crops such as sorghum and (2) land in which we were previously unable to grow maize may become available in the future to alleviate this deficit.

VII. Can Elevated CO_2 Be Beneficial to C_4 Crops?

A. *Response of C_4 Plants to Elevated CO_2 Under Non-limited Water Availability*

In the last 50 years, a significant body of literature has emerged on the response of individual C_3 and C_4 plants to elevated CO_2, as well as the interaction of elevated CO_2 with nutrient supply, water supply and temperature (Wand et al. 1999; Poorter and Navas 2003). It is crucial to comprehensively understand these responses to elevated CO_2 to fully utilise the possible benefits, as it has been predicted that current trends in CO_2

Table 9.2. The predicted change in climatically suitable land for *Z. mays* growth by 2100 in five of the top *Z. mays* producing countries

Country	Estimated change of land with optimum growth conditions for *Z. mays* by 2100
U.S.A.	12.5%
China	16.5%
Brazil	−63.5%
Argentina	0%
Mexico	−46%

Data extracted from Ramirez-Cabral et al. (2017)

concentrations could provide global yield increases of up 1.8% per decade (Lobell and Gourdji 2012). The operation of a CCM during C_4 photosynthesis serves to concentrate CO_2 at the site of its fixation to levels high enough to nearly saturate photosynthesis and suppress photorespiration in air. Consequently, C_4 plants are not expected to show a growth response to rising atmospheric CO_2 (Ghannoum et al. 2000). In potted plants, many C_4 plants grow bigger at elevated CO_2, albeit to a lesser extent than C_3 plants (Allen Jr. et al. 2011; Uddin et al. 2018). In these controlled environment experiments, the vegetative growth stimulation by a doubling of ambient CO_2 averaged 12–33% for C_4 plants (Wand et al. 1999). Under well-watered conditions, elevated CO_2 may enhance the growth of C_4 plants via two routes (Ghannoum et al. 2000, 2005). Firstly, elevated CO_2 can slightly enhance leaf CO_2 assimilation rates when C_4 photosynthesis is not fully CO_2-saturated such as under high light intensities and high soil N supply (Ghannoum et al. 1997; Ghannoum and Conroy 1998). Secondly, elevated CO_2 reduces stomatal conductance to water vapour and leaf transpiration rate (E), which in turn improve shoot water relations (Seneweera et al. 1998, 2001, Knapp et al. 1993) and increase leaf temperature (Kirkham et al. 1991; Wall et al. 2001; Siebke et al. 2003). Both factors tend to enhance plant growth.

With the advent of open top chambers (OTC) and free air CO_2 enrichment (FACE) technologies in the last two decades, more information is becoming available about the long-term (growing season to several years) responses of crop and pasture species in field-like situations. These longer-term and more realistic experiments present a different picture for the response of C_4 plants to elevated CO_2 (Ainsworth and Long 2005; Blumenthal et al. 2018; Reich et al. 2018). Further to this, studies which utilise naturally elevated CO_2 springs exposed to multiple generations of elevated CO_2 now show FACE results appear to hold true (Watson-Lazowski et al. 2016; Saban et al. 2019), enhancing their reliability. Multiple FACE experiments have been conducted using the two C_4 crops maize and sorghum (Conley et al. 2001; Ottman et al. 2001; Cousins et al. 2003; Leakey et al. 2004; Triggs et al. 2004; Hussain et al. 2013; Twine et al. 2013; Manderscheid et al. 2014; Ruiz-Vera et al. 2015). Overall, elevated CO_2 had no significant impact on leaf photosynthesis, aboveground biomass or agronomic yield on field-grown maize or sorghum in the absence of water stress (Fig. 9.5). In contrast, evapotranspiration decreased (Fig. 9.5) and canopy temperature slightly increased (Kimball 2016). Although no FACE studies have been conducted using sugarcane, OTC experiments indicated that sugarcane may be highly responsive to elevated CO_2 (De Souza et al.

2008). Even in the absence of a drought treatment, De Souza et al. (2008) showed elevated CO_2 stimulated a 40% increase in biomass and 29% increase in sucrose content in sugarcane. However, sugarcane is a large crop and maintaining adequate soil moisture throughout the day or the growing season is challenging even under well-watered conditions. This makes it difficult to distinguish between direct leaf-level effects of rising CO_2 on photosynthesis from indirect water-related responses. In a controlled-environment experiment using large pots and a high precision watering system, Stokes et al. (2016) showed that elevated CO_2 reduced whole-plant transpiration by 28% without affecting biomass of sugarcane plants growing with high-frequency, demand-based watering. A simulation model of CO_2 effects, based purely on changes in stomatal conductance (indirect mechanism), showed transpiration was reduced by 30% (initially) to 10% (closed canopy) and yield increased by 3% even in a well-irrigated crop. Taken together, these results indicate that indirect mechanisms dominate CO_2 responses in the C_4 crop sugarcane (Stokes et al. 2016). This is important to consider when determining the future viability of sugarcane as a bioenergy crop, as yield enhancements, due to direct or indirect CO_2 effects, could increase its viability considerably.

B. *Interaction of Water Availability with the Response of C_4 Plants to Elevated CO_2*

In controlled-environment and OTC experiments, the growth response of C_4 plants to elevated CO_2 generally increases as soil water availability decreases, relative to their response under well-watered conditions (Morison and Gifford 1984b; Owensby et al. 1997; Samarakoon and Gifford 1996; Seneweera et al. 1998, 2001). There are two main explanations for this observation. Firstly, elevated CO_2 reduces E which generally translates into reduced plant transpiration because leaf area increases are small in C_4 plants. This may lead to soil water conservation, slowing down the development of water stress and providing more water and time for photosynthesis and growth (Owensby et al. 1997; Seneweera et al. 2001; Serraj et al. 1999; Samarakoon and Gifford 1996). Secondly, elevated CO_2 may directly alleviate the adverse effects of water stress on photosynthesis by alleviating some of the CO_2 limitation caused by reduced stomatal conductance (Lawlor and Cornic 2002), and on growth by improving leaf water relations under soil water deficits (Samarakoon and Gifford 1996; Seneweera et al. 1998; Wall 2001). Few studies have attempted to distinguish between direct (metabolic) effects of elevated CO_2 from indirect (stomatal) ones in C_4 plants. Ghannoum et al. (2003) investigated the contribution of stomatal and non-stomatal factors to photosynthetic inhibition under water stress in four C_4 grasses. CO_2 assimilation rates and quantum yield of PS II of all four grasses decreased rapidly with declining leaf relative water content, while elevated CO_2 (2500μl L^{-1}) had no effect on these parameters at any stage of the drying cycle. This experiment showed that under moderate to severe water stress, inhibition of C_4 photosynthesis depends mainly on biochemical limitations (Ghannoum et al. 2003; Ghannoum 2009).

In FACE experiments, the response of maize and sorghum to elevated CO_2 under water limited conditions is generally in line with controlled environment experiments (Fig. 9.5). In particular, the response of biomass and agricultural yield is significant and larger than that observed under ample water conditions. However, evapotranspiration changes little when water is limiting over a seasonal time frame because plants use all water available (Fig. 9.5). In addition, canopy temperature is not significantly affected by elevated CO_2 under water stress, possibly due to reduced canopy cooling as a result of reduced water uptake and transpiration (Kimball 2016).

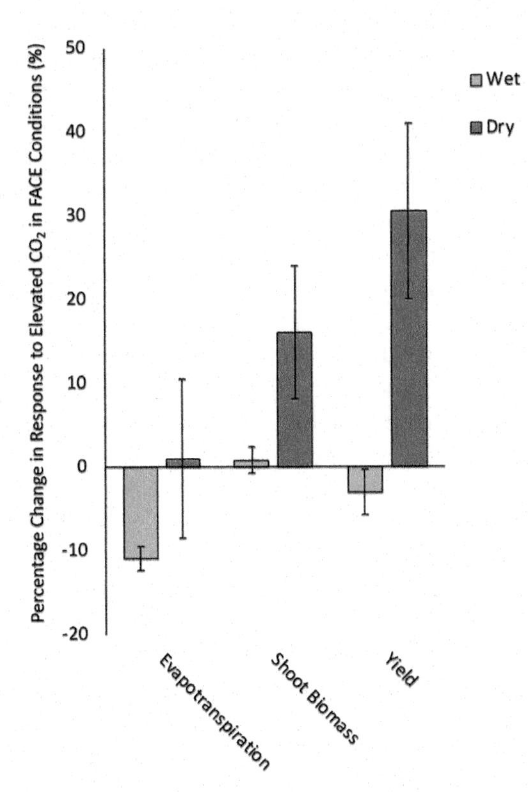

Fig. 9.5. Combined data from FACE experiments using either *Zea mays* or *Sorghum bicolor* where CO_2 concentrations were elevated and conducted using both ample (wet) and limited (dry) watering regimes. Data extracted from Kimball (2016) which utilises raw data from several studies (Ottman et al. 2001; Cousins et al. 2003; Leakey et al. 2004; Triggs et al. 2004; Hussain et al. 2013; Twine et al. 2013; Manderscheid et al. 2014; Ruiz-Vera et al. 2015)

C. *Interaction of Temperature with the Response of C_4 Plants to Elevated CO_2*

Higher air temperatures interact with the growth response to elevated CO_2 by affecting a multitude of processes such as photosynthesis, photorespiration, respiration, transpiration, vegetative growth and reproductive yield (Morison and Lawlor 1999; Dusenge et al. 2019). In general, the growth response to elevated CO_2 increases with temperature for most C_3 plants (Bunce 1998; Kimball et al. 1993; Reddy et al. 1998; Ziska 1998). The interaction of CO_2 and temperature is less well studied for C_4 plants. Due to the lack of apparent photorespiration in C_4 photosynthesis, the short-term response of photosynthesis to increased temperature is unaffected by high CO_2 (Ghannoum et al. 2002). Although it is uncertain whether the relative growth response of C_4 plants to elevated CO_2 may increase at high temperature, there is no theoretical reason why the growth response ought not be sustained as for C_3 plants, within the optimal physiological range for positive growth and photosynthesis of each species.

VIII. Future Outlook for C_4 Crops

A. *Climate Change Is Having Profound Impacts on Crop Yield and Quality Worldwide*

By 2050, global population is expected to rise to 9.7 billion, a 20% increase above today's estimates (United Nations 2019). However, food production needs to increase by more than 60% to meet increasing food demands due to climate change, stagnating yield improvements and limited arable lands (Godfray et al. 2010; Ray et al. 2013). Hence, future food security requires us not only to consider growing population and unfolding environmental change, but also changing human diets, as demands for more unsustainable luxury foods increase with affluence (Tilman and Clark 2014). Projections suggest we currently produce enough food to feed the predicted populations of 2050, but major socio-economic reforms would be required (Berners-Lee et al. 2018). However, if societies do not adapt, over double the amount of edible crops currently grown will be required by 2050 (Berners-Lee et al. 2018).

These projections are without consideration of the impacts of climate change on yields and grain quality. As temperatures and droughts increase, we will likely see decreases in yield in key growing regions of most major C_3 and C_4 crops (Li et al. 2011; Asseng et al. 2015). Although additional yields within C_3 crops may become apparent due to future CO_2 concentrations (Högy et al. 2009), the increased yields have been shown to come with the caveat of less nutritious grain with lower protein content which outweighs the positive effects (Högy et al. 2009; Asseng et al. 2019). Elevated CO_2 reduces the N content in tissues of C_3 plants due to a number of factors, including reduced photorespiration (Bloom et al. 2010, 2014), reduced N acquisition (Feng et al. 2015) and lower transpiration (Sherwin et al. 2013), amongst others. Reduced tissue N leads to reduced grain N content of major C_3 crops such as wheat (Bahrami et al. 2017; Tausz et al. 2017) and rice (Terao et al. 2005; Yang et al. 2007). Other macro (P, S) and micronutrients (Zn and Fe) are also reported to decrease at elevated CO_2, further undermining nutritional grain quality in future grain crops (Myers et al. 2014; Soares et al. 2019). Wheat, which provides about 20% of proteins for humans (Tilman et al. 2011), represents the best studied C_3 crop. Asseng et al. (2019) synthesised results of crop model simulations and observations from outdoor chamber and FACE experiments with increased temperature, heat shocks, and elevated CO_2 combined with increased temperature and drought stress. They predict that between 2014 and 2050, wheat grain protein concentration will decline globally by 1.1%, representing a relative yield change of −8.6%, in spite of the simulated yield increase from elevated CO_2 and the use of adapted genotypes with delayed anthesis and higher rate of grain filling (Asseng et al. 2019). These results suggest we need to consider the ability of C_3 crops to continue to provide the essential calories they currently do, even with the projected increase in production.

B. *What Does the Future Hold for C_4 Crops?*

Given that elevated CO_2 has minimal effects on C_4 photosynthesis, tissue N does not change much under elevated CO_2, however other climate change factors may affect tissue N and hence grain quality of C_4 crops (Myers et al. 2014; Soares et al. 2019). Although the impacts of elevated CO_2 on C_4 crops mostly appear when drought is present, when these two conditions are combined, grain quality has been shown to increase (Erbs et al. 2015). This could create further demand for C_4 crops in future climates, as they begin to provide not only increased yields, but also increased nutrition. It is therefore important to look forward to how environmental stresses can be alleviated within C_4 crops. Here, we briefly consider some possible solutions for improving crop yields and mitigating the effects of climate change.

Expanding and changing agricultural growing regions is feasible but often impractical. This is especially true in continents where climate change (e.g., climate warming, extreme heat and drought events) is reducing environmentally suitable land, and new growing regions with reduced temperatures and increased water are not available. In these cases, alternative genotypes or species need to be considered. Natural diversity has long been utilised to provide beneficial attributes to current germplasm (Hufford et al. 2012), and diversity that has implications for drought tolerance (such as root architecture) are continually being discovered (Li et al. 2018). This natural diversity can be used to improve germplasm to combat the negative impacts of future climates. In particular, there is a need to breed new varieties of crops that can better withstand warmer and extreme temperatures.

Another strategy is to cultivate more tolerant crops which require less irrigation and fertilisation. Switches within C_4 species are already evident in some parts of Africa, where the production of sorghum and millets are increasing (Orr et al. 2016), reducing maize production in these areas (Wang et al. 2018). This is due to the drought and heat tolerance of sorghum and *Pennisetum glaucum* (pearl millet). Planting these species alongside maize is beneficial as it spreads the risk in the case of drought in drought-prone areas such as Africa. The exploration of additional species is becoming more common across the world. For example, in India C_4 species with lower irrigation footprints and higher nutrition than *Oryza sativa* (rice) are being investigated (Davis et al. 2018). However, switches between species can be useful but culturally frowned upon, in particularly in countries where crops are staple within diets and food manufacturing of a country. This has meant the improvement of current germplasm is often pursued but has also meant current techniques are beginning to reach a ceiling for improvement.

The advent of new technologies is enhancing our ability to improve current germplasm. Gene editing has become relatively 'quick and easy' when compared to past methods, the leader of these new technologies currently being CRISPR (clustered regularly interspaced short palindromic)-Cas9. Within model plant species such as *Arabidopsis thaliana* CRISPR-Cas9 can now be completed with on-target efficiencies of up to 92% (Peterson et al. 2016). The technology allows both over expression and knock outs, while also allowing for base pair edits within genetic sequences. In tandem with gene editing technology, RNA sequencing technology has advanced over the past 10 years, producing an array of publicly available genetic information. Prior to the advance of DNA and RNA sequencing, quantitative trait loci (QTL) analysis was used to identify beneficial genetic traits for crops. Although a powerful tool, gene editing now provides the likely path to meeting future food demands. However, this does not make the quantitative genetic method redundant. In fact, gene editing can use the wealth of QTL information to focus efforts and identify beneficial targets for editing. Further to this, RNA-Sequencing data can strengthen this bridge between the two approaches. By extracting associated genes from meta-analyses of trait-based QTLs, we can identify sub-sets of genes which are of heightened interest. An example of this can be seen in Fig. 9.6, where genes associated with maize drought tolerance QTLs were mined via a maize expression database for differential expression in response to drought and tissue specificity (Hoopes et al. 2018). Here, it can be seen that the largest proportion of tissue specific genes are expressed within the seed (12.82%), and only around 25% of genes were significantly differentially expressed. This can give valuable information both when deciding on promoter specificity when introducing a gene to avoid confounding effects, and also whether a gene is likely to have a greater benefit when over-expressed or knocked out, ensuring advances are made efficiently.

Another approach which is currently being pursued is to introduce a C_4 CCM into C_3 crops, such as rice within the C_4 rice project (Lin et al. 2019). Rice is one of the most important crops in the world and is a staple food for more than half of the human population. Asia currently accounts for 90% of global rice production, which will need to increase by 50% within the next 30 years (Lin et al. 2019). If the technological difficulties of introducing a C_4 photosynthetic pathway into rice are overcome, then C_4 rice could potentially increase yields by 50%, double the water-use efficiency and reduce fertiliser use by 40%. These types of projects can be particularly advantageous under future climates where water scarcity and global temperature are predicted to increase (Nordhaus 2018).

We not only need to look towards crops which will feed our future population solely through increased yield and nutrient content, but also to those which ensure resources remain plentiful. With a wealth of new resources available to improve current germplasm, and the data available to predict future climates, we can make strides to ensure C_4 crops are improved and utilised in the correct climates to help ensure future food demands are met, and finite resources are not wasted.

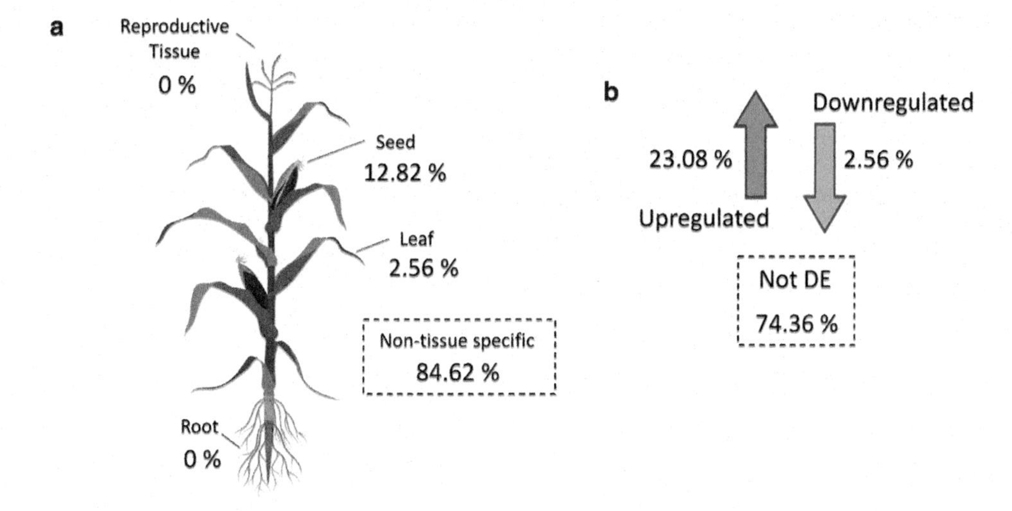

Fig. 9.6. The (**a**) tissue specificity and (**b**) change in transcript expression in response to drought for genes associated with drought tolerance QTLs. Genes associated with drought tolerance QTLs were extracted from a meta-analysis (Hao et al. 2010) then cross-referenced against a *Zea mays* transcriptomic database (Hoopes et al. 2018) to identify whether each gene has tissue specific expression or is significantly differentially expressed (DE) in response to drought. Percentages are out of a total number of 39 genes associated with QTLs. Created with BioRender.com

Acknowledgements

AWL and OG were supported by the ARC Centre of Excellence for Translational Photosynthesis (CE140100015) awarded to OG.

References

Ainsworth EA, Long SP (2005) What have we learned from 15 years of free-air CO_2 enrichment (FACE)? A meta-analytic review of the responses of photosynthesis, canopy properties and plant production to rising CO_2. New Phytol 165:351–372

Allen LH Jr, Kakani VG, Vu JC, Boote KJ (2011) Elevated CO_2 increases water use efficiency by sustaining photosynthesis of water-limited maize and sorghum. J Plant Physiol 168:1909–1918

Amaducci S, Colauzzi M, Battini F, Fracasso A, Perego A (2016) Effect of irrigation and nitrogen fertilization on the production of biogas from maize and sorghum in a water limited environment. Eur J Agron 76:54–65

Anami S, De Block M, Machuka J, Van Lijsebettens M (2009) Molecular improvement of tropical maize for drought stress tolerance in sub-Saharan Africa. Crit Rev Plant Sci 28:16–35

Assefa Y, Staggenborg SA, Prasad VP (2010) Grain sorghum water requirement and responses to drought stress: a review. Crop Manag 9:0–0

Assefa Y, Roozeboom K, Thompson C, Schlegel A, Stone L, Lingenfelser J (2014) Corn and grain sorghum morphology, physiology and phenology. Corn Grain Sorghum Comparison:3–14

Asseng S, Ewert F, Martre P, Rötter RP, Lobell DB, Cammarano D, Kimball BA, …, Zhu Y (2015) Rising temperatures reduce global wheat production. Nat Clim Chang 5:143

Asseng S, Martre P, Maiorano A, Rötter RP, O'leary GJ, Fitzgerald GJ, Girousse C, …, Ewert F (2019) Climate change impact and adaptation for wheat protein. Glob Chang Biol 25:155–173

Bahrami H, De Kok LJ, Armstrong R, Fitzgerald GJ, Bourgault M, Henty S, Tasuz M, Tausz-Posch S (2017) The proportion of nitrate in leaf nitrogen, but not changes in root growth, are associated with decreased grain protein in wheat under elevated [CO_2]. J Plant Physiol 216:44–51

Bathellier C, Tcherkez G, Lorimer GH, Farquhar GD (2018) Rubisco is not really so bad. Plant Cell Environ 41:705–716

Berners-Lee M, Kennelly C, Watson R, Hewitt C (2018) Current global food production is sufficient to meet human nutritional needs in 2050 provided there is radical societal adaptation. Elem Sci Anth 6:52

Berry J, Bjorkman O (1980) Photosynthetic response and adaptation to temperature in higher plants. Annu Rev Plant Physiol 31:491–543

Björkman O (1971) Comparative photosynthetic CO_2 exchange in higher plants, pp 18–32

Bloom AJ, Burger M, Asensio JSR, Cousins AB (2010) Carbon dioxide enrichment inhibits nitrate assimilation in wheat and Arabidopsis. Science 328:899–903

Bloom AJ, Burger M, Kimball BA, Pinter PJ Jr (2014) Nitrate assimilation is inhibited by elevated CO_2 in field-grown wheat. Nat Clim Chang 4:477

Blumenthal DM, Mueller KE, Kray JA, LeCain DR, Pendall E, Duke S, Jane Zelikova T, …, Morgan JA (2018) Warming and elevated CO_2 interact to alter seasonality and reduce variability of soil water in a semiarid grassland. Ecosystems 21:1533–1544

Borrell AK, Mullet JE, George-Jaeggli B, van Oosterom EJ, Hammer GL, Klein PE, Jordan DR (2014) Drought adaptation of stay-green sorghum is associated with canopy development, leaf anatomy, root growth, and water uptake. J Exp Bot 65:6251–6263

Bräutigam A, Schliesky S, Külahoglu C, Osborne CP, Weber AP (2014) Towards an integrative model of C4 photosynthetic subtypes: insights from comparative transcriptome analysis of NAD-ME, NADP-ME, and PEP-CK C4 species. J Exp Bot 65:3579–3593

Brooks A, Farquhar G (1985) Effect of temperature on the CO_2/O_2 specificity of ribulose-1, 5-bisphosphate carboxylase/oxygenase and the rate of respiration in the light. Planta 165:397–406

Brown RH (1999) Agronomic implications of C4 photosynthesis. In: C4 Plant Biology, pp 473–507

Bunce JA (1998) The temperature dependence of the stimulation of photosynthesis by elevated carbon dioxide in wheat and barley. J Exp Bot 49:1555–1561

Burnell JN, Hatch MD (1988) 1Photosynthesis in phosphoenolpyruvate Carboxykinase-type-C4 plants – pathways of C-4 acid decarboxylation in bundle sheath-cells of Urochloa-Panicoides. Arch Biochem Biophys 260:187–199

Burton GW (1983) Breeding pearl millet. In: Plant Breeding Review. Springer, Boston, pp 162–182

Carmo-Silva AE, Soares AS, Marques da Silva J, Bernardes da Silva A, Keys AJ, Arrabaça MC (2007) Photosynthetic responses of three C4 grasses of different metabolic subtypes to water deficit. Funct Plant Biol 34:204–213

Chapman KSR, Hatch MD (1981) Aspartate decarboxylation in bundle sheath-cells of Zea-Mays and its possible contribution to C4 photosynthesis. Aust J Plant Physiol 8:237–248

Chaves MM, Flexas J, Pinheiro C (2009) Photosynthesis under drought and salt stress: regulation mechanisms from whole plant to cell. Ann Bot 103:551–560

Chaves MM, Costa JM, Saibo NJM (2011) Recent advances in photosynthesis under drought and salinity. Adv Bot Res 57:49–104

Choat B, Jansen S, Brodribb TJ, Cochard H, Delzon S, Bhaskar R, Bucci SJ, …, Zanne AE (2012) Global convergence in the vulnerability of forests to drought. Nature 491:752

Christin P-A, Besnard G, Samaritani E, Duvall MR, Hodkinson TR, Savolainen V, Salamin N (2008) Oligocene CO_2 decline promoted C4 photosynthesis in grasses. Curr Biol 18:37–43

Christin P-A, Salamin N, Kellogg EA, Vicentini A, Besnard G (2009) Integrating phylogeny into studies of C_4 variation in the grasses. Plant Physiol 149:82–87

Christin PA, Osborne CP, Chatelet DS, Columbus JT, Besnard G, Hodkinson TR, Garrison LM, …, Edwards EJ (2013) Anatomical enablers and the evolution of C4 photosynthesis in grasses. Proc Natl Acad Sci USA 110:1381–1386

Ciais P, Reichstein M, Viovy N, Granier A, Ogée J, Allard V, Aubinet M, …, Valentini A (2005) Europe-wide reduction in primary productivity caused by the heat and drought in 2003. Nature 437:529

Conley MM, Kimball B, Brooks T, Pinter Jr P, Hunsaker D, Wall G, Adam NR, …, Triggs JM (2001) CO_2 enrichment increases water-use efficiency in sorghum. New Phytol 151:407–412

Cousins A, Adam N, Wall G, Kimball B, Pinter P Jr, Ottman MJ, Leavitt SW, Webber A (2003) Development of C4 photosynthesis in sorghum leaves grown under free-air CO_2 enrichment (FACE). J Exp Bot 54:1969–1975

Crafts-Brandner SJ, Salvucci ME (2002) Sensitivity of photosynthesis in a C4 plant, maize, to heat stress. Plant Physiol 129:1773–1780

Davis KF, Chiarelli DD, Rulli MC, Chhatre A, Richter B, Singh D, DeFries R (2018) Alternative cereals can improve water use and nutrient supply in India. Sci Adv 4:eaao1108

De Souza AP, Gaspar M, Da Silva EA, Ulian EC, Waclawovsky AJ, Nishiyama Jr MY, Santos RVD, …, Buckeridge MS (2008) Elevated CO_2 increases photosynthesis, biomass and productivity, and modifies gene expression in sugarcane. Plant Cell Environ 31:1116–1127

DeLucia EH, Chen S, Guan K, Peng B, Li Y, Gomez-Casanovas N, Kantola IB, …, Ort DR (2019) Are we approaching a water ceiling to maize yields in the United States? Ecosphere 10:e02773

Dengler NG, Dengler RE, Donnelly PM, Hattersley PW (1994) Quantitative leaf anatomy of C3 and C4 grasses (Poaceae): bundle sheath and mesophyll surface area relationships. Ann Bot 73:241–255

Dohleman FG, Long SP (2009) More productive than maize in the Midwest: how does Miscanthus do it? Plant Physiol 150:2104–2115

Dusenge ME, Duarte AG, Way DA (2019) Plant carbon metabolism and climate change: elevated CO_2 and temperature impacts on photosynthesis, photorespiration and respiration. New Phytol 221:32–49

Dwyer SA, Ghannoum O, Nicotra A, von Caemmerer S (2007) High temperature acclimation of C4 photosynthesis is linked to changes in photosynthetic biochemistry. Plant Cell Environ 30:53–66

Edwards G, Walker D (1983) C3, C4: Mechanisms, and Cellular and Environmental Regulation, of Photosynthesis. University of California Press, Berkeley

Edwards EJ, Osborne CP, Strömberg CAE, Smith SA, C4 Grasses Consortium (2010) The origins of C4 grasslands: integrating evolutionary and ecosystem science. Science 328:587–591

Ehleringer J, Pearcy RW (1983) Variation in quantum yield for CO_2 uptake among C_3 and C_4 plants. Plant Physiol 73:555–559

Ehleringer JR, Cerling TE, Helliker BR (1997) C4 photosynthesis, atmospheric CO_2, and climate. Oecologia 112:285–299

Engineer CB, Hashimoto-Sugimoto M, Negi J, Israelsson-Nordström M, Azoulay-Shemer T, Rappel W-J, Iba K, Schroeder JI (2016) CO_2 sensing and CO_2 regulation of stomatal conductance: advances and open questions. Trends Plant Sci 21:16–30

Erbs M, Manderscheid R, Hüther L, Schenderlein A, Wieser H, Dänicke S, Weigel H-J (2015) Free-air CO_2 enrichment modifies maize quality only under drought stress. Agron Sustain Dev 35:203–212

Estrada F, Botzen WW, Tol RS (2017) A global economic assessment of city policies to reduce climate change impacts. Nat Clim Chang 7:403

Evans JR (1989) Photosynthesis and nitrogen relationships in leaves of C_3 plants. Oecologia 78:9–19

FAO (2001) The state of food and agriculture 2001. Food & Agriculture Org

FAO (2019) FAOSTAT. http://www.faoorg/faostat/en/#home (22/02/2019)

Farquhar GD, Sharkey TD (1982) Stomatal conductance and photosynthesis. Annu Rev Plant Physiol 33:317–345

Feng Z, Rütting T, Pleijel H, Wallin G, Reich PB, Kammann CI, Newton PCD, …, Uddling J (2015) Constraints to nitrogen acquisition of terrestrial plants under elevated CO_2. Glob Chang Biol 21:3152–3168

Foster GL, Royer DL, Lunt DJ (2017) Future climate forcing potentially without precedent in the last 420 million years. Nat Commun 8:14845

Gao Y, Gao X, Zhang X (2017) The 2 C global temperature target and the evolution of the long-term goal of addressing climate change – from the United Nations framework convention on climate change to the Paris agreement. Engineering 3:272–278

Ghannoum O (2009) C_4 photosynthesis and water stress. Ann Bot 103:635–644

Ghannoum O (2016) How can we breed for more water use-efficient sugarcane? J Exp Bot 67:557–559

Ghannoum O, Conroy JP (1998) Nitrogen deficiency precludes a growth response to CO2 enrichment in C3 and C4 Panicum grasses. Funct Plant Biol 25:627–636

Ghannoum O, Sv C, Barlow EWR, Conroy JP (1997) The effect of CO2 enrichment and irradiance on the growth, morphology and gas exchange of a C3 (Panicum laxum) and a C4 (Panicum antidotale) grass. Funct Plant Biol 24:227–237

Ghannoum O, Caemmerer SV, Ziska L, Conroy J (2000) The growth response of C_4 plants to rising atmospheric CO_2 partial pressure: a reassessment. Plant Cell Environ 23:931–942

Ghannoum O, Caemmerer SV, Conroy JP (2002) The effect of drought on plant water use efficiency of nine NAD-ME and nine NADP-ME Australian C_4 grasses. Funct Plant Biol 29:1337–1348

Ghannoum O, Conroy JP, Driscoll SP, Paul MJ, Foyer CH, Lawlor DW (2003) Nonstomatal limitations are responsible for drought-induced photosynthetic inhibition in four C(4) grasses. New Phytol 159:599–608

Ghannoum O, Evans JR, Wah Soon C, Andrews TJ, Conroy JP, Susanne von C (2005) Faster Rubisco is the key to superior nitrogen use efficiency in NADP-malic enzyme relative to NAD-malic enzyme C_4 grasses. Plant Physiol 137:638–650

Ghannoum O, Evans JR, Caemmerer S (2011) Nitrogen and water use efficiency of C_4 plants. In: Raghavendra AS, Sage RF (eds) C_4 Photosynthesis and Related CO_2 Concentrating Mechanisms. Springer, Dordrecht, pp 129–146

Godfray HCJ, Beddington JR, Crute IR, Haddad L, Lawrence D, Muir JF, Pretty J, …, Toulmin C (2010) Food security: the challenge of feeding 9 billion people. Science 327:812–818

Grass Phylogeny Working G, II (2012) New grass phylogeny resolves deep evolutionary relationships and discovers C4 origins. New Phytol 193:304–312

Griffiths CA, Sagar R, Geng Y, Primavesi LF, Patel MK, Passarelli MK, Gilmore IS, …, Davis BG

(2016) Chemical intervention in plant sugar signalling increases yield and resilience. Nature 540:574

Hasegawa S, Piñeiro J, Ochoa-Hueso R, Haigh AM, Rymer PD, Barnett KL, Power SA (2018) Elevated CO_2 concentrations reduce C4 cover and decrease diversity of understorey plant community in a Eucalyptus woodland. J Ecol 106:1483–1494

Hatch MD (1987) C_4 photosynthesis: a unique blend of modified biochemistry, anatomy and ultrastructure. Biochim Biophys Acta Rev Bioenergetics 895:81–106

Hattersley PW (1983) The distribution of C_3 and C_4 grasses in Australia in relation to climate. Oecologia 57:113–128

Hattersley P (1992) C_4 photosynthetic pathway variation in grasses (Poaceae): its significance for arid and semi-arid lands. Desertified grasslands: their biology and management. Linn Soc Symp Ser:181–212

Hattersley PW, Watson L (1992) Diversification of photosynthesis. In: Chapman G (ed) Grass Evolution and Domestication. Cambridge University Press, Cambridge, pp 38–116

Henning JC, Brown RH (1986) Effects of irradiance and temperature on photosynthesis in C_3, C_4 and C_3/C_4 *Panicum* species. Photosynth Res 10:101–112

Hernández-Prieto MA, Foster C, Watson-Lazowski A, Ghannoum O, Chen M (2019) Comparative analysis of thylakoid protein complexes in the mesophyll and bundle sheath cells from C3, C4 and C3-C4 Paniceae grasses. Physiol Plant 166:134–147

Hess TM, Sumberg J, Biggs T, Georgescu M, Haro-Monteagudo D, Jewitt G, Ozdogan M, …, Knox JW (2016) A sweet deal? Sugarcane, water and agricultural transformation in Sub-Saharan Africa. Glob Environ Chang 39:181–194

Högy P, Wieser H, Köhler P, Schwadorf K, Breuer J, Franzaring J, Muntifering R, Fangmeier A (2009) Effects of elevated CO_2 on grain yield and quality of wheat: results from a 3-year free-air CO_2 enrichment experiment. Plant Biol 11:60–69

Hoopes GM, Hamilton JP, Wood JC, Esteban E, Pasha A, Vaillancourt B, Provart NJ, Buell CR (2018) An updated gene atlas for maize reveals organ-specific and stress-induced genes. Plant J 97:1154–1167

Hufford MB, Xu X, Van Heerwaarden J, Pyhäjärvi T, Chia J-M, Cartwright RA, Elshire RJ, …, Ross-Ibarra J (2012) Comparative population genomics of maize domestication and improvement. Nat Genet 44:808

Hui P, Tang J, Wang S, Niu X, Zong P, Dong X (2018) Climate change projections over China using regional climate models forced by two CMIP5 global models. Part II: projections of future climate. Int J Climatol 38:e78–e94

Hussain MZ, VanLoocke A, Siebers MH, Ruiz-Vera UM, Cody Markelz R, Leakey AD, Ort DR, Bernacchi CJ (2013) Future carbon dioxide concentration decreases canopy evapotranspiration and soil water depletion by field-grown maize. Glob Chang Biol 19:1572–1584

Jordan DB, Ogren WL (1984) The CO_2/O_2 specificity of ribulose 1,5-bisphosphate carboxylase/oxygenase. Planta 161:308–313

Kanai R, Edwards GE (1999) The biochemistry of C4 photosynthesis. In: C4 Plant Biology, vol 49, p 87

Kimball BA (2016) Crop responses to elevated CO_2 and interactions with H_2O, N, and temperature. Curr Opin Plant Biol 31:36–43

Kimball BA, Mauney JR, Nakayama FS, Idso SB (1993) Effects of increasing atmospheric CO2 on vegetation. Vegetatio 104/105:65–75

Kirkham MB, He H, Bolger TP, Lawlor DJ, Kanemasu ET (1991) Leaf photosynthesis water use of big bluestem under elevated carbon dioxide. Crop Sci 31:1589–1594

Knapp A, Hamerlynck E, Owensby C (1993) Photosynthetic and water relations responses to elevated CO2 in the C4 grass Andropogon gerardii. Int J Plant Sci 154:459–466

Kocacinar F, Sage R (2003) Photosynthetic pathway alters xylem structure and hydraulic function in herbaceous plants. Plant Cell Environ 26:2015–2026

Koteyeva NK, Voznesenskaya EV, Edwards GE (2015) An assessment of the capacity for phosphoenolpyruvate carboxykinase to contribute to C-4 photosynthesis. Plant Sci 235:70–80

Krupa KN, Dalawai N, Shashidhar HE, Harinikumar KM, Manojkumar HB, Bharani S, Turaidar V (2017) Mechanisms of drought tolerance in Sorghum: a review. Int J Pure App Biosci 5:221–237

Kubien DS, Sage RF (2004) Dynamic photo-inhibition and carbon gain in a C_4 and a C_3 grass native to high latitudes. Plant Cell Environ 27:1424–1435

Kubien DS, von Caemmerer S, Furbank RT, Sage RF (2003) C_4 photosynthesis at low temperature. A study using transgenic plants with reduced amounts of Rubisco. Plant Physiol 132:1577–1585

Lawlor DW, Cornic G (2002) Photosynthetic carbon assimilation and associated metabolism in relation to water deficits in higher plants. Plant, Cell and Environment 25:275–294

Lawlor DW, Tezara W (2009) Causes of decreased photosynthetic rate and metabolic capacity in water-deficient leaf cells: a critical evaluation of

mechanisms and integration of processes. Ann Bot 103:561–579
Leakey A, Bernacchi C, Dohleman F, Ort D, Long S (2004) Will photosynthesis of maize (Zea mays) in the US Corn Belt increase in future [CO_2] rich atmospheres? An analysis of diurnal courses of CO_2 uptake under free-air concentration enrichment (FACE). Glob Chang Biol 10:951–962
Leff B, Ramankutty N, Foley JA (2004) Geographic distribution of major crops across the world. Glob Biogeochem Cycles 18
Lhotka O, Kyselý J, Farda A (2018) Climate change scenarios of heat waves in Central Europe and their uncertainties. Theor Appl Climatol 131:1043–1054
Li X, Takahashi T, Suzuki N, Kaiser HM (2011) The impact of climate change on maize yields in the United States and China. Agric Syst 104:348–353
Li P, Pan T, Wang H, Wei J, Chen M, Hu X, Zhao Y, …, Yang Z (2018) Natural variation of ZmHKT1 affects root morphology in maize at the seedling stage. Planta:1–11
Lin HC, Coe RA, Quick WP, Bandyopadhyay A (2019) Climate-resilient future crop: development of C4 Rice. In: Sustainable Solutions for Food Security. Springer, Cham, pp 111–124
Liu H, Osborne CP (2015) Water relations traits of C_4 grasses depend on phylogenetic lineage, photosynthetic pathway, and habitat water availability. J Exp Bot 66:761–773
Liu Y, Li S, Chen F, Yang S, Chen X (2010) Soil water dynamics and water use efficiency in spring maize (Zea mays L.) fields subjected to different water management practices on the Loess Plateau, China. Agric Water Manag 97:769–775
Liu H, Edwards EJ, Freckleton RP, Osborne CP (2012) Phylogenetic niche conservatism in C_4 grasses. Oecologia 170:835–845
Lloyd J, Farquhar GD (1994) ^{13}C discrimination during CO_2 assimilation by the terrestrial biosphere. Oecologia 99:201–215
Lobell DB, Gourdji SM (2012) The influence of climate change on global crop productivity. Plant Physiol 160:1686–1697
Lobell DB, Roberts MJ, Schlenker W, Braun N, Little BB, Rejesus RM, Hammer GL (2014) Greater sensitivity to drought accompanies maize yield increase in the US Midwest. Science 344:516–519
Long SP (1999) Environmental Responses. In: Rowan FS, Russell KM (eds) C_4 Plant Biology. Academic, San Diego, pp 215–249
Long SP, Woolhouse HW (1978) The responses of net photosysthesis to light and temperature in *Spartina townsendii* (sensu lato), a C_4 species from a cool temperate climate. J Exp Bot 29:803–814
Long SP, Zhu XG, Naidu SL, Ort DR (2006) Can improvement in photosynthesis increase crop yields? Plant Cell Environ 29:315–330
Lorimer GH, Badger MR, Andrews TJ (1976) The activation of ribulose-1, 5-bisphosphate carboxylase by carbon dioxide and magnesium ions. Equilibria, kinetics, a suggested mechanism, and physiological implications. Biochemistry 15:529–536
Lunn JE, Furbank RT (1997) Localisation of sucrose-phosphate synthase and starch in leaves of C4 plants. Planta 202:106–111
Lunn JE, Furbank RT (1999) Tansley review no. 105. New Phytol 143:221–237
Makino A, Sakuma H, Sudo E, Mae T (2003) Differences between maize and rice in N-use efficiency for photosynthesis and protein allocation. Plant Cell Physiol 44:952–956
Manderscheid R, Erbs M, Weigel H-J (2014) Interactive effects of free-air CO_2 enrichment and drought stress on maize growth. Eur J Agron 52:11–21
McDowell NG (2011) Mechanisms linking drought, hydraulics, carbon metabolism, and vegetation mortality. Plant Physiol 155:1051–1059
McDowell NG, Sevanto S (2010) The mechanisms of carbon starvation: how, when, or does it even occur at all? New Phytol 186:264–266
Mcdowell NG, Williams A, Xu C, Pockman W, Dickman L, Sevanto S, Pangel R, …, Koven C (2016) Multi-scale predictions of massive conifer mortality due to chronic temperature rise. Nat Clim Chang 6:295
Mitchell TD, Carter TR, Jones PD, Hulme M, New M (2004) A comprehensive set of high-resolution grids of monthly climate for Europe and the globe: the observed record (1901–2000) and 16 scenarios (2001–2100). Tyndall centre for climate change research working paper 55, p 25
Morison JIL, Gifford RM (1984a) Plant growth and water use with limited water supply in high CO2 concentrations. II. Plant dry weight, partitioning and water use efficiency. Aust J Plant Physiol 11:375–384
Morison JIL, Gifford RM (1984b) Plant growth and water use with limited water supply in high CO2 concentrations. I. Leaf area, water use and transpiration. Aust J Plant Physiol 11:361–374
Morison J, Lawlor D (1999) Interactions between increasing CO_2 concentration and temperature on plant growth. Plant Cell Environ 22:659–682
Myers SS, Zanobetti A, Kloog I, Huybers P, Leakey AD, Bloom AJ, Carlisle E, …, Usui Y (2014) Increasing CO_2 threatens human nutrition. Nature 510:139
Nordhaus W (2018) Projections and uncertainties about climate change in an era of minimal climate policies. Am Econ J Econ Pol 10:333–360

Orr A, Mwema C, Gierend A, Nedumaran S (2016) Sorghum and millets in Eastern and Southern Africa: facts, Trends and outlook. Working Paper. ICRISAT, Patancheru, Telangana, India

Osborne CP, Salomaa A, Kluyver TA, Visser V, Kellogg EA, Morrone O, Vorontsova MS, …, Simpson DA (2014) A global database of C4 photosynthesis in grasses. New Phytol 204:441–446

Osmond CB, Winter K, Ziegler H (1982) Functional significance of different pathways of CO_2 fixation in photosynthesis. In: Lange PDOL et al (eds) Physiological Plant Ecology II. Springer, Berlin/Heidelberg, pp 479–547

Oszvald M, Primavesi LF, Griffiths CA, Cohn J, Basu SS, Nuccio ML, Paul MJ (2018) Trehalose 6-phosphate regulates photosynthesis and assimilate partitioning in reproductive tissue. Plant Physiol 176:2623–2638

Ottman M, Kimball B, Pinter Jr P, Wall G, Vanderlip R, Leavitt S, Lamorte R, …, Brooks T (2001) Elevated CO_2 effects on sorghum growth and yield at high and low soil water content. New Phytol 150:261–273

Owensby CE, Ham JM, Knapp AK, Bremer D, Auen LM (1997) Water vapour fluxes and their impact under elevated CO2 in a C4-tallgrass prairie. Glob Chang Biol 3:189–195

Pau S, Edwards EJ, Still CJ (2013) Improving our understanding of environmental controls on the distribution of C-3 and C-4 grasses. Glob Chang Biol 19:184–196

Paul MJ, Gonzalez-Uriarte A, Griffiths CA, Hassani-Pak K (2018) The role of Trehalose 6-phosphate in crop yield and resilience. Plant Physiol 177:12–23

Peacock J (1982) Response and tolerance of sorghum to temperature stress. In: Sorghum in the eighties: proceedings of the international symposium on Sorghum, 2–7 November 1981, Patancheru. A.P. India

Pearcy RW, Ehleringer J (1984) Comparative ecophysiology of C_3 and C_4 plants. Plant Cell Environ 7:1–13

Pengelly JJ, Tan J, Furbank RT, von Caemmerer S (2012) Antisense reduction of NADP-malic enzyme in Flaveria bidentis reduces flow of CO_2 through the C4 cycle. Plant Physiol 160:1070–1080

Pepper G, Pearce R, Mock J (1977) Leaf orientation and yield of maize 1. Crop Sci 17:883–886

Peterson BA, Haak DC, Nishimura MT, Teixeira PJ, James SR, Dangl JL, Nimchuk ZL (2016) Genome-wide assessment of efficiency and specificity in CRISPR/Cas9 mediated multiple site targeting in Arabidopsis. PLoS One 11:e0162169

Pingali PL (2001) CIMMYT 1999/2000 World maize facts and trends. Meeting world maize needs: technological opportunities and priorities for the public sector. CIMMYT, Mexico, DF

Pinto H, Tissue DT, Ghannoum O (2011) Panicum milioides (C-3-C-4) does not have improved water or nitrogen economies relative to C-3 and C-4 congeners exposed to industrial-age climate change. J Exp Bot 62:3223–3234

Pinto H, Sharwood RE, Tissue DT, Ghannoum O (2014) Photosynthesis of C-3, C-3-C-4, and C-4 grasses at glacial CO_2. J Exp Bot 65:3669–3681

Pinto H, Powell JR, Sharwood RE, Tissue DT, Ghannoum O (2016) Variations in nitrogen use efficiency reflect the biochemical subtype while variations in water use efficiency reflect the evolutionary lineage of C4 grasses at inter-glacial CO_2. Plant Cell Environ 39:514–526

Pittermann J, Sage RF (2000) Photosynthetic performance at low temperature of *Bouteloua gracilis* lag., a high-altitude C_4 grass from the Rocky Mountains, USA. Plant Cell Environ 23:811–823

Poorter H, Navas ML (2003) Plant growth and competition at elevated CO_2: on winners, losers and functional groups. New Phytol 157:175–198

Ramirez-Cabral NY, Kumar L, Shabani F (2017) Global alterations in areas of suitability for maize production from climate change and using a mechanistic species distribution model (CLIMEX). Sci Rep 7:5910

Ranum P, Peña-Rosas JP, Garcia-Casal MN (2014) Global maize production, utilization, and consumption. Ann NY Acad Sci 1312:105–112

Ray DK, Mueller ND, West PC, Foley JA (2013) Yield trends are insufficient to double global crop production by 2050. PLoS One 8:e66428

Reddy KR, Hodges H (2000) Climate Change and Global Crop Productivity. CABI, Wallingford

Reddy KR, Robana RR, Hodges HF, Liu XJ, McKinion JM (1998) Interactions of CO2 enrichment and temperature on cotton growth and leaf characteristics. Environ Exp Bot 39:117–129

Regassa TH, Wortmann CS (2014) Sweet sorghum as a bioenergy crop: literature review. Biomass Bioenergy 64:348–355

Reich PB, Hobbie SE, Lee TD, Pastore MA (2018) Unexpected reversal of C-3 versus C-4 grass response to elevated CO_2 during a 20-year field experiment. Science 360:317

Ripley BS, Gilbert ME, Ibrahim DG, Osborne CP (2007) Drought constraints on C4 photosynthesis: stomatal and metabolic limitations in C3 and C4 subspecies of Alloteropsis semialata. J Exp Bot 58:1351–1363

Ripley B, Frole K, Gilbert M (2010) Differences in drought sensitivities and photosynthetic limitations between co-occurring C3 and C4 (NADP-ME) Panicoid grasses. Ann Bot 105:493–503

Roser M, Ritchie H (2013) Yields and land use in agriculture. Our world in data

Roy H, Andrews TJ (2000) Rubisco: assembly and mechanism. In: Photosynthesis. Springer, Dordrecht, pp 53–83

Ruiz-Vera UM, Siebers MH, Drag DW, Ort DR, Bernacchi CJ (2015) Canopy warming caused photosynthetic acclimation and reduced seed yield in maize grown at ambient and elevated [CO_2]. Glob Chang Biol 21:4237–4249

Saban JM, Chapman MA, Taylor G (2019) FACE facts hold for multiple generations; evidence from natural CO_2 springs. Glob Chang Biol 25:1–11

Sage RF (2002) Variation in the k_{cat} of Rubisco in C_3 and C_4 plants and some implications for photosynthetic performance at high and low temperature. J Exp Bot 53:609–620

Sage RF (2004) The evolution of C_4 photosynthesis. New Phytol 161:341–370

Sage RF, Kubien DS (2007) The temperature response of C_3/ and C_4 photosynthesis. Plant Cell Environ 30:1086–1106

Sage RF, Pearcy RW (2000) The physiological ecology of C_4 photosynthesis. In: Leegood RC et al (eds) Photosynthesis. Springer, Dordrecht, pp 497–532

Sage RF, Li M, Monson RK (1999) Chapter 16 – The taxonomic distribution of C_4 photosynthesis. In: Monson RFSK (ed) C4 Plant Biology. Academic, San Diego, pp 551–584

Sage RF, Christin P-A, Edwards EJ (2011) The C4 plant lineages of planet Earth. J Exp Bot 62:3155–3169

Samarakoon AB, Gifford RM (1996) Elevated CO2 effects on water use and growth of maize in wet and drying soil. Aust J Plant Physiol 23:53–62

Schittenhelm S, Schroetter S (2014) Comparison of drought tolerance of maize, sweet sorghum and sorghum-sudangrass hybrids. J Agron Crop Sci 200:46–53

Seneweera SP, Ghannoum O, Conroy J (1998) High vapour pressure deficit and low soil water availability enhance shoot growth responses of a C4 grass (Panicum coloratum cv. Bambatsi) to CO2 enrichment. Funct Plant Biol 25:287–292

Seneweera S, Ghannoum O, Conroy JP (2001) Root and shoot factors contribute to the effect of drought on photosynthesis and growth of the C4 grass Panicum coloratum at elevated CO2 partial pressures. Funct Plant Biol 28:451–460

Serba DD, Yadav RS (2016) Genomic tools in pearl millet breeding for drought tolerance: status and prospects. Front Plant Sci 7:1724

Serraj R, Allen LH, Sinclair TR (1999) Soybean leaf growth and gas exchange response to drought under carbon dioxide enrichment. Glob Chang Biol 5:283–291

Sharma-Natu P, Ghildiyal M (2005) Potential targets for improving photosynthesis and crop yield. Curr Sci:1918–1928

Sharwood RE, Sonawane BV, Ghannoum O (2014) Photosynthetic flexibility in maize exposed to salinity and shade. J Exp Bot 16:3715–3724

Sharwood RE, Ghannoum O, Kapralov MV, Gunn LH, Whitney SM (2016a) Temperature responses of Rubisco from Paniceae grasses provide opportunities for improving C_3 photosynthesis. Nat Plants 2:16186

Sharwood RE, Sonawane BV, Ghannoum O, Whitney SM (2016b) Improved analysis of C4 and C3 photosynthesis via refined in vitro assays of their carbon fixation biochemistry. J Exp Bot 67:3137–3148

Sherwin GL, George L, Kannangara K, Tissue DT, Ghannoum O (2013) Impact of industrial-age climate change on the relationship between water uptake and tissue nitrogen in eucalypt seedlings. Funct Plant Biol 40:201–212

Shiferaw B, Prasanna BM, Hellin J, Bänziger M (2011) Crops that feed the world 6. Past successes and future challenges to the role played by maize in global food security. Food Sec 3:307

Siebers MH, Slattery RA, Yendrek CR, Locke AM, Drag D, Ainsworth EA, Bernacchi CJ, Ort DR (2017) Simulated heat waves during maize reproductive stages alter reproductive growth but have no lasting effect when applied during vegetative stages. Agric Ecosyst Environ 240:162–170

Siebke K, Ghannoum O, Conroy JP, Badger MR, von Caemmerer S (2003) Photosynthetic oxygen exchange in C_4 grasses: the role of oxygen as electron acceptor. Plant Cell Environ 26:1963–1972

Soares JC, Santos CS, Carvalho SM, Pintado MM, Vasconcelos MW (2019) Preserving the nutritional quality of crop plants under a changing climate: importance and strategies. Plant Soil:1–26

Sonawane BV, Sharwood RE, von Caemmerer S, Whitney SM, Ghannoum O (2017) Short-term thermal photosynthetic responses of C-4 grasses are independent of the biochemical subtype. J Exp Bot 68:5583–5597

Sonawane BV, Sharwood RE, Whitney S, Ghannoum O (2018) Shade compromises the photosynthetic efficiency of NADP-ME less than that of PEP-CK and NAD-ME C4 grasses. J Exp Bot 69:3053–3068

Staggenborg SA, Dhuyvetter KC, Gordon W (2008) Grain sorghum and corn comparisons: yield, economic, and environmental responses. Agron J 100:1600–1604

Steffen W, Broadgate W, Deutsch L, Gaffney O, Ludwig C (2015) The trajectory of the Anthropocene: the great acceleration. Anthropocene Rev 2:81–98

Stokes CJ, Inman-Bamber NG, Everingham Y, Sexton J (2016) Measuring and modelling CO_2 effects on sugarcane. Environ Model Softw 78:68–78

Tardieu F, Granier C, Muller B (2011) Water deficit and growth. Co-ordinating processes without an orchestrator? Curr Opin Plant Biol 14:283–289

Taub DR, Lerdau MT (2000) Relationship between leaf nitrogen and photosynthetic rate for three NAD-ME and three NADP-ME C_4 grasses. Am J Bot 87:412–417

Tausz M, Norton R, Tausz-Posch S, Löw M, Seneweera S, O'Leary G, Armstrong R, Fitzgerald G (2017) Can additional N fertiliser ameliorate the elevated CO 2-induced depression in grain and tissue N concentrations of wheat on a high soil N background? J Agron Crop Sci 203:574–583

Taylor SH, Hulme SP, Rees M, Ripley BS, Ian Woodward F, Osborne CP (2010) Ecophysiological traits in C_3 and C_4 grasses: a phylogenetically controlled screening experiment. New Phytol 185:780–791

Taylor SH, Franks PJ, Hulme SP, Spriggs E, Christin PA, Edwards EJ, Woodward FI, Osborne CP (2012) Photosynthetic pathway and ecological adaptation explain stomatal trait diversity amongst grasses. New Phytol 193:387–396

Taylor SH, Aspinwall MJ, Blackman CJ, Choat B, Tissue DT, Ghannoum O (2018) CO_2 availability influences hydraulic function of C3 and C4 grass leaves. J Exp Bot 69:2731–2741

Terao T, Miura S, Yanagihara T, Hirose T, Nagata K, Tabuchi H, Kim H-Y, …, Kobayashi K (2005) Influence of free-air CO_2 enrichment (FACE) on the eating quality of rice. J Sci Food Agric 85:1861–1868

Tezara W, Mitchell VJ, Driscoll SD, Lawlor DW (1999) Water stress inhibits plant photosynthesis by decreasing coupling factor and ATP. Nature 401:914–917

Tilman D, Clark M (2014) Global diets link environmental sustainability and human health. Nature 515:518

Tilman D, Balzer C, Hill J, Befort BL (2011) Global food demand and the sustainable intensification of agriculture. Proc Natl Acad Sci 108: 20260–20264

Tollenaar M, Lee E (2002) Yield potential, yield stability and stress tolerance in maize. Field Crop Res 75:161–169

Tollenaar M, Lee EA (2011) 2 strategies for enhancing grain yield in maize. Plant Breed Rev 34:37–82

Triggs JM, Kimball B, Pinter Jr P, Wall G, Conley M, Brooks T, LaMorte RL, …, Cerveny RS (2004) Free-air CO_2 enrichment effects on the energy balance and evapotranspiration of sorghum. Agric For Meteorol 124:63–79

Tubiello FN, Soussana J-F, Howden SM (2007) Crop and pasture response to climate change. Proc Natl Acad Sci 104:19686–19690

Twine TE, Bryant JJ, Richter KT, Bernacchi CJ, McConnaughay KD, Morris SJ, Leakey AD (2013) Impacts of elevated CO_2 concentration on the productivity and surface energy budget of the soybean and maize agroecosystem in the Midwest USA. Global Chang Biol 19:2838–2852

Uddin S, Löw M, Parvin S, Fitzgerald G, Bahrami H, Tausz-Posch S, Armstrong R, …, Tausz M (2018) Water use and growth responses of dryland wheat grown under elevated [CO_2] are associated with root length in deeper, but not upper soil layer. Field Crop Res 224:170–181

United Nations DoEaSA, Population Division (2019) World population prospects 2019. Online Edition. https://population.un.org/wpp2019/

Varshney RK, Shi C, Thudi M, Mariac C, Wallace J, Qi P, Zhang H, …, Xu X (2017) Pearl millet genome sequence provides a resource to improve agronomic traits in arid environments. Nat Biotechnol 35:969

Ver Sagun J, Badger MR, Chow WS, Ghannoum O (2019) Cyclic electron flow and light partitioning between the two photosystems in leaves of plants with different functional types. Photosynth Res:1–14

Vicentini A, Barber JC, Aliscioni SS, Giussani LM, Kellogg EA (2008) The age of the grasses and clusters of origins of C_4 photosynthesis. Glob Chang Biol 14:2963–2977

von Caemmerer S (2000) Biochemical Models of Leaf Photosynthesis. CSIRO Publishing, Collingwood

von Caemmerer S, Quick PW (2000) Rubisco, physiology *in vivo*. In: Leegood RC et al (eds) Photosynthesis: Physiology and Metabolism. Kluwer Academic Publishers, Dordrecht, pp 85–113

von Caemmerer S, Quick W, Furbank R (2012) The development of C-4 rice: current progress and future challenges. Science 336:1671

von Caemmerer S, Ghannoum O, Pengelly JJL, Cousins AB (2014) Carbon isotope discrimination as a tool to explore C4 photosynthesis. J Exp Bot 65:3459–3470

Wahid A, Gelani S, Ashraf M, Foolad MR (2007) Heat tolerance in plants: an overview. Environ Exp Bot 61:199–223

Wall GW, Brooks TJ, Adam NR, Cousins A, Kimball BA, Pinter PJ Jr, LaMorte RL, Triggs J, Ottman MJ, Leavitt SW, Matthias AD, Williams DG, Webber AN (2001) Elevated atmospheric CO2 improved Sorghum plant water status by ameliorating the adverse effects of drought. New Phytol 152:231–248
Wand SJE, Midgley GF, Jones MH, Curtis PS (1999) Responses of wild C4 and C3 grass (Poaceae) species to elevated atmospheric CO_2 concentration: a meta-analytic test of current theories and perceptions. Glob Chang Biol 5:723–741
Wang Y, Bräutigam A, Weber AP, Zhu X-G (2014) Three distinct biochemical subtypes of C4 photosynthesis? A modelling analysis. J Exp Bot 65:3567–3578
Wang P, Vlad D, A Langdale J (2016) Finding the genes to build C4 rice. Curr Opin Plant Biol 31:44–50
Wang J, Vanga S, Saxena R, Orsat V, Raghavan V (2018) Effect of climate change on the yield of cereal crops: a review. Climate 6:41
Watson-Lazowski A, Lin Y, Miglietta F, Edwards RJ, Chapman MA, Taylor G (2016) Plant adaptation or acclimation to rising CO_2? Insight from first multigenerational RNA-Seq transcriptome. Glob Chang Biol 22:3760–3773
Watson-Lazowski A, Papanicolaou A, Sharwood R, Ghannoum O (2018) Investigating the NAD-ME biochemical pathway within C-4 grasses using transcript and amino acid variation in C-4 photosynthetic genes. Photosynth Res 138:233–248
Watson-Lazowski A, Papanicolaou A, Koller F, Ghannoum O (2020) The transcriptomic responses of C4 grasses to sub-ambient CO_2 and low light are largely species-specific and only refined by photosynthetic subtype. Plant J 101:1170–1184
Weber T, Haensler A, Rechid D, Pfeifer S, Eggert B, Jacob D (2018) Analyzing regional climate change in africa in a 1.5, 2, and 3 C global warming world. Earth's Future 6:643–655
Wheeler T, Von Braun J (2013) Climate change impacts on global food security. Science 341:508–513
Whitney SM, Houtz RL, Alonso H (2011) Advancing our understanding and capacity to engineer nature's CO_2-sequestering enzyme, Rubisco. Plant Physiol 155:27–35
Willis K, McElwain J (2014) The Evolution of Plants. Oxford University Press, Oxford
Wingler A, Walker RP, Chen Z-H, Leegood RC (1999) Phosphoenolpyruvate carboxykinase is involved in the decarboxylation of aspartate in the bundle sheath of maize. Plant Physiol 120:539–546
Yamori W, Hikosaka K, Way DA (2014) Temperature response of photosynthesis in C 3, C 4, and CAM plants: temperature acclimation and temperature adaptation. Photosynth Res 119:101–117
Yan K, Chen P, Shao H, Shao C, Zhao S, Brestic M (2013) Dissection of photosynthetic Electron transport process in sweet Sorghum under heat stress. PLoS One 8:e62100
Yang L, Wang Y, Dong G, Gu H, Huang J, Zhu J, Yang H, …, Han Y (2007) The impact of free-air CO_2 enrichment (FACE) and nitrogen supply on grain quality of rice. Field Crop Res 102:128–140
Zhang YG, Pagani M, Liu Z, Bohaty SM, DeConto R (2013) A 40-million-year history of atmospheric CO_2. Philos Trans R Soc A Math Phys Eng Sci 371:20130096
Zhang P, Wei T, Cai T, Ali S, Han Q, Ren X, Jia Z (2017) Plastic-film mulching for enhanced water-use efficiency and economic returns from maize fields in Semiarid China. Front Plant Sci 8:512
Zhao C, Liu B, Piao S, Wang X, Lobell DB, Huang Y, Huang M, …, Asseng S (2017) Temperature increase reduces global yields of major crops in four independent estimates. Proc Natl Acad Sci 114:9326–9331
Zheng Y, Xu M, Shen R, Qiu S (2013) Effects of artificial warming on the structural, physiological, and biochemical changes of maize (Zea mays L.) leaves in northern China. Acta Physiol Plant 35:2891–2904
Zheng Y, Guo L, Hou R, Zhou H, Hao L, Li F, Cheng D, …, Xu M (2018a) Experimental warming enhances the carbon gain but does not affect the yield of maize (Zea mays L.) in the North China Plain. Flora 240:152–163
Zheng Y, Li R, Guo L, Hao L, Zhou H, Li F, Peng ZP, …, Xu M (2018b) Temperature responses of photosynthesis and respiration of maize (Zea mays) plants to experimental warming. Russ J Plant Physiol 65:524–531
Zhu X-G, Long SP, Ort DR (2008) What is the maximum efficiency with which photosynthesis can convert solar energy into biomass? Curr Opin Biotechnol 19:153–159
Ziska LH (1998) The influence of root zone temperature on photosynthetic acclimation to elevated carbon dioxide concentrations. Ann Bot 81:717 721
Zwart SJ, Bastiaanssen WG (2004) Review of measured crop water productivity values for irrigated wheat, rice, cotton and maize. Agric Water Manag 69:115–133

Chapter 10

Climate Change Responses and Adaptations in Crassulacean Acid Metabolism (CAM) Plants

Paula N. Pereira, Nicholas A. Niechayev, Brittany B. Blair, and John C. Cushman*
Department of Biochemistry and Molecular Biology, University of Nevada, Reno, NV, USA

*Author for correspondence, e-mail: jcushman@unr.edu

K. M. Becklin et al. (eds.), *Photosynthesis, Respiration, and Climate Change*, Advances in Photosynthesis and Respiration 48, https://doi.org/10.1007/978-3-030-64926-5_10

Summary

Global climatic change is predicted to result in certain areas of the earth's surface becoming hotter and drier. These spatially durable changes in climate will likely have negative impacts on the productivity of many plant species that are poorly adapted to increasing air:leaf vapor pressure deficits. However, more that 6% of vascular plants have evolved crassulacean acid metabolism (CAM) as a photosynthetic adaptation to hotter and drier climates. This specialized mode of photosynthesis exploits a temporal CO_2 pump with nocturnal CO_2 uptake and concentration to reduce photorespiration, improve water-use efficiency (WUE), and optimize the adaptability of plants to hotter and drier climates. CAM species have a suite of physiological and anatomical adaptations that make them resilient to extreme heat and high insolation including tissue succulence, water-storage and water-capture strategies to attenuate drought, and thick cuticles, epicuticular wax, low stomatal density, high stomatal responsiveness, UV-light protection, and shallow rectifier-like roots to limit water loss under conditions of water deficit. Various modeling approaches have been developed to estimate the growth potential of CAM species under current and future climatic conditions. Such information suggests that certain CAM species with considerable biomass productivity, such as *Agave* and *Opuntia spp.*, hold great potential as highly water-use efficient, food, forage, and fodder production platforms as well as carbon neutral feedstocks for bioethanol and biogas production to replace fossil petroleum resources. As low-lignin, herbaceous feedstocks, CAM species can be converted to biofuels and biogas for transportation and electricity to displace fossil fuels thereby reducing net greenhouse gas emissions. Given their intrinsic climate-resilience, CAM species are useful for the restoration of abandoned or marginal agricultural lands, to provide ecosystem services, and agroecosystem and urban strategies for terrestrial carbon sequestration. Lastly, we discuss the transformational research area of CAM Biodesign aimed at introducing CAM into C_3 or C_4 photosynthesis food and bioenergy crops to improve their water-use efficiency and drought tolerance.

I. Introduction

Global warming and the associated climate change or weather extremes it produces represents a major threat to global biological diversity and agricultural production systems in the twenty-first century. Rising temperatures result in greater soil drying and vapor-pressure deficits (VPD), which is the humidity gradient between the atmosphere and the water vapor-saturated interior of plant leaves. Drier soils and increased vapor-pressure deficits create greater water-deficit stress, limiting the ability of many plant and crop species to survive or thrive. In this review, we will explore the potential utility of crassulacean acid metabolism (CAM) crops, which can tolerate or attenuate drought stress or have greater intrinsic water-use efficiency (WUE, the unit of CO_2 fixed per unit of water lost), as a pathway to partially offset the negative effects of global warming and potentially make greater use of marginal or degraded agricultural lands (Cushman et al. 2015; Davis et al. 2017, 2019).

A. Global Climate Change

Climate change models predict that the frequency and severity of drought events will increase over the course of the twenty-first century, threatening global food and water security. Global climate change is predicted to increase soil moisture deficits and the severity and duration of drought episodes (Cook et al. 2014, 2015). Approximately 50% of the global land area is already considered arid, semi-arid, or dry sub-humid (Zika and Erb 2009). Models predict that global warming trends coupled with increased terrestrial soil drying in many regions (Treberth 2011; Field et al. 2014) will reduce terrestrial net primary production (NPP), reduce terrestrial carbon sinks, threaten global food security, and curtail biomass feedstock production (Zhao and Running 2009), leading to a global expansion of drylands (Dai 2013; Feng and Fu 2013; Fu and Feng 2014; Huang et al. 2015). Many of these dry areas are expected to experience 5–10-times more frequent droughts with 3 °C warming despite attempts to mitigate greenhouse gas release and to capture or sequester such emissions (Naumann et al. 2018). Modeling of global warming suggests that the resulting increases in surface drying are not likely to be offset by plant's physiological response to elevated CO_2 in the twenty-first century (Dai et al. 2018).

Extreme weather events associated with global climate change are already exerting a drag on the yield growth of major crops (Lobell et al. 2011; Wheeler and Von Braun 2013; Lesk et al. 2016). Cereal grain production has declined 9–10% due to droughts and extreme heat events between 1964 and 2007 (Lesk et al. 2016). Crop productivity is expected to decrease further in response to warming and drying trends in large regions of the world that coincide with countries that already experience food insecurity (Wheeler and Von Braun 2013). Future major crop production gains are predicted to fall well below what is needed to meet projected demands by 2050 (Ray et al. 2013). A meta-analysis of crop yields and other impacts due to global climate change from more than 1700 published reports and simulations indicated that average production losses of 5.4% per °C of warming will occur for a range of crops (Challinor et al. 2014). Furthermore, grain yield losses are predicted to accelerate in the second half of the twenty-first century as temperatures rise (Challinor et al. 2014). A recent ensemble modeling study reported that yield loss risks under drought conditions arising from global warming will increase from 9% to 19% for major crops including wheat, maize, rice and soybeans by the end of the twenty-first century (Leng and Hall 2019).

As the effects of global climate change intensify in the coming decades, the availability and access to freshwater resources will become increasingly difficult. Overall,

Abbreviations: BMP – Best management practices; CAM – Crassulacean acid metabolism; C_i – Leaf intracellular CO_2 concentration; EPI – Environmental productivity index; FBA – Flux balance analysis; GHG – Greenhouse gas; Glu – Glucose; g_m – Mesophyll conductance of CO_2; GREET – Greenhouse gases, regulated emissions, and energy use in transportation; HNT – High nighttime temperature; IAS – Intracellular air space; LCA – Life cycle assessment; LCCA – Life cycle costing assessment; LCI – Life cycle inventories; MDH – Malate dehydrogenase; ME – Malic enzyme; NPP – Net primary production; PAR – Photosynthetically active radiation; PEP – Phosphoenolpyruvate; PEPC – Phosphoenolpyruvate carboxylase; PEPCK – PEP carboxykinase; PPFD – Photosynthetic photon flux densities; PS II – Photosystem II; Rubisco – Ribulose 1,5-bisphosphate carboxylase oxygenase; RUE – Radiation-use efficiency; SLN – Surface leaf nitrogen; Suc – Sucrose; VPD – Vapor-pressure deficits; WUE – Water-use efficiency agricultural production will need to increase by more than 60% to meet the global food demands of a growing human population (Tilman et al. 2011; OECD-FAO 2013), which is expected to reach ~9.6 billion by 2050 (United Nations 2007). In addition to rain-fed crop production, irrigated agriculture contributes to 40% of global crop production, accounts for 20% of total cultivated land area, and results in approximately 70% of global water withdrawals (FAO 2017). However, aquifers in the U.S. (Scanlon et al. 2012; Famiglietti and Rodell 2013) and globally (Moiwo et al. 2012; Voss et al. 2013) are being overdrawn at an alarming rate (United Nations 2007; Gerbens-Leenes et al. 2009). Thus, global water insecurity is not only a looming crisis in agricultural production systems, but also for all uses with approximately 4 billion people experiencing moderate to severe water scarcity at least 1 month per year (Mekeonnen and Hoekstra, 2016).

In a warming world, water availability will become a critically important factor for global agricultural productivity (Leng and Hall 2019). Evidence-based research has linked anthropogenic climate change to the co-occurrence of ground water depletion (Cuthber et al. 2019) and reduced and earlier annual snow pack melt (Bormann et al. 2018). Contemporary estimates predict that the magnitude of droughts will likely double over 30% of the global landmass even with stringent mitigation policies and increase water-demand deficits five-fold within the twenty-first century (Naumann et al. 2018). For each 1 °C of additional warming, an estimated 7% of the global population (~525 million people) could experience a 20% or greater decrease in renewable water resources (FAO 2017). Increased droughts will likely correspond to a 4%, 8%, and 10% increase in the global population exposed to increased water scarcity under 1.5 °C, 2 °C, and 3 °C of global warming, respectively (Koutroulis et al. 2019). These studies all point to a looming global water crisis that can be addressed, in part, through the development of more WUE crops that use less water or use limited water resources more sparingly.

B. *Climate Resilient Water-Use Efficient Crops*

Climate-resilient crops are needed to meet the future food and bioenergy demands of a growing human population. More robust agricultural and forestry sectors could also address the need to increase mitigation and sequestration of CO_2 emissions to limit global temperature rise (Roe et al. 2019). Water availability is the major factor that constrains the cultivation of food and bioenergy crops as these activities consume approximately 80% of water resources (Velasco-Muñoz et al. 2018). To meet the food and feed demands of a growing human population, agricultural water consumption is estimated to increase 70–90% by 2050, depending on actual population growth and changes in dietary preferences (De Fraiture and Wichelns 2010). Urban water demand is expected to increase 80% by 2050 further threatening limited freshwater resources (Flörke et al. 2018). Increasing conventional bioenergy crop production in an attempt to mitigate greenhouse gas (GHG) emissions would use 7% more available water resources in the U.S. (King et al. 2010) and 5% globally by 2030 (Gerbens-Leenes et al. 2012; Gerbens-Leenes and Hoekstra 2012). Unfortunately, using traditional bioenergy crops to mitigate GHG emissions is predicted to increase water deficits more than if no mitigation effort were attempted (Hejazi et al. 2015). Traditional bioenergy crops, such as sugar cane and maize, carry large water footprints for the production of sweeteners and bioethanol (Gerbens-Leenes et al. 2009; Gerbens-Leenes and Hoekstra 2012). Thus, there are clear and present needs to diversify our food and biomass feedstock portfolio and adopt food and bioenergy crops that will be resilient to global warming and use limited water resources more efficiently.

Several options exist to improve WUE and access to water resources associated with agricultural production systems. One option is to increase the efficiency of irrigation systems through technological innovation and improved irrigation practices, such as shifting from flood irrigation to mini-sprinkle, trickle, sub-surface drip, and regulated deficit irrigation methods to reduce water demand (Levidow et al. 2014). Another useful option is to expand the use of brackish or saline water to grow halophytes in arid regions with limited freshwater resources (Cheeseman 2016). A third option is to expand the development and use of agriculturally important CAM crop species, which possess extremely high water-use efficiencies and would use far less water than traditional C_3 and C_4 crops (Davis et al. 2011, 2017, 2019; Cushman et al. 2015; Mason et al. 2015; Yang et al. 2015a, b; Owen et al. 2016). Lastly, an alternative option is to explore the potential for engineering CAM into C_3 and C_4 photosynthetic species as a means of improving plant WUE (Borland et al. 2014; DePaoli et al. 2014; Borland et al. 2015; Yang et al. 2015a, b; Lim et al. 2019).

II. Crassulacean Acid Metabolism

CAM is an elaboration of C_3 photosynthesis that improves WUE. Water-saving is achieved by temporal shifting of primary CO_2 acquisition from the day to the night when air:leaf water VPDs are usually reduced relative to the day along with corresponding "inverted" stomatal behaviors (Griffiths 1989a, b). CO_2 uptake and fixation is aligned with the use of two separated carbon fixing pathways: (1) primary nighttime carbon fixation by phospho*enol*pyruvate (PEP) carboxylase (PEPC) using HCO_3^- (Phase I) and (2) secondary daytime (re)fixation by ribulose-1, 5-bisphosphate carboxylase/oxygenase (Rubisco) using CO_2 (Phase III) (Osmond 1978). Interestingly, RNAi-mediated knock-down of the CAM PEPC isogene (PPC1) in *Kalanchoe laxiflora* resulted in nearly completed disruption of dark period CO_2 fixation and stomatal conductance and altered the temporal phasing of expression of genes involved in controlling stomatal movements demonstrating that inverse stomatal behavior is dependent upon the activity of the primary carboxylation reaction (Boxall et al. 2020). During phase I, the oxaloacetate (OAA) formed by PEPC is then converted to malate by NAD(P)-malate dehydrogenase (NAD(P)-MDH). In strongly obligate CAM plants, only these two phases occur. However, in other CAM plants, the two phases are separated by a transitional phase at dawn from nighttime carboxylation to daytime carboxylation (Phase II) and a second transitional phase near dusk from the daytime carboxylation to nighttime carboxylation (Phase IV) (Osmond 1978). Phase I and III are linked by the nocturnal accumulation and storage of organic acid storage intermediates, such as malate in the vacuole, and the reciprocal, daytime accumulation of storage carbohydrates (such as starch).

A. *Metabolic Plasticity of CAM*

In most species within families of CAM species (e.g., Agavaceae, Aizoaceae, Cactaceae, Crassulaceae, Nolindoideae, and Orchidaceae), the malate that accumulates overnight is subsequently decarboxylated during the day by NAD(P)-malic enzyme (ME) to release pyruvate and CO_2 with pyruvate being regenerated to PEP by pyruvate orthophosphate dikinase (PPDK) to fuel carbohydrate synthesis via gluconeogenesis (Osmond 1978; Christopher and Holtum 1996; Winter and Smith 1996). However, in most species of other CAM families (e.g., Apocynaceae, Asphodelaceae, Bromeliaceae, Clusiaceae, Euphorbiaceae, and Liliaceae), PEP carboxykinase (PEPCK) serves as the predominant decarboxylase, wherein NAD(P)-malate dehydrogenase (MDH) converts malate to oxaloacetate, which is then decarboxylated by PEPCK to release CO_2 and regenerate PEP (Osmond 1978; Christopher and Holtum 1996; Winter and Smith 1996). However, designations of familial enzyme types are generalizations and exceptions are either known to exist or are likely to be discovered. The CO_2 that is (re)fixed by chloroplastic Rubisco (Phase III) leads to carbohydrate production *via* the C_3 Calvin–Benson cycle. This intracellular release of CO_2 in the vicinity of Rubisco can elevate CO_2 concentrations within the leaf 2- to 60-fold higher than atmospheric CO_2 levels (Lüttge 2002). The result of this 'CO_2 pump' is to promote Rubisco's carboxylase activity over its oxygenase activity, which reduces photorespiration, which can reduce the efficiency of photosynthesis up to 40% in C_3 plants (Ehleringer and Monson 1993). CAM can result in a 3- to 6-fold improvement in WUE compared with C_4 and C_3 photosynthesis species, respectively. Nocturnal net CO_2 fixation can also enhance the duration and overall magnitude of net CO_2 uptake over a 24-h cycle in resource-limited environments (Winter et al. 2005; Borland et al. 2009, 2011a, b).

In addition to the two major decarboxylation systems, CAM species can be categorized by the types of storage carbohydrates used to supply substrates for nocturnal C_4 acid accumulation. For example, for CAM eudicot species within the Apocynaceae, Aizoaceae, Cactaceae, Crassulaceae, Cuburbitaceae, and Orchidaceae, starch accumulation in the chloroplast serves as the primary daytime storage carbohydrate, whereas species within the Clusiaceae accumulate soluble sugars (Christopher and Holtum 1998). However, designations of familial carbohydrate storage forms are generalizations and exceptions are either known to exist or are likely to be discovered. For example, monocot CAM species within the Bromeliaceae accumulate either starch or soluble sugars, whereas species within the Agavaceae accumulate fructans (Christopher and Holtum 1996, 1998). Still others, such as species within the Asphodelaceae accumulate either starch/glucose or glucomannan, whereas species within the Nolindoideae accumulate sucrose (Christopher and Holtum 1996). This diversification is thought to reflect the evolutionary histories of these families. The selection of storage carbohydrate systems might also be driven by the energetic requirements of the specific environments in which each family evolved (Shameer et al. 2018).

B. *Flexible Engagement of CAM*

CAM plants also show great flexibility in the extent to which they engage in net nocturnal CO_2 uptake and fixation over any given 24-h period (Winter et al. 2015; Winter 2019). The degree of net nocturnal CO_2 fixation extends along a continuum from 0% to 100% with obvious bimodal distributions depending on variations in stomata behavioral dynamics and daytime accumulation of storage carbohydrates available to supply C_3 substrate for nocturnal CO_2 fixation (Winter et al. 2015; Edwards 2019). Proximate environmental conditions (mainly availability of water supply) govern the relative degree to which CAM is engaged. CAM engagement can also be reversed under conditions of adequate irrigation or rainfall. The extent to which CAM is engaged can be measured by the classic carbon-isotope ratio $\delta^{13}C$ (a measure of the ratio of ^{13}C:^{12}C isotopes), which is based on the observation that PEPC discriminates less against ^{13}C than does Rubisco. Thus, CAM plants can be readily distinguished from C_3 photosynthesis species, which have $\delta^{13}C$ values in the range of −33 to −22‰, from CAM species, which have less negative $\delta^{13}C$ values in the range of −20‰ to −8‰ (Winter et al. 2015). Large isotopic surveys of the Bromeliaceae, Crassulaceae, Euphorbiaceae, and Orchidaceae with mixed distributions of both C_3 photosynthesis and CAM species have shown striking bimodal distribution of values with most species showing C_3 photosynthesis or strong CAM values (Teeri et al. 1981; Crayn et al. 2004, 2015; Silvera et al. 2009, 2010a, b; Horn et al. 2014; Winter et al. 2015; Edwards 2019). A minimum frequency distribution of values occurs around −20‰. This isotopic ratio value represents an approximately 40% contribution of net nocturnal CO_2 fixation to total daily carbon gain (Winter and Holtum 2002). CAM species with >70% CO_2 uptake during the night are considered strong CAM, whereas species with less than 33% nocturnal CO_2 fixation are considered weak CAM (Winter and Holtum 2002).

Discrete modes of CAM engagement along the CAM spectrum have been described. For example, CAM cycling is used to designate the ability to detect organic acid accumulation in the absence of any detectable net nocturnal CO_2 fixation (Sipes and Ting 1985). While CAM cycling does not contribute directly to net nocturnal CO_2 uptake, it is thought to maintain a positive carbon balance by reducing CO_2 loss through the recapture of respired CO_2 at night, thereby promoting plant growth and reproductive fitness (Herrera 2008). Weak CAM (or C_3-CAM) is used to describe detectable nocturnal organic acid accumulation reflected in $\delta^{13}C$ values that range from −26‰ to −20‰ (Winter and Holtum 2002). Weak CAM is thought to contribute to overall carbon balance, thereby enabling plant fitness during periods of water-deficit stress (Pikart et al. 2018). Strong or obligate CAM is used to describe prominent nocturnal organic acid accumulation reflected in $\delta^{13}C$ values that range from −20‰ to −8‰. Between the two distribution maxima for C_3 photosynthesis and CAM species lies a transitional zone referred to as the Winter–Holtum zone (Winter et al. 2015; Males 2018) encompassing $\delta^{13}C$ values that range from −23 to −19‰ (Winter et al. 2015). Within this transitional and weak CAM zone, species that display facultative or inducible CAM are often found. CAM induction can occur in many different species in response to water-deficit stress or related treatments such as exposure to high salinity or high light intensity (Herrera et al. 1991; Winter and Holtum 2005, 2011a, b, 2017; Cushman et al. 2008; Heyduk et al. 2015, 2018; Brilhaus et al. 2016). For example, the defining features of facultative CAM have been described for a number of terrestrial CAM species including *Calandrinia polyandra*, *Clusia pratensis*, *Mesembryanthemum crystallinum*, *Portulaca oleracea*, and *Talinum triangulare* (Winter and Holtum 2014). Alternatively, CAM can be induced by exposure to abscisic acid (Taybi and Cushman 1999, 2002; Minardi et al. 2014; Brilhaus et al. 2016) and can be reversed following the return to well-watered conditions as observed in *Mesembryanthemum crystallinum, Portulaca oleracea*, and *Talinum triangulare* (Winter and Holtum 2011a, b, 2014). CAM can also be under developmental control in some species (Winter et al. 2008, 2015). Lastly, CAM idling is used to describe a low-level of CAM that occurs under extreme water-deficit stress conditions, which is characterized by sustained diel fluctuations in organic acids behind closed stomata wherein essentially all CO_2 fixed into organic acids is derived from internally recycled respiratory CO_2 (Szarek et al. 1973).

C. *Habitat Diversity of CAM Plants*

While many CAM species occupy seasonally warm and dry terrestrial or epiphytic habitats (Niechayev et al. 2019a, b), one exception to this tendency was the discovery of CAM in aquatic species. Members of the submerged aquatic macrophytes within the *Isoetes* genus are known to perform CAM (Keeley 1981, 1996, 1998). CAM likely aids in improving the uptake of limited inorganic carbon availability while the plants are submerged (Keeley and Rundel 2003). In contrast to terrestrial plants that switch from C_3 to CAM under conditions of water-deficit stress, the aquatic species, *I. howelii* and

Crassula aquatica switch from CAM to C_3 when water levels subside exposing the leaves to the atmosphere (Keeley 1981; Keeley and Sandquist 1991). When leaves are submerged, CAM can contribute up to 40% of total carbon gain via passive CO_2 diffusion in the water, whereas when exposed, CAM contributes to <1% of total carbon gain (Keeley and Sandquist 1991; Keeley 1998). The switch from CAM to C_3 photosynthesis is partially reversible and is thought to be a response to the increased CO_2 availability under non-submerged conditions, which would eliminate the ecological advantage of nocturnal CO_2 fixation (Keeley 1983, 1998). However, the loss of CAM can be prevented under conditions of high humidity, suggesting that water status of the exposed tissues might also influence the return of C_3 photosynthesis (Aulio 1986; Keeley 1998).

Another exception to the propensity of CAM species to occupy warm and dry climates was the discovery of some CAM species (e.g., *Opuntia fragilis*) that are able to survive seasonally freezing temperatures as low as −50 °C when cold acclimated (Ishikawa and Gusta 1996; Gorelick et al. 2015). Cold acclimation and survival appears to be associated with substantial water loss from cladodes (Loik and Nobel 1993; Ishikawa and Gusta 1996; Cota-Sánchez 2002; Gorelick et al. 2015) and prostrate growth to ensure snow coverage, which ameliorates low temperatures and avoids freeze-thaw cycles that can contribute to plant death (Cota-Sánchez 2002; Gorelick et al. 2015). Freezing tolerance also likely involves mechanisms that slow down and limit dehydration and prevent intracellular and extracellular ice formation such as the accumulation of cryoprotectant sugars, sugar alcohols (e.g., mannitol, glycerol), and extracellular polysaccharides (Goldstein and Nobel 1991, 1994). The accumulation of cold-shock proteins might also aid in adaptation to colder temperatures in less cold-tolerant species such as *O. ficus-indica*, which are able to survive to about −9 °C (Nobel 1988). In general, most large agronomically important *Opuntia* spp. used for fruit, vegetable, or fodder production are susceptible to freezing temperatures (−9 to 11 °C), but there are exceptions (Russell and Felker 1987; Valdez-Cepeda et al. 2002). The identification of cold-tolerant varieties would be useful for expanding the production ranges in more northerly latitudes as tropical and subtropical regions become too warm for successful commercial cultivation.

III. Co-Adaptive Traits of CAM Plants

In addition to the intrinsic water-use efficiency afforded by CAM itself, CAM is often accompanied by a suite of morphological adaptations, such as tissue succulence and water-storage and water-capture strategies, that also help to attenuate drought and allow these plants to survive and thrive in hot and dry habitats. Other associated traits include thickened cuticles and epicuticular waxes, reduced stomatal guard cell density, increased stomatal dynamics or responsiveness, high-light and UV-light protection, and shallow rectifier-like roots, which provide adaptive advantages to the increasingly dry and hot conditions associated with global warming.

A. *Tissue Succulence*

Tissue succulence is defined by the ability of a plant to store sufficient water within its living tissues and is an important adaptation that permits plants to grow temporarily without access to an external water supply (Eggli and Nyffeler 2009). Tissue succulence is present in 3–5% of flowering plants including C_3 and C_4 photosynthesis and CAM species (Eggli and Nyffeler 2009; Griffiths and Males 2017). Succulent plants avoid the damaging effects of drought often associated with epiphytic, epilithic, semi-arid, and arid environments and display remarkable taxonomic diversity (Grace 2019). For example, the mucilaginous cells in the fruits and cladodes of cactus have the capacity to imbibe and retain water (Saag et al. 1975). Highly succulent species not only slow water loss during times of prolonged drought, but also possess three-dimensional venation patterns with moderate hydraulic path lengths, which allow for rapid water uptake when water becomes available again (Table 10.1) (Griffiths 2013; Ogburn and Edwards 2013). While succulence is not unique to CAM species, some degree of succulence is typically correlated with strong CAM, but less so in weak (C_3-CAM) CAM species (de Santo et al. 1983; Silvera et al. 2005, 2010a, b; Winter et al. 2015; Griffiths and Males 2017; Males 2018). For example, columnar stem succulents are typically obligate CAM species and are well known for their ability to survive under extremely hot and dry conditions for extended time periods (Williams et al. 2014; Hultine et al. 2019). CAM plants

Table 10.1. Co-adaptive traits in crassulacean acid metabolism (CAM) species conferring climate resilience

Trait	Description	Major benefit	Citations
Tissue succulence	Water storage within metabolically active tissues	Drought attenuation, rapid water uptake, high thermal capacity, biomass production, salinity tolerance	Eggli and Nyffeler (2009)
Water-capture organs	Dense epidermal trichomes	Water capture and absorption	Males (2016)
Water-storage organs	Bromeliad tanks, orchid pseudobulbs	Drought attenuation, nutrient absorption	Rodrigues et al. (2013), Yang et al. (2016), Males (2016), Freschi et al. (2010)
Thickened cuticle	Barrier against water loss, protective shield against high-light and UV-light	Reduced water loss, reflection and attenuation of UV-light, heat reduction	Yeats and Rose (2013), Yang et al. (2016), Fernández et al. (2017), Heredia-Guerrero et al. (2018)
Epicuticular waxes	Barrier against water loss, non-wettable (hydrophobic) and self-cleaning surface	Protection against microbial and insect pathogens, channel rainfall to roots	Rykaczewski et al. (2016), Bargel et al. (2006), Domínguez et al. (2011)
Epidermal trichomes, root velamen	Contain light-absorbing pigments to limit light damage	High-light and UV-light reflection and protection, limit water loss, facilitate water re-uptake	Chomicki et al., 2015
Reduced stomatal density	Lower density of stomata resulting in lower stomatal conductance	Reduce water loss due to evapotranspiration	Barrera-Zambrano et al. (2014), Yang et al. (2016), Males and Griffiths (2017)
Enhanced stomatal responsiveness	Rapid response to environmental signals and ABA; less responsiveness to light signals	Greater water-use efficiency	Osmond (1978), Griffiths et al. (2007), von Caemmerer and Griffiths (2009), Males and Griffiths (2017)
Rectifier-like roots	Rapid loss of hydraulic conductivity following drought; rapid increase in hydraulic conductivity following rewatering	Limit water loss to dry soils; maximize water uptake after rainfall events	North and Nobel (1991), Nobel and Cui (1992), Graham and Nobel (1999)

typically have larger cell size with low surface area to volume (SA:V) ratios to accommodate nocturnal vacuolar storage of organic acids (mainly malate) necessary for the performance of CAM. However, malate accumulation might also assist in osmotically driven water uptake and remobilization within the chlorenchyma (Lüttge and Nobel 1984; Smith et al. 1987).

CAM species display a wide variety of anatomical configurations associated with tissue succulence (Griffiths and Males 2017; Males 2017). Leaf and stem anatomies can consist of 'all-cell succulents', which are defined by water storage as part of an expanded chlorenchyma. Alternatively, 'storage succulents' have a centralized water-storing hydrenchyma accompanied by well-defined zones of chlorenchyma. Various configurations of these two anatomies can also exist. The enlarged cells reduce intracellular air space (IAS). The enlarged cells also reduce the mesophyll cell surface area that is exposed to the IAS, which results in a reduction in mesophyll conductance of CO_2 (g_m) and a decrease in CO_2 efflux from the leaf, which in turn improves overall carbon capture and CAM performance (Nelson et al. 2005; Nelson and Sage 2008). However, while reduced g_m is advantageous for the performance of CAM, restricted access to CO_2 can restrict the performance of C_3 photosynthesis (Nelson and Sage 2008). This functional trade off might explain why large $\delta^{13}C$ surveys of species within genera performing both C_3 photosynthesis and CAM photosynthesis have revealed bimodal distributions centered around −20‰ (Winter et al. 2015). Examination of diverse *Clusia* species with varying degrees of succulence have shown that some leaf anatomical configurations with enlarged and densely packed palisade mesophyll cells, as well as large vacuoles, can accommodate greater nocturnal C_4 carboxylation and organic acid storage capacity than less succulent species (Barrera-Zambrano et al. 2014). This study suggested that a balance between relatively well-aerated spongy mesophyll cells and enlarged palisade mesophyll cells might be considered an optimal leaf anatomy for C_3-CAM intermediates or bioengineered CAM.

In addition to providing protection against intermittent drought and slowing diffusion of CO_2 for efficient fixation into C_4 acids, succulent tissues typically display high thermal capacity allowing them to tolerate high tissue temperatures during the day (Griffiths and Males 2017). This trait also slows the rate of radiative heat loss during the night to allow the plants to better protect sensitive tissues against chilling or freezing damage that might occur during cold nights (Griffiths and Males 2017; Males 2017).

B. *Water Capture and Storage Strategies*

In addition to succulence, many CAM species, particularly those occupying epiphytic or epilithic habitats have developed specialized anatomical features to facilitate the capture and storage of water. Perhaps the best example of water capture is present in the 'atmospheric' or 'fog-catching' nebulophytic CAM epiphytes within the *Tillandsia* genus (Males 2016). These species possess dense epidermal trichomes to maximize capture and absorb transiently available water (Table 10.1). In addition to capturing water, this dense trichome layer can also reduce CO_2 exchange. Thus, such species often perform CAM to recycle CO_2 from respiration and to increase CO_2 uptake at night, as well as improving WUE during the dry season (Pierce et al. 2002).

Species within the Orchidaceae, particularly within the Oncidiinae (Neubig et al. 2012), have evolved specialized water storage and conservation organs called 'pseudobulbs' (Table 10.1). Pseudobulbs are enlarged stem segments that lack stomata and are comprised of large, tightly packed chlorenchyma cells and aerenchyma tissue capable of storing water as well as performing weak CAM for recycling respiratory CO_2 under water-deficit stress conditions (Rodrigues et al. 2013; Yang et al. 2016). In addition to lacking stomata, pseudobulbs have a lignified exodermis to reduce water loss. Orchid roots can be encased in a spongy epidermal layer called a 'velamen', which not only helps to limit water loss during periods of drought, but can also accumulate water and minerals within internal spaces to facilitate reuptake (Chomicki et al. 2015).

Some species in the Bromeliaceae family have evolved tank structures for the storage and conservation of water as well as nutrients (Table 10.1) (Males 2016). Tank structures are comprised of overlapping basal portions of the leaves, where trichomes form a hydrophobic liner for retaining water within the tank. Nutrients are absorbed by epidermal trichomes present at a higher density towards the base compared with the apical portion (Freschi et al. 2010). In CAM facultative epiphytic tank bromeliad species, such as *Guzmania monostachia*, the base of the leaves that form the tank contains more hydrenchymal cells to buffer the water status in the more apical regions of the leaves. In contrast, the apical portions of leaves have a greater stomatal density and exhibit stronger CAM activity when plants are exposed to water-deficit stress (Freschi et al. 2010).

C. *Thickened Cuticles and Epicuticular Waxes*

Plant cuticles act as a key interface between the plant and its environment and protect the plant against microbial pathogens, insect pests, xenobiotics, mechanical damage, UV radiation, and water loss (Yeats and Rose 2013; Fernández et al. 2017; Heredia-Guerrero et al. 2018). In many CAM species, cuticular thickness has been associated with reduced water loss (Table 10.1). For example, a study of four *Dendrobium* species showed that species with a thicker upper leaf cuticle (*D. chrysotoxum* (C_3 species) and *D. officinale* (facultative CAM species)) showed lower rates of water loss than two C_3 species (*D. chrysanthum* and *D. crystallinum*) with thinner leaf cuticles (Yang et al. 2016). In fact, the CAM species had the thickest cuticle, which presumably would limit water loss under dry conditions. Interestingly, the C_3 species with thinner cuticles also had pseudobulbs, which would be expected to partially offset the more rapid water loss in these species (Yang et al. 2016).

Many CAM species have very thick cuticles and extensive epicuticular wax deposition compared with C_3 and C_4 photosynthesis species, which reduce water loss (Ranjan et al., 2016). For example, the cuticles of the obligate CAM species, *O. ficus-indica,* is hundreds of times thicker than those found in *Arabidopsis thaliana* (North et al. 1995). Thicker cuticles along with wax accumulation provide a major barrier against water loss in semi-arid and arid environments, with the intra-cuticular wax (not the epicuticular wax layer) providing the more important barrier against transpirational water loss (Zeisler-Diehl et al. 2018).

In addition to limiting water loss, the cuticle and epicuticular wax accumulation perform other protective functions (Table 10.1). For example, epicuticular surfaces are typically highly hydrophobic, non-wetting and self-cleaning, which protects against pathogens and insect predation (Bargel et al. 2006). An excellent illustration of extreme non-wettability is the surface of juvenile *Opuntia* pads, which allows water to transit quickly from the juvenile pads to mature base pads, which are more wettable, thereby facilitating the movement of rainfall to the older pads and roots for absorption (Rykaczewski et al. 2016). Key functions of the cuticular layer are protection against UV-irradiation and heat dissipation. Many CAM species occupy semi-arid and arid environments with high insolation rates. A thick cuticle can effectively absorb short-wave radiation to limit heating of the leaf surface in desert-adapted *Agave* species (Nobel and Smith 1983). The production and accumulation of flavonoids in the cuticle or epicuticular wax layers help to reflect and attenuate UV-light (Bargel et al. 2006; Domínguez et al. 2011). In addition, polymeric components in the cuticle have a higher specific heat relative to other plant polymers, which helps to limit the heating of the aerial surfaces of leaves and stems thereby allowing plants to occupy hotter and drier environments (Domínguez et al. 2011).

Many CAM species currently occupy terrestrial environments that resemble the hotter and drier climate of the future. Thus, CAM species arguably have cuticular and epicuticular characteristics that will likely reflect future changes expected in plant species that currently occupy less harsh climates. Increased heating and drying arising from global warming will likely change the components of the cuticle including increasing cutin and polysaccharides, wax loading, and alter the wax composition via the accumulation of longer aliphatic compounds (Heredia-Guerrero et al. 2018). These compositional changes are predicted to increase the stiffness, thickness, and hydrophobicity, while decreasing the permeability of plant cuticles (Heredia-Guerrero et al. 2018). Increasing the rigidity of the cuticle during growth might curtail leaf or stem expansion and limit yield.

D. Reduced Stomatal Density

Although surveys are limited, CAM species occupying xerophytic and epiphytic habitats typically display lower rates of stomatal conductance due in part to lower stomatal densities (Table 10.1) (Sayed 1998; Barrera-Zambrano et al. 2014; Yang et al. 2016; Males and Griffiths 2017). In *Clusia* spp. thicker leaves were correlated with larger stomata and lower density relative to species with thinner leaves (Barrera-Zambrano et al. 2014). Fewer, larger stomata might help decrease stomatal metabolic machinery costs and lower guard cell respiration rates (Franks et al. 2009). However, quantitative data on the true metabolic costs of guard cell respiration are limited and additional investigations are warranted. This adaptation might be useful for plants that occupy water-limited environments especially when the additional energetic costs of performing CAM are considered (Borland et al. 2018; Shameer et al. 2018). Given the presumed high energetic costs of stomatal movement, a lower density would decrease the cost for CAM plants when compared to plants that have higher stomatal density. In addition, CAM requires large cells with large vacuoles to store C_4 acids overnight, so low stomatal density may serve as a favorable trade-off for these plants (Males and Griffiths 2017). Several CAM plant families, such as the Agavaceae, also have deeply sunken stomata, which is thought to reduce water vapor loss (Davis et al. 2011).

E. Enhanced Stomatal Responsiveness

Stomatal behavior is driven by various environmental stimuli and internal signals and sensitivity, and responses to such cues can vary widely among species (Lawson and Blatt 2014). In most plants, stomatal opening is stimulated by light, low [CO_2], high temperatures, and low VPD, whereas stomatal closing is driven by no or low light, high [CO_2], and high VPD (Lawson 2008). Water-use efficiency gains might be attainable by exploiting the natural variation in stomatal responsiveness (Lawson et al. 2012; Lawson and Blatt 2014; Faralli et al. 2019). In CAM species, stomata are not responsive to white- or blue-light-activated opening (Lee and Assmann 1992; Tallman et al. 1997). Key blue-light photoreceptors (e.g., *CRY1*, *PHOT1*) might not be involved in stomatal opening in *Agave americana*, an obligate CAM species (Abraham et al. 2016). However, nocturnal stomata opening is controlled by a decline in leaf intracellular [CO_2] (C_i) driven by PEPC activation and malate production (Griffiths et al. 2007; von Caemmerer and Griffiths 2009; Males and Griffiths 2017). Nocturnal stomatal conductance is also controlled by the presence of an active and functional nighttime PEPC (Boxall et al. 2020). Recent evidence suggests that lower C_i appears to drive stomatal opening along with nocturnal malate accumulation, whereas increased C_i appeared to drive daytime stomatal closing accompanied by malate depletion (Lim et al. 2019). CAM plant guard cells are thought to be more sensitive to changes in ABA and leaf-air vapor pressure differences than C_3 photosynthesis plants (Osmond 1978). This hyper- responsiveness is consistent with the supposition that succulent plants tend to attenuate water-deficit stress by avoiding water loss (Males and Griffiths 2017). Lastly, stomatal responsiveness in CAM plants is modulated by the ideal nighttime temperature range (15–25 °C) within which stomatal open, which predictably corresponds with the optimal performance range of CAM (Yamori et al. 2014).

F. High-Light and UV-Light Protection

CAM plants typically occupy hot, sunny regions and have developed several adaptations to protect against high light and UV-light damage (Luttge 2000; Borland et al. 2009). The presence of spines and dense trichomes reduce transpiration by lowering surface area and play a role in reflecting light in many cactus species (Nobel 1978, 1994; Rebman and Pinkava 2001; Ranjan et al. 2016). The orientation of barrel, columnar, and platyopuntia cacti with respect to the sun can be important for reducing surface area that is exposed to direct sunlight and avoid high tissue temperature often associated with high light desert environments (Nobel 1978; Rebman and Pinkava 2001; Ranjan et al. 2016).

High light stress is thought to act on CAM photosynthesis through increasing VPD (Lüttge 2000); however, direct effects have been documented. For example, well-watered *Clusia minor* (Clusiaceae) subjected to high light treatment showed increased malate and citrate accumulation (Franco et al. 1992). High light exposure was observed to increase CAM in the facultative CAM species *G. monostachia* (Bromeliaceae) (Maxwell et al. 1997). The high light exposed plants also rapidly acclimated to a high light regime by increasing violaxanthin, antheraxanthin, and zeaxanthin levels as well as by the extent of zeaxanthin conversion, which served to down-regulate PS II and prevent long-term damage. In this instance, CAM induction might alleviate photoinhibition by maintaining internal CO_2 through refixation of respiratory CO_2 (Maxwell et al. 1994). In *C. minor*, a facultative CAM species, UV-A/blue light was the most effective light range for CAM induction (Grams and Thiel 2002). In its natural habitat, *C. minor* is subject to high light and water shortage, which makes CAM an important mechanism to cope with excess radiational energy and to avoid photoinhibition (Grams and Thiel 2002). Exposure of *Rosularia elymaitica* (Crassulaceae) plants to UV-light treatment resulted in increased nocturnal organic acid accumulation, but not significantly higher CAM enzyme activities when compared to the control treatments (Habibi and Hajiboland 2011). However, reactive oxygen scavenging enzyme activities (e.g., ascorbate peroxidase (APX) and catalase (CAT)) increased under UV radiation compared to the control (Habibi and Hajiboland 2011). Mechanistically, greater CAM activity is thought to alleviate the oxidative burden in the chloroplast, mitochondrial, and cytosolic compartments (Borland et al. 2006). This observation was consistent with multiple reports in CAM species that daytime malate decarboxylation leading to CO_2 release and recycling of respiratory CO_2 can protect against PS II photoinhibition (Adams and Osmond 1988; Griffiths 1989a, b; Griffiths et al. 1989).

The thickened cuticle and epicuticular wax layer of CAM species contain a variety of sunscreen protectants to reflect or absorb light. For example, high light induces the accumulation of a thick layer of epicuticular wax that reflects light in *Cotyledon orbiculata* (Crassulaceae), a South African obligate CAM species (Robinson and Osmond 1994; Barker et al. 1997). This species also accumulates xanthophyll cycle pigments, lutein, and β-carotene in young leaves to dissipate excess light energy and anthocyanins in the upper and lower epidermis to reflect excess incident light (Barker et al. 1997). High solar irradiance treatment (containing UV-A and UV-B components) of *Sedum album* (Crassulaceae), a CAM cycling species, resulted in the accumulation of anthocyanins (e.g., cyanidine-3-glucose) (Bachereau et al. 1998). While not specific to CAM species, leaf trichomes reflect light and reduce its absorption under high photosynthetic photon flux densities (PPFD), and protect against the damaging effects of UV-B radiation via the accumulation of flavonoids within trichomes and the leaf cuticle (Bickford 2016). Lastly, epiphytic orchid roots are protected from UV radiation damage by the presence of UV-B-absorbing flavonoids in the velamen (Chomicki et al. 2015).

G. *Rectifier-Like Roots*

CAM species, such as *Agave deserti* (Asparagaceae), occupy semi-arid regions with limited seasonal rainfall. To take advantage of such limited and unpredictable precipitation, these species have shallow, rectifier-like root systems that maximize water uptake after rainfall events, while preventing water loss during periods of drought (Nobel and Sanderson 1984; North and Nobel 1991; Nobel and Cui 1992). Lateral or 'rain' and nodal roots of *Agave deserti* exhibit rapid declines in hydraulic conductivity following episodes of drought to curtail water loss to the dry soil, while young nodal roots will recover 50% of their hydraulic conductivity after just 2 days of rewatering (North and Nobel 1991). Curtailment of water loss can be extremely rapid with a ten-fold decrease in radial water flow after just 3–6 h of drying and a 200-fold decrease after 72 h (Nobel and Sanderson 1984). After 4 days of rewetting, both water uptake and new root growth contributed equally to water uptake (Nobel and Sanderson 1984). After 7 days of rehydration of the entire root system, the leaf storage capacity of the plant was restored to pre-drought levels (Graham and Nobel 1999). Similar capture of limited, wet-season rainfall occurs in shallow-rooted members of the Aizoaceae and Crassulaceae in the Namibian desert (February et al. 2013). These shallow rooting systems maximize water uptake, while tissue succulence permits water to be stored, thus extending plant survival and reproduction well into the dry season. These adaptations, coupled with the performance of facultative CAM, minimizes competition with non-Aizoaceae species and partially explains the spectacular diversification of these species (February et al. 2013).

IV. Environmental Effects on CAM Photosynthesis

Global warming and associated changes in climate are driven by increasing greenhouse gas emissions and temperatures. Although the responses to these environmental perturbations have been well studied in C_3 and C_4 photosynthesis plants (Kirschbaum 2004), fewer studies have been performed using CAM species (Drennan and Nobel 2000; Poorter and Navas 2003; Ceusters and Borland 2010). On average, CO_2 uptake over the diel cycle increases about 35–40% in response to elevated CO_2 conditions (Ceusters and Borland 2010). However, the effects can differ greatly depending on local conditions, the species in question, anatomical configurations, developmental status, and growth rates of the plants (Poorter and Navas 2003; Ceusters and Borland 2010). In general, the nocturnal CO_2 uptake of CAM species appear to be far from saturating, likely due to the effects of tissue succulence, which is characterized by tight cell packing and low intercellular air spaces, resulting in limited internal CO_2 diffusion and CO_2 concentrations during phase III of CAM when Rubisco is likely to become saturated (Maxwell et al. 1997; Nelson and Sage 2008). Thus, increases in atmospheric CO_2 concentrations are expected to have growth stimulation effects for many CAM species. A survey of nine species revealed that the biomass enrichment ratios (BER, a ratio between total plant mass of high CO_2-grown plants and plants grown at control concentrations) of CAM species was >23%, which was intermediate to the ratios of C_3 and C_4 photosynthesis species (Poorter and Navas 2003). Increasing temperatures result in increasing VPD with a concomitant increase in evapotranspiration rates across the leaf canopy and increased soil drying, which restrict water availability. Under such conditions of water limitation, the high WUEs of CAM species are expected to make them more climate-resilient than C_3 and C_4 photosynthesis plants. However, without significant daytime transpirational cooling, CAM species must be able to buffer daytime heating of leaves or tolerate the heat stress effects more effectively than other plant species. Fortunately, with acclimation, many desert CAM species, such as those within the Agavoideae and Cactaceae, can tolerate tissue temperatures up to 60 °C and beyond (Nobel and Smith 1983; Nobel 1984; Smith et al. 1984; Nobel et al. 1986; Garcia-Moya et al. 2011). Lastly, the WUE of CAM plants helps to maintain their productivity under hot and dry conditions.

A. Responses to Atmospheric CO_2 Enrichment

Several studies have investigated the responses of CAM plants to elevated atmospheric CO_2 concentrations (Drennan and Nobel 2000; Ceusters and Borland 2010). In a majority of CAM species, increased CO_2 leads to increased CO_2 uptake, increased malate accumulation indicating increase nocturnal CO_2 uptake, increased leaf thickness (mainly associated with increased chlorenchyma thickness), and increases in biomass productivity (Drennan and Nobel 2000; Ceusters and Borland 2010). CAM plants averaged an increase of 35% dry biomass after 3 months exposure to elevated CO_2 concentrations of 650–750 μmol mol^{-1} (Drennan and Nobel 2000).

Several *Agave* spp. have been investigated for response to CO_2 enrichment (Table 10.2). For example, CO_2 enrichment from 350 to 650 mmol mol^{-1} for 1 year increased daily CO_2 uptake by 28% resulting in an 28% increase in shoot dry biomass of *Agave deserti* (Nobel and Hartsock 1986a, b). The effects of longer term (17 months) CO_2 enrichment from 370 to 750 mmol mol^{-1} were even more dramatic resulting in an 88% increase in total dry biomass of *Agave deserti* (Asparagaceae) (Graham and Nobel 1996). Treatment of *Agave vilmoriniana* ('octopus' Agave) plants with up to 560–885 mmol mol^{-1} for 6 months increased total dry biomass by 28% (Idso et al. 1986).

Within Bromeliaceae, 4 months of CO_2 enrichment from 330 to 730 mmol mol^{-1} resulted in a 5% increase in leaf area, an 11% increase in leaf thickness, and a 19% increase in dry biomass of *A. comosus* (Bromeliaceae) with the enhancement resulting from nocturnal CO_2 fixation (Zhu et al. 1997). However, an earlier study reported negative growth for *A. comosus* as a result of atmospheric CO_2 enrichment (Ziska et al. 1991). In the terrestrial obligate CAM species *Aechmea magdalenae* (Bromeliaceae), CO_2 enrichment from 354 to 712 mmol mol^{-1} for 3 months resulted in a 43% increase in leaf area and a 36% increase in total dry biomass (Ziska et al. 1991). In contrast to other studies, CO_2 enrichment from 380 to 750 mmol mol^{-1} for 8.5 months resulted in a 41% decrease in leaf area, a 9% decrease in leaf thickness, and a 25% decrease in dry leaf biomass for *Aechmea fasciata* cv. 'Primera'

Table 10.2. Effects of long-term, elevated CO_2 concentrations on crassulacean acid metabolism (CAM) species

Species	CO_2 Treatment	Major effects	Biomass change	Citations
Aechmea fasciata	560 μmol mol^{-1} (280 μmol mol^{-1} control)	Mean 24% increase in relative growth rate grown under low/high nutrients and low/high PFD	–	Monteiro et al. (2009)
Aechmea fasciata cv. 'Primera'	750 μmol mol^{-1} (380 μmol mol^{-1} control)	41% decrease in leaf area; 9% decrease in leaf thickness	25% decrease in leaf dry biomass	Croonenborghs et al. (2009)
Aechmea hybrid cv. 'Maya'	750 μmol mol^{-1} (380 μmol mol^{-1} control)	4% increase in leaf thickness[a]; no change in leaf area	7% decrease in leaf dry biomass	Croonenborghs et al. (2009)
Aechmea magdalenae	712 mmol mol^{-1} (354 mmol mol^{-1} control)	43% increase in leaf area	32% increase in shoot dry biomass, and a 53% increase in root dry biomass	Ziska et al. (1991)
Agave deserti	650 μmol mol^{-1} (350 μmol mol^{-1} control)	28% increase in daytime net CO_2 uptake	28% increase in shoot and 29% root mass (dry weight) over 12 months	Nobel and Hartsock (1986a, b)
Agave deserti	750 μmol mol^{-1} (370 μmol mol^{-1} control)	49% increase in daytime net CO_2 uptake; 110% increase in WUE; 17% longer and 11% thicker leaves	88% increase in total mass (dry weight) over 17 months	Graham and Nobel, (1996)
Agave vilmoriniana	560–885 μmol mol^{-1} (350 μmol mol^{-1} control)	Growth enhancement observed under dry conditions.	28% increase in total mass (dry weight) over 6 months	Idso et al. (1986)
Ananas comosus	712 mmol mol^{-1} (354 mmol mol^{-1} control)	6% decrease in leaf area	9% decrease in shoot and 12% root dry biomass	Ziska et al. (1991)
Ananas comosus	730 mmol mol^{-1} (330 mmol mol^{-1} control)	5% increase in leaf area; 11% increase in leaf thickness	19% increase in dry biomass	Zhu et al. (1997)
Bulbophyllum longissimum	560 μmol mol^{-1} (280 μmol mol^{-1} control)	Mean 19% increase in relative growth rate grown under low/high nutrients and low/high PFD	–	Monteiro et al. (2009)
Ferocactus acanthodes	650 μmol mol^{-1} (350 μmol mol^{-1} control)	30% increase in daytime net CO_2 uptake	30% increase shoot and 28% root mass (dry weight) over 12 months	Nobel and Hartsock (1986a, b)
Hylocereus undatus	740 μmol mol^{-1} (370 μmol mol^{-1} control)	34% increase in total daily net CO_2 uptake	–	Raveh et al. (1995)

(continued)

Table 10.2. (continued)

Species	CO_2 Treatment	Major effects	Biomass change	Citations
Hylocereus undatus	1000 μmol mol^{-1} (380 μmol mol^{-1} control)	52% increase in total daily net CO_2 uptake, 22%, increase in shoot elongation; 175% increase in number of reproductive buds; 7% increase in fresh fruit mass	18% increase in shoot dry biomass over 12 months	Weiss et al. (2010)
Kalanchoë blossfeldiana	700 μmol mol^{-1} (300 μmol mol^{-1} control)	Increased leaf number and flower stem length, and plant height	46% increase in inflorescence fresh weight, and a 36% increase in dry biomass	Mortensen and Moe (1992)
Kalanchoë pinnata	700 μmol mol^{-1} (350 μmol mol^{-1} control)	Increased plant height; 38–42% increase in leaf area	42–51% increase in dry biomass (over 33–72 days)	Winter et al. (1997)
Opuntia ficus-indica	720 μmol mol^{-1} (360 μmol mol^{-1} control)	–	7% increase in daughter cladode dry biomass over 63 days	Luo and Nobel (1993)
Opuntia ficus-indica	720 μmol mol^{-1} (370 μmol mol^{-1} control)	98% increase in daily net CO_2 uptake in basal cladodes; 49% increase in daily net CO_2 uptake in daughter cladodes after 9 weeks	55% increase in shoot dry biomass over 23 weeks	Cui et al. (1993)
Opuntia ficus-indica	750 μmol mol^{-1} (370 μmol mol^{-1} control)	96% increase in daily net CO_2 uptake in daughter cladodes after 8 weeks	32% increase in shoot dry biomass over 12 weeks	Cui and Nobel (1994)
Opuntia ficus-indica	720 μmol mol^{-1} (360 μmol mol^{-1} control)	Cladodes were 11–16% thicker after 12 months	37–40% increase in shoot dry biomass over 12 months	Nobel and Israel (1994)
Opuntia ficus-indica	720 μmol mol^{-1} (370 μmol mol^{-1} control)	30% higher rate of growth after 20 days; cladodes were 27% thicker after 4 months; 31% thicker chlorenchyma	–	North et al. (1995)
Selenicereus megalanthus	1000 μmol mol^{-1} (380 μmol mol^{-1} control)	129% increase in total daily net CO_2 uptake, shoot, 73%, increase in shoot elongation; 233% increase in number of reproductive buds; 63% increase in fresh fruit mass	68% increase in shoot dry biomass over 12 months	Weiss et al. (2010)
Tillandsia fasciculata	560 μmol mol^{-1} (280 μmol mol^{-1} control)	Mean – 8% decrease in relative growth rate grown under low/high nutrients and low/high PFD	–	Monteiro et al. (2009)

[a]Not significant

(Croonenborghs et al. 2009). The *Aechmea* hybrid cv. 'Maya' showed no significant changes in leaf area or thickness and only a 7% decrease in dry leaf biomass (Croonenborghs et al. 2009). When grown under low/high nutrient conditions with low/high photosynthetic flux densities under natural humidity conditions, *Aechmea fasciata* exhibited a mean 24% increase in relative growth rate with CO_2 enrichment from 280 to 560 mmol mol^{-1} for 6 months (Monteiro et al. 2009) (Table 10.2). In contrast, *Tillandsia fasciculata*, despite growing well under these same conditions, showed a mean 8% decrease in relative growth rate (Monteiro et al. 2009). Another tropical epiphyte, *Bulbophyllum longissimum* (Orchidaceae), responded positively to elevated CO_2 concentrations with a mean 19% increase in relative growth rate under the same conditions (Table 10.2).

For the horticulturally important species of *Kalanchoë*, *Kalanchoë blossfeldiana*, CO_2 enrichment from 300 to 700 mmol mol^{-1} resulted in a 36% increase in dry biomass when plants were grown under 23 °C day/14 °C night temperatures (Mortensen and Moe 1992) (Table 10.2). CO_2 enrichment from 350 to 700 mmol mol^{-1} over 72 days resulted in increased plant height, a 38–42% increase in leaf area, and a 42–51% increase in dry biomass *K. pinnata* (Winter et al. 1997) (Table 10.2). The growth stimulation was attributed to daytime CO_2 uptake increase and had little effect on CAM cycle activity.

Early studies of the agronomically important CAM cacti, such as 'red pitaya' cactus (*Hylocereus undatus*) showed a 34% increase in total daily net CO_2 uptake when exposed to CO_2 enrichment from 370 to 740 mmol mol^{-1} (Raveh et al. 1995). More detailed studies showed that treatment of *H. undatus* with CO_2 enrichment from 380 to 1000 mmol mol^{-1} for 12 months resulted in 52%, 22%, 18%, and 175% increases in total daily net CO_2 uptake, shoot elongation, shoot dry mass, and number of reproductive buds, respectively (Weiss et al. 2010). Even more dramatic growth stimulation effects were obtained with 'yellow pitaya' cactus (*Selenicereus megalanthus*), which displayed 129%, 73%, 68%, and 233% increases in total daily net CO_2 uptake, shoot elongation, shoot dry mass, and number of reproductive buds, respectively (Weiss et al. 2010). Notably, *S. megalanthus* responded to CO_2 treatment with much greater fresh fruit mass production than did *H. undatus* (Table 10.2).

Among other species within the Cactaceae, exposure of *Ferocactus acanthodes* to CO_2 enrichment from 350 to 650 mmol mol^{-1} for 12 months resulted in a 30% increase in daytime net CO_2 uptake and a 30% and 28% increase in shoot and root dry biomass, respectively (Nobel and Hartsock 1986a, b) (Table 10.2). Probably the most extensively studied cactus species, the prickly pear cactus (*Opuntia ficus-indica*), showed significant growth stimulation in response to CO_2 enrichment from 360 to 720 mmol mol^{-1} for ~2 months (Luo and Nobel 1993). The dry biomass of new daughter cladodes increased 7% mmol mol^{-1} after 63 days of treatment (Table 10.2). *O. ficus-indica* exposed to CO_2 enrichment from 370 to 720 mmol mol^{-1} showed dramatic increases in net daily CO_2 uptake, WUE, and biomass production. After 9 weeks, daily net CO_2 uptake increased 98% for basal cladodes and 49% for daughter cladodes (Cui et al. 1993). WUE was 88% higher for basal cladodes and 57% higher for daughter cladodes. Plant biomass production increased 55% after 23 weeks (Cui et al. 1993). In a related study, nocturnal CO_2 uptake in daughter pads was 96% higher under CO_2 concentrations that were double that of atmospheric levels after 8 weeks (Cui and Nobel 1994). CO_2 enrichment from 370 to 750 mmol mol^{-1} resulted in a 32% increase in dry cladode biomass after 12 weeks (Cui and Nobel 1994). In addition, daughter cladodes showed a 96% higher nocturnal CO_2 uptake at 720 mmol mol^{-1} CO_2 compared with 360 mmol mol^{-1}, showing that nocturnal CO_2 fixation in CAM plants was not saturated under elevated atmospheric CO_2 (Table 10.2). CO_2 enrichment from 360 to 720 mmol mol^{-1} for 12 months showed a 40% increase in biomass (Nobel and Israel 1994). In addition, an increased number of daughter cladodes grew on the basal cladodes and were 11–16% thicker after 1 year of exposure to elevated atmospheric CO_2 concentrations (Nobel and Israel 1994). CO_2 enrichment from 360 to 720 mmol mol^{-1} for 4 months showed a 30% higher rate of growth, a 27% increase in cladode thickness with a 37% thicker chlorenchyma after 4 months (North et al. 1995). Although these responses might be subjected to nutrient limitation, these results suggest that *O. ficus-indica* might serve as a useful species for terrestrial CO_2 sequestration. In summary, while limited in scope and having inconsistent results compared to C_3 and C_4 photosynthesis species, these studies generally point to an advantage for CAM species to flexibly engage CO_2 uptake during both day and night photoperiods while displaying growth stimulation in response to rising atmospheric CO_2 levels.

B. Responses to Increasing Temperatures

Photosynthetic organisms are sensitive to high temperatures (above 40 °C) with PS II and Rubisco activase being among the most thermosensitive components (Mathur et al. 2014; Feller 2016; Kaushal et al. 2016). Indeed, thermostable forms of Rubsico activase can improve growth and yield stability under conditions of mild heat stress in *Arabidopsis* (Kurek et al. 2007; Kumar et al. 2009). The heat sensitivity of Rubisco activase varies among higher plants supporting its limiting role in photosynthesis under high temperature conditions (Salvucci and Crafts-Brandner 2004). In the CAM species *Agave tequilana*, the Rubisco activase isoforms were 10 ° C more thermostable *in vitro* than isoforms from rice, a C_3 photosythetic species (Shivhare and Mueller-Cajar 2017). This report suggests that CAM plants might serve as a source of highly active and thermostable forms of Rubisco activase for improving thermostability in other crop species. However, more research is needed to investigate the thermostability of Rubisco activase in other CAM species occupying extremely hot desert environments.

Model CAM species, such as *Kalanchoë daigremontiana* and *K. pinnata*, typically display day/night air temperature optima that facilitate substantial nocturnal acid accumulation (25 °C/15 °C) (Nobel and Hartsock 1984; Yamori et al. 2014). Increasing acclimation temperatures from 20 °C to 30 °C resulted in increased optimal nighttime CO_2 fixation rates from 16.1 °C and 21.1 °C and 15.3 °C and 19.6 °C for *K. daigremontiana* and *K. pinnata*, respectively (Yamori et al. 2014). Plants adapted to 30 °C also displayed greater daytime electron transport rates. Increasing temperatures above 34 °C disrupted the endogenous rhythm of nocturnal CO_2 uptake in *K. daigremontiana* (Grams et al. 1995).

Temperature response studies using *Clusia rosea* (obligate CAM), *Clusia pratensis* (facultative CAM), and *Agave angustifolia* (obligate CAM) showed that the highest ratios of maximum chlorophyll fluorescence (F_v/F_m), an indicator of the potential efficiency of PS II, were observed at 46 °C, 48 °C, and 54 °C, respectively, during the day (Table 10.3). During the night, higher temperatures decreased the ratios of maximum chlorophyll fluorescence in all three species (Krause et al. 2016). When photosynthetic tissues are subjected to a high daytime temperature, acids released from the vacuoles are likely metabolized, minimizing or preventing damage and allowing these species to establish in novel habitats. Additionally, high nighttime temperatures are associated with an over acidification of the cytoplasm, which contributes to cellular damage (Krause et al. 2016).

When *Ananas comosus* and *Kalanchoë pinnata* were kept under a high nighttime

Table 10.3. Effects of elevated temperatures on crassulacean acid metabolism (CAM) species

Species	Heat stability threshold	Thermal stability assay	Citations
Agave americana, A. angustifolia, A. bovicornuta, A. deserti, A. lecheguilla, A. mutlifilifera, A. palmeri, A. parryi, A. patonii, A. pedunculifera, A. rhodacantha, A. schottii, A. shawii, A. utahensis	57.2–64.7 °C (mean 61.9 °C)	Neutral red dye uptake for cell viability	Nobel and Smith (1983)
Agave deserti, Ferocactus. acanthodes (seedlings)	54.8 °C (60.7 °C with acclimation at 50 °C); 56.2 °C (64.8 °C with acclimation at 50 °C)	Neutral red dye uptake for cell viability	Nobel (1984)
Agave deserti	56.7 °C (62.9 °C with acclimation at 50 °C)	Neutral red dye uptake for cell viability	Kee and Nobel (1986)
Agave tequilana	55 °C	Percentage of leaf area damage after recovery	Nobel et al. (1998), Lujan et al. (2009)
Aloe vera	53.2 °C	Electrolyte leakage	Huerta et al. (2013)
Carnegiea gigantea, Ferocactus covillei, Lemaireocereus thurberi, Lophocereus schottii, Opuntia engelmannii, Echinocactus polvcephalus, O. basilaris, O. echinocarpa, O. fulgida	49.8–57.9 (mean 54.7 °C)	Chlorophyll fluorescence increase	Downton et al. (1984)
Carnegiea gigantea	55.3 °C; (2.8 °C with acclimation at 50 °C)	Neutral red dye uptake for cell viability	Kee and Nobel (1986)
Clusia rosea, Clusia pratensis, and Agave angustifolia	46 °C, 48 °C, and 54 °C, respectively	Chlorophyll fluorescence increase	Krause et al. (2016)
Epithelantha bolcei, Mammillaria lasiacantha, Ariocarpus fissuratus	56 °C, 57 °C, and 58 °C (with acclimation at 50 °C), respectively	Neutral red dye uptake for cell viability	Nobel et al. (1986)
Ferocactus. acanthodes	58.9 °C (66.1 °C with acclimation at 50 °C)	Neutral red dye uptake for cell viability	Kee and Nobel (1986)
Kalanchoe daigremontiana, Mamillaria woodsia, O. vulgaris	45 °C	Net nocturnal acid accumulation	Winter et al. (1986)
Opuntia bigelovii	52 °C (59 °C with acclimation at 50 °C)	Neutral red dye uptake for cell viability; electrolyte leakage	Didden-Zofpy and Nobel (1982)
Opuntia ficus-indica	55 °C (60 °C with acclimation at 45 °C)	PS I and PS II electron transport assays using isolated chloroplasts; net CO_2 uptake rates	Chetty and Nobel (1988)

temperature (HNT) (37 °C), a decrease in PEPC activity was observed, because HNT increases the sensitivity of the enzyme to malate (Lin et al. 2006). Increasing the night temperature raised the rate of malate exported from the vacuole and caused a rapid decline of PEPCK activity, thus accelerating the PEPC inhibition by L-malate (Lin et al. 2006). In *A. comosus*, a slight CO_2 uptake at night was still observed under HNT, while in *K. pinnata* nocturnal CO_2 uptake was completely blocked. The higher observed OAA levels in *A. comosus* indicated that PEPC was activated under HNT, and OAA generated by carboxylation by PEPC exceeded the requirements for malate synthesis during the night. In *K. pinnata*, lower levels of OAA may be generated by the TCA cycle rather than carboxylation by PEPC, indicating that this enzyme may be inactive in this species under HNT (Lin et al. 2006). This study showed that the sensitivity to high temperature varies among different CAM species.

Heat stress studies in *Aloe barbadensis* (*Aloe vera*) showed that the lethal temperature LT_{50} was 52.3 °C as estimated by an electrolyte leakage assay (Table 10.3) (Huerta et al. 2013). Heat acclimation was associated with the accumulation of transcripts for *Hsp*70, *Hsp*100, and Ubiquitin genes.

For cacti and agaves, temperature is usually not a limiting factor for CO_2 uptake over the course of a year, except during episodes of low temperatures (Nobel 1996). However, CAM cactus species from the Mohave and Sonoran deserts are exposed to extremely high tissue temperatures and did not experience damage to the photosynthetic apparatus until temperatures reached a mean of 54.7 °C during the day (Table 10.3) (Downton et al. 1984). This mean temperature is about 5 °C greater than other warm-temperature species tested. *Opuntia bigelovii* cladodes could tolerate temperatures of 52 °C and up to 59 °C when heat acclimated (Didden-Zopfy and Nobel 1982). Small desert cacti tolerated temperatures as high as 56–58 °C when heat acclimated at 50 °C (Table 10.3) (Nobel et al. 1986). *Opuntia ficus-indica* cladodes could tolerate temperatures as high as 55 °C and up to 60 °C when heat acclimated (Chetti and Nobel 1988). Seedlings of *Agave deserti* and *Ferocactus acanthodes* show temperature limits of 60.7 °C and 64.8 °C with heat acclimation of 50/40 °C (day/night) (Nobel 1984). Heat acclimation of desert cactus species typically increases the high temperature limit by 6–8 °C (Table 10.3), takes up to 10 days to achieve, and is associated with the accumulation of species-specific heat shock proteins (Kee and Nobel 1986).

The central spike and unfolded peripheral leaves of *Agave tequilana*, which are the most heat-tolerant parts of the plant, can tolerate temperatures up to 55 °C (Table 10.3) (Nobel et al. 1998; Lujan et al. 2009). This notable heat tolerance is likely due, in part, to the accumulation of small and large heat shock proteins (e.g., HSP17.7, HSP18.4B, and HSP90A). A survey of *Agave* spp. from the Chihuahuan and Sonoran deserts showed heat tolerance to temperatures ranging from 57.2–64.7 °C after heat acclimation at 50/40 °C (day/night) (Nobel and Smith 1983). Lastly, transcriptomic and proteomic analyses have been performed on *A. tequilana* plants exposed to high temperatures (45 °C, 55 °C, and 65 °C) (Shakee et al. 2013).

In summary, under high daytime temperatures (particularly when acclimated), CAM plants seem to be far more tolerant, maintain higher photosynthetic rates, and display lower photorespiration rates than C_3 and C_4 photosynthetic plants, suggesting that CAM plants might be more resilient to global warming.

C. *Water-Use Efficiency and Productivity of CAM Plants*

As a result of global warming, improving the water use-efficiency of crops will become an increasingly critical factor for global agricultural productivity (Leng and Hall 2019). The high WUE of CAM plants compared with C_3 and C_4 photosynthetic plants clearly demonstrates their potential for use in sustainable agricultural production systems on arid and marginal lands (Borland et al. 2009, 2011a, b). The WUE of CAM plants increases with the enhancement of CAM expression, as shown by the relationship between transpiration ratio and $\delta^{13}C$ value (Winter et al. 2005). Several highly WUE CAM crops show great potential due to their high productivity, economic importance, climate-resilience, and overall utility. For example, *Agave* and *Opuntia* spp. serve as a source of food, forage and fodder, fiber, biofuel, various high-value secondary products, and alcohol-containing beverages in arid and semi-arid regions (Borland et al. 2009; Owen and Griffiths 2014; Cushman et al. 2015; Yang et al. 2015a, b; Owen et al. 2016; Ranjan et al. 2016; Davis et al. 2017, 2019).

In relation to the biomass productivity of CAM crops, under ambient precipitation conditions, *Agave* spp. showed dry-weight productivity values from <1 to 34 Mg ha^{-1} year^{-1}, while under well-watered conditions, biomass productivity can be as high as 38 to 44 Mg ha^{-1} year^{-1} (Nobel et al. 1992; Davis et al. 2011; Garcia-Moya et al. 2011; Yan et al. 2011). For *Opuntia* spp., under arid conditions (200–400 mm annual precipitation), dry-weight productivity values ranged from 3 to 15 Mg ha^{-1} year^{-1}, while under well-watered conditions, the productivity values ranged from 40 to 50 Mg ha^{-1} year^{-1} (Garcia de Cortázar and Nobel 1991; Nobel 1991a, b; Garcia de Cortazar and Nobel 1992). *Agave* spp. and *Opuntia* spp. can both exceed yields of commodity crops that use C_3 or C_4 photosynthetic pathways. In comparison, the productivities of C_4 photosynthetic species grown for biofuels such as maize, switchgrass, and sugarcane range from 5 to 26 Mg ha^{-1} year^{-1}, and C_3 photosynthetic species grown for biofuels such as oil palm, poplar, and willow produce between 2 and 14 Mg ha^{-1} year^{-1} (Somerville et al. 2010).

Simulated production modeling on low-grade land showed that *O. ficus-indica* is more resilient to climate changes than *Agave tequilana* (Owen et al. 2016). A deviation of 5 °C from the optimum temperature reduced the yield by 2 to 12% in *O. ficus-indica*, while *A. tequilana* yield was reduced by 20 to 33%. In a simulated worst-case scenario prediction for 2070, *A. tequilana* yield was predicted to fall by 11%, whereas *O. ficus-indica* yield will remain unchanged in 2070 (Owen et al. 2016). The development of biomass productivity models can help identify geographical regions where CAM crops might be more successful and can predict how they will respond under global climate change (Cushman et al. 2015).

When grown under arid conditions (e.g., 200 mm year^{-1}), the superior WUE of CAM species can result in theoretical biomass yield potentials that are 147% greater than those from C_4 photosynthetic species (Davis et al. 2014). *Agave americana* cultivated under 330 mm year^{-1} showed a biomass production of 2.0 to 4.0 Mg ha^{-1} year^{-1}, proving that commercially viable yields in arid regions with minimal irrigation are possible (Davis et al. 2017). In contrast, cotton, a C_3 plant, one of the most important crops in Arizona required 1046 mm year^{-1} of irrigation and yields only 1.46 Mg ha^{-1} year^{-1} (U.S. Department of Agriculture (USDA) 2012). *A. americana* is an alternative crop for climate change that can be cultivated in dry lands with far less water input in the southwestern U.S.A. (Davis et al. 2017).

Dry mass and transpiration ratios were reported for C_3, C_4, and CAM plants growing under tropical conditions in the Republic of Panama (Winter et al. 2005). The dry mass was from 2.2 to 5.5 times higher for CAM plants when compared to C_3 photosynthesis plants, while for C_4 photosynthesis plants it was from 2.0 to 3.5 times lower when compared to CAM plants. The transpiration ratio was from 2.3 to 7.0 times higher in C_3 photosynthesis plants when compared to CAM, and from 1.2 to 2.5 times higher in C_4 photosynthesis species than in CAM species at the same site (Winter et al. 2005). The highest WUE was observed in *Aloe vera*, the median in *A. comosus*, and the lowest for *Kalanchoë daigremontiana*.

V. Productivity Modeling of Major CAM Crop Species

Seasonal variation in environmental factors determine the overall productivity of a plant species. Understanding how new or alternative agricultural crop species respond to variation in seasonal climates allows for comparative estimates of productivity across geographic regions without the need for labor intensive trial and error (Jones et al. 2017). For C_3 photosynthesis plants, a physiological model has been established that takes into account light, ambient CO_2 levels, temperature-dependent kinetics of Rubisco, and electron transport rates to calculate CO_2 assimilation at the leaf level (Farquhar et al. 1980). A similar model was also developed for C_4 photosynthesis plants using the same principles with additional parameters for the initial fixation of CO_2 by PEPC, as well as parameters addressing the specialized Kranz anatomy that allows for CO_2 saturation in bundle-sheath cells (von Caemmerer and Furbank 1999). While the scale of these models are at the leaf level, they can be used as a basis for field-scale models that can relate limitations on overall yield to specific biochemical reactions (Wu et al. 2016). The development of physiological models of carbon assimilation in CAM have been lacking until recently, as C_3 and C_4 photosynthesis models were developed, in part, by measuring changes in carbon assimilation with light intensity, which is a unique challenge in CAM plants as CO_2 is mainly assimilated when light is not available (Owen and Griffiths 2013; Bartlett et al. 2014; Cheung et al. 2014; Shameer et al. 2018). Below we summarize the Environmental Productivity Index (EPI) model that has allowed for empirical estimates without the need of physiological models, as well as contemporary physiological, biochemical, and *in situ* models of CAM plant productivity that might prove useful for agricultural predictions in the near future.

A. Environmental Productivity Index Modeling

The EPI model is a quantitative method for evaluating primary environmental factor effects on net CO_2 uptake, and therefore, overall productivity for large agricultural CAM species, such as *Opuntia* and *Agave* spp. as well as other crop plants (Nobel 1988). An EPI model is typically developed using empirical data from laboratory studies that measure changes in the nocturnal accumulation of organic acids, a proxy of productivity in CAM plants as CO_2 is first fixed into organic acids, mainly malic acid, before being decarboxylated during the day (Osmond 1978; Ting 1985).

The EPI model's strength lies in its simplicity. The underlying experimentation to determine productivity responses is not laborious or expensive, as it consists of collecting tissue samples for titratable acidity analysis (Nobel 1988). Moreover, having multiplicative indices that represent individual abiotic factors allows for the addition of more parameters on productivity without the need for major adjustment to the core equation. Once generated, an EPI model can be validated at any field site where the crop in question is grown by inputting climate variables and comparing the estimated yield with the observed yield. The validated EPI model can then be used to make yield predictions on a global scale.

EPI modeling is based upon nocturnal measurements of organic acid accumulation and is only accurate for obligate CAM species as it does not include CO_2 that is fixed during Phase II and IV of CAM that occurs in facultative and other obligate CAM species. For facultative species, EPI equations can instead be based upon measurements of diurnal gas exchange and biomass with changes to environmental conditions as these measurements will reveal total carbon assimilation regardless of fixation by C_3 or CAM photosynthesis (Niechayev et al. 2019a, b). Changes in nocturnal organic acid accumulation and/or diurnal gas exchange in response to changes in soil water, photosynthetically active radiation (PAR), and temperature are each measured independently while all other conditions are held constant (Nobel 1988). Separate index equations are then derived from the experimentally found responses to available water, PAR, and temperature. For each index, a value of 1.00 corresponds to an optimal condition, and 0.00–0.99 is the proportional percent decrease of malic acid accumulation or diurnal gas exchange and biomass accumulation due to deviations from the experimentally identified optimal condition (Nobel 1988). The product of the water (I_w), PAR (I_P), and temperature (I_T) indices is equal to the EPI:

$$I_W * I_P * I_T = \text{EPI}$$

The productivity in a field site can then be estimated as:

$$\text{Maximum theoretical Yield} * \text{EPI} = \text{Estimated Yield}$$

Once an EPI model is developed, monthly I_w, I_P, and I_T can be calculated for field sites by inputting the average monthly soil moisture content, PAR, and minimum and maximum temperatures, respectively. These monthly EPI values can be averaged for the entire year to determine the annual EPI value. Additional indices can also be added to the EPI model in order to account for responses to variation in soil nutrients (Nobel 1989), plant spacing (de Cortázar et al. 1986), elevation (Nobel and Hartsock 1986a, b), ambient atmospheric CO_2 concentration (Nobel 1991a, b), as well as biotic factors, such as rhizosphere composition that can influence nutrient availability and drought tolerance.

EPI values have been shown to correlate strongly with actual field growth measurements of *Agave lechuguilla* (Nobel and Quero 1986), *A. tequilana* (Nobel and Valenzuela 1987), *A. deserti* (Nobel and Hartsock 1986a, b), *A. salmiana* (Nobel and Meyer 1985), *O. ficus-indica* (Nobel 1991a, b), and *A. americana* (Niechayev et al. 2019a, b). Constructed EPI models have been used to calculate predictive productivity estimates at specific geological locations allowing for spatial comparisons of species productivity (Nobel and Hartsock 1986a, b; Quero and Nobel 1987).

Recent studies have implemented Global Information Systems (GIS) software to increase the resolution of mapping potential productivity, while also using ethanol conversion values from the literature to estimate ethanol yield by species across vast geographical ranges (Sánchez et al. 2012; Owen and Griffiths 2014). Yield simulations under predicted changes in climate due to global warming suggest that productive ranges of *O. ficus-indica* and *A. tequilana* are likely to expand in the near future as described above.

However, the EPI model is often criticized for being an over simplification as it is empirical and does not include a physiological/biochemical basis for calculations (Owen and Griffiths 2013; Bartlett et al. 2014; Hartzell et al. 2018). The EPI model is not capable of elucidating specific enzymatic steps within the CAM cycle that are limiting to productivity or potential targets for genetic modification in the future. Another drawback is that the traditional EPI model calculates at a timescale of 1 month, and fails to take into account environmental variability at daily and weekly timescales (Hartzell et al. 2018). However, daily index values can be calculated using the same equations with daily climate inputs, and therefore, daily changes can be implemented to calculate overall productivity using the EPI model (Niechayev et al. 2019b). The EPI model may also be limited in that it does not account for the interdependency between factors considered. For example, the PAR response of a given species might change when water is limiting or *vice versa* in ways not considered in the current EPI modeling approach.

B. Biochemical Models of CAM Productivity

The biochemical pathway of CAM has been well established and is typically described in four phases that take place over a 24-hour period (Osmond 1978; Ting 1985; Owen and Griffiths 2013). To summarize, phase I begins when sunlight is not available, stomata are open, and CO_2 is fixed by PEPC that converts PEP into OAA, which is ultimately stored in vacuoles as organic acids. Phase II occurs during sunrise as stomata are still open, and the combined fixation of CO_2 by PEPC and photoactivated Rubisco causes a sharp spike in the carbon assimilation rate. Phase III begins when stomata close during the day. Malic acid is released from the vacuoles, decarboxylated by PEPCK or ME depending on the species, and the released CO_2 is fixed by Rubisco in the C_3 photosynthetic pathway. Lastly, phase IV takes place as stomata begin to open before the sun has set, causing another carbon assimilation spike from the combined fixing of CO_2 by PEPC and Rubisco.

While the underlying reactions of CAM photosynthesis have long been understood and continually clarified, models that predict carbon assimilation rates given abiotic inputs using biochemical and physiological processes as a basis have only been published within the last decade. The System Dynamics (SD) model developed by Owen and Griffiths (2013) expands upon a computational model of the metabolic processes in CAM (Nungesser et al. 1984). The SD model incorporates carbon flow through stomatal and mesophyll conductance, and malic acid transport across the tonoplast with consideration of the feedback effects that stomatal aperture, malic acid inhibition of PEPC, and enzyme kinetics have upon these flow rates. The SD model uses temperature, PAR, Rubisco activation factor, and PEPC activation factor (as activities of these two enzymes are unique across species) as inputs to predict the carbon flow over the diel period in each of the four phases. The SD model simulations of gas exchange and malic acid accumulation showed a strong correlation ($R^2 > 0.9$) with actual gas exchange measurements in *K. daigremontiana* and *A. tequilana* (Owen and Griffiths 2013).

Bartlett et al. (2014) published another physiological model of CAM that takes the Farquhar model of C_3 photosynthesis (Farquhar et al. 1980) and adjusted it to account for the CAM circadian rhythm. This model incorporates a soil-plant-atmosphere continuum model that includes the effect of soil water availability, stomatal conductance, plant water capacitance, and vapor-pressure deficit on whole plant carbon fluxes, allowing for an estimation of photosynthesis given a specific ecohydrology region. Hartzell et al. (2018) further improved upon Bartlett et al. (2014) by creating a unified representation of C_3, C_4, and CAM using Python coding to create the Photo3 model. The strength of the Photo3 model is that it uses the same modeling assumptions for all three types of photosynthesis and provides a basis for a robust comparison of productivity of traditional crop species with that of agricultural CAM plants under the same environmental conditions. The Photo3 model simulations were validated by prediction of diurnal carbon uptake in *O. ficus-indica* (CAM), *Sorghum bicolor* (C_4), and *Triticum aestivum* (C_3) (Hartzell et al. 2018).

Similar to the SD modeling approaches, flux balance analysis (FBA) has been used to predict fluxes in large metabolic networks using known photosynthetic flux dynamics (Sweetlove and Ratcliffe 2011). FBA is a user defined, constraints-based modeling approach in which steady-state fluxes in a metabolic network are predicted by applying mass-balance constraints to a model of the network based on the matrix of reaction stoichiometries. Typically, simple and easy to measure mass-balance information, such as growth rate, biomass composition, and substrate-consumption rate, is used to place boundaries on the flux solution space (Sweetlove and Ratcliffe 2011). FBA models are constructed under the premise that nature has optimized efficiency of metabolic networks, and therefore, optimization principles can be applied to make flux predictions without knowing the specific enzyme-kinetics of a pathway. FBA models take into consideration the cost of making the enzyme machinery and use genome-scale metabolic maps to characterize all biochemical reactions occurring within a system. Only a subset of capable metabolic pathways is used at any one time depending on stage of development, resource availability, time of day, cell/tissue type, and environment. Several optimal flux solutions can be generated for each objective, and objectives will have unique flux solutions. For example, flux solutions for a maximized growth rate objective would be unique when compared to the flux solutions for minimized substrate consumption (Sweetlove and Ratcliffe 2011).

An FBA model that describes the diel pattern of CAM was constructed from the C_3 genome scale model from *Arabidopsis thaliana,* as a genome scale-model of a CAM species was not yet available (Cheung et al. 2014). The model places constraints on the genome-scale model of *A. thaliana* metabolism such that storage compounds synthesized during the day were available for use in the dark and *vice versa* as in the CAM pathway in a single optimization problem. The model includes different options of carbon storage (e.g., starch, glucose, fructose, malate, fumarate, and citrate), and amino acid accumulation (only during day) without any direct constraints on which compounds were stored, allowing the model to choose the most optimized pathway. This computational model allowed for a full accounting of ATP, PPi, and NADPH production and expenditure across the entire CAM network and shows that there is unlikely to be a substantial benefit involved with the reduced amount of photorespiration than that of C_3 photosynthesis due to the carbon concentrating mechanism of CAM (Cheung et al. 2014). The same model has been refined, and now uses only the core *A. thaliana* metabolism genes that are highly conserved across C_3, C_4, and CAM plant species. In addition, this refined model includes proton-balancing of the vacuole at night, which is dependent upon the activity of energy-dependent ATP or PP_i proton pumps including those in the tonoplast, as it effects overall flux (Shameer et al. 2018). Predictions made by the newly refined model were able to show that energy consumption is three-fold higher at night in CAM plants compared to C_3 species, but that the energy cost might be entirely offset by the daytime decarboxylation of malate (Shameer et al. 2018). Therefore, the *in situ* FBA CAM models demonstrate that plants using CAM may not be at an energetic disadvantage when compared to that of species using C_3 photosynthesis pathways (Shameer et al. 2018). However, photon uptake for this model was constrained to 200 $\mu mol\, m^{-2}\, s^{-1}$, which is sufficient for *A. thaliana*, but low for large CAM species, which are typically found in full sun conditions (900–1500 $\mu mol\, m^{-2}\, s^{-1}$) (Nobel 1988).

C. *Coupling Biochemical Models to Field Predictions*

While biochemical models provide a fundamental framework to identify key limiting factors to photosynthesis in an environment, they do so at the leaf or plant level (Owen et al. 2016; Hartzell et al. 2018; Shameer et al. 2018). In order to make use of these insights at the canopy or crop level, connections must be made between biochemical models and crop-level growth and developmental dynamics (Wu et al. 2016). The biochemical models presented previously could be especially useful when combined with canopy radiation-use efficiency (RUE) and surface leaf nitrogen (SLN) crop models (Leuning et al. 1995; Sinclair and Muchow 1999). Biochemical models can also assess the changes in productivity due to variation in air temperature and ambient CO_2 levels more reliably than empirical crop models. Furthermore, whole crop models, which include biochemical models as a basis, will allow for the elucidation of limiting traits within the photosynthetic pathway that could be targeted by genetic engineering or genome editing to enhance yields (Wu et al. 2016).

Crop yield is directly linked to the capacity of plants to intercept solar radiation (Sinclair and Muchow 1999). RUE measurements provide the efficiency at which a species uses PAR at the field level, which is dependent upon the three-dimensional structure of the canopy as not all leaves intercept the same amount of light due to shading. Cacti species provide a unique challenge for modeling RUE, as they mostly rely upon photosynthetic stems or stem segments instead of leaves for PAR interception (de Cortazar et al. 1985, 1986). Interestingly, cactus spines have not only been shown to mainly reduce the amount of PAR that reaches the stem surface thereby decreasing productivity, but also to reduce high temperatures on the stem surface in desert ecosystems (Nobel 1983). A systems-level understanding of CAM productivity in *Agave* or *Opuntia* spp. requires that the arrangement of photosynthetic stems and leaves is estimated in such a way to allow for the maximal absorption of PAR (de Cortázar et al. 1986; Davis et al. 2015). RUE has been studied in *Opuntia ficus-indica* (Ochoa and Uhart 2004; Franck et al. 2013), *Opuntia basilaris, Opuntia chlorotica, Stenocereus gummossus* (Nobel 1980), *Ferocactus acanthodes, Opuntia bigelovii* (Nobel 1983), and *Agave americana* (Davis et al. 2015).

D. Assessing Environmental and Economic Potential

Once a validated productivity model has been established for making field-level predictions, assessments can be made to compare the environmental and economic viability of crops to one another across vast regions. The environmental impact of a crop can vary depending upon the pollutant release associated with required land-use changes, conversion efficiencies to desired products, transportation, and other factors. Life cycle assessment (LCA) models are a common tool for assessing the overall emissions produced from initial planting to final production and transport to market (Yan et al. 2011). Similarly, a life cycle costing assessment (LCCA) can be used to assess the cost of producing a crop-based product from cradle to grave as well as to consider the detailed costs associated with the same set of steps considered in an LCA (Swarr et al. 2011).

An LCA model considers the amount of pollutants gained or lost from each production process step to determine the net emissions of an energy source. A common issue with LCAs of biomass feedstocks in the past has been the lack of standardized data that are included in LCAs or life cycle inventories (LCI) (Stork and Singh 1995; Swarr et al. 2011). The absence of a defined LCI presents challenges when comparing LCAs from one species to another. Greenhouse Gases, Regulated Emissions, and Energy Use in Transportation (GREET) modeling software has been developed in an attempt to standardize inventories to calculate costs (Wang 1999). GREET includes over 100 energy production pathways, including petroleum fuels, natural gas fuels, biofuels, hydrogen, and electricity produced from various energy sources.

Similarly, LCCA estimates of the overall cost for energy pathways have not been standardized for biofuels (Swarr et al. 2011). While several studies have simply used the same life cycle inventory from developed LCAs to calculate LCCAs, a common issue with LCCA models is that they often fail to incorporate economic uncertainties including variation in resource price, biomass yield, and a market value within their estimates. These economic uncertainties are not trivial as the economic viability of biofuel sources is interlinked with an ever-changing market. New LCCA models allow for estimates of net biofuel production costs by inclusion of cost uncertainties of individual steps in the life cycle inventory (Ren et al. 2015). This model incorporates the costs of a biofuel supply chain design that includes all the essential steps from cropping to market. The strength of this model is that it represents steps in the supply chain with uncertain costs such as multiple agriculture zones, multiple transportation modes, multiple biofuel plants, and multiple market centers as interval numbers instead of constants. Determining the LCA and LCCA estimates for highly productive CAM species is still in its infancy, but such assessment models will prove critical in the future for estimating the GHG emissions and economic viability of using these species for bioenergy feedstocks in arid and semi-arid regions around the world. In addition, LCA will be expected to include single feedstock and mixed digestion systems such as *O. ficus-indica* cladodes and livestock manure (Ramírez-Arpide et al. 2018).

VI. CAM Species as Bioenergy Feedstocks

The expanded use of bioenergy feedstocks represents one critical strategy to partially offset the effects of global warming brought about by GHG emissions. As the demand for alternative energy resources has increased, the need for novel bioenergy feedstocks that utilize nontraditional, marginal, abandoned, semi-arid, and arid agricultural lands will also likely expand. CAM species like *Agave* and *Opuntia* spp. have been considered for their potential to act as economically viable bioenergy crops (Cushman et al. 2015; Yang et al. 2015a, b; Davis et al. 2019). The extreme WUE characteristics, low lignin content, and high biomass potential of *Agave* and *Opuntia* spp. make these crops highly desirable second-generation biofuel sources (Li et al. 2012, 2014; Cushman et al. 2015; Yang et al. 2015a, b).

A. *Low Lignin Content Herbaceous Feedstocks*

Lignin, the complex polymer that increases rigidity of cell walls, is a major impediment to the efficient conversion of biomass into bioenergy (Trajano et al. 2013; Zeng et al. 2014). Higher lignin content increases hydrolysis hinderance, called recalcitrance, during biofuel feedstock processing and can negatively impact saccharification by interfering with the breakdown of sugars in the cell walls (Trajano et al. 2013; Zeng et al. 2014). Additionally, lignin associated with herbaceous rather than woody crops, tends to be less inhibiting to enzymatic degradation of the biomass (Zeng et al. 2014). Typically, second-generation bioenergy crops such as switchgrass, poplar, and corn stover are high in lignocellulose and need additional processing such as chemical acidification and alkylation to complete the biofuel production process (Escamilla-Treviño 2012).

Corn stover typically has an average lignin content of 26% biomass, switchgrass 12%, poplar 17.7%, and sugarcane 19.1% lignin, whereas CAM species, such as *Agave spp.* vary in lignin content from 4% to 16% and *Opuntia* spp. vary between 5% and 16% (Escamilla-Treviño 2012; Li et al. 2012; Yang et al. 2015a, b). Pineapple leaf wastes can also be used for the production of fermentable sugars and subsequent use in biofuel production (Banerjee et al. 2017; Gil and Maupoey 2018).

B. *Pretreatment and Digestion Strategies*

Many bioenergy production processes involve a combined use of mechanical, thermal, chemical, and biological degradation processes to effectively release fermentable monosaccharides and lipids from biomass products to form useful solid, liquid, or gaseous fuels. Chemical acidification and alkylation pretreatments include the use of concentrated acids and alkali agents that have additional costs of corroding equipment, generating toxic byproducts, and are expensive to use (Maurya et al. 2015). Other delignifying pretreatments include milling, steam explosion, liquid hot water, ammonia fiber expansion, wet oxidation, ozonolysis, organosolv treatments, and biological treatments, all of which besides biological treatments are expensive due to either high energy demands or high cost of materials (Maurya et al. 2015). CAM plants like *Agave* and *Opuntia* spp. are also considered to be lignocellulosic bioenergy crops; however, their relatively low lignin contents may decrease the additional processing costs typical for second-generation bioenergy crops (Yang et al. 2015a, b).

VII. Mechanical and Thermal Degradation

The mechanical degradation of biomass is a combination of natural degradation due to harvesting and storing conditions and physical processes such as milling, grinding, and cutting of the biomass products in order to increase the surface area to volume ratio and to release lipids, sugars, and other fermentable compounds into solution (Lindner et al. 2015; Graham et al. 2017). Mechanical degradation is frequently the first step in biofuel feedstock processing where the raw biomass is milled, ground, or cut into a biomass slurry (Voloshin et al. 2016). Mechanical degradation of post-anaerobically treated biomass can triple the increase of methane yields, increasing the biomass-to-energy yields (Lindner et al. 2015).

Thermal degradation is often paired with mechanical degradation through either a series of heating steps using increasingly hotter conditions or maintaining a set boiling temperature throughout the degradation process (Ye et al. 2018). Pyrolysis is a thermal degradation process that can reprocess waste products such as liquor bran waste into products such as bio-oil, syngas, and biochar (Ye et al. 2018). Mechanical degradation is present in most biofuel feedstock processes and must be optimized to ensure lower biofuel production costs. For biogas production, sonification of 50% of the biomass ensures optimal methane production, whereas mechanical milling of wood products can increase total sugar yield from 65% to 95% (Divyalakshmi et al. 2017; Liu et al. 2017).

In ethanol and methane production trials of *O. ficus-indica*, each experimental design included a mechanical degradation step that involved either mechanical or manual shredding of the cladodes into pieces averaging 1 cm^3 (Kuloyo et al. 2014; Myovela et al. 2019). In an ethanol production analysis of *O. ficus-indica*, thermal treatment of the biomass was performed concurrently with a dilute acid pretreatment rather than as a separate treatment (Kuloyo et al. 2014).

VIII. Chemical Degradation

Chemical degradation of biomass products involves the use of acids, bases, and organic compounds to degrade biomolecules like cellulose and hemi-cellulose and interact with naturally present alkali and alkaline earth metals such as Si, Na, K, Mg, and Ca that can decrease biofuel yields and cause the production of adverse secondary products (Rodríguez-Machín et al. 2018, 2019). Pretreatment leaching processes typically use sulfuric acid, hydrochloric acid, hydrogen fluoride, nitric acid, and phosphoric acids, all of which have increased costs for the biofuel production process and lead to environment concerns such as increased leachates in water systems (Rodríguez-Machín et al. 2018). In *O. ficus-indica* ethanol trials, using an acid pretreatment with sulfuric acid at 121 °C is a common method to create a biomass slurry (Kuloyo et al. 2014; Pérez-Cadena et al. 2018). Given that the acid pretreatments were neutralized before the fermentation process, there was little effect of the pretreatment on microbial growth and only hinders the biofuel process through the additional cost and time spent degrading the biomass slurry.

Chemical acidification is typically used in lignocellulosic biofuel production and is often faster and more effective in lignocellulosic degradation in short-term pretreatments, whereas biological degradation of lignin and cellulose is more efficient in long-term pretreatment of biomass (Taherzadeh and Karimi 2008; Ranjithkumar et al. 2017). Ammonia fiber expansion (AFEX), followed by enzymatic carbohydrate degradation, has also been used experimentally to degrade *Agave* leaf and bagasse residues to use for bioethanol production (Flores-Gómez et al. 2018). Alkaline hydrolysis of *O. ficus-indica* and *N. cochenillifera* fruit provided the best enzymatic saccharification yields using a simultaneous saccharification and fermentation process (de Souza Filho et al. 2016).

IX. Biological Degradation

Biological degradation of biofuels presents a cheaper and often more effective way to process unfermentable polysaccharides and lipids from bioenergy crops into smaller, easier to process molecules needed to complete the biofuel process. Biological degradation primarily relies upon enzymatic activity that has increased specificity for plant proteins and other compounds (Maurya et al. 2015; Lima et al. 2016). While chemical degradation can degrade complex polymers like pectin, cellulose, and lignin into simpler monomers, enzymes derived from microbial species are more specific, cheaper, and more effective at degrading plant polymers (Benoit et al. 2012, 2015; Maurya et al. 2015; Lima et al. 2016). Often enzymatic degradation of biomass prevents the production of secondary products, like microbial growth inhibitors, that are detrimental to downstream processes typical in acid-base treatments used in second-generation biofuel production (Benoit et al. 2012, 2015; Lima et al. 2016).

Microbial species can contain enzymes such as pectinase and cellulase that can deconstruct biomass more efficiently than acidic or mechanical degradation (Jayani et al. 2010). Microbial species such as *Pectobacterium* can process the complex sugar pectin into easily fermentable sugars such as glucose and fructose (Valenzuela-Soto et al. 2015), whereas complex organisms such as the Black Soldier fly (*Hermetia illucens*) larvae can further degrade lignocellulosic and cellulosic material into biological wastes that are easier to saccharify using anaerobic fermentation (Li et al. 2018).

Microbial species can be optimized to overcome ethanol production limitations of *Opuntia* spp. by hydrolyzing pectin and other carbohydrates not easily degraded by chemical acidification (Benoit et al. 2012, 2015; Pérez-Cadena et al. 2018; Myovela et al. 2019). Combinations of chemical pretreatment for lignocellulosic products and biological degradation of carbohydrate products may be the future for CAM biofuels, where the optimization of enzymatic degradation minimizes the degree of chemical pretreatment needed to complete the biofuel production process.

X. Conversion to Ethanol

Ethanol production from biomass involves the conversion of soluble sugars such as sucrose, glucose, galactose, and fructose into ethanol using microbial species that can undergo anaerobic fermentation. Typically, yeast species such as *Kluyveromyces marxianus* and *Saccharomyces cerevisiae* have been used in the industrial fermentation process, with unique strains and blends of species being proprietary (Kuloyo et al. 2014).

Current attempts to convert *Opuntia* spp. into ethanol have resulted in low ethanol yields despite the high pectin and polysaccharide concentration in the cladodes, where the yields remain around 2.6% w/v ethanol and need to be above 4% w/v to be economically viable (Kuloyo et al. 2014). Additional studies using yeast strains isolated from *O. ficus-indica* cladodes achieved a maximum ethanol production of 2.5 g/L with the yeast strain *Candida intermedia*, compared to a more recent study that reported a maximum ethanol production of 19.6 g/L using *S. cerevisiae* and 19.5 g/L using *K. marxianus* (Kuloyo et al. 2014; Pérez-Cadena et al. 2018).

Notably, trials using acid-pretreatments with enzymatic degradation of the biomass slurry resulted in ethanol yields of 34.9 g/L (Alencar et al. 2018). Others studies have conducted simultaneous saccharification and fermentation for ethanol production using *O. ficus-indica* and *N. cochenillifera* fruit (de Souza Filho et al. 2016). Further studies need to be conducted on *Opuntia* spp. in order to increase the fermentable carbohydrate concentration of the biomass slurry. Ethanol production studies on *Agave* spp. leaf and bagasse reported ethanol yields ranging from 0.154 to 0.198 g ethanol per gram biomass, which corresponds to ethanol g/L yields of between 35 and 40 g/L (Flores-Gómez et al. 2018).

XI. Conversion to Biogas

Beyond liquid biofuels, CAM bioenergy crops can be processed into biogas, mainly methane products, for use in electricity production and transportation. The use of biogas as a replacement of transportation fossil fuels could result in a 49–84% reduction in the carbon dioxide footprint (Budzianowski and Postawa 2017). Biogas is produced from the anaerobic degradation of organic material. Biogas production typically involves the use of waste organic material such as corncob, rice straw, corn stover, and mixtures of grasses and animal manure to produce biogas composed of ~50% methane (Lindner et al. 2015; Li et al. 2018). The high lignin content in many lignocellulosic biofuel crops hampers the biogas conversion process due to the low degradation rates of lignin by anaerobic microbes (Taherzadeh and Karimi 2008; Ranjithkumar et al. 2017).

CAM species with low lignin content such as *Agave* and *Opuntia* spp. can overcome these limitations by using the appropriate pretreatment techniques and feedstock or mixtures of feedstock for anaerobic digestion (Myovela et al. 2019). Short-term (9 h) aerobic digestion of sisal (*A. sisalana*) pulp using an activated sludge mixed culture increased the biomethane yield compared with no pretreatment (Mshandete et al. 2005). Acidic pretreatment of *O. ficus-indica* resulted in the greatest methane yields compared with alkaline or thermal pretreatments (Calabrò et al. 2018). When *Opuntia* spp. biomass was used in anaerobic digestion, approximately 325 liters of biogas was produced per 1 kg of biomass (Mason et al. 2015). With the addition of aerobic pretreatments, methane yields from bioenergy crops can be increased. Nine-hours of aerobic pre-treatment of *O. ficus-indica* cladodes using cow rumen fluids caused methane yields to increase 123% with a maximum yield of 0.72 l/kg volatile solid (Myovela et al. 2019). Further studies need to be conducted to assess the economic viability of *Opuntia* spp. as a biogas resource, but initial estimates seem to hold great promise. For example, annual electrical power generated globally from methane gas (about 5 PW) could be generated from CAM plants (e.g., *O. ficus-indica*) on 100–380 Mha of semi-arid land or between 4% and 15% of such currently available land area (Mason et al. 2015). An evaluation of methane production from representative species of five different CAM genera (e.g., *Ananas comosus*, *Agave angustifolia*, *Euphorbia virosa*, *Kalanchoe daigremontiana*, and *Opuntia fragilis*) showed that *A. angustifolia* produced the highest methane yields and that biomethane yields were similar to those obtained from maize biomass (Lueangwattanapong et al. 2020). The CAM species *Euphorbia tirucalli* and *O. ficus-indica* are able to produce 1791 m^3 and 1860 m^3 of methane from 1 ha of land planted at high planting densities, respectively (Krümpel et al. 2020). These studies suggest that low-lignin CAM species can serve as highly attractive bioenergy feedstocks for biorefinery industries.

XII. Carbon Sequestration Using CAM Species

A. Terrestrial Carbon Sequestration Strategies

Terrestrial carbon sequestration strategies are a means to remove carbon from the atmosphere to reverse the damaging effects of GHG emissions. Terrestrial carbon sequestration primarily focuses on either soil carbon sequestration or land-based phytosequestration. Soil carbon sequestration is aided by increasing soil carbon content through no-till farming, addition of biochar to the soil, perennial crop growth, nitrogen fixation, bioenergy crop production, wood burial, and the activity of various plant products (Nogia et al. 2016). Phytosequestration is carried out by terrestrial plants conducting C_3, C_4, and CAM photosynthesis, and aquatic or marine cyanobacterial carboxysomes, and algal pyrenoids (Nogia et al. 2016). Overall, the carbon storage capacity of soil is more than the vegetative capacity and presents a significant alternative carbon sink to nonbiological systems like oceanic and geological sequestration (Nogia et al. 2016; Roe et al. 2019).

Soil carbon sequestration is especially beneficial from perennial bioenergy crop production due to the sequestration of carbon via expansive and long-lived root systems (Lal 2008). The above ground leaf litter of perennial CAM crops such as *Agave* spp. and *Opuntia* spp. can reduce soil evaporation and increase soil organic matter (Nobel 1994). The production of biochar from bioenergy crops can enhance the soil carbon sequestration abilities of soil while also enhancing the soil health and fertility (Qambrani et al. 2017).

Best management practices (BMPs) utilized on various agricultural and non-agricultural lands contribute to a global average carbon sequestration rate of 0.05 to 1 Mg C ha^{-1} $year^{-1}$ (Lal 2008). Global soil carbon reservoirs to 1-m depth are estimated to account for 1462 to 1548 Pg of carbon, which is approximately 2.68 times the amount of biotic carbon reservoirs and 1.71 times the amount of atmospheric carbon (Lal 2008).

Adoption of BMPs, changes in land management, and the development of alternative agricultural and bioenergy crops have the potential to increase global soil organic carbon reservoirs. Given that semi-arid and arid lands cover 42% of the earth's surface and that desert regions typically have very low carbon sequestration capacities due to soil quality and degradation (Zika and Erb 2009), these areas also provide the greatest potential for future expansion of terrestrial and soil carbon sequestration projects by increasing net primary productivity (NPP) through managed agroecosystems (Daily 1995).

B. Agroecosystems to Combat Climate Change

Arid land crops present an opportunity to improve the use of marginal lands, which show low rates of NPP due to lack of adequate precipitation and are more vulnerable to erosion and degradation than areas receiving more rainfall (Daily 1995; Zika and Erb 2009). If managed carefully, arid land perennial crops such as *Opuntia* spp. and *Agave* spp., can provide important ecosystem services to marginal lands, such as preventing soil erosion, improving soil properties, provisioning food and forage, improving water quality and quantity, and providing wildlife habitats (Le Houérou 1996; Rodrígueza et al. 2006; Davis et al. 2017, 2019).

As of 2014, renewable energy comprised 30% of the global energy supply. Notable countries and associations such as the United States, the European Union, Brazil, and Australia have adopted policies to support renewable energy expansion through the development of biofuels (Acheampong et al. 2017). However, along with global initiatives to increase renewable energy use and promote bioenergy production, adverse environmental effects have been seen with unchecked bioenergy crop production. Combinations of livestock grazing and sugarcane expansion in critical areas of the rainforests offset the benefits of biofuel production through decreased viability of the local ecosystem and degradation of local carbon reservoirs typically held by the previously forested land (Otto et al. 2016). Additionally, the expansion of sugarcane in tropical regions caused a loss of 40% of the diversity of soil macrofauna groups and an overall soil species abundance decrease of approximately 90% (Franco et al. 2016).

Large-scale bioenergy production is estimated to reach 300 EJ/year by 2095 and use 490 Mha per year under well-irrigated conditions and increase to 690 Mha under rainfed conditions (Bonsch et al. 2016). In both scenarios, bioenergy crop production expands by appropriating forest- and pasture- land, where good water practices increase the ecosystem water stability at the expense of more land use. In contrast, the development of alternative CAM bioenergy crop production would avoid the use of forest- and pasture-land and utilize species that are far more water-use efficient.

Conservation agriculture aims to minimize the negative effects that agriculture has on the soil and local ecosystems through encouraging minimum soil disturbance tillage practices, retention of crop residues to replenish the soil nutrients, and diversification of agricultural products grown on the land to reduce soil degradation and nutrient depletion (Powlson et al. 2016).

Agroecosystem management also relies upon the stability, resilience, diversity, complexity, efficiency, and equality of the agricultural land (Zhu et al. 2018). In this context, CAM species present a new dimension to semi-arid land agricultural diversity, stability, and resilience for abandoned or degraded lands that deserve further investigation. For example, *O. ficus-indica* helped maintain soil organic carbon content and microbial growth similar to that of the previously forested conditions, whereas maize systems caused decreases in soil organic carbon content and microbial activity (Bautista-Cruz et al. 2018). Cactus canopies have the ability to trap fine soil materials and plant detritus from nearby lands and deposit them into the soil to increase soil carbon content as well as decrease soil erosion that typically occurs with annual crops (Bautista-Cruz et al. 2018). Soil management practices associated with cactus growth contribute to soil carbon content through the absence of ploughing that destroys soil aggregates that contribute to soil carbon stability (Bautista-Cruz et al. 2018).

However, in Mediterranean climates, the benefits of growing *O. ficus-indica* are not nearly as dramatic nor as impactful as the natural maquis shrubland (Novara et al. 2014). Moving from maquis to *Opuntia* spp. cultivation, soil carbon content decreased 65% after 28 years; from *Opuntia* spp. to olive grove cultivation, soil carbon content decreased a further 14% in 7 years (Novara et al. 2014). Soil mineralization and tillage of cactus soil contributed to the loss of soil carbon content (Novara et al. 2014; Bautista-Cruz et al. 2018). These observations would argue for low- or no-till strategies to reduce soil carbon losses in perennial production systems.

While CAM species are not always as beneficial to soil content retention and soil fertility, CAM species like *Opuntia* spp. are more beneficial to the land than other traditional annual C_3 photosynthesis crops (Bautista-Cruz et al. 2018). The preserved soil organic carbon content and microbial activity from *Opuntia* spp. cultivation contributes to environmental stability and has the potential to reverse soil damages caused by desertification and excessive soil usage. Additionally, CAM species can occupy abandoned pasture and farmlands and unutilized arid and semi-arid landscapes with minimal inputs (Borland et al. 2009; Cushman et al. 2015; Yang et al. 2015a, b).

C. *Urban Strategies for Carbon Sequestration*

In contrast to agronomically important CAM species, few studies have been performed on CAM species beyond small-scale production. Photosynthetic and transpiration rates were evaluated for three different facultative CAM *Sedum* spp., to determine their cooling effect on green roofs and potential use as alternative, low-water-input green roof plants (Kuronuma and Watanabe 2017). Under well-watered conditions, *Sedum* spp. showed similar photosynthetic and transpiration rates compared to the green roof plants *Zoysia matrella* and *Ophiopogon japonicus,* revealing that *Sedum* spp. may be a useful green roof alternative, especially in areas prone to droughts and high temperatures or for green roofs without irrigation systems (Kuronuma and Watanabe 2017; Kuronuma et al. 2018).

Green roofs offer an urban alternative means of capturing CO_2 and other pollutants, while also decreasing the costs of air conditioning through shading and evapotranspiration cooling effects (Dimitrijević et al. 2018; Kuronuma et al. 2018). For example, in Los Angeles, 2000 m^2 of uncut grass of a green roof could remove 4000 kg of air particulate matter, the equivalent of a year's emissions from one car (Dimitrijević et al. 2018). *Sedum* spp. have the potential to store on average 168 g C/m^2 in above-ground plant matter and root biomass (Dimitrijević et al. 2018). Expanding the use of CAM species for green roof applications is a key opportunity for CO_2 sequestration that deserves greater investment.

XIII. CAM Biodesign

An alternative to the expanded use of CAM crops is to transfer the CAM molecular machinery to C_3 and C_4 photosynthesis models or crop species, a process called CAM Biodesign (Borland et al. 2014, 2015). This approach, which leverages our improved understanding of CAM genomics, is expected to improve the water-use efficiency of the target host plant, thereby conferring greater climate resiliency (Yang et al. 2015a, b; Liu et al. 2018). Recently, progress towards CAM Biodesign has been reported by the overexpression and subcellular localization of individual carboxylation and decarboxylation module enzymes in *A. thaliana* (Lim et al. 2019). Notably, this report provided evidence that the expression of the carboxylation enzymes promotes malate accumulation and is associated with lower C_i, which appears to drive stomatal opening. Conversely, overexpression of decarboxylating enzymes reduced malate accumulation and likely increased C_i, which appears to drive stomatal closing.

While this is a useful foundation, assembly and installation of multi-gene circuits is needed to demonstrate that coordinated expression of multiple module enzymes can work together to improve the efficiency of the core C_4 cycle carboxylation and decarboxylation modules. Modeling the effects of each enzyme and suite of enzymes and transporters within a given gene circuit will inform each design-build-test-learn iteration of the synthetic biology cycle (Cheung et al. 2014; Shameer et al. 2018). Combining engineered tissue succulence, with CAM Biodesign can result in increased biomass and seed production (Lim et al. 2018; Lim et al. 2020) and salinity tolerance (Han et al. 2013).

XIV. Outlook

Current models of climate changes over the course of this century predict that the frequency and severity of drought events will increase as the atmosphere becomes warmer. A warmer and drier climate for much of the earth's surface will lead to drier soil conditions and the expansion of semi-arid and arid regions, resulting in lower terrestrial net primary productivity and reduced carbon sinks that threaten global food security and our ability to expand biomass production for bioenergy to offset future GHG emissions from the burning of fossil fuels. However, some CAM species offer greater climate resiliency due to greater WUE and a suite of anatomical adaptations that have co-evolved with CAM to improve their ability to occupy hotter and drier climates.

Increasing atmospheric CO_2 concentrations generally increases the CO_2 uptake and biomass productivity of CAM species. Desert-adapted CAM species also display considerable heat tolerance and selected *Agave* and *Opuntia* spp. exhibit annual productivities that rival those of traditional C_3 and C_4 photosynthesis crops. Understanding how these CAM species perform under current and future field conditions will require extensive field trials and the application of various productivity modeling approaches.

In addition, life cycle and life cycle costing assessments will be necessary to better estimate the economical viability and carbon sequestration potential of these water-use efficient bioenergy feedstocks for transportation fuels and electricity production. Although most CAM feedstocks are low in lignin, improving the methods for their efficient degradation and conversion into biofuels and biogas will require substantial research investments. Transfer of the water-use efficiency of CAM into non-CAM crops (CAM Biodesign) should be feasible in the near future; however, continued research investments are needed to optimize and disseminate this climate-resilience approach to a wide variety of crops.

Acknowledgements

This work was supported, in part, by the U.S. Department of Energy, Office of Science, Genomic Science Program under award number DE-SC0008834, the U.S. National Institute of Food and Agriculture (NIFA)/U.S. Department of Agriculture (USDA) under award number 2018-68005-27924, and the Nevada Agricultural Experiment Station (NEV-00377 and NEV-00380). The authors would like to thank Lisa Petrusa for critical reading of the manuscript.

References

Abraham PE, Yin H, Borland AM, Weighill D, Lim SD, De Paoli HC, Engle N, …, Yang X (2016) Transcript, protein and metabolite temporal dynamics in the CAM plant *Agave*. Nat Plants 2:16178

Acheampong M, Ertem FC, Kappler B, Neubauer P (2017) In pursuit of Sustainable Development Goal (SDG) number 7: will biofuels be reliable? Renew Sust Energ Rev 75:927–937

Adams WW, Osmond CB (1988) Internal CO_2 supply during photosynthesis of sun and shade grown CAM plants in relation to photoinhibition. Plant Physiol 86:117–123

Alencar BR, Dutra ED, Sampaio EV, Menezes RS, Morais JMA (2018) Enzymatic hydrolysis of cactus pear varieties with high solids loading for bioethanol production. Bioresour Technol 250:273–280

Aulio K (1986) CAM-like photosynthesis in *Littorella uniflora* (L.) Aschers.: the role of humidity. Ann Bot 58:273–275

Bachereau F, Marigo G, Asta J (1998) Effect of solar radiation (UV and visible) at high altitude on CAM-cycling and phenolic compound biosynthesis in *Sedum album*. Physiol Plant 104:203–210

Banerjee R, Chintagunta A, Ray S (2017) A cleaner and eco-friendly bioprocess for enhancing reducing sugar production from pineapple leaf waste. J Clean Prod 149:387–395

Bargel H, Koch K, Cerman Z, Neinhuis C (2006) Evans review no. 3: structure–function relationships of the plant cuticle and cuticular waxes – a smart material? Funct Plant Biol 33:893–910

Barker DH, Seaton GG, Robinson SA (1997) Internal and external photoprotection in developing leaves of the CAM plant *Cotyledon orbiculata*. Plant Cell Environ 20:617–624

Barrera-Zambrano VA, Lawson T, Olmos E, Fernández-García N, Borland AM (2014) Leaf anatomical traits which accommodate the facultative engagement of crassulacean acid metabolism in tropical trees of the genus *Clusia*. J Exp Bot 65:3513–3523

Bartlett MS, Vico G, Porporato A (2014) Coupled carbon and water fluxes in CAM photosynthesis: modeling quantification of water use efficiency and productivity. Plant Soil 383:111–138

Bautista-Cruz A, Leyva-Pablo T, de León-González F, Zornoza R, Martínez-Gallegos V, Fuentes-Ponce M, Rodríguez-Sánchez L (2018) Cultivation of *Opuntia ficus-indica* under different soil management practices: a possible sustainable agricultural system to promote soil carbon sequestration and increase soil microbial biomass and activity. Land Degrad Dev 29:38–46

Benoit I, Coutinho P, Schols H, Gerlach J, Henrissat B, de Vries R (2012) Degradation of different pectins by fungi: correlations and contrasts between the pectinolytic enzyme sets identified in genomes and the growth on pectins of different origin. BNC Genomics 13:321

Benoit I, Culleton H, Zhou M, DiFalco M, Aguilar-Osorio G, Battaglia E, Bouzid O, …, Gruben BS (2015) Closely related fungi employ diverse enzymatic strategies to degrade plant biomass. Biotechnol Biofuels 8:107

Bickford CP (2016) Ecophysiology of leaf trichomes. Funct Plant Biol 43:807–814

Bonsch M, Humpenöder F, Popp A, Bodirsky B, Dietrich JP, Rolinski S, Biewald A, …, Stevanovic M (2016) Trade-offs between land and water requirements for large-scale bioenergy production. Glob Change Biol Bioenergy 8:11–24

Borland A, Elliott S, Patterson S, Taybi T, Cushman J, Pater B, Barnes J (2006) Are the metabolic components of crassulacean acid metabolism up-regulated in response to an increase in oxidative burden? J Exp Bot 57:319–328

Borland A, Griffiths H, Hartwell J, Smith J (2009) Exploiting the potential of plants with crassulacean acid metabolism for bioenergy production on marginal lands. J Exp Bot 60:2879–2896

Borland A, Zambrano V, Ceusters J, Shorrock K (2011a) The photosynthetic plasticity of crassulacean acid metabolism: an evolutionary innovation for sustainable productivity in a changing world. New Phytol 191:619–633

Borland AM, Barerra-Zambrano VA, Ceusters J, Shorrock K (2011b) The photosynthetic plasticity of crassulacean acid metabolism: an evolutionary innovation for sustainable productivity in a changing world. New Phytol 191:619–633

Borland AM, Hartwell J, Weston DJ, Schlauch KA, Tschaplinski TJ, Tuskan GA, Yang X, Cushman JC (2014) Engineering crassulacean acid metabolism to improve water-use efficiency. Trends Plant Sci 15:327–338

Borland AM, Wullschleger SD, Weston DJ, Hartwell J, Tuskan GA, Yang X, Cushman JC (2015) Climate-resilient agroforestry: physiological responses to climate change and engineering of crassulacean acid metabolism (CAM) as a mitigation strategy. Plant Cell Environ 38:1833–1849

Borland AM, Leverett A, Hurtado-Castano N, Hu R, Yang X (2018) Functional anatomical traits of the photosynthetic organs of plants with crassulacean acid metabolism. In: Adams WW III, Terashima I (eds) The Leaf: A Platform for Performing Photosynthesis, vol 44. Springer, Cham, pp 281–306

Bormann KJ, Brown RD, Derksen C, Painter TH (2018) Estimating snow-cover trends from space. Nat Clim Chang 8:924

Boxall SF, Kadu N, Dever LV, Kneřová J, Waller JL, Gould PJ, Hartwell J (2020) Kalanchoë PPC1 is essential for crassulacean acid metabolism and the regulation of core circadian clock and guard cell signaling genes. Plant Cell 32:1136–1160

Brilhaus D, Bräutigam A, Mettler-Altmann T, Winter K, Weber AP (2016) Reversible burst of transcriptional changes during induction of crassulacean acid metabolism in *Talinum triangulare*. Plant Physiol 170:102–122

Budzianowski WM, Postawa K (2017) Renewable energy from biogas with reduced carbon dioxide footprint: implications of applying different plant configurations and operating pressures. Renew Sust Energ Rev 68:852–868

Calabrò PS, Catalán E, Folino A, Sánchez A, Komilis D (2018) Effect of three pretreatment techniques on the chemical composition and on the methane yields of *Opuntia ficus-indica* (prickly pear) biomass. Waste Manag Res 36:17–29

Ceusters J, Borland AM (2010) Impacts of elevated CO_2 on the growth and physiology of plants with crassulacean acid metabolism. Prog Bot 72:163–181

Challinor AJ, Watson J, Lobell DB, Howden SM, Smith DR, Chhetri N (2014) A meta-analysis of crop yield under climate change and adaptation. Nat Clim Chang 4:287–291

Cheeseman J (2016) Food security in the face of salinity, drought, climate change, and population growth. In: Khan MA, Ozturk M, Gul B, Ahmed MZ (eds) Halophytes for Food Security in Dry Lands. Academic, Amsterdam, pp 111–123

Chetti MB, Nobel PS (1988) Recovery of photosynthetic reactions after high-temperature treatments of a heat-tolerant cactus. Photosynth Res 18:277–286

Cheung CYM, Poolman MG, Fell DA, Ratcliffe RG, Sweetlove LJ (2014) A diel flux balance model captures interactions between light and dark metabolism during day-night cycles in C_3 and crassulacean acid metabolism leaves. Plant Physiol 165:917–929

Chomicki G, Bidel LP, Ming F, Coiro M, Zhang X, Wang Y, Baissac Y, …, Renner SS (2015) The velamen protects photosynthetic orchid roots against UV-B damage, and a large dated phylogeny implies multiple gains and losses of this function during the Cenozoic. New Phytol 205:1330–1341

Christopher JT, Holtum JAM (1996) Patterns of carbohydrate partitioning in the leaves of Crassulacean acid metabolism species during deacidification. Plant Physiol 112:393–399

Christopher JT, Holtum JAM (1998) Carbohydrate partitioning in the leaves of Bromeliaceae performing C_3 photosynthesis or Crassulacean acid metabolism. Aust J Plant Physiol 25:371–376

Cook BI, Smerdon JE, Seager R, Coats S (2014) Global warming and 21st century drying. Clim Dyn 43:2607–2627

Cook B, Ault T, Smerdon J (2015) Unprecedented 21st century drought risk in the American Southwest and Central Plains. Sci Adv 1:e1400082

Cota-Sánchez JH (2002) Taxonomy, distribution, rarity status and uses of Canadian cacti. Haseltonia 9:17–25

Crayn DM, Smith JAC, Winter K (2004) Multiple origins of Crassulacan acid metabolism and the epiphytic habit in the neotropical family Bromeliaceae. Proc Natl Acad Sci USA 101:3703–3708

Crayn DM, Winter K, Schulte K, Smith JA (2015) Photosynthetic pathways in Bromeliaceae: phylogenetic and ecological significance of CAM and C_3 based on carbon isotope ratios for 1893 species. Bot J Linn Soc 178:169–221

Croonenborghs S, Ceusters J, Londers E, De Proft MP (2009) Effects of elevated CO_2 on growth and morphological characteristics of ornamental bromeliads. Sci Hortic 121:192–198

Cui M, Nobel PS (1994) Gas exchange and growth responses to elevated CO_2 and light levels in the CAM species *Opuntia ficus-indica*. Plant Cell Environ 17:935–944

Cui M, Miller PM, Nobel PS (1993) CO_2 exchange and growth of the crassulacean acid metabolism plant *Opuntia ficus-indica* under elevated CO_2 in open-top chambers. Plant Physiol 103:519–524

Cushman JC, Tillett RL, Wood JA, Branco JA, Schlauch KA (2008) Large-scale mRNA expression profiling in the common ice plant, *Mesembryanthemum crystallinum*, performing C_3 photosynthesis and Crassulacean acid metabolism (CAM). J Exp Bot 59:1875–1894

Cushman JC, Davis SC, Yang X, Borland AM (2015) Development and use of bioenergy feedstocks for semi-arid and arid lands. J Exp Bot 66:4177–4193

Cuthber MO, Gleeson T, Moosdorf N, Befus KM, Schneider A, Hartmann J, Lehner B (2019) Global patterns and dynamics of climate–groundwater interactions. Nat Clim Chang 9:137

Dai A (2013) Increasing drought under global warming in observations and models. Nat Clim Chang 3:52–58

Dai A, Zhao T, Chen J (2018) Climate change and drought: a precipitation and evaporation perspective. Curr Climate Change Rep 4:301–312

Daily GC (1995) Restoring value to the world's degraded lands. Science 269:350–354

Davis SC, Dohleman FG, Long SP (2011) The global potential for *Agave* as a biofuel feedstock. Glob Change Biol Bioenergy 3:68–78

Davis SC, LeBauer DS, Long SP (2014) Light to liquid fuel: theoretical and realized energy conversion efficiency of plants using Crassulacean Acid Metabolism (CAM) in arid conditions. J Exp Bot 65:3471–3478

Davis SC, Ming R, LeBauer DS, Long SP (2015) Toward systems-level analysis of agricultural production from crassulacean acid metabolism (CAM): scaling from cell to commercial production. New Phytol 208:66–72

Davis SC, Kuzmick ER, Niechayev N, Hunsake DJ (2017) Productivity and water use efficiency of

Agave americana in the first field trial as bioenergy feedstock on arid lands. Glob Change Biol Bioenergy 9:314–325
Davis SC, Simpson J, Vega G, Del Carmen K, Niechayev NA, Tongerlo EV, Castano NH, …, Búrque A (2019) Undervalued potential of crassulacean acid metabolism for current and future agricultural production. J Exp Bot 70:6521–6537
de Cortazar VG, Acevedo E, Nobel PS (1985) Modeling of PAR interception and productivity by *Opuntia ficus-indica*. Agric For Meteorol 34:145–162
de Cortázar VG, Acevedo R, Nobel PS (1986) Modeling of PAR interception and productivity of a prickly pear cactus, *Opuntia ficus-indica* L., at various spacings. Agron J 78:80–85
De Fraiture C, Wichelns D (2010) Satisfying future water demands for agriculture. Agric Water Manag 97:502–511
de Santo AV, Alfani A, Russo G, Fioretto A (1983) Relationship between CAM and succulence in some species of Vitaceae and Piperaceae. Bot Gaz 144:342–346
de Souza Filho PF, Ribeiro VT, dos Santos ES, de Macedo GR (2016) Simultaneous saccharification and fermentation of cactus pear biomass – evaluation of using different pretreatments. Ind Crop Prod 89:425–433
DePaoli HC, Borland AM, Tuskan GA, Cushman JC, Yang X (2014) Synthetic biology as it relates to CAM photosynthesis: challenges and opportunities. J Exp Bot 65:3381–3393
Didden-Zopfy B, Nobel PS (1982) High temperature tolerance and heat acclimation of *Opuntia bigelovii*. Oecologia 52:176–180
Dimitrijević D, Živković P, Branković J, Dobrnjac M, Stevanović Ž (2018) Air pollution removal and control by green living roof systems. Acta Tech Corviniensis Bull Eng 11:47–50
Divyalakshmi P, Murugan D, Sivarajan M, Sivasamy A, Saravanan P, Rai CL (2017) Optimization and biokinetic studies on pretreatment of sludge for enhancing biogas production. Int J Environ Sci Technol 14:813–822
Domínguez E, Heredia-Guerrero JA, Heredia A (2011) The biophysical design of plant cuticles: an overview. New Phytol 189:938–949
Downton WJ, Berry J, Seemann JR (1984) Tolerance of photosynthesis to high temperature in desert plants. Plant Physiol 74:786–790
Drennan PM, Nobel PS (2000) Responses of CAM species to increasing atmospheric CO_2 concentrations. Plant Cell Environ 23:767–781
Edwards EJ (2019) Evolutionary trajectories, accessibility and other metaphors: the case of C_4 and CAM photosynthesis. New Phytol 223:1742–1755
Eggli U, Nyffeler R (2009) Living under temporarily arid conditions–succulence as an adaptive strategy. Bradleya 27:13–36
Ehleringer JR, Monson RK (1993) Evolutionary and ecological aspects of photosynthetic pathway variation. Annu Rev Ecol Syst 24:411–439
Escamilla-Treviño LL (2012) Potential of plants from the genus *Agave* as bioenergy crops. Bioenergy Res 5:1–9
Famiglietti J, Rodell M (2013) Water in the balance. Science 340:1300–1301
FAO (2017) Coping with water scarcity in agriculture: a global framework for actions in a changing climate. In: Food and Agriculture Organization of the United Nations
Faralli M, Matthews J, Lawson T (2019) Exploiting natural variation and genetic manipulation of stomatal conductance for crop improvement. Curr Opin Plant Biol 49:1–7
Farquhar GD, von Caemmerer S, Berry JA (1980) A biochemical model of photosynthetic CO_2 assimilation in leaves of C_3 species. Planta 149:78–90
February EC, Matimati I, Hedderson TA, Musil CF (2013) Root niche partitioning between shallow rooted succulents in a South African semi desert: implications for diversity. Plant Ecol 214:1181–1187
Feller U (2016) Drought stress and carbon assimilation in a warming climate: reversible and irreversible impacts. J Plant Physiol 203:84–94
Feng S, Fu Q (2013) Expansion of global drylands under a warming climate. Atmos Chem Phys 13:10081–10094
Fernández V, Bahamonde HA, Peguero-Pina JJ, Gil-Pelegrín E, Sancho-Knapik D, Gil L, Goldbach HE, Eichert T (2017) Physico-chemical properties of plant cuticles and their functional and ecological significance. J Exp Bot 68:5293–5306
Field C, Barros V, Mach K, Mastrandrea M, van Aalst M, Adger W, Arent D, …, Yohe G (2014) Climate Change 2014: Impacts, Adaptation, and Vulnerability. Part A: Global and Sectoral Aspects. Cambridge University Press, Cambridge/New York
Flores-Gómez CA, Silva EM, Zhong C, Dale BE, da Costa SL, Balan V (2018) Conversion of lignocellulosic agave residues into liquid biofuels using an AFEX™-based biorefinery. Biotechnol Biofuels 11:7
Flörke M, Schneider C, McDonald RI (2018) Water competition between cities and agriculture driven by climate change and urban growth. Nat Sustain 1:51–58
Franck N, Muñoz V, Alfaro F, Arancibia D, Pérez-Quezada J (2013) Estimating the carbon assimilation of growing cactus pear cladodes through different methods. Acta Hortic 19:157–164

Franks PJ, Drake PL, Beerling DJ (2009) Plasticity in maximum stomatal conductance constrained by negative correlation between stomatal size and density: an analysis using *Eucalyptus globulus*. Plant Cell Environ 32:1737–1748

Franco AC, Ball E, Lüttge U (1992) Differential effects of drought and light levels on accumulation of citric and malic acids during CAM in Clusia. Plant Cell Environ 15:821–829

Franco ALC, Bartz MLC, Cherubin MR, Baretta D, Cerri CEP, Feigl BJ, Wall DH, …, Cerri CC (2016) Loss of soil (macro) fauna due to the expansion of Brazilian sugarcane acreage. Sci Total Environ 563:160–168

Freschi L, Takahashi CA, Cambui CA, Semprebom TR, Cruz AB, Mioto PT, de Melo Versieux L, …, Mercier H (2010) Specific leaf areas of the tank bromeliad *Guzmania monostachia* perform distinct functions in response to water shortage. J Plant Physiol 167:526–533

Fu Q, Feng S (2014) Responses of terrestrial aridity to global warming. J Geophys Res Atmos 119:7863–7875

Garcia de Cortázar V, Nobel P (1991) Prediction and measurement of high annual productivity for *Opuntia ficus-indica*. Agric For Meterol 56:261–272

Garcia de Cortazar V, Nobel PS (1992) Biomass and fruit production for the prickly pear cactus *Opuntia ficus-indica*. J Am Soc Hortic Sci 117:568–562

Garcia-Moya E, Romero-Mananares A, Nobel PS (2011) Highlights of *Agave* productivty. Glob Change Biol Bioenergy 3:4–14

Gerbens-Leenes W, Hoekstra AY (2012) The water footprint of sweeteners and bio-ethanol. Environ Int 40:202–211

Gerbens-Leenes W, Hoekstra A, van der Meer T (2009) The water footprint of bioenergy. Proc Natl Acad Sci USA 106:10219–10223

Gerbens-Leenes PW, Van Lienden AR, Hoekstra AY, Van der Meer TH (2012) Biofuel scenarios in a water perspective: the global blue and green water footprint of road transport in 2030. Glob Environ Chang 22:764–775

Gil LS, Maupoey PF (2018) An integrated approach for pineapple waste valorisation. Bioethanol production and bromelain extraction from pineapple residues. J Clean Prod 172:1224–1231

Goldstein G, Nobel PS (1991) Changes in osmotic pressure and mucilage during low-temperature acclimation of *Opuntia ficus-indica*. Plant Physiol 97:954–961

Goldstein G, Nobel PS (1994) Water relations and low-temperature acclimation for cactus species varying in freezing tolerance. Plant Physiol 104:675–681

Gorelick R, Drezner TD, Hancock K (2015) Freeze-tolerance of cacti (Cactaceae) in Ottawa, Ontario, Canada. Madrono 62:33–46

Grace OM (2019) Succulent plant diversity as natural capital. Plants People Planet 1:336–345

Graham EA, Nobel PS (1996) Long-term effects of a doubled atmospheric CO_2 concentration on the CAM species *Agave deserti*. J Exp Bot 47:61–69

Graham EA, Nobel PS (1999) Root water uptake, leaf water storage and gas exchange of a desert succulent: implications for root system redundancy. Ann Bot 84:213–223

Graham S, Eastwick C, Snape C, Quick W (2017) Mechanical degradation of biomass wood pellets during long term stockpile storage. Fuel Process Technol 160

Grams TE, Thiel S (2002) High light-induced switch from C_3-photosynthesis to Crassulacean acid metabolism is mediated by UV-A/blue light. J Exp Bot 53:1475–1483

Grams TE, Kluge M, Lüttge U (1995) High temperature-adapted plants of *Kalanchoë daigremontiana* show changes in temperature dependence of the endogenous CAM rhythm. J Exp Bot 46:1927–1929

Griffiths H (1989a) Carbon dioxide concentrating mechanisms and the evolution of CAM in vascular epiphytes. In: Lüttge U (ed) Vascular Plants as Epiphytes: Evolution and Ecophysiology. Springer, Berlin, pp 42–86

Griffiths H (1989b) Crassulacean acid metabolism: a re-appraisal of physiological plasticity in form and function. Adv Bot Res 15:43–92

Griffiths H (2013) Plant venation: from succulence to succulents. Curr Biol 23:R340–R341

Griffiths H, Males J (2017) Succulent plants. Curr Biol 27:R890–R896

Griffiths H, Ong BL, Avadhani PN, Goh CJ (1989) Recycling of respiratory CO_2 during Crassulacean acid metabolism: alleviation of photoinhibition in *Pyrrosia piloselloides*. Planta 179:115–122

Griffiths H, Cousins AB, Badger MR, von Caemmerer S (2007) Discrimination in the dark. Resolving the interplay between metabolic and physical constraints to phospho*enol*pyruvate carboxylase activity during the crassulacean acid metabolism cycle. Plant Physiol 143:1055–1067

Habibi G, Hajiboland R (2011) Comparison of water stress and UV radiation effects on induction of CAM and antioxidative defense in the succulent *Rosularia elymaitica* (Crassulaceae). Acta Biol Cracov Ser Bot 53:15–24

Han Y, Wang W, Sun J, Ding M, Zhao R, Deng S, Wang F, …, Chen S (2013) *Populus euphratica* XTH overexpression enhances salinity tolerance by the development of leaf succulence in transgenic tobacco plants. J Exp Bot 64:4225–4238

Hartzell S, Bartlett MS, Porporato A (2018) Unified representation of the C_3, C_4, and CAM photosynthetic pathways with the Photo3 model. Ecol Model 384:173–187

Hejazi M, Voisin N, Liu L, Bramer L, Fortin D, Hathaway J, Huang M, …, Zhou Y (2015) 21st century United States emissions mitigation could increase water stress more than the climate change it is mitigating. Proc Natl Acad Sci USA 112:10635–10640

Heredia-Guerrero JA, Guzman-Puyol S, Benítez JJ, Athanassiou A, Heredia A, Domínguez E (2018) Plant cuticle under global change: biophysical implications. Glob Chang Biol 24:2749–2751

Herrera A (2008) Crassulacean acid metabolism and fitness under water deficit stress: if not for carbon gain, what is facultative CAM good for? Ann Bot 103:645–653

Herrera A, Delgado J, Paraguatey I (1991) Occurrence of inducible crassulacean acid metabolism in leaves of *Talinum triangulare* (Portulacaceae). J Exp Bot 42:493–499

Heyduk K, Burrell N, Lalani F, Leebens-Mack J (2015) Gas exchange and leaf anatomy of a C3–CAM hybrid, *Yucca gloriosa* (Asparagaceae). J Exp Bot 67:1369–1379

Heyduk K, Ray JN, Ayyampalayam S, Leebens-Mack J (2018) Shifts in gene expression profiles are associated with weak and strong crassulacean acid metabolism. Am J Bot 105:587–601

Horn JW, Xi Z, Riina R, Peirson JA, Yang Y, Dorsey BL, Berry PE, …, Wurdack KJ (2014) Evolutionary bursts in *Euphorbia* (Euphorbiaceae) are linked with photosynthetic pathway. Evolution 68:3485–3504

Huang J, Yu H, Guan X, Wang G, Guo R (2015) Accelerated dryland expansion under climate change. Nat Clim Chang 6:166–171

Huerta C, Freire M, Cardemil L (2013) Expression of hsp70, hsp100 and ubiquitin in *Aloe barbadensis* Miller under direct heat stress and under temperature acclimation conditions. Plant Cell Rep 32:293–307

Hultine KR, Dettman DL, English NB, Williams DG (2019) Giant cacti: isotopic recorders of climate variation in warm deserts of the Americas. J Exp Bot 70:6509–6519

Idso SB, Kimball BA, Anderson MG, Szarek SR (1986) Growth response of a succulent plant, *Agave vilmoriniana*, to elevated CO_2. Plant Physiol 80:796–797

Ishikawa M, Gusta LV (1996) Freezing and heat tolerance of *Opuntia* cacti native to the Canadian prairie provinces. Can J Bot 74:1890–1895

Jayani RS, Shukla SK, Gupta R (2010) Screening of bacterial strains for polygalacturonase activity: its production by *Bacillus sphaericus* (MTCC 7542). Enzyme Res 2010

Jones JW, Antle JM, Basso B, Boote KJ, Conant RT, Foster I, Godfray HCJ, …, Wheeler TR (2017) Brief history of agricultural systems modeling. Agric Syst 155:240–254

Kaushal N, Bhandari K, Siddique KH, Nayyar H (2016) Food crops face rising temperatures: an overview of responses, adaptive mechanisms, and approaches to improve heat tolerance. Cogent Food Agric 2:1134380

Kee SC, Nobel PS (1986) Concomitant changes in high temperature tolerance and heat-shock proteins in desert succulents. Plant Physiol 80:596–598

Keeley JE (1981) *Isoetes howellii*: a submerged aquatic CAM plant? Am J Bot 68:420–424

Keeley JE (1983) Crassulacean acid metabolism in the seasonally submerged aquatic *Isoetes howellii*. Oecologia 58:57–62

Keeley JE (1996) Aquatic CAM photosynthesis. In: Winter K, Smith J (eds) Crassulacean Acid Metabolism. Springer, Berlin, pp 281–295

Keeley JE (1998) CAM photosynthesis in submerged aquatic plants. Bot Rev 64:121–175

Keeley JE, Rundel PW (2003) Evolution of CAM and C_4 carbon-concentrating mechanisms. Int J Plant Sci 164:S55–S77

Keeley JE, Sandquist DR (1991) Diurnal photosynthesis cycle in CAM and non-CAM seasonal-pool aquatic macrophytes. Ecology 72:716–727

King C, Webber M, Duncan I (2010) The water needs for LDV transportation in the United States. Energy Policy 38:1157–1167

Kirschbaum MU (2004) Direct and indirect climate change effects on photosynthesis and transpiration. Plant Biol 6:242–253

Koutroulis AG, Papadimitriou LV, Grillakis MG, Tsanis IK, Warren R, Betts RA (2019) Global water availability under high-end climate change: a vulnerability based assessment. Glob Planet Chang 175:52–63

Krause GH, Winter K, Krause B, Virgo A (2016) Protection by light against heat stress in leaves of tropical crassulacean acid metabolism plants containing high acid levels. Funct Plant Biol 43:1061–1069

Krümpel J, George T, Gasston B, Francis G, Lemmer A (2020) Suitability of *Opuntia ficus-indica* (L) mill. And *Euphorbia tirucalli* L. as energy crops for anaerobic digestion. J Arid Environ 174:104047

Kuloyo OO, du Preez JC, del Prado García-Aparicio M, Kilian SG, Steyn L, Görgens J (2014) *Opuntia*

ficus-indica cladodes as feedstock for ethanol production by *Kluyveromyces marxianus* and *Saccharomyces cerevisiae*. World J Microbiol Biotechnol 30:3173–3183
Kumar A, Li C, Portis AR (2009) *Arabidopsis thaliana* expressing a thermostable chimeric Rubisco activase exhibits enhanced growth and higher rates of photosynthesis at moderately high temperatures. Photosynth Res 100:143–153
Kurek I, Chang TK, Bertain SM, Madrigal A, Liu L, Lassner MW, Zhu G (2007) Enhanced thermostability of *Arabidopsis* Rubisco activase improves photosynthesis and growth rates under moderate heat stress. Plant Cell 19:3230–3241
Kuronuma T, Watanabe H (2017) Photosynthetic and transpiration rates of three *Sedum* species used for green roofs. Environ Control Biol 55:137–141
Kuronuma T, Watanabe H, Ishihara T, Kou D, Toushima K, Ando M, Shindo S (2018) CO_2 payoff of extensive green roofs with different vegetation species. Sustainability 10:2256
Lal R (2008) Carbon sequestration. Philos Trans R Soc London B Biol Sci 363:815–830
Lawson T (2008) Guard cell photosynthesis and stomatal function. New Phytol 181:13–34
Lawson T, Blatt MR (2014) Stomatal size, speed, and responsiveness impact on photosynthesis and water use efficiency. Plant Physiol 164:1556–15570
Lawson T, Kramer DM, Raines CA (2012) Improving yield by exploiting mechanisms underlying natural variation of photosynthesis. Curr Opin Biotechnol 23:215–220
Le Houérou HN (1996) The role of cacti (*Opuntia* spp.) in erosion control, land reclamation, rehabilitation and agricultural development in the Mediterranean Basin. J Arid Environ 33:135–159
Lee DM, Assmann SM (1992) Stomatal responses to light in the facultative crassulacean acid metabolism species, *Portulacaria afra*. Physiol Plant 85:35–42
Leng G, Hall J (2019) Crop yield sensitivity of global major agricultural countries to droughts and the projected changes in the future. Sci Total Environ 654:811–821
Lesk C, Rowhani P, Ramankutty N (2016) Influence of extreme weather disasters on global crop production. Nature 529:84–87
Leuning R, Kelliher FM, De Pury DG, Schulze ED (1995) Leaf nitrogen, photosynthesis, conductance and transpiration: scaling from leaves to canopies. Plant Cell Environ 18:1183–1200
Levidow L, Zaccaria D, Maia R, Vivas E, Todorovic M, Scardigno A (2014) Improving water-efficient irrigation: prospects and difficulties of innovative practices. Agric Water Manag 146:84–94
Li H, Foston MB, Kumar R, Samuel R, Gao X, Hu F, Ragauskas AJ, Wyman CE (2012) Chemical composition and characterization of cellulose for *Agave* as a fast-growing, drought-tolerant biofuels feedstock. RSC Adv 2:4951–4958
Li H, Pattathil S, Foston MB, Ding SY, Kumar R, Gao X, Mittal A, …, Wyman CE (2014) Agave proves to be a low recalcitrant lignocellulosic feedstock for biofuels production on semi-arid lands. Biotechnol Biofuels 7:50
Li W, Li Q, Wang Y, Zheng L, Zhang Y, Yu Z, Chen H, Zhang J (2018) Efficient bioconversion of organic wastes to value-added chemicals by soaking, black soldier fly (*Hermetia illucens* L.) and anaerobic fermentation. J Environ Manag 227:267–276
Lim SD, Yim WC, Liu D, Hu R, Yang X, Cushman JC (2018) A *Vitis vinifera* basic helix–loop–helix transcription factor enhances plant cell size, vegetative biomass and reproductive yield. Plant Biotechnol J 16:1595–1615
Lim SD, Lee S, Yim WC, Choi W-G, Cushman JC (2019) Laying the foundation for crassulacean acid metabolism (CAM) biodesign: expression of the C_4 metabolism cycle genes of CAM in *Arabidopsis*. Front Plant Sci 10:101
Lim SD, Mayer JA, Yim WC, Cushman JC. (2020) Plant tissue succulence engineering improves water-use efficiency, water-deficit stress attenuation and salinity tolerance in Arabidopsis. The Plant Journal. 103: 1049–1072
Lima MS, Damasio AR, Crnkovic PM, Pinto MR, da Silva AM, da Silva JC, Segato F, …, Polizeli MD (2016) Co-cultivation of *Aspergillus nidulans* recombinant strains produces an enzymatic cocktail as alternative to alkaline sugarcane bagasse pretreatment. Front Microbiol 7:583
Lin Q, Abe S, Nose A, Sunami A, Kawamitsu Y (2006) Effects of high night temperature on crassulacean acid metabolism (CAM) photosynthesis of *Kalanchoë pinnata* and *Ananas comosus*. Plant Prod Sci 9:10–19
Lindner J, Zielonka S, Oechsner H, Lemmer A (2015) Effects of mechanical treatment of digestate after anaerobic digestion on the degree of degradation. Bioresour Technol 178:194–200
Liu Y, Wang J, Wolcott MP (2017) Evaluating the effect of wood ultrastructural changes from mechanical treatment on kinetics of monomeric sugars and chemicals production in acid bisulfite treatment. Bioresour Technol 226:24–30
Liu D, Palla KJ, Hu R, Moseley RC, Mendoza C, Chen M, Abraham PE, …, Yang X (2018) Perspectives on the basic and applied aspects of crassulacean acid metabolism (CAM) research. Plant Sci 274:394–401

Lobell D, Schlenker W, Costa-Roberts J (2011) Climate trends and global crop production since 1980. Science 333:616–620

Loik ME, Nobel PS (1993) Freezing tolerance and water relations of *Opuntia fragilis* from Canada and the United States. Ecology 74:1722–1732

Lueangwattanapong K, Ammam F, Mason PM, Whitehead C, McQueen-Mason SJ, Gomez LD, Smith JA, Thompson IP (2020) Anaerobic digestion of Crassulacean acid metabolism plants: exploring alternative feedstocks for semi-arid lands. Bioresour Technol 297:122262

Lujan R, Lledias F, MartÍnez LM, Barreto R, Cassab GI, Nieto-Sotelo JO (2009) Small heat-shock proteins and leaf cooling capacity account for the unusual heat tolerance of the central spike leaves in *Agave tequilana* var. Weber Plant Cell Environ 32:1791–1803

Luo Y, Nobel PS (1993) Growth characteristics of newly initiated cladodes of *Opuntia ficus-indica* as affected by shading, drought and elevated CO_2. Physiol Plant 87:467–474

Luttge U (2000) Light-stress and crassulacean acid metabolism. Phyton-Horn 40:65–82

Lüttge U (2002) CO_2-concentrating: consequences in crassulacean acid metabolism. J Exp Bot 53:2131–2142

Lüttge U, Nobel PS (1984) Day-night variations in malate concentration, osmotic pressure, and hydrostatic pressure in *Cereus validus*. Plant Physiol 5:804–807

Males J (2016) Think tank: water relations of Bromeliaceae in their evolutionary context. Bot J Linn Soc 181:415–440

Males J (2017) Secrets of succulence. J Exp Bot 68:2121–2134

Males J (2018) Concerted anatomical change associated with crassulacean acid metabolism in the Bromeliaceae. Funct Plant Biol 45:681–695

Males J, Griffiths H (2017) Stomatal biology of CAM plants. Plant Physiol 174:550–560

Mason PM, Glover K, Smith JAC, Willis KJ, Woods J, Thompson IP (2015) The potential of CAM crops as a globally significant bioenergy resource: moving from 'fuel or food' to 'fuel and more food'. Energy Enivon Sci 8:2320–2329

Mathur S, Agrawal D, Jajoo A (2014) Photosynthesis: response to high temperature stress. J Photochem Photobiol B Biol 137:116–126

Maurya DP, Singla A, Negi S (2015) An overview of key pretreatment processes for biological conversion of lignocellulosic biomass to bioethanol. 3Biotech 5:597–609

Maxwell K, von Caemmerer S, Evans JR (1997) Is a low internal conductance to CO_2 diffusion a consequence of succulence in plants with crassulacean acid metabolism? Aust J Plant Physiol 24:777–786

Maxwell C, Griffiths H, Young AJ (1994) Photosynthetic acclimation to light regime and water stress by the C -CAM epiphyte *Guzmania monostachia*: gas-exchange characteristics, photochemical efficiency and the xanthophyll cycle. Funct Ecol 8:746–754

Mekeonnen MM, Hoekstra AY (2016) Four billion people facing severe water scarcity. Sci Adv 2:21500323

Minardi BD, Voytena AP, Santos M, Randi ÁM (2014) Water stress and abscisic acid treatments induce the CAM pathway in the epiphytic fern *Vittaria lineata* (L.) Smith. Photosynthetica 52:404–412

Moiwo J, Lu W, Tao F (2012) GRACE, GLDAS and measured groundwater data products show water storage loss in Western Jilin, China. Water Sci Technol 65:1606–1614

Monteiro JAF, Zotz G, Körner C (2009) Tropical epiphytes in a CO_2-rich atmosphere. Acta Oecol 35:60–68

Mortensen LM, Moe R (1992) Effects of CO_2 enrichment and different day/night temperature combinations on growth and flowering of *Rosa* L. and *Kalanchoe blossfeldiana* v. poelln. Sci Hortic 51:145–153

Mshandete A, Björnsson L, Kivaisi AK, Rubindamayugi ST, Mattiasson B (2005) Enhancement of anaerobic batch digestion of sisal pulp waste by mesophilic aerobic pre-treatment. Water Res 39:1569–1575

Myovela H, Mshandete AM, Imathiu S (2019) Enhancement of anaerobic batch digestion of spineless cacti (*Opuntia ficus indica*) feedstock by aerobic pre-treatment. Afr J Biotechnol 18:12–22

Naumann G, Alfieri L, Wyser K, Mentaschi L, Betts RA, Carrao H, Spinoni J, …, Feyen L (2018) Global changes in drought conditions under different levels of warming. Geophys Res Lett 45:3285–3296

Nelson EA, Sage RF (2008) Functional constraints of CAM leaf anatomy: tight cell packing is associated with increased CAM function across a gradient of CAM expression. J Exp Bot 59:1841–1850

Nelson EA, Sage TL, Sage RF (2005) Functional leaf anatomy of plants with crassulacean acid metabolism. Funct Plant Biol 32:409–419

Neubig KM, Whitten WM, Williams NH, Blanco MA, Endara L, Burleigh G, Silvera K, …, Chase MW (2012) Generic recircumscriptions of Oncidiinae (Orchidaceae: Cymbidieae) based on maximum likelihood analysis of combined DNA datasets. Bot J Linn Soc 168:117–146

Niechayev NA, Jones AM, Rosenthal DM, Davis SC (2019a) A model of environmental limitations on production of *Agave americana* L. grown as a biofuel crop in semi-arid regions. J Exp Bot 70:6549–6559

Niechayev NA, Pereira PN, Cushman JC (2019b) Understanding trait diversity associated with crassulacean acid metabolism (CAM). Curr Opin Plant Biol 49:74–85

Nobel PS (1978) Surface temperatures of cacti: influences of environmental and morphological factors. Ecology 59:986–996

Nobel PS (1980) Interception of photosynthetically active radiation by cacti of different morphology. Oecologia 45:160–166

Nobel PS (1983) Spine influences on PAR interception, stem temperature, and nocturnal acid accumulation by cacti. Plant Cell Environ 6:153–159

Nobel PS (1984) Extreme temperatures and thermal tolerances for seedlings of desert succulents. Oecologia 62:310–317

Nobel PS (1988) Environmental Biology of Agaves and Cacti. Cambridge University Press, New York

Nobel PS (1989) A nutrient index quantifying productivity of agaves and cacti. J Appl Ecol 26:635–645

Nobel PS (1991a) Achievable productivities of certain CAM plants: basis for high values compared with C_3 and C_4 plants. New Phytol 119:183–205

Nobel PS (1991b) Environmental productivity indices and productivity for *Opuntia ficus-indica* under current and elevated atmospheric CO_2 levels. Plant Cell Environ 14:637–646

Nobel PS (1994) Remarkable Agaves and Cacti. Oxford University Press, New York

Nobel PS (1996) High productivity of certain agronomic CAM species. In: Winter K, Smith JAC (eds) Crassulacean Acid Metabolism: Biochemistry, Ecophysiology and evolution. Springer, Berlin, pp 255–265

Nobel PS, Cui M (1992) Hydraulic conductances of the soil, the root-soil air gap, and the root: changes for desert succulents in drying soil. J Exp Bot 43:319–326

Nobel PS, Hartsock TL (1984) Physiological responses of *Opuntia ficus-indica* to growth temperature. Physiol Plant 60:98–105

Nobel PS, Hartsock TL (1986a) Short-term and long-term responses of crassulacean acid metabolism plants to elevated CO_2. Plant Physiol 82:604–606

Nobel PS, Hartsock TL (1986b) Temperature, water, and PAR influences on predicted and measured productivity of *Agave deserti* at various elevations. Oecologia 68:181–185

Nobel PS, Israel AA (1994) Cladode development, environmental responses of CO_2 uptake, and productivity for *Opuntia ficus-indica* under elevated CO_2. J Exp Bot 45:295–303

Nobel PS, Meyer SE (1985) Field productivity of a CAM plant, *Agave salmiana*, estimated using daily acidity changes under various environmental conditions. Physiol Plant 65:397–404

Nobel PS, Quero E (1986) Environmental productivity indices for a Chihuahuan Desert CAM plant, *Agave lechuguilla*. Ecology 67:1–11

Nobel PS, Sanderson J (1984) Rectifier-like activities of roots of two desert succulents. J Exp Bot 35:727–737

Nobel PS, Smith SD (1983) High and low temperature tolerances and their relationships to distribution of agaves. Plant Cell Environ 6:711–719

Nobel PS, Valenzuela AG (1987) Environmental responses and productivity of the CAM plant, *Agave tequilana*. Agric For Meterol 39:319–334

Nobel PS, Geller GN, Kee SC, Zimmerman AD (1986) Temperatures and thermal tolerances for cacti exposed to high temperatures near the soil surface. Plant Cell Environ 9:279–287

Nobel PS, Garciamoya E, Quero E (1992) High annual productivity of certain agaves and cacti under cultivation. Plant Cell Environ 15:329–335

Nobel PS, Castañeda M, North G, Pimienta-Barrios E, Ruiz A (1998) Temperature influences on leaf CO_2 exchange, cell viability and cultivation range for *Agave tequilana*. J Arid Environ 39:1–9

Nogia P, Sidhu GK, Mehrotra R, Mehrotra S (2016) Capturing atmospheric carbon: biological and nonbiological methods. Int J Low-Carbon Technol 11:266–274

North GB, Nobel PS (1991) Changes in hydraulic conductivity and anatomy caused by drying and rewetting roots of *Agave deserti* (Agavaceae). Am J Bot 78:906–915

North GB, Moore L, Nobel PS (1995) Cladode development for Opuntia ficus-indica (Cactaceae) under current and doubled CO_2 concentrations. Am J Bot 82:159–166

Novara A, Pereira P, Santoro A, Kuzyakov Y, La Mantia T (2014) Effect of cactus pear cultivation after Mediterranean maquis on soil carbon stock, δ13C spatial distribution and root turnover. Catena 118:84–90

Nungesser D, Kluge M, Tolle H, Oppelt W (1984) A dynamic computer model of the metabolic and regulatory processes in Crassulacean acid metabolism. Planta 162:204–214

Ochoa MJ, Uhart S (2004) Nitrogen availability and fruit yield generation in cactus pear (*Opuntia ficus-indica*): II. Effects on solar radiation use effi-

ciency and dry matter accumulation. Acta Hortic 728:125–130
OECD-FAO (2013) Agricultural outlook 2013–2022 highlights. In. Food and Agricultural Organization (FAO) of the United Nations, pp 1–116
Ogburn RM, Edwards EJ (2013) Repeated origin of three-demensional leaf venation releases constraints of the evolution of succulence in plants. Curr Biol 23:722–726
Osmond CB (1978) Crassulacean acid metabolism: a curiosity in context. Annu Rev Plant Physiol 29:379–414
Otto R, Castro SA, Mariano E, Castro SG, Franco HC, Trivelin PC (2016) Nitrogen use efficiency for sugarcane-biofuel production: what is next? Bioenergy Res 9:1272–1289
Owen NA, Griffiths H (2013) A system dynamics model integrating physiology and biochemical regulation predicts extent of crassulacean acid metabolism (CAM) phases. New Phytol 200:1116–1131
Owen NA, Griffiths H (2014) Marginal land bioethanol yield potential of four crassulacean acid metabolism candidates (*Agave fourcroydes*, *Agave salmiana*, *Agave tequilana* and *Opuntia ficus-indica*) in Australia. Glob Change Biol Bioenergy 6:687–703
Owen NA, Fahy KF, Griffiths H (2016) Crassulacean acid metabolism (CAM) offers sustainable bioenergy production and resilience to climate change. Glob Clim Chang Bioenergy 8:737–749
Pérez-Cadena R, Espinosa Solares T, Medina-Moreno SA, Martínez A, Lizardi-Jiménez MA, Téllez-Jurado A (2018) Production of ethanol by three yeasts in defined media and hydrolyzed cladodes of *Opuntia ficus-indica* var. Atlixco. Int J Agric For 8:26–34
Pierce S, Winter K, Griffiths H (2002) The role of CAM in high rainfall cloud forests: an *in situ* comparison of photosynthetic pathways in Bromeliaceae. Plant Cell Environ 25:1181–1189
Pikart FC, Marabesi MA, Mioto PT, Gonçalves AZ, Matiz A, Alves FR, Mercier H, Aidar MP (2018) The contribution of weak CAM to the photosynthetic metabolic activities of a bromeliad species under water deficit. Plant Physiol Biochem 123:297–303
Poorter H, Navas ML (2003) Plant growth and competition at elevated CO_2: on winners, losers and functional groups. New Phytol 157:175–198
Powlson DS, Stirling CM, Thierfelder C, White RP, Jat ML (2016) Does conservation agriculture deliver climate change mitigation through soil carbon sequestration in tropical agro-ecosystems? Agric Ecosyst Environ 220:164–174
Qambrani NA, Rahman MM, Won S, Shim S, Ra C (2017) Biochar properties and eco-friendly applications for climate change mitigation, waste management, and wastewater treatment: a review. Renew Sust Energ Rev 79:255–273
Quero E, Nobel PS (1987) Predictions of field productivity for *Agave lechuguilla*. J Appl Ecol 24:1053–1062
Ramírez-Arpide FR, Demirer GN, Gallegos-Vázquez C, Hernández-Eugenio G, Santoy-Cortés VH, Espinosa-Solares T (2018) Life cycle assessment of biogas production through anaerobic co-digestion of nopal cladodes and dairy cow manure. J Clean Prod 172:2313–2322
Ranjan P, Ranjan JK, Misra RL, Dutta M, Singh B (2016) Cacti: notes on their uses and potential for climate change mitigation. Genet Resour Crop Evol 63:901–917
Ranjithkumar M, Ravikumar R, Sankar MK, Kumar MN, Thanabal V (2017) An effective conversion of cotton waste biomass to ethanol: a critical review on pretreatment processes. Waste Biomass Valoriz 8:57–68
Raveh E, Gersani M, Nobel PS (1995) CO_2 uptake and fluorescence responses for a shade-tolerant cactus *Hylocereus undatus* under current and doubled CO_2 concentrations. Physiol Plant 93:505–511
Ray D, Mueller N, West P, Foley J (2013) Yield trends are insufficient to double global crop production by 2050. PLoS One 8:e66428
Rebman JP, Pinkava DJ (2001) Opuntia cacti of North America: an overview. Fla Entomol 84:475–483
Ren J, Dong L, Sun L, Goodsite ME, Tan S, Dong L (2015) Life cycle cost optimization of biofuel supply chains under uncertainties based on interval linear programming. Bioresour Technol 187:6–13
Robinson SA, Osmond CB (1994) Internal gradients of chlorophyll and carotenoid pigments in relation to photoprotection in thick leaves of plants with crassulacean acid metabolism. Funct Plant Biol 21:497–506
Rodrigues MA, Matiz A, Cruz AB, Matsumura AT, Takahashi CA, Hamachi L, Félix LM, …, Kerbauy GB (2013) Spatial patterns of photosynthesis in thin-and thick-leaved epiphytic orchids: unravelling C_3–CAM plasticity in an organ-compartmented way. Ann Bot 112:17–29
Rodrígueza L, Unai Pascualb U, Niemeyer HM (2006) Local identification and valuation of ecosystem goods and services from *Opuntia* scrublands of Ayacucho, Peru. Ecol Econ 57:30–44
Rodríguez-Machín L, Arteaga-Pérez LE, Pérez-Bermúdez RA, Casas-Ledón Y, Prins W, Ronsse F

(2018) Effect of citric acid leaching on the demineralization and thermal degradation behavior of sugarcane trash and bagasse. Biomass Bioenergy 108:371–380
Rodríguez-Machín L, Arteaga-Pérez LE, Pala M, Herregods-Van De Pontseele K, Pére Bermúdez RA, Feys J, Prins W, Ronsse F (2019) Influence of citric acid leaching on the yield and quality of pyrolytic bio-oils from sugarcane residues. J Anal Appl Pyrolysis 137:43–53
Roe S, Streck C, Obersteiner M, Frank S, Griscom B, Drouet L, Fricko O, …, Lawrence D (2019) Contribution of the land sector to a 1.5° C world. Nat Clim Chang 9:817–828
Russell C, Felker P (1987) Comparative cold-hardiness of *Opuntia* spp. and cvs grown for fruit, vegetable and fodder production. J Hortic Sci 62:545–550
Rykaczewski K, Jordan JS, Linder R, Woods ET, Sun X, Kemme N, Manning KC, …, Majure LC (2016) Microscale mechanism of age dependent wetting properties of prickly pear cacti (*Opuntia*). Langmuir 32:9335–9341
Saag KML, Sanderson G, Moyna P, Ramos G (1975) Cactaceae mucilage composition. J Sci Food Agric 26:993–1000
Salvucci ME, Crafts-Brandner SJ (2004) Relationship between the heat tolerance of photosynthesis and the thermal stability of Rubisco activase in plants from contrasting thermal environments. Plant Physiol 134:1460–1470
Sánchez J, Sánchez F, Curt MD, Fernández J (2012) Assessment of the bioethanol potential of prickly pear (*Opuntia ficus-indica* (L.) Mill.) biomass obtained from regular crops in the province of Almeria (SE Spain). Israel J Plant Sci 60:301–318
Sayed OH (1998) Phenomorphology and ecophysiology of desert succulents in eastern Arabia. J Arid Environ 40:177–189
Scanlon B, Faunt C, Longuevergne L, Reedy R, Alley W, McGuire V, McMahon P (2012) Groundwater depletion and sustainability of irrigation in the US High Plains and Central Valley. Proc Natl Acad Sci USA 109:9320–9325
Shakee SN, Aman S, Haq NU, Heckathorn SA, Luthe D (2013) Proteomic and transcriptomic analyses of *Agave americana* in response to heat stress. Plant Mol Biol Report 31:840–851
Shameer S, Baghalian K, Cheung CM, Ratcliffe RG, Sweetlove LJ (2018) Computational analysis of the productivity potential of CAM. Nat Plants 4:165–171
Shivhare D, Mueller-Cajar O (2017) In vitro characterization of thermostable CAM Rubisco activase reveals a Rubisco interacting surface loop. Plant Physiol 174:1505–1516
Silvera K, Santiago LS, Winter K (2005) Distribution of crassulacean acid metabolism in orchids of Panama: evidence of selection of weak and strong modes. Funct Plant Biol 32:397–407
Silvera K, Santiago LS, Cushman JC, Winter K (2009) Crassulacean acid metabolism and epiphytism linked to adaptive radiations in the Orchidaceae. Plant Physiol 149:1838–1847
Silvera K, Neubig KM, Whitten WM, Williams NH, Winter K, Cushman JC (2010a) Evolution along the crassulacan acid metabolism continuum. Funct Plant Biol 37:995–1010
Silvera K, Santiago LS, Cushman JC, Winter K (2010b) Incidence of crassulacean acid metabolism in the Orchidaceae derived from carbon isotope ratios: a checklist of the flora of Panama and Costa Rica. Bot J Linn Soc 163:194–222
Sinclair TR, Muchow RC (1999) Radiation use efficiency. Adv Agron 65:215–265
Sipes DL, Ting IP (1985) Crassulacean acid metabolism and crassulacean acid metabolism modifications in *Peperomia camptotricha*. Plant Physiol 77:59–63
Smith SD, Didden-Zopfy B, Nobel PS (1984) High-temperature responses of North American cacti. Ecology 65:643–651
Smith JAC, Schulte PJ, Nobel PS (1987) Water flow and water storage in *Agave deserti*: osmotic implications of crassulacean acid metabolism. Plant Cell Environ 10:639–648
Somerville C, Youngs H, Taylor C, Davis S, Long SP (2010) Feedstocks for lignocellulosic biofuels. Science 329:790–792
Stork KC, Singh MK (1995) Impact of the renewable oxygenate standard for reformulated gasoline on ethanol demand, energy use, and greenhouse gas emissions
Swarr TE, Hunkeler D, Klöpffer W, Pesonen H-L, Ciroth A, Brent AC, Pagan R (2011) Environmental life-cycle costing: a code of practice. Int J Life Cycle Assess 16:389–391
Sweetlove LJ, Ratcliffe RG (2011) Flux-balance modelling of plant metabolism. Front Plant Sci 2:38
Szarek SR, Johnson HB, Ting IP (1973) Drought adaptation in *Opuntia basilaris*. Significance of recycling carbon through crassulacean acid metabolism. Plant Physiol 52:539–541
Taherzadeh M, Karimi K (2008) Pretreatment of lignocellulosic wastes to improve ethanol and biogas production: a review. Int J Mol Sci 9:1621–1651
Tallman G, Zhu JX, Mawson BT, Amodeo G, Nouhi Z, Levy K, Zeiger E (1997) Induction of CAM in

Mesembryanthemum crystallinum abolishes the stomatal response to blue light and light-dependent zeaxanthin formation in guard cell chloroplasts. Plant Cell Physiol 38:236–242

Taybi T, Cushman JC (1999) Signaling events leading to Crassulacean acid metabolism (CAM) induction in the common ice plant, *Mesembryanthemum crystallinum*. Plant Physiol 121:545–555

Taybi T, Cushman JC (2002) Abscisic acid signaling and protein synthesis requirements for CAM induction in the common ice plant. J Plant Physiol 159:1235–1243

Teeri JA, Tonsor SJ, Turner M (1981) Leaf thickness and carbon isotope composition in the Crassulaceae. Oecologia 50:367–369

Tilman D, Balzer C, Hill J, Befort BL (2011) Global food demand and the sustainable intensification of agriculture. Proc Natl Acad Sci USA 108:20260–20264

Ting I (1985) Crassulacean acid metabolism. Annu Rev Plant Physiol 36:595–622

Trajano HL, Engle NL, Foston M, Ragauskas AJ, Tschaplinski TJ, Wyman CE (2013) The fate of lignin during hydrothermal pretreatment. Biotechnol Biofuels 6:110–125

Treberth K (2011) Changes in precipitation with climate change. Clim Res 47:123–138

U.S. Department of Agriculture (USDA) (2012) Pima cotton 2012 yield per har- vested acre by county for selected states. In: USDo Agriculture (ed) National Agricultural Statistics Service

United Nations (2007) World population prospects: the 2006 revision, highlights. In: PD Department of Economic and Social Affairs (ed) Vol working paper no. ESA/P/WP. 202. United Nations, New York

Valdez-Cepeda RD, Blanco-Macías F, Gallegos-Vázquez C, Salinas-García GE, Vázquez-Alvarado RE (2002) Freezing tolerance of *Opuntia* spp. J Prof Assoc Cactus Dev 4:105–116

Valenzuela-Soto J, Maldonado-Bonilla L, Hernández-Guzmán G, Rincón-Enríquez G, Martínez-Gallardo N, Ramírez-Chávez E, Hernández I, …, Délano-Frier J (2015) Infection by a coronatine-producing strain of *Pectobacterium cacticidum* isolated from sunflower plants in Mexico is characterized by soft rot and chlorosis. J Gen Plant Pathol 81:368–381

Velasco-Muñoz J, Aznar-Sánchez J, Belmonte-Ureña L, Román-Sánchez I (2018) Sustainable water use in agriculture: a review of worldwide research. Sustainability 10:1084

Voloshin RA, Rodionova MV, Zharmukhamedov SK, Veziroglu TN, Allakhverdiev SI (2016) Biofuel production from plant and algal biomass. Int J Hydrog Energy 41:17257–17273

von Caemmerer S, Furbank R (1999) Modeling C_4 photosynthesis. In: Sage RF, Monson RK (eds) C4 Plant Biology. Academic Press, San Diego, pp 173–211

von Caemmerer S, Griffiths H (2009) Stomatal responses to CO_2 during a diel crassulacean acid metabolism cycle in *Kalanchoe daigremontiana* and *Kalanchoe pinnata*. Plant Cell Environ 32:567–576

Voss KA, Famiglietti JS, Lo M, Linage C, Rodell M, Swenson SC (2013) Groundwater depletion in the Middle East from GRACE with implications for transboundary water management in the Tigris-Euphrates-Western Iran region. Water Resour Res 49:904–914

Wang MQ (1999) GREET 1.5-transportation fuel-cycle model-Vol. 1: methodology, development, use, and results. Argonne National Lab, IL, USA

Weiss I, Mizrahi Y, Raveh E (2010) Effect of elevated CO_2 on vegetative and reproductive growth characteristics of the CAM plants *Hylocereus undatus* and *Selenicereus megalanthus*. Sci Hortic 123:531–536

Wheeler T, Von Braun J (2013) Climate change impacts on global food security. Science 341:508–513

Williams DG, Hultine KR, Dettman DL (2014) Functional trade-offs in succulent stems predict responses to climate change in columnar cacti. J Exp Bot 65:3405–3413

Winter K (2019) Ecophysiology of constitutive and facultative CAM photosynthesis. J Exp Bot 70:6495–6508

Winter K, Holtum JAM (2002) How closely do the δ13C values of Crassulacean acid metabolism plants reflect the proportion of CO_2 fixed during day and night? Plant Physiol 129:1843–1851

Winter K, Holtum JAM (2005) The effects of salinity, crassulacean acid metabolism and plant age on the carbon isotope composition of *Mesembryanthemum crystallinum* L., a halophytic C3-CAM species. Planta 222:201–209

Winter K, Holtum JAM (2011a) Drought-stress-induced up-regulation of CAM in seedlings of a tropical cactus, *Opuntia elatior*, operating predominantly in the C_3 mode. J Exp Bot 62:4037–4042

Winter K, Holtum JAM (2011b) Induction and reversal of crassulacean acid metabolism in *Calandrinia polyandra*: effects of soil moisture and nutrients. Funct Plant Biol 38:576–582

Winter K, Holtum JAM (2014) Facultative crassulacean acid metabolism (CAM) plants: powerful tools for unravelling the functional elements of CAM photosynthesis. J Exp Bot 65:3425–3441

Winter K, Holtum JAM (2017) Facultative CAM photosynthesis (crassulacean acid metabolism) in four small C_3 and C_4 leaf-succulents. Aust J Bot 65:103–108

Winter K, Smith JAC (1996) Crassulacean acid metabolism. Current status and perspectives. In: Winter K, Smith JAC (eds) Crassulacean Acid Metabolism. Biochemistry, Ecophysiology and Evolution. Springer, Berlin, pp 389–426

Winter K, Schröppel-Meier G, Caldwell MM (1986) Respiratory CO_2 as carbon source for nocturnal acid synthesis at high temperatures in three species exhibiting crassulacean acid metabolism. Plant Physiol 81:390–394

Winter K, Richter A, Engelbrecht B, Posada J, Virgo A, Popp M (1997) Effect of elevated CO_2 on growth and crassulacean acid metabolism activity of *Kalanchoe pinnata* under tropical conditions. Planta 201:389–396

Winter K, Aranda J, Holtum JA (2005) Carbon isotope composition and water-use efficiency in plants with crassulacean acid metabolism. Funct Plant Biol 32:381–388

Winter K, Garcia M, Holtum JAM (2008) On the nature of facultative and constitutive CAM: environmental and developmental control of CAM expression during early growth of *Clusia, Kalanchoë,* and *Opuntia*. J Exp Bot 59:1829–1840

Winter K, Holtum JAM, Smith JAC (2015) Crassulacean acid metabolism: a continuous or discrete trait? New Phytol 208:73–78

Wu A, Song Y, van Oosterom EJ, Hammer GL (2016) Connecting biochemical photosynthesis models with crop models to support crop improvement. Front Plant Sci 7:1518

Yamori W, Hikosaka K, Way DA (2014) Temperature response of photosynthesis in C_3, C_4, and CAM plants: temperature acclimation and temperature adaptation. Photosynth Res 119:101–117

Yan X, Tan DKY, Inderwildi OR, Smith JAC, King DA (2011) Life cycle energy and greenhouse gas analysis for agave-derived bioethanol. Energy Environ Sci 4:3110–3121

Yang L, Carl S, Lu M, Mayer JA, Cushman JC, Tian E, Lin H (2015a) Biomass characterization of *Agave* and *Opuntia* as potential biofuel feedstocks. Biomass Bioenergy 76:43–53

Yang X, Cushman JC, Borland AM, Edwards EJ, Wullschleger SD, Tuskan GA, Owen NA, …, Holtum JAM (2015b) A roadmap for research on crassulacean acid metabolism to enhance sustainable food and bioenergy production in a hotter, drier world. New Phytol 207:491–504

Yang SJ, Sun M, Yang QY, Ma RY, Zhang JL, Zhang SB (2016) Two strategies by epiphytic orchids for maintaining water balance: thick cuticles in leaves and water storage in pseudobulbs. AoB Plants 8:plw046

Ye G, Luo H, Ren Z, Ahmad MS, Liu CG, Tawab A, Al-Ghafari AB, …, Mehmood MA (2018) Evaluating the bioenergy potential of Chinese liquor-industry waste through pyrolysis, thermogravimetric, kinetics and evolved gas analyses. Energy Convers Manag 163:13–21

Yeats TH, Rose JK (2013) The formation and function of plant cuticles. Plant Physiol 163:5–20

Zeisler-Diehl V, Müller Y, Schreiber L (2018) Epicuticular wax on leaf cuticles does not establish the transpiration barrier, which is essentially formed by intracuticular wax. J Plant Physiol 227:66–74

Zeng Y, Zhao S, Yang S, Ding S-Y (2014) Lignin plays a negative role in the biochemical process for producing lignocellulosic biofuels. Curr Opin Biotechnol 27:38–45

Zhao M, Running SW (2009) Drought-induced reduction in global terrestrial net primary production from 2000 through 2009. Science 329:940–943

Zhu J, Bartholomew DP, Goldstein G (1997) Effect of elevated carbon dioxide on the growth and physiological responses of pineapple, a species with crassulacean acid metabolism. J Am Soc Hortic Sci 122:233–237

Zhu X, Liang C, Masters MD, Kantola IB, DeLucia EH (2018) The impacts of four potential bioenergy crops on soil carbon dynamics as shown by biomarker analyses and DRIFT spectroscopy. Glob Clim Change Bioenergy 10:489–500

Zika M, Erb K (2009) The global loss of net primary production resulting from human-induced soil degradation in drylands. Ecol Econ 69:310–318

Ziska LH, Hogan KP, Smith AP, Drake BG (1991) Growth and photosynthetic response of nine tropical species with long-term exposure to elevated carbon dioxide. Oecologia 86:383–389

Part V

Engineering Photosynthesis for Climate Change

Chapter 11

Engineering Photosynthetic CO_2 Assimilation to Develop New Crop Varieties to Cope with Future Climates

Robert E. Sharwood*

ARC Centre of Excellence for Translational Photosynthesis, Australian National University, Canberra, ACT, Australia

Hawkesbury Institute for the Environment, Western Sydney University, Richmond, NSW, Canberra, ACT, Australia

and

Benedict M. Long

ARC Centre of Excellence for Translational Photosynthesis, Australian National University, Canberra, ACT, Australia

*Author for correspondence, e-mail: robert.sharwood@anu.edu.au

K. M. Becklin et al. (eds.), *Photosynthesis, Respiration, and Climate Change*, Advances in Photosynthesis and Respiration 48, https://doi.org/10.1007/978-3-030-64926-5_11

Summary

Agricultural crop production must significantly increase in the next 30 years to ensure supply of enough nutritious food for the burgeoning global population. However, increasing variability in global climates and reductions in arable land are placing significant pressure on crop production. Most of the key food production crops such as wheat, rice, soybean and barley operate C_3 photosynthetic biochemistry that is often limited by the efficiency of CO_2 fixation, underpinned by the enzyme Rubisco (ribulose-1,5-bisphosphate carboxylase/oxygenase). Rubisco has long been the target enzyme for plant engineers to improve the catalytic efficiency of CO_2 fixation. Kinetic screens have offered promising solutions to improve CO_2 fixation by sifting through the natural diversity of Rubisco catalysis and comparing catalytic constants [including catalytic turnover speed (k_{cat}^c), Michaelis constant for CO_2 (K_c), relative specificity for CO_2 over O_2 ($S_{c/o}$)] and their associated thermal dependencies. New advances in directed evolution and assembly of higher plant Rubisco in *E. coli* has provided the long sought-after Synthetic Biology (SynBio) tools to rapidly explore Rubisco sequence space for opportunities to improve CO_2 fixation by creating catalytic variation not yet observed in nature. These recent advances will enable new frontiers to improve CO_2 fixation either through direct manipulation of Rubisco or a combination of bioengineering strategies. Other strategies are gaining significant momentum, which include installing CO_2 concentrating mechanisms (CCMs) into crop chloroplasts. Significant steps have been achieved in this approach, through engineering carboxysomes and their cognate Rubisco within C_3 plants with an aim to increase CO_2 fixation by saturation of Rubisco with CO_2 and minimizing oxygenation and photorespiratory flux. In addition, the discovery of liquid-liquid phase separation mediated by unstructured proteins in photosynthetic bacteria and algae enable new possibilities to incorporate pyrenoid structures into chloroplasts. New generation solutions will require new advances in SynBio and possible expansion of the availability for chloroplast transformation among crops.

I. Threat of Climate Change to Agricultural Production

The global challenge facing plant scientists is to ensure world crop production increases by more than 60% over the coming decades as the growing global population will require adequate calories for nourishment (Ray et al. 2013). As annual yield progress in key food crops such as rice and wheat are declining, the challenge is to restore positive gain in crop yields (Zhu et al. 2010; Reynolds et al. 2012; Long et al. 2015). However, this challenge is being frustrated by the accelerating increase of anthropogenic CO_2 emissions in the earth's atmosphere, causing major consequences in climate variability, particularly rising temperatures and increases in extreme weather events (IPCC 2012). Rapid changes in the climate of crop growth seasons has required new solutions to ensure crops are responsive to this variability. Atmospheric CO_2 concentrations first reached 400 ppm in 2014 and is currently >415 ppm with a record-breaking 12 month increase of 3.5 ppm since 2018 (www.research.noaa.gov). Already in Australia, the growth temperature of crops has risen by an average of 1 °C with some regions the anomaly being >2 °C (www.bom.gov.au/state-of-the-climate). This has resulted in shorter crop duration and farm management strategies are being altered to include early planting. Variability in rainfall is increasing alongside the rising intensity and frequency of drought (www.bom.gov.au/state-of-the-climate). Recent analysis of wheat yields and subsequent crop model testing at elevated temperatures indicates that temperature is already reducing yield (Hochman et al. 2017) and is predicted to decrease by 6% for every 1 °C rise in temperature (Asseng et al. 2014). Globally, climate change is associated with >30% variability in crop yields (Ray et al. 2015).

As traditional plant breeding techniques are reaching their theoretical maximum for harvest index, new solutions to increase plant biomass must be sought (Furbank et al. 2019). A major research direction, substantially reviewed in recent times, is improvement to the key processes of photosynthesis (Ort et al. 2015). Several modelling efforts have indicated that the next step-change required to improve productivity could come from optimizing the light-dependent and so-called dark reactions of photosynthesis. This chapter will focus on efforts to improve photosynthetic CO_2 assimilation with recent advances in SynBio tools.

II. Rubisco Catalytic and Structural Diversity and the Requirement for Rubisco Activase

A. Rubisco Bifunctional Catalysis Underpins Plant Growth and Yield

The assimilation of CO_2 into the earth's biosphere is crucial for the synthesis of carbohydrates to maintain plant growth and ultimately influence crop yield (Fig. 11.1a). The key rate-limiting enzyme in this process is Rubisco (ribulose-1,5-bisphosphate carboxylase/oxygenase). Rubisco forms an integral part of the photosynthetic carbon reduction cycle [Calvin-Benson-Bassham (CBB) cycle] and has complex catalytic properties that offer prospects to improve CO_2 assimilation (Sharwood 2017b). Carboxylation of the substrate ribulose-1,5-bisphosphate (RuBP) by Rubisco underpins plant growth through the production of two molecules of 3-phosphoglycerate (PGA). This product is cycled through the CBB cycle to produce carbohydrate precursors for plant growth and yield (Raines 2011). Remarkably, Rubisco is bifunctional and competitively inhibited by oxygen, which leads to one molecule of PGA and one molecule of 2-phosphoglycolate (2-PG) when fixed (Andrews et al. 1973). The molecule 2-PG is regarded as toxic and is exported from the chloroplast then cycled through the photorespiratory pathway to regenerate PGA (Fig. 11.1a; Zelitch et al. 2009; Bauwe et al. 2010). This process requires energy and further results in release of previously fixed CO_2 (Sharkey 1988; Walker et al. 2016). Rubisco oxygenation increases in conditions of drought, heatwaves and high seasonal growing temperatures. Therefore, improving the propensity for carboxylation over oxygenation is a key target to mitigate declines in yield potential (Fig. 11.1a).

B. Rubisco Structural Properties

In nature, the Rubisco superfamily is comprised of different structural forms (Form I–IV) that occur among different organisms.

Abbreviations: 2-PG – 2-phosphoglycolate; CA1P – carboxyarabinitol-1-phosphate; CBB cycle – Calvin-Benson-Bassham cycle; CCM – CO_2 concentrating mechanism; k_{cat}^c – Rubisco catalytic speed per second; K_c – Michaelis constant for CO_2; $K_c^{21\%O_2}$ – Michaelis constant for CO_2 measured in air (21% O_2); L subunit – Rubisco large subunit; NAD-ME – NAD malic enzyme C_4 sub-type; NADP-ME – NADP malic enzyme C_4 sub-type; PCK – phosphoenolpyruvate carboxykinase; PEPC – phosphoenolpyruvate carboxylase; PGA – 3-phosphoglycerate; *rbc*L – Rubisco large subunit gene; *rbc*S – Rubisco small subunit gene; RCA – Rubisco activase; RDE – Rubisco dependent *E. coli*; Rubisco – ribulose-1,5-bisphosphate carboxylase/oxygenase; RuBP – ribulose-1,5-bisphosphate; S subunit – Rubisco small subunit; $S_{c/o}$ – relative specificity for CO_2 as opposed to O_2; SynBio – Synthetic Biology; XuBP – xylulose-1,5-bisphosphate

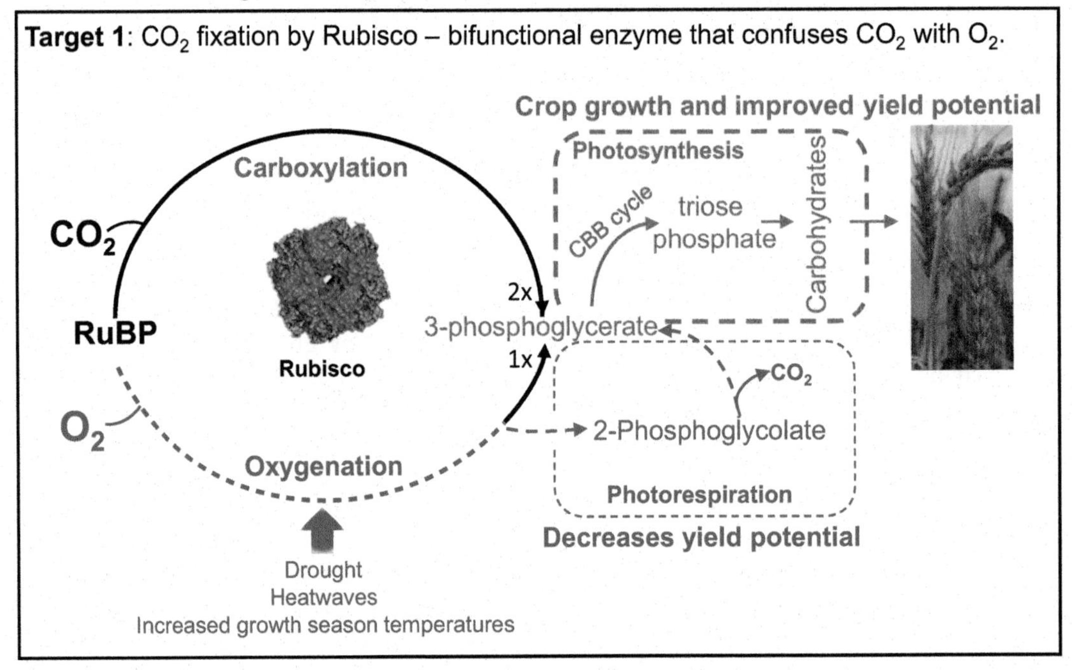

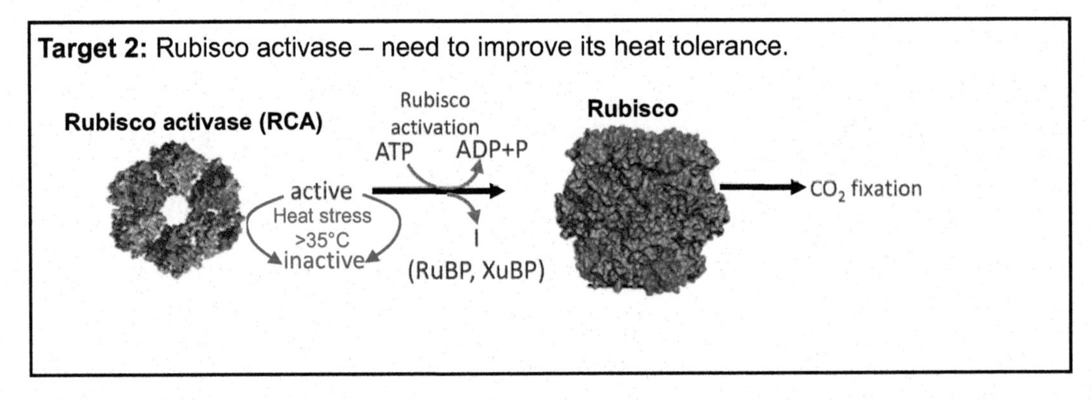

Fig. 11.1. **Targeting biochemical components of CO_2 assimilation to improve plant productivity under future climates.** (**a**) The bifunctional catalytic reactions of photosynthesis. The fixation of CO_2 to substrate ribulose-1,5-bisphosphate (RuBP) results in the production of two molecules of 3-phosphoglycerate. These are utilized by the Calvin, Benson and Bassham (CBB) cycle to synthesize the triose phosphates that are either utilized for the regeneration of substrate RuBP for the synthesis of carbohydrates that are requisite for plant growth and productive grain yield (Raines 2011). CO_2 fixation is competitively inhibited by oxygen and results in the production of only one molecule of 3-phosphoglycerate and 2-phosphoglycolate. Oxygenation of RuBP is exacerbated under conditions of drought, heatwaves and increases in seasonal growing temperatures and thus flux is increased through the photorespiratory cycle that consumes energy and releases previously fixed CO_2 (Bauwe et al. 2010). This contributes to significant declines in yield potential. (**b**) Rubisco activase is the second biochemical target and is responsible for maintaining Rubisco activation by remodeling the active site powered by the hydrolysis of ATP to remove active site inhibitors such as RuBP or carboxyarabinitol-1-phosphate (CA1P) and xylose-1,5-bisphosphate (XuBP; Parry et al. 2008). Remarkably, Rubisco activase in C_3 crops is thermolabile and is rendered inactive at elevated temperatures resulting in Rubisco deactivation (Crafts-Brandner and Salvucci 2000). Alternative RCA isoforms are being sought that remain active at temperatures >35 °C. (Shivhare and Mueller-Cajar 2017)

Form I Rubisco is the predominant structure found within this superfamily and is comprised of eight large (L) and eight small (S) subunits. Within the Form I Rubisco superfamily exists a major phylogenetic divide between Form I 'green' type Rubisco present in plants, freshwater cyanobacteria, and algae and Form I 'red' type that exists in non- green algae (Tabita et al. 2008; Whitney et al. 2011a).

The Rubisco L subunits are arranged as antiparallel dimers that form the L_8 core (Wilson and Hayer-Hartl 2018). Each dimer harbours two active sites that are formed by the N-terminal domain of one L subunit and the C-terminal domain of an adjacent L subunit (Bracher et al. 2017). The S subunits are required for competent catalysis and four subunits assemble at each end of the L_8 core (Sharwood and Whitney 2008). Although many crystal structures exist for the diverse forms of Rubisco, the active sites are superimposable from each enzyme, making rational design of better performing Rubisco difficult since it would appear that catalytic diversity is not necessarily coupled to changes in active site amino acid residues (Andersson and Backlund 2008). For example, amino acids outside of the conserved active site have been shown to influence catalysis (Whitney et al. 2011b). Unlike the cyanobacterial ancestor, terrestrial plant Rubisco is encoded in two disparate genetic locations. The L subunit is encoded in the chloroplast by *rbc*L, while the S subunit is encoded in the nucleus by *rbc*S as multigene family (Sharwood 2017b). Therefore, coordination between the nucleus and chloroplast is important for the synthesis and assembly of plant Rubisco. Understanding the molecular regulation of *rbc*S and *rbc*L expression will be crucial for future efforts to transplant Rubisco from foreign sources into crops (Sharwood 2017b).

C. *Rubisco Activation, Complex Catalysis and the Requirement for Rubisco Activase*

For Rubisco to become catalytically competent, the conserved active site residue K201 binds non-substrate CO_2 to form a carbamate (Lorimer and Miziorko 1980). The carbamate is then stabilized through the rapid binding of Mg^{2+} that enables the active site to be catalytically active (Cleland et al. 1998). Once active, RuBP binds within the active site. Enolization of RuBP occurs through the abstraction of a proton from the C3 carbon and creates a highly reactive enediol transition state intermediate that can either react with CO_2 or O_2 (Cleland et al. 1998). Reaction with CO_2 produces a carboxyketone that further goes through three partial reactions to produce two molecules of PGA (Sharwood 2017b). However, the fixation of O_2 produces a peroxyketone intermediate and the subsequent three partial reactions produce one molecule of phosphoglycerate and 2-phosphoglycolate (Cleland et al. 1998). This active site chemistry is consistent across all Rubisco forms.

For Form I 'green' Rubisco present in land plants, the CO_2 fixing catalytic cycle varies from 2 to 6 reactions per second at each catalytic site (i.e., the turnover number; (Sharwood 2017b). However, turnover numbers are consistently higher in cyanobacterial Form IA and IB Rubiscos, although at the expense of K_c and specificity for CO_2 (Figs. 11.2 and 11.3; (Tcherkez et al. 2006). Land plants have adjusted for the poor performance of Rubisco by investing significant amounts of nitrogen to synthesize large amounts of the enzyme (Evans 1989). This contributes to Rubisco being one of the most abundant enzymes on the planet (Ellis 1979; Bar-On and Milo 2019). A further catalytic complication for Rubisco is the catalytic

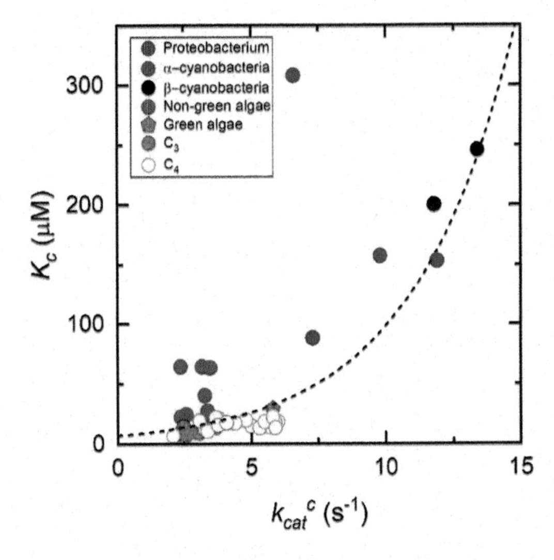

Fig. 11.2. **Diversity in Rubisco catalysis among different organisms**
Within the Rubisco superfamily large diversity exists in the Michaelis constant for CO_2 (K_c, μM) and catalytic speed (turnover number; k_{cat}^c, s^{-1}). The dashed line is represented by $y = 7.003e^{0.2641x}$, $r_2 = 0.53$. The trade-off between K_c and k_{cat}^c in Form I "Green-type" Rubisco is linear but this contrasts to non-green algae and cyanobacteria

misfiring reactions, which produces the inhibitor xylulose-1,5-bisphosphate (XuBP) which must be removed from the active site by Rubisco activase (Pearce 2006).

Rubisco activase (RCA) modulates the activation state of Rubisco by removing sugar phosphate inhibitors from the active site (Fig. 11.1b). It is a member of the AAA+ protein superfamily and utilizes the energy from ATP hydrolysis to remodel the active site to enable the inhibitors to disassociate (Mueller-Cajar 2017). In most plants RCA has two isoforms. An α isoform that has two proximal C-terminal cysteine residues that regulate activity by redox state and a shorter β isoform (Portis 2003). In wheat, a second β isoform is present that is upregulated under heat stress and exhibits thermostable properties (Scafaro et al. 2019).

In addition to XuBP, the primary substrate RuBP can also inhibit catalysis if it binds to non-carbamylated active sites and is largely responsible for Rubisco inhibition in illuminated leaves (Parry et al. 2008; Sharwood

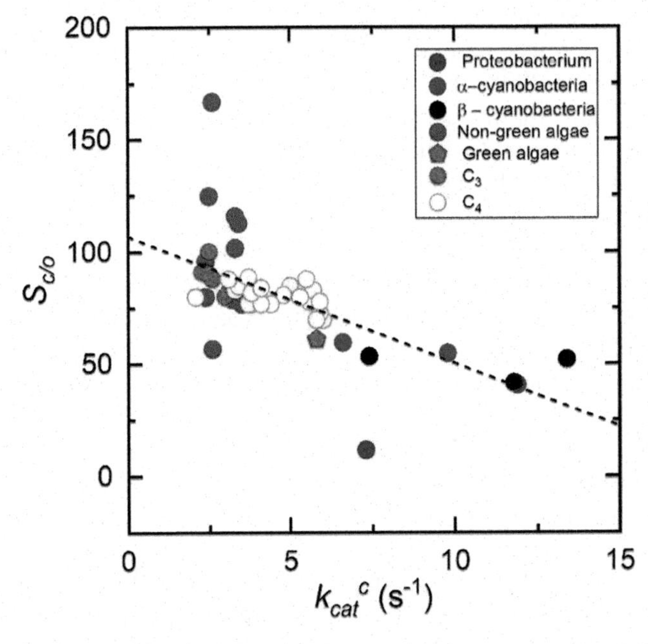

Fig. 11.3. **Catalytic trade-off between Rubisco specificity ($S_{c/o}$) and catalytic speed (k_{cat}^c)**
Diversity in Rubisco specificity for CO_2 as opposed to O_2 ($S_{c/o}$) within the Rubisco superfamily exists. The negative relationship with catalytic speed is represented by the dashed line which is a linear regression fitted to the data $y = -5.6555x + 106.94$, $r^2 = 0.41$. Substantial diversity away from this relationship is observed in non-green algae

et al. 2016c). The removal of RuBP by RCA allows non-substrate CO_2 to bind, followed by Mg^{2+}, and the enzyme is re-primed for catalysis. However, the nocturnal inhibitor carboxyarabinitol-1-phosphate (CA1P) binds to Rubisco in the dark, presumably to protect the enzyme from hydrolysis and to keep a carbamylated pool of Rubisco competent for catalysis after sunrise (Seemann et al. 1990; Khan et al. 1999). Once leaves become illuminated, RCA is activated by ATP from the light reactions and CA1P is removed from the active site, leaving Rubisco primed for catalysis. CA1P is subsequently dephosphorylated by CA1P phosphatase to prevent rebinding of the inhibitor (Andralojc et al. 2012).

D. The Thermal Sensitivity of Rubisco Activase Limits Photosynthetic CO_2 Assimilation Under Elevated Temperatures and Is a Target for Improvement

At elevated temperatures, it has been demonstrated that the decline in photosynthesis is associated with the reduction in Rubisco activity (Crafts-Brandner and Salvucci 2000; Fig. 11.1b). This decline has been linked, at least in part, to the thermolability of RCA (Kurek et al. 2007; Scafaro et al. 2016; Scafaro et al. 2018). This is an unfortunate feature of C_3 crop RCA and at temperatures slightly above (+4–5 °C) the photosynthetic thermal optimum, RCA activity declines and eventually denatures, becoming associated with heat-shock proteins (Salvucci 2008; Scafaro et al. 2019). The importance of improving RCA thermotolerance was further verified through transforming thermostable Arabidopsis RCA isoforms generated through directed evolution in *E. coli* into an Arabidopsis background deficient in RCA (Kurek et al. 2007).

Transgenic plants displayed improvements in CO_2 assimilation and biomass when grown above optimal temperatures, which correlated to improved heat tolerance of RCA (Kurek et al. 2007). Further evidence supporting the improvement of RCA thermotolerance as a target for increasing photosynthetic CO_2 assimilation under future climates is the identification of thermotolerant RCA isoforms from wild relatives of crops and alternate plant species originating from hot and dry climates (Fig. 11.1b). The β RCA isoform from *Oryza australensis* was demonstrated to maintain activity until 42 °C whereas the *Oryza sativa* (rice) counterpart only maintained activity until 36 °C (Scafaro et al. 2016). Transformation of this isoform into rice resulted in increased plant biomass and seed yield when exposed to 40–45 °C heat stress throughout the vegetative growth stage (Scafaro et al. 2018). However, it is important to note there were no significant changes the instantaneous measurements of CO_2 assimilation, suggesting that overall diurnal accumulation of carbohydrate is driving the improvement plant productivity.

Analysis of ATPase activity and activation of Rubisco at elevated temperatures has also identified β isoforms of RCA from hot/dry adapted *Agave tequiliana* (a CAM plant) and Creosote (C_3) that have increased thermotolerance and display higher ATPase activity compared to more widespread C_3 isoforms (Salvucci and Crafts-Brandner 2004; Shivhare and Mueller-Cajar 2017). Unsurprisingly, plant species originating from hot and dry environments offer novel opportunities to improve RCA function at elevated temperatures and represents natural diversity that remains to be mined for additional thermal biochemical traits.

Recently, a thermotolerant wheat β isoform of RCA (TaRCA1-β) was found to be induced when plants were exposed to heat stress (Scafaro et al. 2019). This isoform displayed a thermal midpoint at 42 °C compared to 35 °C for the TaRCA2 isoforms, however the activity of this isoform was reduced at 25 °C. Comparison of the TaRCA2-β amino sequence with the thermotolerant versions of rice and wheat identified 11 amino acids conferring thermotolerance of TaRCA1-β (Scafaro et al. 2019). In the future, it will be important to determine how universally these substitutions can confer thermotolerance to other thermolabile RCA isoforms and to improve the understanding of the mechanism underpinning the heat responsive regulation of the TaRCA2-β promoter.

E. Natural Diversity in Rubisco Kinetics

Exploring the catalytic performance within the Rubisco superfamily has identified substantial diversity among different photosynthetic organisms (Fig. 11.2a). Analysis of structure-function relationships has identified amino acid residues responsible for changes in performance and are termed 'catalytic switches' (Whitney et al. 2011b; Orr et al. 2016; Sharwood et al. 2016a). The identification of catalytic switches has reinforced that improvements to catalytic properties are possible and there is still an impetus to continue to explore photosynthetic organisms to identify Rubisco isoforms that are better suited to future climates.

The interrogation of Rubisco catalytic properties relies on determination of catalytic speed (k_{cat}^{c}), the Michaelis constants for CO_2 in 0% O_2 (K_c), and 21% O_2 ($K_c^{21\%O_2}$) and the ratio between carboxylation and oxygenation ($S_{c/o}$). The relationship between k_{cat}^{c} and K_c in C_3 and C_4 plants is linear, with increasing speed causing a catalytic trade-off with decreased affinity for CO_2 (Fig. 11.2 and Table 11.1; (Tcherkez et al. 2006). C_4 plant Rubisco typically has faster catalytic speeds compared to C_3 counterparts, which is most pronounced from NADP-ME and PCK photosynthetic subtypes (Table 11.1) compared to Rubisco from NAD-ME plants (Sharwood et al. 2016a). In addition, the faster catalytic speeds of NADP-ME Rubisco is linked to superior nitrogen use efficiency compared to NAD-ME counterparts as less investment in Rubisco is required within NADP-ME plants (Ghannoum et al. 2005; Sharwood et al. 2016c). Interestingly, Rubisco from non-green algae and cyanobacteria do not share this linear relationship between k_{cat}^{c} and K_c (Fig. 11.2). The second established catalytic trade-off is the negative relationship observed with k_{cat}^{c} and $S_{c/o}$ (Fig. 11.3; Tcherkez et al. 2006; Shih et al. 2016). Importantly, diversity exists in this relationship (r^2 = 0.4) offering promise to discover Rubisco enzymes with high specificity without severely compromising k_{cat}^{c} (Fig. 11.3).

Determining the thermal dependence of Rubisco catalysis is crucial to identify versions of Rubisco that are better suited to future climates consisting of elevated CO_2 and temperature. High resolution determination of the temperature response of Rubisco catalysis is required to properly assess the catalytic response to temperature (10–35 °C; Sharwood et al. 2016a). Within C_4 grasses, PCK and NADP-ME Rubisco have been modelled to provide improvements to CO_2 assimilation under future climates if transplanted into C_3 plants (Sharwood et al. 2016a).

F. Modelling the Impacts of Rubisco Catalytic Diversity for C_3 Chloroplasts and High CO_2 Environments

The impact of changing Rubisco kinetics on leaf CO_2 assimilation in C_3 plants can be determined using the C_3 photosynthetic model that is represented by equations for Rubisco and light-limited regeneration of RuBP (Farquhar et al. 1980). Comparing the catalytic properties of Rubisco from tobacco, *Griffithsia monilis, Urochloa mosambicensis, Cyanobium marinum PCC7001, Synechococcus elongatus PCC6301* and *Rhodobacter sphaeroides* using the C_3 model demonstrates that the superior carboxylation efficiency ($k_{cat}^{c} / K_c^{21\%O_2}$) from *G. monilis* and *U. mosambicensis* would offer improvements to tobacco CO_2 assimilation up to 330 μbar *Cc* compared to *R. sphaeroides, Cyanobium PCC7001* and *S. elongatus PCC6301* (Fig. 11.4a). However, the high

Table 11.1. Rubisco kinetic parameters from diverse photosynthetic organisms

Organism	c^{-1} k_{cat} ^{c}(s^{-1})	K_c (μM)	$S_{c/o}$	References
Purple bacteria				
R. rubrum	7.3	89	12	Whitney et al. (2001)
Cyanobacteria				
α-Cyanobacteria				
Prochloroccous marinus	6.6	309	59.9	Whitney et al. (2011a)
Cyanobium PCC7001	9.8	158	55	Long et al. (2018)
Synechococcus WH8102	11.9	154	41	This report.
β-Cyanobacteria				
Synechococcus 6301	11.8	200	41.5	Mueller-Cajar and Whitney (2008)
Synechococcus 7002	13.4	246	52	Wilson et al. (2018)
T. elongatus	7.4	n.m	53.3	Wilson et al. (2018)
Non-green				
G. monilis	2.6	9.3	167	Whitney et al. (2001)
P. tricornutum	3.4	27.9	113	Whitney et al. (2001)
P. tricornutum (CS-29)	3.3	41	116	Heureux et al. (2017)
T. weissflogii	3.2	65	79	Young et al. (2016)
T. oceania	2.4	65	80	Young et al. (2016)
F. cyldindrus	3.5	64	77	Heureux et al. (2017)
C. calcitrans	2.6	25	57	Young et al. (2016)
C. muelleri	2.4	23	96	Young et al. (2016)
P. lutheri	2.5	15	125	Heureux et al. (2017)
P. carterae	3.3	18	102	Heureux et al. (2017)
Green algae				
C. reinhardtii	5.8	29	61	Savir et al. (2010)
C_3 plants				
Tobacco	3.1	9.7	82	Sharwood et al. (2016a)
Arabidopsis	3.0	9.8	80	Whitney et al. (2015)
Helianthus annuus	3.3	12.4	84	Sharwood et al. (2008a)
F. pringlei	3.5	13.7	81	Whitney et al. (2011b)
S. oleracea	3.7	14	82	Young et al. (2016)
Glycine max	2.5	13.2	100	Orr et al. (2016)
Triticum aestivum	2.5	14	90	Savir et al. (2010)
S. laxa	2.3	7.7	91	Sharwood et al. (2016a)
P. bisulcatum	2.6	7.8	88	Sharwood et al. (2016a)
C_4 plants				
Paniceae				
C. ciliaris (NADP-ME)	6.0	19.0	70	Sharwood et al. (2016a)
S. viridis (NADP-ME)	5.9	18.1	73	Sharwood et al. (2016a)
P. virgatum (NAD-ME)	3.3	12.7	83	Sharwood et al. (2016a)
P. milliaceum (NAD-ME)	2.1	7.2	80	Sharwood et al. (2016a)
P. coloratum (NAD-ME)	3.4	11.1	85	Sharwood et al. (2016a)
U. panicoides (PCK)	5.6	15.4	78	Sharwood et al. (2016a)
U. mosambicensis (PCK)	5.7	14.8	83	Sharwood et al. (2016a)
P. deustum (PCK)	5.0	15.4	85	Sharwood et al. (2016a)
M. maximus (PCK)	5.3	13.9	80	Sharwood et al. (2016a)
Andropogoneae				
Z. mays (NADP-ME)	5.5	18.9	88	Sharwood et al. (2016b)
S. bicolor (NADP-ME)	5.8	22.9	70	Ogren (1984) and Sharwood et al. (2016b)

(continued)

Table 11.1. (continued)

Organism	c^{-1} $k_{cat}{}^{c}$(s^{-1})	K_c (μM)	$S_{c/o}$	References
Paspaleae				
P. dilatatum (NADP-ME)	3.4	19.9	88	Carmo-Silva et al. (2010)
Chloridoideae				
C. dactylon (NAD-ME)	3.7	21	89	Whitney et al. (2011a)
Tageteae				
F. bidentis (NADP-ME)	4.8	20.4	81	Whitney et al. (2011b)
F. trinervia (NADP-ME)	4.4	17.9	77	Kubien et al. (2008)
F. australasica (NADP-ME)	3.8	22.0	77	Kubien et al. (2008)
F. kochiana (NADP-ME)	3.7	22.7	77	Kubien et al. (2008)
Amaranthaceae				
A. edulis (NAD-ME)	4.1	18.2	77	Kubien et al. (2008)
A. hybridus (NAD-ME)	3.8	16	82	Whitney et al. (2011a)
Other species				
P. oleracea (C_4/CAM)	5.9	14	78	Whitney et al. (2011a)
Z. japonica (PCK)	4.1	18.5	84	Carmo-Silva et al. (2010)

$S_{c/o}$ of *G. monilis* increases the RuBP regeneration limited region of the curve, making this enzyme highly desirable for C_3 plants at ambient CO_2 (Fig. 11.4b; Whitney et al., 2001). However, when modelled for a high CO_2 environment, the $k_{cat}{}^{c}$ and $k_{cat}{}^{c} / K_c{}^{21\%O_2}$ parameters are largely driving the differences CO_2 assimilation response which corresponds to *U. mosambicensis* showing improved CO_2 assimilation <5000 μbar, while cyanobacterial Rubiscos become favoured at CO_2 concentrations >5000 μbar (Fig. 11.4b), which might be expected in the high CO_2 environment of a CO_2 concentrating mechanism (CCM).

Future modelling efforts can also include the influence of RCA under different environmental variables (i.e., elevated CO_2 and temperature). Measuring the Rubisco activation status under these conditions will provide information to what percentage of Rubisco catalytic sites are active for catalysis. This will allow comparisons of CO_2 assimilation under different abiotic stress conditions and will aid to the understanding of the influence of RCA on plant productivity.

III. Synthetic Biology (SynBio) Approaches to Improve Carbon Assimilation

A. *Plant Rubisco Assembly in E. coli*

A major SynBio advance that will provide new avenues in altering and screening for new variants of Rubisco exhibiting desirable catalytic parameters was the recent assembly of plant Rubisco in *E. coli* (Aigner et al. 2017). This required the transplantation of the chloroplast Rubisco biogenesis pathway into the bacterium using a three-plasmid system (Fig. 11.5).

Rubisco biogenesis is a complicated pathway that involves chloroplast chaperonin (Cpn60αβ, Cpn20/10) and four ancillary folding proteins RbcX, Raf1, Raf2 and BSDII (Wilson et al. 2019). The role of the ancillary protein RbcX was the first identified in 2007 as a molecular staple to stabilize the L_8 core of cyanobacterial Rubisco preceding the binding of Rubisco S subunits (Saschenbrecker et al. 2007). Additional ancillary proteins Raf1 and Raf2 were both identified subsequently in the maize photo-

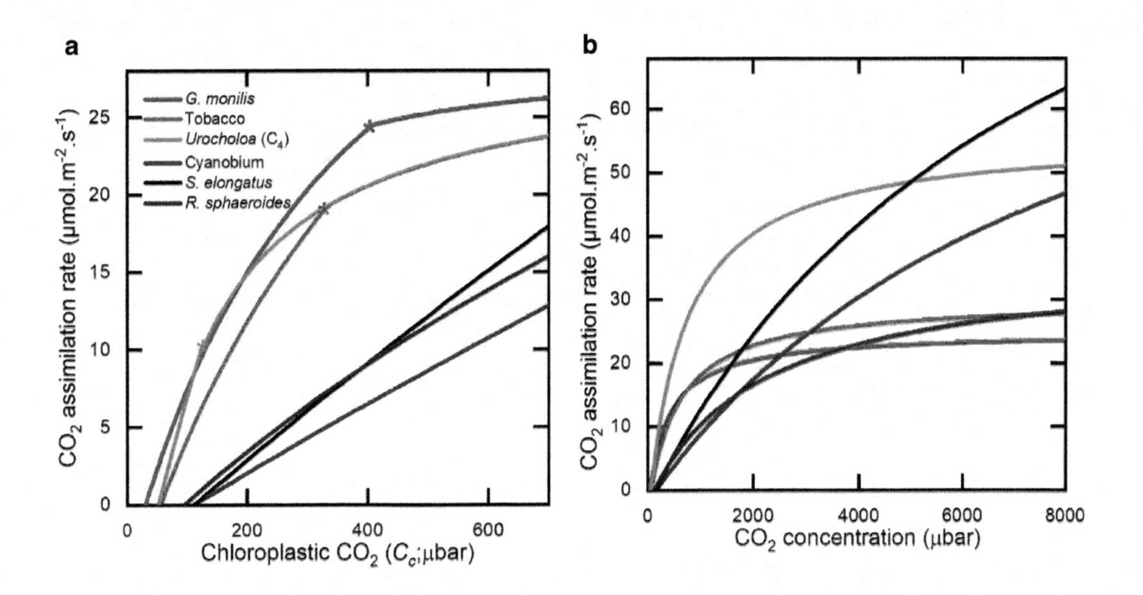

c

Organism	k_{cat}^c (s^{-1})	K_c (µM)	$K_c^{21\%O2}$(µM)	$S_{c/o}$	Reference:
Tobacco	3.1	9.7	18.3	82	(Young et al. 2016)
U. mosambicensis	5.7	14.8	22.8	82.5	(Sharwood et al. 2016a)
G. monilis	2.6	9.3	12.6	167	(Whitney et al. 2001)
Cyanobium PCC7001	9.8	169	275	55	(Long et al. 2018)
S. elongatus PCC6301	12.6	169	251	45.9	(Whitehead et al. 2014; Wilson et al. 2018)
R. sphaeroides	3.7	60	68.8	58	(Zhou and Whitney 2019)

Fig. 11.4. **Comparing Rubisco performance within a C_3 chloroplast and at elevated CO_2 experienced within a higher plant operating a CO_2 concentrating mechanism**
Modelling CO_2 assimilation in two environments: (**a**) C_3 chloroplast and, (**b**) a high CO_2 environment with a higher plant operating a CO_2 concentrating mechanism using Rubisco catalytic parameters from tobacco (C_3), *Urochloa mosambicensis* (C_4, PCK) *Griffithsia monilis* (non-green alga), *Cyanobium marinum PCC7001* (α-cyanobacterium), *Synechococcus elongatus PCC6301* (β-cyanobacterium at 25 °C) and *Rhodobacter sphaeroides* ('red-type' Form 1C). C_3 chloroplast CO_2 assimilation was modelled in response to chloroplastic CO_2 partial pressures (C_C) under Rubisco limiting activity (Eq. 1) and under electron transport limiting (Eq. 2) conditions. CO_2 conditions experienced by plants operating a Kranz-type C_4-photosynthetic pathway or a bacterial microcompartment under Rubisco activity limiting conditions (Eq. 1 where *CC* is substituted with *C* (CO_2 concentration around Rubisco). Coloured lines represent the modelled CO_2 assimilation rates according to (Farquhar et al. 1980; von Caemmerer 2000) using the equations:

$$A = A = \frac{B \cdot \left(C_C \cdot s_C - 0.5 \, {}^{O}\!/_{S_{C/O}} \right) k_{cat}^{C}}{C_C \cdot s_C + K_C^{21\%O_2}} - R_d$$

$$A = A = \frac{\left(C_C \cdot s_C - 0.5 \, {}^{O}\!/_{S_{C/O}} \right) J}{4 \left(C_C \cdot s_C + {}^{O}\!/_{S_{C/O}} \right)} - R_d$$

using the Rubisco parameters listed in panel C and using the CO_2 solubility in H_2O (s_c) of 0.0334 857 M·bar^{-1}, ambient O_2 concentrations (O) of 253 µM, a non-photorespiratory CO_2 release (R_d) of 1 µmol m^{-2} s^{-1} and Rubisco contents (B) of either 20 µmol catalytic sites.m^{-2} (panel A) or 10 µmol catalytic sites•m^{-2} (panel B). For Eq. 2, an electron transport rate was set to (J) of 120 µmol m^{-2} s^{-1}. (Sharwood 2017a, 2017b)

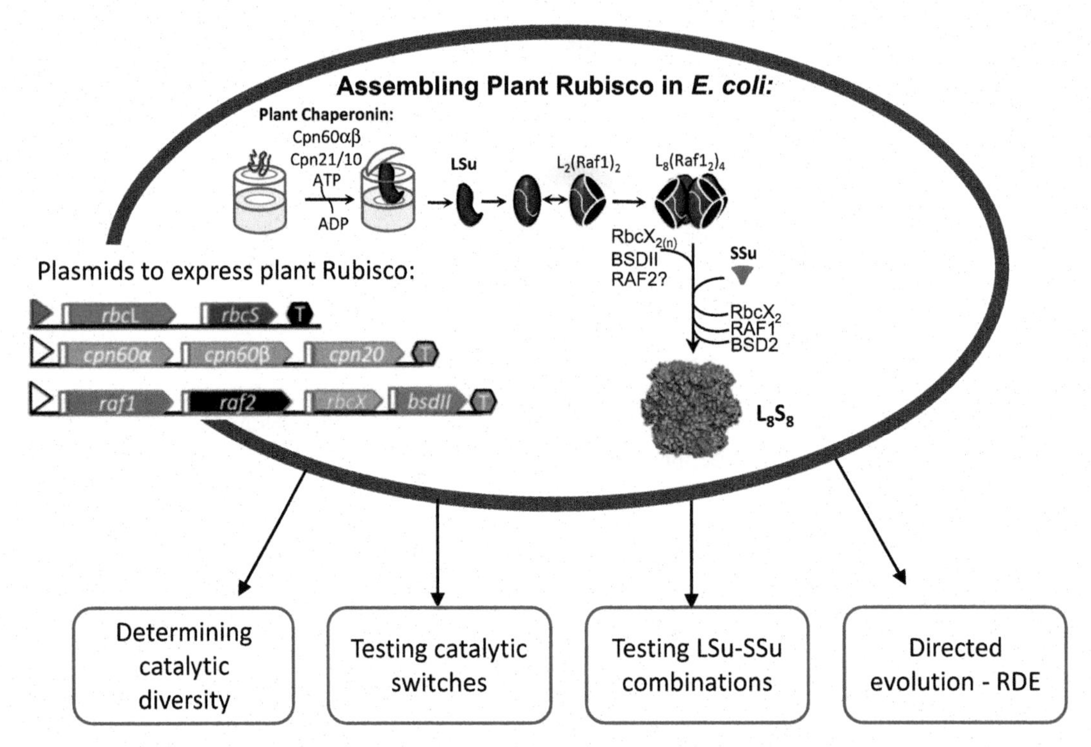

Fig. 11.5. **Transplanting the higher plant Rubisco assembly pathway into *E. coli* to provide new synthetic biology opportunities to improve Rubisco catalysis.** Rubisco assembly in plant chloroplasts is a multistep process requiring the correct folding of nascent large subunit (LSu) chains within the Cpn60αβ cage and then stabilization by Rubisco ancillary folding factors including Raf1, RbcX, BSDII and Raf2 (Wilson and Hayer-Hartl 2018). These factors are displaced by the binding of Rubisco small subunits (SSu) to form a stable functional L_8S_8 enzyme. This pathway has been successfully transplanted into *E. coli* which resulted in the synthesis and assembly of Arabidopsis Rubisco using a three component plasmid system as depicted (Aigner et al. 2017). This synthetic biology advancement will allow testing of catalytic diversity from plants that are recalcitrant to protein purification, rapid testing of catalytic residues responsible for catalytic diversity, the testing of subunit combinations to determine the catalytic phenotype and to reach new catalytic varation through directed evolution using the Rubisco dependent *E. coli* version 2. (Wilson et al. 2018)

synthetic mutant library (PML; Feiz et al. 2012, 2014). Raf1 is crucial for the stabilization of the L subunit antiparallel dimer and demonstrated to have coevolved with its cognate Rubisco making it species specific (Hauser et al. 2015; Whitney et al. 2015). Co-expression of Raf1 in tobacco chloroplasts expressing Arabidopsis Rubisco L subunits containing tobacco S subunits resulted in substantial improvement in the content of the hybrid Arabidopsis Rubisco (Whitney et al. 2015). Raf2 has been shown to be important for assembly but its precise role in biogenesis is still yet to be determined (Feiz et al. 2014). Recently, the role of BSDII has also been demonstrated to stabilize the end-state intermediate of the L_8 core prior to the assembly with Rubisco S subunits (Aigner et al. 2017). Unlike Raf1, BSDII interactions are not limited to cognate Rubisco L subunit, which suggests that its interactions are less specific for surface L subunit amino acid residues (Conlan et al. 2019).

Recently, it was confirmed that chloroplast chaperonin is essential and specific for folding the nascent plant Rubisco L subunit peptides (Aigner et al. 2017). This chloroplast protein folding machinery is homologous to the Group I chaperonin GroEL (Cpn60) that forms a two-ring tetradecamer and hydrolyses ATP to fold client proteins (Wilson and Hayer-Hartl 2018).

Chloroplast chaperonin is comprised of Cpn60α and Cpn60β subunits in a 1:1 ratio (Cpn60α7β7) and cooperates with Cpn20 either as a tetramer or a heterooligomer with Cpn10 (Wilson and Hayer-Hartl 2018). These components bind to the ends of the chaperonin cage that is regulated by the Cpn60 ATPase.

B. Utilizing the E. coli Plant Rubisco Platform

Confirmation that the catalytic measurements from recombinantly expressed Arabidopsis Rubisco mirror the plant form (Aigner et al. 2017) provides the confidence that recombinant Rubisco protein can be used for the kinetic screening of plant Rubisco with substantially increased throughput. Furthermore, this suggests that Rubisco catalysis is not influenced by the absence of chloroplast specific post-translation modifications. Figure 11.5 outlines how the Rubisco *E. coli* platform can now be used to interrogate Rubisco kinetics and alter them either by site-directed mutagenesis and/or directed evolution.

The traditional 'grind and find' approach to determine the diversity in Rubisco catalysis has been utilized to demonstrate the natural diversity in Rubisco from many different photosynthetic organisms. However, extracts can be difficult to assay as secondary metabolites can inhibit Rubisco catalysis or directly impact the extractable Rubisco protein from leaves such as those from *Eucalyptus* species (Sharwood et al. 2017; Crous et al. 2018). The Rubisco *E. coli* platform requires gene sequence information of the five key proteins involved in the chloroplast chaperone assisted folding pathway and the cognate *rbc*L and *Rbc*S genes. Whether the folding and assembly requirements for Rubisco from closely related species in this system are properly met is still yet to be fully determined. Redundancy with folding diverse Rubisco will reduce the number of genes required for expression in *E. coli.*

Once the Rubisco of the target species is successfully expressed in *E. coli*, site directed mutagenesis can be utilised to test the influence of associated catalytic residues in both the L and S subunits on enzyme catalysis. Furthermore, testing the catalytic influence of each S subunit of the multi-gene family can also be achieved. This will allow catalytic determination of the role of different small subunits on Rubisco catalysis within species and from foreign sources.

The development of Rubisco-dependent *E. coli* (RDE) has enabled directed evolution to alter catalysis and solubility of cyanobacterial Rubisco enzymes (Mueller-Cajar and Whitney 2008). The recent development of RDE-2 through the fusion of phosphoribulokinase (PRK) to neomycin phosphotransferase (Kan^R), enabled reductions in false positives (Wilson et al. 2018). Mutations generated in *Thermosynechococcus elongatus* BP1 L and S subunits resulted in improvements in catalytic turnover, carboxylation efficiency and specificity for CO_2 (Wilson et al. 2018).

Importantly, this system can now be used as a platform for directed evolution of higher plant Rubisco to improve catalytic efficiency and specificity for CO_2 by exploring novel fitness landscapes. However, thought needs to be given to the mutant library generation, whether it be random mutagenesis through error-prone PCR or gene shuffling (Wilson et al. 2018). Gene shuffling for example between plant and non-green algae might not yield Rubisco amenable for folding in *E. coli.*

C. Transplanting Bacterial Microcompartments into Higher Plant Chloroplasts

Another major synthetic biology challenge to improve CO_2 assimilation in higher plants is the transplantation of Rubisco-containing bacterial microcompartments (carboxysomes) into plant chloroplasts, along with a system of transporters to elevate chloroplastic bicarbonate concentrations (Fig. 11.6). Carboxysomes form an integral part of the cyanobacterial CCM by condensing Rubisco within a protein capsule that is resistant to leakage of CO_2 (Long et al. 2016). The cyanobacterial CCM is also comprised of membrane transporters for active uptake of HCO_3^- and conversion of CO_2 to HCO_3^- inside the cell, to generate a large cytosolic HCO_3^- pool (5–20 mM; Fig. 11.6a). The HCO_3^- that diffuses into the carboxysome is converted to CO_2 by a localised carbonic anhydrase, thus concentrating CO_2 around Rubisco and enhancing CO_2 fixation (Fig. 11.6a).

Within cyanobacterial phylogenies exists a divergence of Rubiscos between α- and β-subtypes. The α-cyanobacteria are generally marine species and harbor α-carboxysomes containing Form IA Rubisco, like that found in many γ-proteobacteria (Badger and Bek 2008). β-cyanobacteria are primarily freshwater and contain β-carboxysomes, which encapsulate the higher plant-like Form IB Rubisco (Badger and Bek 2008).

Initial steps to incorporate Rubisco containing β carboxysomes in plant chloroplasts paved the way for expressing cyanobacterial Form IB Rubisco from *S. elongatus* within tobacco chloroplasts (Lin et al. 2014b). Limited by Rubisco content and the poor kinetics of *S. elongatus* Rubisco in atmospheric CO_2 conditions, the transgenic plants required high CO_2 for growth (Lin et al. 2014b). However, co-expression with the carboxysomal Rubisco-binding protein CcmM35 formed large protein structures. Efforts to express the genetically complex β carboxysome shell proteins through transient expression assays in tobacco provided early indication

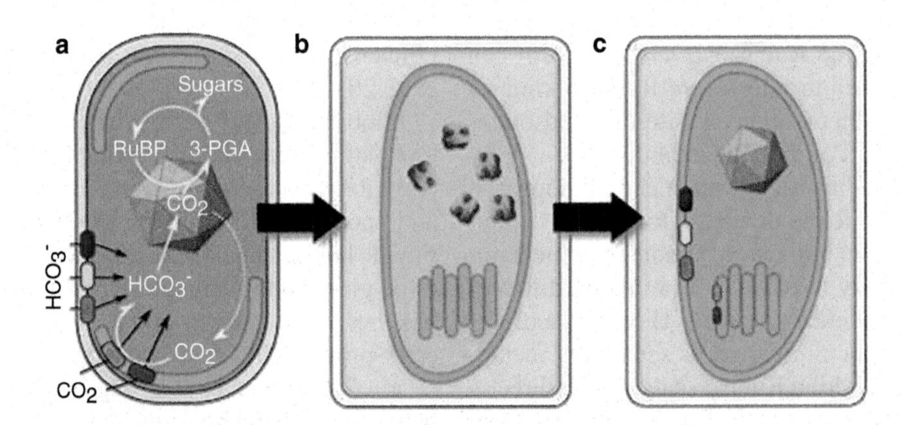

Fig. 11.6. **Transplanting the cyanobacterial CO_2 concentrating mechanism into higher plant chloroplasts** (**a**) The CO_2 concentrating mechanism (CCM) in cyanobacteria relies on the active transport of HCO_3^- via transporters located in the plasma membrane, and CO_2 to HCO_3^- conversion systems on the thylakoid membranes, to accumulate inorganic carbon primarily as HCO_3^- in the cytoplasm (Rae et al. 2017). The large HCO_3^- pool is then utilized by the carboxysome where a co-localized carbonic anhydrase converts it back to CO_2 inside the structure (Long et al. 2018). CO_2 saturates the active sites of Rubisco due to resistance to efflux and the enzyme can operate near its maximal catalytic rate. (**b**) A recipient plant cell that has its cognate Rubisco localised with the chloroplast stroma is reliant on the diffusion of CO_2 from the atmosphere to fuel carbon fixation. Here there is a draw-down effect which results in CO_2 concentrations significantly low to limit Rubisco catalysis. (**c**) Successful transplantation of the simple α-carboxysomes which encapsulate cyanobacterial Rubisco (Long et al. 2018), β-carboxysomal Rubisco (Lin et al. 2014b), and independently β-carboxysomal shell proteins (Lin et al. 2014a), in tobacco chloroplasts has been achieved. However, the incoporation of bicarbonate transporters into the chloroplast membrane and removal of the chloroplastic carbonic anhydrase to maintain an inorganic carbon pool as HCO_3^- are future requirements for success (Pengelly et al. 2014; Rolland et al. 2016; Rae et al. 2017). Atlhough the targeting of the transporters to the chloroplast envelope has been achieved (Rolland et al. 2016ss there is still a requirement to decipher the mechanism of activation within the chloroplast envelope

of possible microcompartment assembly without Rubisco (Lin et al. 2014a). However, further effort is required to understand the process of carboxysome assembly with Rubisco and the appropriate stoichiometry of shell components. Knowledge in this field has recently been expanded by determining the binding of CcmM to Rubisco to form a protein liquid condensate (Wang et al. 2019).

Remarkably, the genetically simpler α-carboxysome from *Cyanobium marinum PCC7001* assembled into icosahedral structures within tobacco chloroplasts using just two carboxysomal components: CsoS1A (outer shell protein) and CsoS2 (disordered peptide that binds/links Rubisco) that were reminiscent of their bacterial form (Long et al. 2018). The cyanobacterial Rubisco replaced the tobacco counterpart and plants required an atmosphere supplemented with CO_2 for growth, again due to relatively low content and poor kinetics of the *Cyanobium* Rubisco at ambient CO_2. However, the addition of the HCO_3^- transporters, which are required for a fully functional CCM, should remedy the poor catalytic performance of *Cyanobium* by providing elevated levels of substrate CO_2.

New efforts have begun to explore the kinetic variation of Rubiscos from α-cyanobacteria. This will provide further understanding of the diversity and to explore natures landscape to find α- cyanobacterial Rubisco that would outperform *Cyanobium* Rubisco predominately for catalytic speed. An example is *Synechococcus* PCC8102 (Table 11.1) which has an improved k_{cat}^c and K_c compared to *Cyanobium*. Some of the β-cyanobacteria have improved catalytic properties that provide opportunities to engineer these improvements into *Cyanobium* carboxysomes (Fig. 11.4b). While there is a growing dataset of Rubisco kinetic properties from plant and β-cyanobacterial systems, the paucity of data for Form IA cyanobacterial Rubiscos warrants greater exploration in this area.

Further research is needed to explore the temperature sensitivity of carboxysome-based systems to understand the responses of plants harbouring a cyanobacterial CCM to a warming world. While the expectation is that a carboxysome will minimize energetic losses to photorespiration, normally exacerbated by elevated growth temperatures in wild-type plants, little is known about how this potential benefit might translate to crops.

D. Prospects for Transplanting an Algal CCM into Higher Plants

Advances in the understanding of pyrenoid formation and function of the CCM in algae have enabled new prospects for engineering pyrenoids into plant chloroplasts (Sharwood 2017a). The discovery of the Essential Pyrenoid Component 1 (EPYC1) protein has been crucial for understanding how Rubisco is brought together within a pyrenoid (Mackinder et al. 2016). EPYC1 causes liquid-liquid phase separation (LLPS) of catalytically active Rubisco (Wunder et al. 2018; Wunder et al. 2019). Analogous to EPYC1, microcompartment protein CcmM (from β carboxysomes) also links Rubisco causing LLPS, which is a requirement for microcompartment assembly (Wang et al. 2019).

Recently, the bestrophin-like proteins have been implicated for the transport of HCO_3^- into the thylakoid lumen that traverse the pyrenoid in *Chlamydomonas reinhardtii* (Mukherjee et al. 2019). HCO_3^- transported into the lumen is subsequently converted to CO_2 by a localized CA and enables the buildup of CO_2 around Rubisco in a manner analogous to the carboxysome system (Rae et al. 2017), therefore improving CO_2 fixation at the expense of oxygenation. As more components of the algal CCM are revealed, this will provide new opportunities to incorporate algal CCM's into higher plants to improve the efficiency of CO_2 assimilation particularly under future hot and dry climates where photorespiration is exacerbated.

E. Transplanting the NADP-ME Subtype CCM into Higher Plants

To mitigate poor catalytic properties of Rubisco, C_4 plants evolved a CCM which has occurred multiple times across >70 plant lineages through convergent evolution (Edwards et al. 2010; Sage 2017). All the genes required for the CCM exist in C_3 plants but rewiring of expression, localisation and activity of CCM components was required (Niklaus and Kelly 2019). The C_4 plant CCM is comprised of biochemical and anatomical features that separate the two carboxylases PEPC (phosphoenolpyruvate carboxylase) and Rubisco (Kajala et al. 2011). To incorporate this CCM into C_3 plants, chloroplast number must to be increased within C_3 bundle sheath cells along with compartmentalization of Rubisco and the reduction in vein spacing between vascular bundles (von Caemmerer et al. 2012). Furthermore, enzymes and metabolite transporters need to be installed with appropriate regulatory mechanisms for cell specific expression. This is further reviewed by (Furbank et al. 2015; Furbank 2017; Ermakova et al. 2019).

IV. Engineering Rubisco into Key Crops

Ultimately, the use of SynBio will provide new opportunities to improve crop productivity through altering metabolic pathways such as photosynthetic carbon assimilation. There are several engineering strategies available for altering photosynthetic genes in higher plants. These include plastid transformation (Sharwood 2017b), nucleus transformation (Rodermel et al. 1988) and CRISPR CAS9 mediated gene editing. However, plastid transformation is limited by the number of amenable dicot species and no available technology is currently present for monocots (Bock 2015). Exploiting nucleus transformation mediated by *Agrobacterium tumefaciens* remains viable, however, the random nature of transgene insertion into the nucleus can limit successful expression of transgenes (Gelvin 2003). Therefore, there is a requirement to analyse many T_0 transgenic plants to ensure the correct expression line is identified. The burgeoning field of CRISPR CAS9 has new opportunities for editing gene coding sequences to change enzyme function and used for hyper-expressing genes of interest (Agarwal et al. 2018). The technology has already been implemented in key crops such as rice, wheat and cotton and C_4 plants such as Maize and Sorghum (Belhaj et al. 2013; Belhaj et al. 2015; Bortesi and Fischer 2015). In addition, new cloning tools such as Golden Gate have now made it simpler to clone multiple genes into a single construct equipped with promoter and terminator elements for plant transformation (Engler et al. 2014).

A. Strategies for Manipulating Rubisco and Rubisco Activase

Rubisco content and catalytic properties can be altered in higher plants through transgenic technologies. Rubisco content can be altered through RNAi of the S subunit gene and crucial assembly factors. Plastid transformation has been successfully utilized for site directed mutagenesis of *rbc*L (Whitney et al. 1999), the partial and full replacement of *rbc*L (Sharwood et al. 2008b; Whitney and Sharwood 2008).

The development of tobacco masterline, which has *rbc*L replaced with *rbc*M (codes for *R. rubrum* L_2 Rubisco), has enabled more efficient site-directed mutagenesis because of the low sequence homology between the two genes (Whitney et al. 1999; Whitney and Sharwood 2008). Furthermore, it has increased the efficiency of generating homoplasmic plants as *R. rubrum* Rubisco cannot support growth in air (Whitney and Sharwood 2008). Therefore, plastomes encoding Rubisco that are suited for ambient conditions of CO_2 will be selected for, thus providing additional selection alongside the antibiotic resistance marker *aad*A.

Within higher plants, new strategies are required to determine the influence of S subunits on Rubisco catalysis. S subunits are encoded in the nucleus as multigene families and evidence suggests that different S subunits may be induced and assembled into Rubisco depending on environmental conditions (Cavanagh and Kubien 2014). Recombinant expression of higher plant Rubisco will enable the rapid analysis of individual S subunits expressed with cognate L subunits (Aigner et al. 2017). Nuclear expression and or RNAi is required to make modifications to existing plant systems (Rodermel et al. 1988; Dhingra et al. 2004; Ishikawa et al. 2011). The competition for assembly between nucleus encoded Sorghum S subunits and rice S subunits reduced the capacity of Sorghum smalls to assemble with rice Rubisco L subunits within the chloroplast (Ishikawa et al. 2011).

Nucleus insertion of codon modified maize Rubisco L and S subunit genes and the Rubisco assembly factor Raf1 within *Z. mays* revealed Rubisco L subunits could be targeted successfully to the chloroplast for assembly into Rubisco (Salesse-Smith et al. 2018). The transgenic maize plants were found to overexpress Rubisco, which resulted in increased rates of CO_2 assimilation and increased plant biomass (Salesse-Smith et al. 2018). Nucleus expression of Rubisco L subunits will provide new opportunities to transplant altered Rubisco into monocots such as maize and wheat that are not yet amenable for chloroplast transformation.

Rubisco activase is nucleus encoded, therefore altering specific amino acid residues of the endogenous isoforms is possible through CRISPR-CAS9 technology. However, complete replacement of RCA β isoforms with thermotolerant versions is viable through *Agrobacterium* mediated nucleus transformation but there is a requirement to determine whether foreign RCA is capable of interacting with the endogenous Rubisco to modulate activation.

V. Conclusions

The significant advances in SynBio technologies will enable future high throughput testing of Rubisco performance. This technology will be utilised to identify naturally occurring or artificially evolved forms of Rubisco that will provide improvements in CO_2 assimilation under future climates. Exploiting natural diversity in species that have already adapted to hot and dry environments are now providing access to heat tolerant proteins involved in photosynthetic CO_2 assimilation such as RCA. Heat tolerant RCA isoforms have been demonstrated to improve plant productivity when tested under non-optimal conditions. The rapid advance in cloning and fast DNA synthesis technologies is set to significantly advance throughput and rapidly produce vast libraries for screening using newly developed methods. Application of transposon screening methods, for example, are now being used to identify important elements in multicomponent systems, such as the classification of DAB bicarbonate accumulation factors within the CCM of *Halothiobacillus neapolitanus* (Desmarais et al. 2019). While the SynBio technologies are under the umbrella of GMO technologies, we can no longer ignore this technology to achieve new strategies for plants to cope with future climates.

Acknowledgements

RES research is funded by ARC Centre of Excellence for Translational Photosynthesis (CE140100015), the NSW Environmental Trust (2016/RD/0006), Australian Academy of Sciences – Thomas Davies Award and the Cotton Research and Development Corporation (CRDC). RES acknowledges Spencer Whitney for the measurement of *Synechococcus* WH8102 Rubisco kinetics.

References

Agarwal A, Yadava P, Kumar K, Singh I, Kaul T, Pattanayak A, Agrawal PK (2018) Insights into maize genome editing via CRISPR/Cas9. Physiol Molec Biol Plants 24:175–515

Aigner H, Wilson RH, Bracher A, Calisse L, Bhat JY, Hartl FU, Hayer-Hartl M (2017) Plant RuBisCo assembly in E. coli with five chloroplast chaperones including BSD2. Science 358:1272–1278

Andersson I, Backlund A (2008) Structure and function of Rubisco. Plant Physiol Biochem 46:275–291

Andralojc PJ, Madgwick PJ, Tao Y, Keys A, Ward JL, Beale MH, Loveland JE … Parry MAJ (2012) 2-Carboxy-D-arabinitol 1-phosphate (CA1P) phosphatase: evidence for a wider role in plant Rubisco regulation. Biochem J 442: 733–742

Andrews TJ, Lorimer GH, Tolbert NE (1973) Ribulose diphosphate oxygenase. I. Synthesis of phosphoglycolate by fraction-1 protein of leaves. Biochemistry 12:11–18

Asseng S, Ewert F, Martre P, Rötter RP, Lobell DB, Cammarano D, … Zhu Y (2014) Rising temperatures reduce global wheat production. Nat Clim Chang 5: 143

Badger MR, Bek EJ (2008) Multiple Rubisco forms in proteobacteria: their functional significance in relation to CO2 acquisition by the CBB cycle. J Exp Bot 59:1525–1541

Bar-On YM, Milo R (2019) The global mass and average rate of rubisco. Proc Natl Acad Sci USA 116:4738–4743

Bauwe H, Hagemann M, Fernie AR (2010) Photorespiration: players, partners and origin. Trends Plant Sci 15:330–336

Belhaj K, Chaparro-Garcia A, Kamoun S, Nekrasov V (2013) Plant genome editing made easy: targeted mutagenesis in model and crop plants using the CRISPR/Cas system. Plant Methods 9:39

Belhaj K, Chaparro-Garcia A, Kamoun S, Patron NJ, Nekrasov V (2015) Editing plant genomes with CRISPR/Cas9. Curr Opin Biotechnol 32:76–84

Bock R (2015) Engineering plastid genomes: methods, tools, and applications in basic research and biotechnology. Annu Rev Plant Biol 66:211–241

Bortesi L, Fischer R (2015) The CRISPR/Cas9 system for plant genome editing and beyond. Biotechnol Adv 33:41–52

Bracher A, Whitney SM, Hartl FU, Hayer-Hartl M (2017) Biogenesis and metabolic maintenance of Rubisco. Annu Rev Plant Biol 68:29–60

Carmo-Silva AE, Keys AJ, Andralojc PJ, Powers SJ, Arrabaca MC, Parry MA (2010) Rubisco activities, properties, and regulation in three different C_4 grasses under drought. J Exp Bot 61:2355–2366

Cavanagh AP, Kubien DS (2014) Can phenotypic plasticity in Rubisco performance contribute to photosynthetic acclimation? Photosynth Res 119:203–214

Cleland WW, Andrews TJ, Gutteridge S, Hartman FC, Lorimer GH (1998) Mechanism of Rubisco – the carbamate as general base [review]. Chem Rev 98:549–561

Conlan B, Birch R, Kelso C, Holland S, De Souza AP, Long SP, Beck JL … Whitney SM (2019) BSD2 is a Rubisco-specific assembly chaperone, forms intermediary hetero-oligomeric complexes, and is nonlimiting to growth in tobacco. Plant Cell Environ 42: 1287–1301

Crafts-Brandner SJ, Salvucci ME (2000) Rubisco activase constrains the photosynthetic potential of leaves at high temperature and CO2. Proc Natl Acad Sci U S A 97:13430–13435

Crous KY, Drake JE, Aspinwall MJ, Sharwood RE, Tjoelker MG, Ghannoum O (2018) Photosynthetic capacity and leaf nitrogen decline along a controlled climate gradient in provenances of two widely distributed Eucalyptus species. Glob Chang Biol 24:4626–4644

Desmarais JJ, Flamholz AI, Blikstad C, Dugan EJ, Laughlin TG, Oltrogge LM, Chen AW … Savage DF (2019) DABs are inorganic carbon pumps found throughout prokaryotic phyla. Nat Microbiol 4: 2204–2215

Dhingra A, Portis AR, Daniell H (2004) Enhanced translation of a chloroplast-expressed RbcS gene restores small subunit levels and photosynthesis in nuclear RbcS antisense plants. Proc Natl Acad Sci U S A 101:6315–6320

Edwards EJ, Osborne CP, Stromberg CA, Smith SA, Consortium CG, Bond WJ, Christin PA … Tipple B (2010) The origins of C4 grasslands: integrating evolutionary and ecosystem science. Science 328: 568 587–591

Ellis RJ (1979) The most abundant protein in the world. Trends Biochem Sci 4:241–244

Engler C, Youles M, Gruetzner R, Ehnert TM, Werner S, Jones JD, Patron NJ, Marillonnet S (2014) A golden gate modular cloning toolbox for plants. ACS Synth Biol 3:839–843

Ermakova M, Danila FR, Furbank RT, von Caemmerer S (2019) On the road to C4 rice: advances and perspectives. Plant J 101:940–950

Evans JR (1989) Photosynthesis and nitrogen relationships in leaves of C_3 plants. Oecologia 78:9–19

Farquhar GD, von Caemmerer S, Berry JA (1980) A biochemical model of photosynthetic CO_2 assimilation in leaves of C_3 species. Planta 149:78–90

Feiz L, Williams-Carrier R, Wostrikoff K, Belcher S, Barkan A, Stern DB (2012) Ribulose-1,5-bisphosphate carboxylase/oxygenase accumulation factor1 is required for holoenzyme assembly in maize. Plant Cell 24:3435–3446

Feiz L, Williams-Carrier R, Belcher S, Montano M, Barkan A, Stern DB (2014) A protein with an inactive pterin-4a-carbinolamine dehydratase domain is required for Rubisco biogenesis in plants. Plant J 80:862–869

Furbank RT (2017) Walking the C4 pathway: past, present, and future. J Exp Bot 68:4057–4066

Furbank RT, Quick WP, Sirault XRR (2015) Improving photosynthesis and yield potential in cereal crops by targeted genetic manipulation: prospects, progress and challenges. Field Crops Res 182:19–29

Furbank RT, Jimenez-Berni JA, George-Jaeggli B, Potgieter AB, Deery DM (2019) Field crop phenomics: enabling breeding for radiation use efficiency and biomass in cereal crops. New Phytol 223:1714–1727

Gelvin SB (2003) Agobacterium-mediated plant transformation: the biology behind the "gene-jockeying" tool [review]. Microbiol Molec Biol Rev 67:16–37

Ghannoum O, Evans JR, Chow WS, Andrews TJ, Conroy JP, von Caemmerer S (2005) Faster rubisco is the key to superior nitrogen-use efficiency in NADP-malic enzyme relative to NAD-malic enzyme C_4 grasses. Plant Physiol 137:638–650

Hauser T, Bhat JY, Miličić G, Wendler P, Hartl FU, Bracher A, Hayer-Hartl M (2015) Structure and mechanism of the Rubisco assembly chaperone Raf1. Nat Struct Mol Biol 22:720–728

Heureux AMC, Young JN, Whitney SM, Eason-Hubbard MR, Lee RBY, Sharwood RE, Rickaby REM (2017) The role of Rubisco kinetics and pyrenoid morphology in shaping the CCM of haptophyte microalgae. J Exp Bot 68:3959–3969

Hochman Z, Gobbett DL, Horan H (2017) Climate trends account for stalled wheat yields in Australia since 1990. Glob Chang Biol 23:2071–2081

IPCC (2012) Managing the risks of extreme events and disasters to advance climate change adaptation. In: Field CB et al (eds) A Special Report of Working Groups I and II of the 606 Intergovernmental Panel on Climate Change. IPCC, Cambridge/New York, p 582

Ishikawa C, Hatanaka T, Misoo S, Miyake C, Fukayama H (2011) Functional incorporation of sorghum small subunit increases the catalytic turnover rate of Rubisco in transgenic rice. Plant Physiol 156:1603–1611

Kajala K, Covshoff S, Karki S, Woodfield H, Tolley BJ, Dionora MJ, Mogul RT … Quick WP (2011) Strategies for engineering a two-celled C(4) photosynthetic pathway into rice. J Exp Bot 62: 3001–3010

Khan S, Andralojc PJ, Lea PJ, Parry MA (1999) 2′-carboxy-D-arabitinol 1-phosphate protects ribulose 1, 5-bisphosphate carboxylase/oxygenase against proteolytic breakdown. Eur J Biochem 616:840–847

Kubien DS, Whitney SM, Moore PV, Jesson LK (2008) The biochemistry of Rubisco in Flaveria. J Exp Bot 59:1767–1777

Kurek I, Chang TK, Bertain SM, Madrigal A, Liu L, Lassner MW, Zhu G (2007) Enhanced Thermostability of Arabidopsis Rubisco activase improves photosynthesis and growth rates under moderate heat stress. Plant Cell 19:3230–3241

Lin MT, Occhialini A, Andralojc PJ, Devonshire J, Hines KM, Parry MA, Hanson MR (2014a) Beta-Carboxysomal proteins assemble into highly organized structures in Nicotiana chloroplasts. Plant J 79:1–12

Lin MT, Occhialini A, Andralojc PJ, Parry MA, Hanson MR (2014b) A faster Rubisco with potential to increase photosynthesis in crops. Nature 513:547–550

Long Stephen P, Marshall-Colon A, Zhu X-G (2015) Meeting the global food demand of the future by engineering crop photosynthesis and yield potential. Cell 161:56–66

Long BM, Rae BD, Rolland V, Forster B, Price GD (2016) Cyanobacterial CO_2-concentrating mechanism components: function and prospects for plant metabolic engineering. Curr Opin Plant Biol 31:1–8

Long BM, Hee WY, Sharwood RE, Rae BD, Kaines S, Lim YL, Nguyen ND … Price GD (2018) Carboxysome encapsulation of the CO_2-fixing enzyme Rubisco in tobacco chloroplasts. Nat Commun 9: 3570

Lorimer GH, Miziorko HM (1980) Carbamate formation on the epsilon-amino group of a lysyl residue as the basis for the activation of ribulosebisphosphate carboxylase by carbon dioxide and magnesium(2+). Biochemistry 19:5321–5328

Mackinder LC, Meyer MT, Mettler-Altmann T, Chen VK, Mitchell MC, Caspari O, Rosenzweig F … Jonikas MC (2016) A repeat protein links Rubisco to form the eukaryotic carbon-concentrating organelle. Proc Natl Acad Sci U S A 113: 5958–5963

Mueller-Cajar O (2017) The diverse AAA+ machines that repair inhibited Rubisco active sites. Front Mol Biosci 4:31

Mueller-Cajar O, Whitney SM (2008) Evolving improved *Synechococcus* Rubisco functional expression in *Escherichia coli*. Biochem J 414:205–214

Mukherjee A, Lau CS, Walker CE, Rai AK, Prejean CI, Yates G, Emrich-Mills T … Moroney JV (2019) Thylakoid localized bestrophin-like proteins are essential for the CO_2 concentrating mechanism of *Chlamydomonas reinhardtii*. Proc Natl Acad Sci USA, 116 16915, 16920

Niklaus M, Kelly S (2019) The molecular evolution of C4 photosynthesis: opportunities for understanding and improving the world's most productive plants. J Exp Bot 70:795–804

Ogren WL (1984) Photorespiration: pathways, regulation and modification. Annu Rev Plant Physiol 35:415–442

Orr D, Alcantara A, Kapralov MV, Andralojc J, Carmo-Silva E, Parry MA (2016) Surveying Rubisco diversity and temperature response to improve crop photosynthetic efficiency. Plant Physiol 172:707–717

Ort DR, Merchant SS, Alric J, Barkan A, Blankenship RE, Bock R, … Zhu XG (2015) Redesigning photosynthesis to sustainably meet global food and bioenergy demand. Proc Natl Acad Sci U S A 112: 8529–8536

Parry MA, Keys AJ, Madgwick PJ, Carmo-Silva AE, Andralojc PJ (2008) Rubisco regulation: a role for inhibitors. J Exp Bot 59:1569–1580

Pearce FG (2006) Catalytic by-product formation and ligand binding by ribulose bisphosphate carboxylases from different phylogenies. Biochem J 399:525–534

Pengelly JJ, Forster B, von Caemmerer S, Badger MR, Price GD, Whitney SM (2014) Transplastomic integration of a cyanobacterial bicarbonate transporter into tobacco chloroplasts. J Exp Bot 65:3071–3080

Portis AR (2003) Rubisco activase – Rubisco's catalytic chaperone [Review]. Photosyn Res 75:11–27

Rae BD, Long BM, Forster B, Nguyen ND, Velanis CN, Atkinson N, … McCormick AJ (2017) Progress and challenges of engineering a biophysical CO_2-concentrating mechanism into higher plants. J Exp Bot 68: 3717–3737

Raines CA (2011) Increasing photosynthetic carbon assimilation in C_3 plants to improve crop yield: current and future strategies. Plant Physiol 155:36–42

Ray DK, Mueller ND, West PC, Foley JA (2013) Yield trends are insufficient to double global crop production by 2050. PLoS One 8:e66428

Ray DK, Gerber JS, MacDonald GK, West PC (2015) Climate variation explains a third of global crop yield variability. Nat Commun 6:5989

Reynolds M, Foulkes J, Furbank R, Griffiths S, King J, Murchie E, Parry M … Slafer G (2012) Achieving yield gains in wheat. Plant Cell Environ 35: 1799–1823

Rodermel SR, Abbott MS, Bogorad L (1988) Nuclear-organelle interactions: nuclear antisense gene inhibits ribulose bisphosphate carboxylase enzyme levels in transformed tobacco plants. Cell 55:673–681

Rolland V, Badger MR, Price GD (2016) Redirecting the cyanobacterial bicarbonate transporters BicA and SbtA to the chloroplast envelope: soluble and membrane cargos need different chloroplast targeting signals in plants. Front Plant Sci 7:185

Sage RF (2017) A portrait of the C4 photosynthetic family on the 50th anniversary of its discovery: species number, evolutionary lineages, and hall of fame. J Exp Bot 68:4039–4056

Salesse-Smith CE, Sharwood RE, Busch FA, Kromdijk J, Bardal V, Stern DB (2018) Overexpression of Rubisco subunits with RAF1 increases Rubisco content in maize. Nat Plants 4:802–810

Salvucci ME (2008) Association of Rubisco activase with chaperonin-60beta: a possible mechanism for protecting photosynthesis during heat stress. J Exp Bot 59:1923–1933

Salvucci ME, Crafts-Brandner SJ (2004) Inhibition of photosynthesis by heat stress: the activation state of Rubisco as a limiting factor in photosynthesis. Physiologia Plant 120:179–186

Saschenbrecker S, Bracher A, Rao KV, Rao BV, Hartl FU, Hayer-Hartl M (2007) Structure and function of RbcX, an assembly chaperone for hexadecameric Rubisco. Cell 129:1189–1200

Savir Y, Noor E, Milo R, Tlusty T (2010) Cross-species analysis traces adaptation of Rubisco toward optimality in a low-dimensional landscape. Proc Natl Acad Sci U S A 107:3475–3480

Scafaro AP, Galle A, Van Rie J, Carmo-Silva E, Salvucci ME, Atwell BJ (2016) Heat tolerance in a wild Oryza species is attributed to maintenance of Rubisco activation by a thermally stable Rubisco activase ortholog. New Phytol 211:899–911

Scafaro AP, Atwell BJ, Muylaert S, Reusel BV, Ruiz GA, Rie JV, Galle A (2018) A thermotolerant variant of rubisco activase from a wild relative improves growth and seed yield in rice under heat stress. Front Plant Sci 9:1663

Scafaro AP, Bautsoens N, den Boer B, Van Rie J, Galle A (2019) A conserved sequence from heat-adapted species improves Rubisco activase thermostability in wheat. Plant Physiol 181:43–54

Seemann JR, Kobza J, Moore BD (1990) Metabolism of 2-carboxyarabinitol 1-phosphate and regulation of ribulose-1,5-bisphosphate carboxylase activity. Photosyn Res 23:119–130

Sharkey TD (1988) Estimating the rate of photorespiration in leaves. Physiologia Plant 73:147–152

Sharwood RE (2017a) A step forward to building an algal pyrenoid in higher plants. New Phytol 214:496–499

Sharwood RE (2017b) Engineering chloroplasts to improve Rubisco catalysis: prospects for translating improvements into food and fiber crops. New Phytol 213:494–510

Sharwood R, Whitney SM (2008) Engineering the sunflower Rubisco subunits into tobacco chloroplasts: new considerations. In: Rebeiz C (ed) The Chloroplast: Basics And Applications. Springer Science, Dordrecht, pp 285–306

Sharwood R, von Caemmerer S, Maliga P, Whitney S (2008a) The catalytic properties of hybrid Rubisco comprising tobacco small and sunflower large subunits mirror the kinetically equivalent source Rubiscos and can support tobacco growth. Plant Physiol 146:83–96

Sharwood RE, von Caemmerer S, Maliga P, Whitney SM (2008b) The catalytic properties of hybrid Rubisco comprising tobacco small and sunflower large subunits mirror the kinetically equivalent source Rubiscos and can support tobacco growth. Plant Physiol 146:83–96

Sharwood RE, Ghannoum O, Kapralov MV, Gunn LH, Whitney SM (2016a) Temperature responses of Rubisco from Paniceae grasses provide opportunities for improving C_3 photosynthesis. Nat Plants 2:16186

Sharwood RE, Ghannoum O, Whitney SM (2016b) Prospects for improving CO_2 fixation in C_3 crops through understanding C_4 Rubisco biogenesis and catalytic diversity. Curr Opin Plant Biol 31:135–142

Sharwood RE, Sonawane BV, Ghannoum O, Whitney SM (2016c) Improved analysis of C4 and C3 photosynthesis via refined in vitro assays of their carbon fixation biochemistry. J Exp Bot 67:3137–3148

Sharwood RE, Crous KY, Whitney SM, Ellsworth DS, Ghannoum O (2017) Linking photosynthesis and leaf N allocation under future elevated CO2 and climate warming in Eucalyptus globulus. J Exp Bot 68:1157–1167

Shih PM, Occhialini A, Cameron JC, Andralojc PJ, Parry MA, Kerfeld CA (2016) Biochemical characterization of predicted Precambrian RuBisCO. Nat Commun 7:10382

Shivhare D, Mueller-Cajar O (2017) In vitro characterization of thermostable CAM rubisco activase reveals a rubisco interacting surface loop. Plant Physiol 174:1505–1516

Tabita FR, Satagopan S, Hanson TE, Kreel NE, Scott SS (2008) Distinct form I, II, III, and IV Rubisco proteins from the three kingdoms of life provide clues about Rubisco evolution and structure/function relationships. J Exp Bot 59:1515–1524

Tcherkez GG, Farquhar GD, Andrews TJ (2006) Despite slow catalysis and confused substrate specificity, all ribulose bisphosphate carboxylases may be nearly perfectly optimized. Proc Natl Acad Sci U S A 103:7246–7251

von Caemmerer S (2000) Biochemical Models of Leaf Photosynthesis. CSIRO Publishing, Collingwood

von Caemmerer S, Quick WP, Furbank RT (2012) The development of C(4)rice: current progress and future challenges. Science 336:1671–1672

Walker BJ, Van Loocke A, Bernacchi CJ, Ort DR (2016) The costs of photorespiration to food production now and in the future. Annu Rev Plant Biol 67:107–129

Wang H, Yan X, Aigner H, Bracher A, Nguyen ND, Hee WY, … Hayer-Hartl M (2019) Rubisco condensate formation by CcmM in beta-carboxysome biogenesis. Nature 566: 131–135

Whitehead L, Long BM, Price GD, Badger MR (2014) Comparing the in vivo function of alpha- carboxysomes and beta-carboxysomes in two model cyanobacteria. Plant Physiol 165:398–411

Whitney SM, Sharwood RE (2008) Construction of a tobacco master line to improve Rubisco engineering in chloroplasts. J Exp Bot 59:1909–1921

Whitney SM, von Caemmerer S, Hudson GS, Andrews TJ (1999) Directed mutation of the Rubisco large subunit of tobacco influences photorespiration and growth. Plant Physiol 121:579–588

Whitney SM, Baldet P, Hudson GS, Andrews TJ (2001) Form I Rubiscos from non-green algae are expressed abundantly but not assembled in tobacco chloroplasts. Plant J 26:535–547

Whitney SM, Houtz RL, Alonso H (2011a) Advancing our understanding and capacity to engineer nature's CO2-sequestering enzyme, Rubisco. Plant Physiol 155:27–35

Whitney SM, Sharwood RE, Orr D, White SJ, Alonso H, Galmes J (2011b) Isoleucine 309 acts as a C4 catalytic switch that increases ribulose-1,5-bisphosphate carboxylase/oxygenase (rubisco) carboxylation rate in Flaveria. Proc Natl Acad Sci U S A 108:14688–14693

Whitney SM, Birch R, Kelso C, Beck JL, Kapralov MV (2015) Improving recombinant Rubisco biogenesis, plant photosynthesis and growth by coexpressing its ancillary RAF1 chaperone. Proc Natl Acad Sci USA 112:3564–3569

Wilson RH, Hayer-Hartl M (2018) Complex chaperone dependence of rubisco biogenesis. Biochemistry 57:3210–3216

Wilson RH, Martin-Avila E, Conlan C, Whitney SM (2018) An improved Escherichia coli screen for Rubisco identifies a protein-protein interface that can enhance CO2-fixation kinetics. J Biol Chem 293:18–27

Wilson RH, Thieulin-Pardo G, Hartl FU, Hayer-Hartl M (2019) Improved recombinant expression and purification of functional plant Rubisco. FEBS Lett 593:611–621

Wunder T, Cheng SLH, Lai SK, Li HY, Mueller-Cajar O (2018) The phase separation underlying the pyrenoid-based microalgal Rubisco supercharger. Nat Commun 9:5076

Wunder T, Oh ZG, Mueller-Cajar O (2019) CO_2 -fixing liquid droplets: towards a dissection of the microalgal pyrenoid. Traffic 20:380–389

www.bom.gov.au/state-of-the-climate

Young JN, Heureux AM, Sharwood RE, Rickaby RE, Morel FM, Whitney SM (2016) Large variation in the Rubisco kinetics of diatoms reveals diversity among their carbon-concentrating mechanisms. J Exp Bot 67:3445–3456

Zelitch I, Schultes NP, Peterson RB, Brown P, Brutnell TP (2009) High glycolate oxidase activity is required for survival of maize in normal air. Plant Physiol 149:195–204

Zhou Y, Whitney S (2019) Directed evolution of an improved Rubisco; in vitro analyses to decipher fact from fiction. Int J Mol Sci 20

Zhu XG, Long SP, Ort DR (2010) Improving photosynthetic efficiency for greater yield. Annu Rev Plant Biol 61:235–261

Chapter 12

With a Little Help from My Friends: The Central Role of Photorespiration and Related Metabolic Processes in the Acclimation and Adaptation of Plants to Oxygen and to Low-CO_2 Stress

Hermann Bauwe
Plant Physiology Department, University of Rostock, Rostock, Germany

and

Alisdair R. Fernie*
Max Planck Institute of Molecular Plant Physiology, Potsdam, Golm, Germany

*Author for correspondence, e-mail: fernie@mpimp-golm.mpg.de

K. M. Becklin et al. (eds.), *Photosynthesis, Respiration, and Climate Change*, Advances in Photosynthesis and Respiration 48, https://doi.org/10.1007/978-3-030-64926-5_12

Summary

The photorespiratory repair pathway (photorespiration) is a supplement to the Calvin-Benson cycle that allows photosynthetic fixation of CO_2 to be based on the photochemical splitting of water, that is, in the presence of oxygen. Photorespiration is necessary because oxygen can replace CO_2 at the CO_2 fixation enzyme ribulose-1,5-bisphosphate carboxylase (Rubisco) and oxidize the CO_2 acceptor molecule ribulose-1,5-bisphosphate (RuBP), generating the powerful enzyme inhibitor 2-phosphoglycolate (2PG). 2PG must be removed because its accumulation would block operation of the Calvin-Benson cycle and in turn halt photosynthesis. Photorespiratory removal and detoxification of 2PG occurs by recycling to 3-phosphoglycerate which can be used to generate fresh RuBP. This repair process efficiently recovers three out of four misdirected carbon atoms while one carbon is oxidized and lost as CO_2. Photorespiration is a high-flux bearing pathway in most land plants that involves more enzymes than the Calvin-Benson cycle itself and in plants requires cooperation of different cellular compartments. It thereby represents one of the most complex examples of metabolic organisation and in many ways interacts with other metabolism, such as respiration, oxidative phosphorylation, nitrogen and sulphur assimilation. Photorespiration is also a very ancient pathway that likely evolved in the dual-photosystem bearing, oxygen-evolving photoautotrophic cyanobacteria at least 2.5–2.6 billion years ago. In this chapter, we will focus on recent advances concerning the metabolic integration and the regulation of photorespiration, its response to past and predicted future climate changes, and its potential for engineering more productive crops.

I. Introduction

The photorespiratory pathway (in short photorespiration) is best understood as a set of enzyme reactions that collectively prevent accumulation of the metabolic inhibitor 2-phosphoglycolate (2PG) produced in an idle side-reaction of the enzyme at the interface between inorganic and organic carbon, ribulose 1,5-bisphosphate (RuBP) carboxylase. Thereby, photorespiration is a repair pathway that is integral and essential to photosynthetic CO_2 fixation in all plants, algae and cyanobacteria. Historically, this section of metabolism was known as the 'glycolate pathway', while the term 'photorespiration'was initially used to describe the associated loss of CO_2 from photosynthesising leaves.

RuBP carboxylase is commonly known by the acronym Rubisco for RuBP carboxylase/oxygenase, indicating the promiscuous chemistry of the enzyme. The enzyme reaction starts with the reversible enolization of RuBP to produce a reactive 2,3-enediolate intermediate (Fig. 12.1). In fact, Rubisco evolved from a primordial enolase (Ashida et al. 2005). While the enediolate normally binds CO_2 to form two molecules of 3-phosphoglycerate (3PGA), it can also react with oxygen to produce one molecule each of 2PG and 3PGA (Bowes et al. 1971). The oxidation ('metabolite damage') of RuBP to 2PG and 3PGA is the most important of several side reactions catalysed by Rubisco. Its rate is determined by a number of factors including Rubisco amount and kinetic properties and the concentrations of RuBP, oxygen and CO_2 (Laing et al. 1974; Peisker 1974). Even though Rubisco strongly favours CO_2 over oxygen, the 2PG biosynthetic rate is enormous particularly in most land plants. This is because the concentration of oxygen very much exceeds that of CO_2 in the atmo-

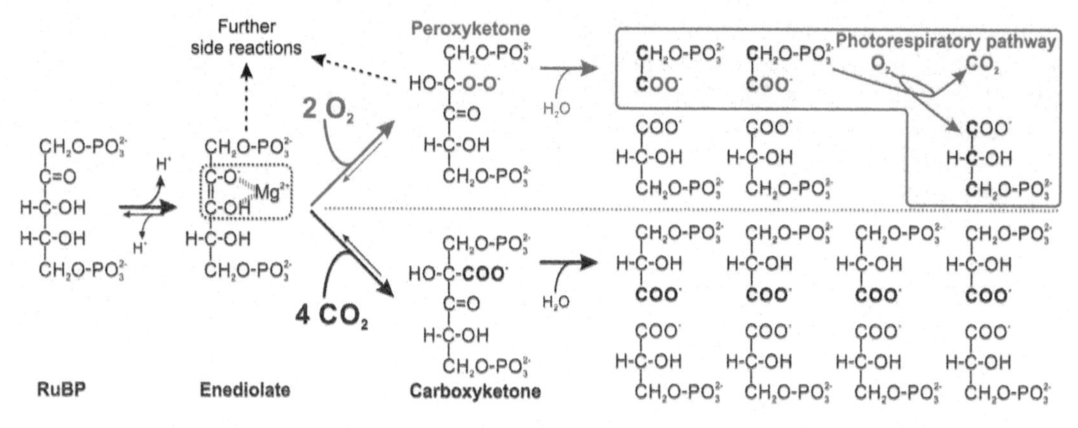

Fig. 12.1. CO_2 fixation by Rubisco. The enediolate form of RuBP can be carboxylated leading to the formation of two molecules 3PGA, or it can be oxidized to form one molecule each of 2PG. Under present atmospheric conditions, about every third RuBP would become oxidized, producing two molecules of 2PG carrying four previously fixed carbon atoms (upper part). At the same time, carboxylation of RuBP can only fix about four CO_2 molecules as 3PGA (lower part). In sum, all freshly fixed carbon becomes locked in 2PG and can be released only by operation of the photorespiratory pathway. (Modified after Bauwe et al. (2012)

sphere (21% oxygen versus 0.04% CO_2 in present air) and particularly in photosynthesizing cells. Thus, RuBP oxidation has three very direct implications for photosynthesis and hence plant growth: competitive inhibition by oxygen of CO_2 fixation, non-productive consumption of RuBP, and the inhibition of RuBP regeneration in the Calvin-Benson by 2PG. This corresponds to the fact that plants cannot thrive in air and require artificial conditions to survive when 2PG degradation is blocked (Somerville and Ogren 1979; Schwarte and Bauwe 2007). Removal of 2PG in the photorespiratory pathway involves that two molecules of 2PG are converted to one molecule of 3PGA, scavenging 3 out of four carbon atoms that were misdirected to 2PG, while only one carbon atom becomes reoxidized to CO_2 (Tolbert 1997). Only about half of the re-liberated CO_2 can be reassimilated particularly in C_3 plants, whereas a very considerable fraction is lost to the atmosphere. This feature explains why photorespiration was long misunderstood as wasteful process that should be eliminated in order to improve photosynthesis and crop yields.

Overall, the photosynthetic-photorespiratory supercycle formed from the Calvin-Benson cycle and its photorespiratory supplement belongs to the key inventions of evolution, being responsible for nearly all biological CO_2 fixation on Earth. It is also a very ancient metabolic process that

Abbreviations: 2OG – 2-oxoglutarate; 2OGDH – 2OG dehydrogenase; 2PG – 2-phosphoglycolate; 3HP – Hydroxypyruvate; 3PGA – 3-phosphoglycerate; AOX – Alternative oxidase; BASS6 – bBile acid sodium symporter 6; CAT – Catalase; CCM – Carbon dioxide concentrating mechanism; Complex II – Succinate dehydrogenase; DiT1 – Dicarboxylate translocator (chloroplastic); Fd-GOGAT – Glutamate synthase (ferredoxin-dependent); F MN – Flavin mononucleotide; GCS – Glycine cleavage system; GGAT – Glutamate-glyoxylate aminotransferase; GLYK – Glycerate 3-kinase; GOE – Great Oxidation Event; GOX – Glycolate oxidase; GS2 – Glutamine synthetase (chloroplastic); HPR1 – NADH-dependent hydroxypyruvate reductase (peroxisomal); HPR2 – NADPH-dependent hydroxypyruvate reductase (cytosolic); MDH – Malate dehydrogenase; OA – Oxaloacetate; PAL – Present atmospheric level; PGLP – 2PG phosphatase; PLGG1 – Plastidal glycolate glycerate translocator 1; PLP – Pyridoxal 5-phosphate; Rubisco – RuBP carboxylase/oxygenase; RuBP – Ribulose 1,5-bisphosphate; SBPase – Sedoheptulose 1,7-bisphosphatase; SGAT – Serine-glyoxylate aminotransferase; SHMT – Serine hydroxymethyltransferase; THF – Tetrahydrofolate; TPI – Triosephosphate isomerase; UCP – Uncoupling protein (aspartate-glutamate carrier) evolved together with the ability to use water, instead of hydrogen sulphide, as the electron donor for CO_2 fixation in photosynthetic cyanobacteria about 2.5–2.6 and possibly much earlier. Looking into the future, given the sheer magnitude of photorespiration in C_3 plants and its multiple interactions particularly with the Calvin-Benson cycle but also with mitochondrial energy metabolism and many other processes, it is not surprising that photorespiration has become a key target for breeders in order to generate crops with better photosynthetic performance and higher yield.

In this chapter, we will focus on recent advances concerning the metabolic integration and the regulation of photorespiration, the response of photosynthetic-photorespiratory metabolism to past and predicted future climate changes, and its potential for engineering more productive crops. This article is in part based on two earlier reviews (Hagemann and Bauwe 2016; Bauwe 2018). We additionally recommend reviews on Rubisco (Bracher et al. 2017), metabolite transport (Eisenhut et al. 2015), metabolic interactions (Abadie et al. 2017; Heyneke and Fernie 2018), and methods and protocols relevant for photorespiration research (Fernie et al. 2017) for more information on individual topics.

II. A Bird's-Eye View at the Core Pathway

Plant photorespiration requires ten different enzyme reactions in the core pathway and a number of auxiliary enzymes for additional processes, such as ammonia refixation. In land plants, these enzymes are distributed over four subcellular compartments: the chloroplast, the peroxisome (lost in green algae), the mitochondrion and the cytosol (Fig. 12.2). Rubisco triggers but, by definition, is not part of the repair pathway. When assessed by the number of participating enzymes and subcellular compartments, photorespiration is more complex than the Calvin-Benson cycle and indeed represents one of the most complex examples of metabolic organisation in eukaryotes. Central to photorespiration is the conversion of the two-carbon skeleton of 2PG into a three-carbon skeleton, 3PGA, which is compatible with the Calvin-Benson cycle.

A. 2-Phosphoglycolate Is Dephosphorylated in the Chloroplast

First, still within the chloroplast, 2PG phosphatase (PGLP) removes the phosphate, producing glycolate. This tetrameric enzyme belongs to the large family of haloacid dehalogenase-type phosphatases and requires Mg^{2+} for activity. An additional, cytosolic PGLP is not involved in photorespiration (Schwarte and Bauwe 2007) but serves to destroy toxic side products from mainline carbon metabolism (Collard et al. 2016).

Glycolate moves out of the chloroplast through two recently identified transporters, the glycolate-glycerate antiporter PLGG1 (Pick et al. 2013) and the glycolate transporter BASS6 (South et al. 2017). Partnering of the two transporters, with and without co-transport of glycerate in the opposite direction, balances photorespiratory carbon flow in a way that two glycolate molecules are exported per one imported glycerate. Plants deficient in both transporters are still viable in air (South et al. 2017), which confirms earlier reports that glycolic acid and glyceric acid to some extent can diffuse through the chloroplast envelope.

B. Glycolate Is Converted to Glycine in the Peroxisome

Glycolate enters the peroxisome through a protein pore, likely PMP22, (Reumann 2011), and becomes oxidised to glyoxylate followed by transamination to glycine. This amino acid is central to the photorespiratory pathway

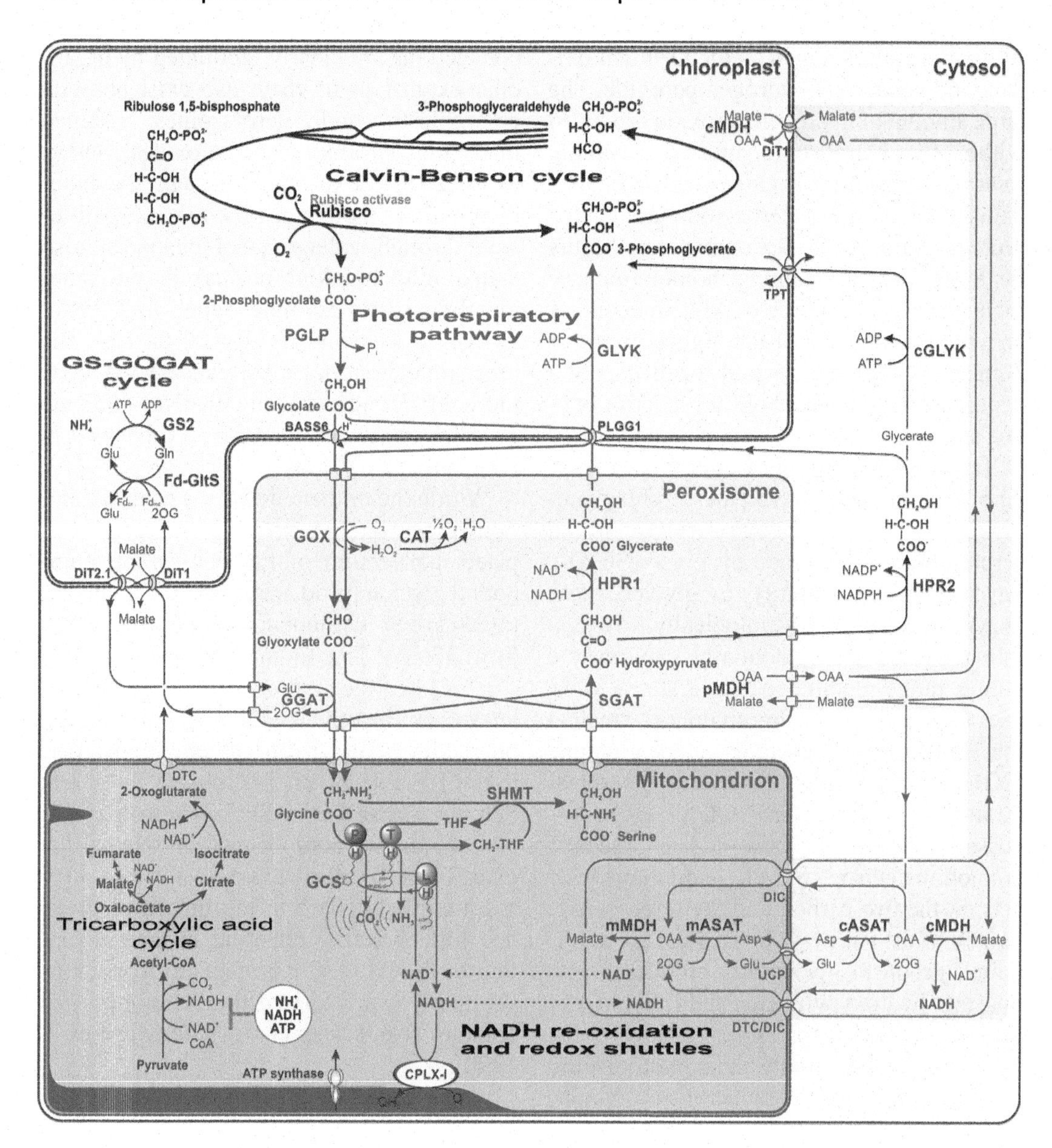

Fig. 12.2. The plant photorespiration core pathway (light blue) spans three organelles: the chloroplast, the peroxisome, and the mitochondrion. The enzymes are PGLP, GOX, GGAT, SGAT, GCS (comprising three enzymes), SHMT, HPR1, and GLYK. Known transporters are PLGG1 and BASS6. CAT detoxifies the generated hydrogen peroxide. The GS-GOGAT cycle (green) re-fixates the ammonia released by GCS. Complex I and several shuttle systems (light grey underlay) keep the redox balance. At least two enzymes of the core pathway can be circumvented in specific conditions. First, the cytosolic HPR2 bypass supports HPR1 when the peroxisomal MDH does not provide NADH rapidly enough for hydroxypyruvate reduction. Second, ATP consumption by cGLYK in the cytosol of shade-grown plants helps alleviating photoinhibition of chloroplasts. Note this scheme is adapted from Fernie and Bauwe, Plant Journal submitted

because, later in the mitochondria, it can serve as a one-carbon donor and as a one-carbon acceptor to produce one molecule of serine from two molecules of glycine.

First, the flavin mononucleotide (FMN)-dependent enzyme glycolate oxidase (GOX; also known as short-chain L-2-hydroxy acid oxidase)) oxidizes glycolate to glyoxylate, using molecular oxygen as the terminal oxidant and generating hydrogen peroxide. The large amount of hydrogen peroxide produced within the peroxisome must be rapidly removed because it can cause oxidative DNA damage and trigger peroxisome-associated protein degradation. Decomposition occurs by a Fe heme-containing homotetrameric monofunctional catalase (CAT), an ancillary enzyme that has one of the highest turnover numbers of all enzymes and forms part of a general scavenging network for reactive oxygen species.

Transamination of glyoxylate to form glycine requires two enzymes, glutamate-glyoxylate aminotransferase (GGAT) and serine-glyoxylate aminotransferase (SGAT). Given their low affinity to glycine these enzymes catalyse physiologically irreversible reactions and potentially can operate with a range of different substrates. GGAT uses glutamate as the amino donor, which at the same time generates 2-oxoglutarate (2OG) needed for the refixation of photorespiratory NH_3 in the GS-GOGAT cycle discussed further below. The second aminotransferase, SGAT, simultaneously acts on the two-carbon and the three-carbon branch of the pathway: It transfers the serine amino group to glyoxylate, producing glycine and hydroxypyruvate (3HP). SGAT can also work with asparagine (Zhang et al. 2013), which potentially links photorespiration with general N metabolism (Modde et al. 2017). It is hypothesized that glyoxylate escaping transamination under stress conditions is scavenged by a pair of high-affinity glyoxylate reductases, the cytosolic GLYR1 and the plastidial GLYR2 (Allan et al. 2009; Zarei et al. 2017).

C. *Mitochondrial Enzymes Convert Glycine to a Three-Carbon Compound, Serine*

The path on which photorespiratory metabolites enter and exit the mitochondrion is currently unknown, but inter-organellar connections are clearly facilitated by organellar extensions as they also exist between chloroplasts and peroxisomes (termed matrixules, stromules, and peroxules, Mathur et al. 2012). Exchange through the outer membrane is thought to rely on passive diffusion through voltage-gated channels (also called mitochondrial porins) in the outer membrane (for example Schell and Rutter 2013). Movement of glycine through the inner mitochondrial membrane seems to be more restrictive and likely involves diffusion and carrier mediated transport (reviewed in Eisenhut et al. 2013a).

Within the mitochondria, the central function of the photorespiratory pathway takes place: conversion of the two-carbon compound glycine (produced from 2PG) into the three-carbon compound serine (to finally form 3PGA). This requires the multienzyme glycine cleavage system (GCS, in plants also known as the glycine decarboxylase complex) and serine hydroxymethyltransferase (SHMT; reviews in Kikuchi et al. 2008; Schirch and Szebenyi 2005). These enzymes are ubiquitous components of one-carbon metabolism in all plant cells; however, they are highly abundant in the matrix of green leaf mitochondria, resulting in the association of the four GCS proteins to form an as yet poorly characterized fragile multiprotein complex (Neuburger et al. 1986; Oliver et al. 1990).

The GCS reaction cycle comprises three reactions, which are catalysed by the pyridoxal 5-phosphate (PLP)-dependent enzyme P-protein (glycine decarboxylase), the polyglutamyl tetrahydrofolate (THF)-dependent enzyme T-protein (aminomethyltransferase), and the NAD^+-dependent enzyme L-protein (dihydrolipoamide dehydrogenase). The lipoic acid-containing H-protein (hydrogen carrier protein), the fourth GCS protein, interacts as a shared substrate successively with the P-, T-, or L-protein to transfer reaction intermediates and reducing equivalents bound to its lipoyllysine arm from one enzyme to the other and finally to

NAD^+. The entire process requires one molecule each of glycine, tetrahydrofolate (THF), and NAD^+ to produce one molecule of methylenetetrahydrofolate (CH_2THF). In addition, one molecule each of CO_2 and ammonia are released (by P- and T-protein, respectively) and NADH is produced (by L-protein). The mitochondrial glutamate transporter BOUT DE SOUFFLE (BOU) could be important for THF polyglutamylation (Porcelli et al. 2018). To provide the GCS with sufficient fresh NAD^+, much of the generated NADH is rapidly reoxidized in a process that is, to some extent though not completely, coupled to adenosine triphosphate (ATP) synthesis (Gardeström and Igamberdiev 2016). That way photorespiration is a major supplier of NADH to daytime oxidative phosphorylation. The remaining NADH is exported to the cytosol via malate shuttles (Selinski and Scheibe 2019). For these reasons, photorespiratory metabolism has a strong effect on the cellular NADH/NAD^+ balance and influences other cellular processes such as sucrose synthesis, the tricarboxylic acid (TCA or Krebs) cycle and nitrate assimilation.

Next, the PLP-dependent enzyme SHMT combines CH_2THF with a second molecule of glycine to make serine and regenerate THF for the GCS. It was found that SHMT also produces considerable amounts of 5-formyl THF. This compound strongly inhibits SHMT and must therefore continually be detoxified (Collakova et al. 2008), adding a second-level metabolic repair system on top of photorespiration (Fig. 12.3).

D. *Back in the Peroxisome, Hydroxypyruvate Is Produced from Serine and Becomes Oxidized to Glycerate*

Serine goes back into the peroxisomes, where it delivers its amino group to glyoxylate to produce glycine and hydroxypyruvate, the latter of which becomes reduced to glycerate by NADH-dependent hydroxypyruvate reductase (HPR1; Givan and Kleczkowski 1992). This enzyme, together with CAT, GOX, and NAD^+-malate dehydrogenase

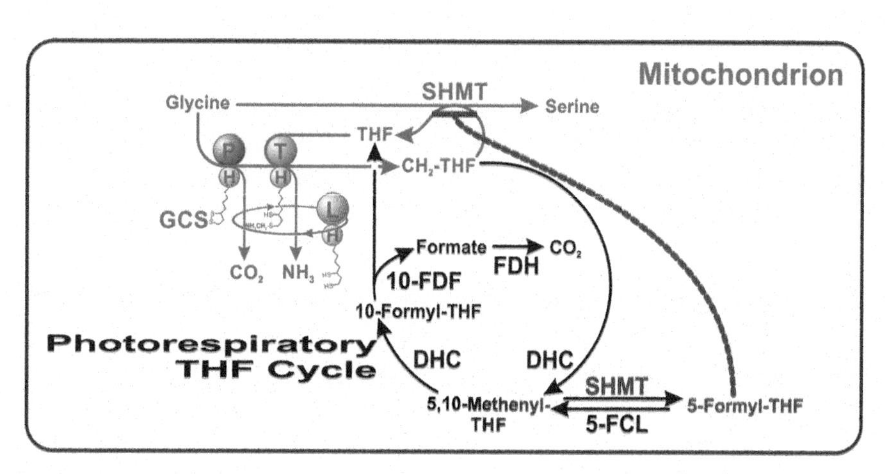

Fig. 12.3. Photorespiration requires several secondary-level repair pathways. For example, SHMT and other enzymes are inhibited by 5-formyl THF, which is produced in considerable amounts by SHMT itself (shown in red). Detoxification and recycling to THF (black route) requires four folate-interconverting enzymes, 5,10-CH_2THF dehydrogenase combined with 5,10-methenyl THF cyclohydrolase in a bifunctional enzyme (DHC), 5-formyl THF cycloligase (5-FCL) and 10-formyl THF deformylase (10-FDF). Formate dehydrogenase (FDH) oxidizes the produced formate to CO_2. If the photorespiratory THF cycle is blocked, plants accumulate massive amounts of glycine and cannot survive in normal air (Collakova et al. 2008). Note this scheme is adapted from Bauwe (2019)

(MDH), is one of the four major proteins in leaf peroxisomes. NADH cannot permeate the peroxisomal membrane and must be provided by peroxisomal malate dehydrogenase (pMDH), which oxidizes malate imported from the cytosol.

In contrast to the deleterious effect of blocks in other reactions of the pathway, neither the deletion of HPR1 (Murray et al. 1989; Timm et al. 2008) nor that of pMDH (Cousins et al. 2008) strongly impairs plant growth. This is because hydroxypyruvate can exit the peroxisome and alternatively be reduced to glycerate in the cytosol by the auxiliary enzyme NADPH-dependent hydroxypyruvate reductase 2 (HPR2; Timm et al. 2011; Ye et al. 2014). Hydroxypyruvate flux through the cytosol is small in moderate environments and likely defined by the rate of NADH supply to HPR1. It is speculated that the bypass allows more flexibility in the short-term adaptation of photorespiratory metabolism to changing environmental conditions (Givan and Kleczkowski 1992; Timm et al. 2011).

E. *Phosphorylation of Glycerate to 3PGA Completes the Photorespiratory Pathway*

Glycerate enters the chloroplast in exchange for glycolate through PLGG1 and to some extent possibly by diffusion of glyceric acid. In the chloroplasts, and in some conditions also the cytosol (Ushijima et al. 2017), D-glycerate 3-kinase (GLYK; Kleczkowski and Randall 1983; Boldt et al. 2005) completes the photorespiration pathway by returning three out of four 2PG carbon atoms that were initially misdirected to 2PG back to the Calvin-Benson cycle in the form of 3PGA. GLYK is the only glycerate kinase that produces 3PGA. Most bacteria and animals do not use GLYK but a glycerate 2-kinase, which produces 2PGA (Bartsch et al. 2008; Kern et al. 2011).

III. Photorespiration Interacts with Other Metabolism and Requires Secondary-Level Repair Pathways

A. *Regulatory Interaction with the Calvin-Benson Cycle*

The photorespiratory pathway obviously is most intimately connected to the Calvin-Benson cycle; however, the interaction goes beyond the mere recycling of 2PG to 3PGA. For example, studies with mutants and chemical inhibitors have consistently shown that any impairment of the photorespiratory pathway compromises photosynthesis (discussed in Timm et al. 2016), but we do not yet fully understand the molecular mechanisms of how photosynthesis is inhibited in this circumstances. Related to this and even more important is the fact that we do not know much about the significance of such interaction for the regulation of photosynthesis in natural conditions.

In cyanobacteria, the prokaryotic inventors of photorespiration, 2PG acts as a low-CO_2 signal that serves to coordinate carbon and nitrogen metabolism (Jiang et al. 2018). At low CO_2, more 2PG binds to two transcriptional regulator proteins: the transcriptional activator CmpR and the transcriptional repressor NdhR. The resulting conformational changes promote dissociation of NdhR from and association of CmpR with their target DNA, increasing the expression of photosynthesis-related genes. NdhR can also bind 2OG, which acts as a corepressor. Therefore, 2PG (sensing CO_2 starvation) and 2OG (sensing nitrogen starvation) function together as fine sensors for the coordination of carbon and nitrogen metabolism in cyanobacteria (Jiang et al. 2018; Zhang et al. 2018). Because of the restriction of 2PG to the chloroplast, it is not likely this kind of transcriptional control occurs in eukaryotic phototrophs.

2PG can however also signal changes of the CO_2 concentration at the biochemical level. This is possible because 2PG very efficiently inhibits the activity of SBPase and TPI, already at μM levels (Anderson 1971; Flügel et al. 2017). Varying inhibition of these enzymes in response to short-term changes in the environment in turn modulates the rate of RuBP regeneration and hence the activity of the Calvin-Benson cycle as a whole, including RuBP oxygenation (that is, photorespiration) and starch synthesis. This type of biochemical regulation by 2PG indeed occurs in plants as demonstrated by the better photosynthetic performance and faster growth observed after PGLP overexpression (Flügel et al. 2017). Notably, inhibition of the chloroplastidal TPI can to some extent be bypassed by the export of glyceraldehyde 3-phosphate to the cytosol where it is converted to glucose 6-phosphate (G6P). The produced G6P is re-imported into the chloroplast and refills the Calvin-Benson cycle, either by direct conversion to fructose 6-phosphate or by oxidative decarboxylation to ribulose 5-phosphate, the glucose 6-phosphate shunt (Sharkey and Weise 2016; Li et al. 2019).

While there is little direct evidence yet, it appears that photorespiratory metabolites which in contrast to 2PG equilibrate between different subcellular compartments, such as glycine and serine (Timm et al. 2013; Modde et al. 2017) or glyoxylate and glycolate (Leegood et al. 1995), are also involved in the regulation of photosynthesis. For example, overexpression in Arabidopsis (*Arabidopsis thaliana*) of the GCS H-protein alone (Timm et al. 2012a) or in combination with the chloroplastidal enzyme SBPase (Simkin et al. 2017; tobacco, Lopez-Calcagno et al. 2018) produced plants with increased rates of net-CO_2 assimilation and plant growth. Similar growth effects were observed with plants overexpressing the GCS L-protein (Timm et al. 2015). All this shows that the facilitation of carbon flux through the mitochondrial section of the photorespiratory pathway stimulates photosynthetic CO_2 assimilation in the Calvin-Benson cycle. The molecular players in this inter-organellar crosstalk remain to be identified. In addition to the metabolite-level control of enzyme activities, it appears that transcriptional, post-transcriptional and post-translational processes are likewise important.

B. *Photorespiration and Photoinhibition*

The interplay between photorespiration, photosynthetic electron transport and photoinhibition of PS II has long been controversially debated. In hindsight, the magnitude of the contribution of photorespiration to photoprotection may have been overemphasized as the result of the aspiration to ascribe a reasonable function to the supposedly 'wasteful' process of photorespiration. At present, there is a consensus that plants have a number of much more efficient mechanisms by which to protect their photosynthetic apparatus from photoinhibition under natural conditions (reviewed in Derks et al. 2015), but photorespiration contributes in specific circumstances.

First, photorespiration consumes ATP and reducing power, providing an alternative electron sink when CO_2 is not sufficiently available. This feature can gain some relevance when plants suffer from a severe water deficit and the stomata of the leaves close, which at the same time restricts the entry of CO_2 (for example Osmond et al. 1997).

Second, the rapid photoinhibition that can be observed in many mutants of the photorespiratory pathway at low CO_2 has been attributed to translational suppression of the de novo synthesis of the D1 protein, which in turn inhibits the repair of photodamaged PS II (Takahashi et al. 2007). Repair of PS II requires large amounts of ATP (Murata and Nishiyama 2018), which links this finding to the unexpected demonstration that photorespiration at least in some conditions does not decrease but increase the chloroplastidal ATP level (Ushijima et al. 2017). These authors found that, due to phytochrome-mediated alternative promotor selection, shade-grown Arabidopsis plants accumulate a cytoplasmic isoform of GLYK, cytGLYK. This isoform alone – without the plastidic GLYK – is able to some extent drive photorespiration and represents another cytosolic photorespiratory bypass. Photorespiratory ATP consumption on the cytosolic route lowers the ATP consumption by photorespiration in the chloroplast and alleviates photodamage induced by fluctuating high light, which, by contrast to constant light used in many studies (Apelt et al. 2017), is the standard day condition for plants or individual leaves in many ecosystems (Slattery et al. 2018).

Finally, photorespiration can also interfere with photosynthetic electron transport. One example is the better light-use efficiency and the significantly higher leaf ATP levels of GCS L-protein overexpressors (Timm et al. 2015), which points to the possibility of crosstalk between photorespiration and cyclic electron flow around photosystem I. Whilst this hypothesis still needs to be tested, it was shown that glycolate can replace bicarbonate from the non-heme iron at the acceptor side of photosystem II, lowering the production of singlet oxygen species and downregulating photosynthetic electron transport and maybe the synthesis of RuBP synthesis in the Calvin-Benson cycle (Messant et al. 2018).

C. Photorespiration and Stomatal Regulation

A further interaction between photorespiration and photosynthesis of note is the early observation that α-hydroxysulfonates which act as competitive inhibitors of GOX prevent the opening of stomata in the light in a manner that is consistent with the levels of glycolate in tobacco (Zelitch and Walker 1964). Furthermore, the authors demonstrated in the same manuscript that CO_2-induced closing of stomata can be reversed on the provision of glycolate to the leaf discs. In support of this a recent study identified that a mutant allelic to *plgg1-1* demonstrated the involvement of this transporter in abscisic acid (ABA)-mediated stomatal movement, however, mechanistic details underlying this phenomena are yet to be resolved (Dong et al. 2018). Moreover, it is currently unclear to what extent such a regulation will interact with the influence on stomatal movement known to be exerted by sugars and organic and fatty acids (Daloso et al. 2017).

D. Metabolite Shuttles

Photorespiration is a major highway of plant metabolism and likely tightly regulated on multiple levels (Leegood et al. 1995; Bauwe et al. 2010; Keech et al. 2017). Particularly, photorespiration produces and consumes in equimolar amounts large amounts of NADH, but this occurs in different organelles and is not directly coupled. Moreover, an estimated two-thirds of the NADH generated during glycine decarboxylation is immediately reoxidized in the mitochondrial oxidative respiratory chain, which ensures rapid regeneration of NAD^+ to drive rapid glycine decarboxylation and produces ATP for sucrose synthesis and nitrate reduction in the cytosol (reviewed in Gardeström and Igamberdiev 2016).

The remaining third of reducing equivalents, if not dissipated in the alternative respiratory pathway, is exported to the cytosol via two shuttle systems (Fig. 12.2). One shuttle, the mitochondrial 'malate valve', exchanges malate for oxaloacetate (OA), which requires mitochondrial and cytosolic isoforms of malate dehydrogenase (MDH) and one or several dicarboxylate translocators (DIC; Hanning et al. 1999; Palmieri et al. 2008). The second system is a malate-aspartate shuttle, which keeps OA levels low and involves an exchange of glutamate for

2OG to ensure the nitrogen equilibrium between the two compartments. It requires compartment-specific isoforms of MDH and aspartate-glutamate aminotransferase (ASAT) and one or two mitochondrial transporter(s) of aspartate, glutamate, and dicarboxylates that were recently identified (Monné et al. 2018) and previously known as mitochondrial uncoupling proteins (UCP) 1 and 2 (Sweetlove et al. 2006).

In any case, the exported fraction of reducing equivalents does not match the needs of photorespiratory HP reduction in the peroxisome and is complemented from the chloroplast (Voon et al. 2018). It was long thought that this occurs by malate export from the chloroplast (the chloroplast 'malate valve'; reviewed in Selinski and Scheibe 2019) with the vacuole serving to temporarily store excess malate (Szecowka et al. 2013), but the actual situation is less clear. For example, full inactivation of neither the peroxisomal pMDH (Cousins et al. 2008), nor the chloroplastidal malate valve enzyme NADP-MDH (Hebbelmann et al. 2012) had any substantial effect on photorespiration. Recent work has meanwhile proven that the plastidial NAD-dependent MDH has a moonlighting role in early chloroplast development through its interaction with an FtsH12-FtsHi protease complex that is more important than any role in redox regulation (Schreier et al. 2018). Moreover, the combined inactivation of HPR1 and pMDH also did not produce the characteristic phenotype of a photorespiratory mutant (Pracharoenwattana et al. 2010; Cousins et al. 2011). Therefore, it appears that not a single process but rather a multi-pathway network including proteins of as yet unknown function (Wang et al. 2018) provides NADH to the cytosol at daytime. Recent surveys appear to have provided comprehensive information concerning the plastidial complement of metabolite transporters (Facchinelli and Weber 2011) and functional and evolutionary analysis of these proteins will likely provide highly insightful in the identification of other proteins mediating inter-organellar redox transfer. Actually, it appears that indeed several bypasses complement the core photorespiratory pathway, including a 2PG-triggered glucose 6-phosphate shunt that releases additional "photorespiratory" CO_2, explaining the extraordinary effect of oxygen on the CO_2 compensation point of plants without peroxisomal hydroxypyruvate reductase (Li et al. 2019).

Recent evidence for the role of two further transporters in, or associated with, photorespiratory metabolism has been provided (Eisenhut et al. 2013b). The loss-of-function of the first of these – an endoplasmic reticulum ATP antiporter – was demonstrated to exhibit a photorespiratory phenotype although the exact mechanistic reason underlying this remains to be uncovered (Hoffmann et al. 2013). Loss of function of the mitochondrial glutamate transporter A BOUT DE SOUFFLE (BOU), which is necessary for the generation of polyglutamylated THF, is also associated with defective photorespiratory metabolism and is required for meristem growth at ambient CO_2 (Eisenhut et al. 2013b; Porcelli et al. 2018) as well as being important for sulphur assimilation and its cross talk with carbon and nitrogen metabolism in Arabidopsis (Samuilov et al. 2018).

E. Nitrogen Metabolism

Photorespiration provides glycine and serine for a number of biosynthetic pathways in photosynthesising cells, including the synthesis of cysteine and glutathione, proteins and the many compounds which require one-carbon units for their biosynthesis. This efflux is small relative to total photorespiratory flux though (Abadie et al. 2016). By contrast, reassimilation of the NH_3 released in GCS T-protein reaction happens in a high-flux bearing pathway that operates with very high efficiency: not more than about 0.01% of photorespiratory NH_3 is lost from the leaf (Mattsson et al.

1997). The process requires ammonia transport across organellar membranes to the chloroplast, possibly through aquaammoniaporins (Jahn et al. 2004; Li et al. 2017). Import of 2OG into the chloroplasts and the export of glutamate is accomplished by two translocators (Woo et al. 1987). The bifunctional dicarboxylate translocator (DiT1), which is identical with the chloroplastic malate valve mentioned above, exchanges 2OG or oxaloacetate for malate (Kinoshita et al. 2011). DiT2.1 exchanges Glu for malate in the opposite direction (Renné et al. 2003), which at the same time balances the opposed malate flow through DiT1. Once in the chloroplast, the ATP-dependent glutamine synthetase (GS2) binds the ammonia to glutamate forming glutamine. Next, the ferredoxin-dependent glutamate synthase (Fd-GOGAT) transfers the amide-nitrogen from glutamine to 2-oxoglutarate to yield two molecules of glutamate, which can be used as a general amino donor including glycine synthesis by GGAT or enters a new round of ammonia (re-)fixation. This GS-GOGAT cycle is a secondary-level repair pathway concerning the reassimilation of photorespiratory ammonia, but it is also essential for nitrate assimilation and the central link between carbon and nitrogen metabolism (Zhang et al. 2018).

The high interdependence of the processes of photorespiration, nitrate assimilation, and mitochondrial respiration was yet further illustrated in the evaluation of transgenic tomato (*Solanum lycopersicum*) plants deficient in the expression of the mitochondrial isoforms of citrate synthase and isocitrate dehydrogenase, where both sets of plants displayed compromised nitrate assimilation and altered rates of photorespiration (Sienkiewicz-Porzucek et al. 2008; Sienkiewicz-Porzucek et al. 2010). Corresponding to these experimental findings, elementary flux mode analysis suggests that photorespiratory metabolism exists in multiple forms, a subset of which is stoichiometrically coupled to nitrate reduction. (Huma et al. 2018).

Some controversy remains though. Firstly, a report suggesting dual targeting of GS2 to both chloroplasts and mitochondria (Taira et al. 2004), with the accumulation in mitochondria having a regulatory rather than enzymatic function, provoked the postulation of several functional modes of photorespiratory ammonia refixation (Linka and Weber 2005). More convincing was the demonstration that photorespiratory SHMT activity requires the mitochondrial accumulation of Fd-GOGAT (Voll et al. 2006), but, secondly, it is not known why this is necessary and how it occurs. Conversely, antisense reduction of SHMT activity resulted in a shift in the diurnal ammonium metabolism in potato (*Solanum tuberosum*) leaves indicating that internal accumulation of post-photorespiratory ammonia leads to nocturnal activation of GS2 and Fd-GOGAT (Schjoerring et al. 2006). Thirdly, it is controversially debated whether, or how much and by which mechanisms, lower photorespiration as induced by elevated atmospheric CO_2 interferes with nitrate assimilation (Bloom and Lancaster 2018; Busch et al. 2018; Huma et al. 2018; Andrews et al. 2019; Bloom et al. 2019).

F. TCA Cycle, Respiratory Electron Transport Chain, Oxidative Phosphorylation

The links between photorespiration and day respiration are supported by considerable cumulative evidence and highly important for several reasons (Obata et al. 2016; Tcherkez et al. 2017). Firstly, photorespiration is the highest flux bearing process in mitochondria under photosynthetic conditions. Secondly, the photorespiratory influx of reducing equivalents via glycine very much exceeds any realistic rate of daytime ATP production by oxidative phosphorylation, but the NADH produced during glycine decarboxylation nevertheless must be rapidly re-oxidized to NAD^+ in order to provide further oxidant for glycine decarboxylation and to maintain the mitochondrial redox balance. Thirdly, glycine oxidation and the TCA cycle potentially compete for available NAD^+. These conflicts are balanced by various mechanisms operating on the levels of (i) downregulation of the TCA cycle as a major competitor for NAD^+, (ii) the mitochondrial electron transport chain and oxidative phosphorylation, (iii) outsourcing NADH oxidation to other cellular compartments by shuttle mechanisms as discussed above.

Downregulation of the TCA cycle occurs by the phosphorylation of the mitochondrial PDH (Budde and Randall 1990), redox regulation of other TCA cycle enzymes (Daloso et al. 2015) and metabolite-level regulation particularly by ammonia and NADH (Nunes-Nesi et al. 2013). Collectively, these changes not only downregulate but also re-organize carbon flow through the TCA cycle at daytime in a way that it is not a cyclic process anymore because the conversion of 2OG to succinate becomes drastically reduced or even stops in the light, while malate and fumarate pools build up (Tcherkez et al. 2009; Tcherkez et al. 2012; Lee and Millar 2016). The resulting 'horseshoe" structure of the TCA 'cycle' at daytime comprises a reductive branch (producing malate and fumarate, consuming NADH) and an oxidative branch (producing 2OG, generating NADH). Together, this helps to balance the consumption and production of NADH in the photorespiring mitochondrion, alleviating competition with glycine oxidation. The daytime mode of the TCA cycle also reminds to the evolutionary origin of the pathway, where the addition of just one enzyme, succinyl-CoA synthetase, may have combined two different linear pathways (one for amino acid and one for heme synthesis) into a cyclic, potentially reversible, multi-functional process (Meléndez-Hevia et al. 1996).

It is not yet known how the 2OG dehydrogenase (2OGDH) and/or succinyl-CoA synthetase and/or the succinate dehydrogenase (Complex II) are down-regulated in the light. Recently, thioredoxin o-mediated redox regulation of Complex II and possibly also of succinyl-CoA synthetase was demonstrated (Daloso et al. 2015). By these not yet fully understood changes, 2OG synthesis is retained to adequately serve nitrate assimilation and dynamic changes in the photorespiratory nitrogen flux but becomes separated from malate metabolism in the light.

Indeed, photorespiratory mutants are generally characterized as exhibiting altered levels of TCA cycle intermediates. Intriguingly, succinate and γ-aminobutyric acid (GABA) levels displayed very similar changes in mutants of the core photorespiratory pathway (Florian et al. 2013) including single and multiple mutants of HPR1, 2 and 3 (Timm et al. 2008; Timm et al. 2011; Timm et al. 2012b) and PGLP1 (Timm et al. 2012b) but not in mutants indirectly related to transport activities effecting photorespiration such as mitochondrial UCP1 (Sweetlove et al. 2006; Monné et al. 2018), plastidial MDH (Cousins et al. 2008) and a GOX overexpressing line (Fahnenstich et al. 2008). Similarly, although the function of the carbonic anhydrase subunit of NADH dehydrogenase (Complex I) is still under debate, a recent study of an Arabidopsis line deficient in the expression two carbonic anhydrase like gene also indicates the relationship between respiration and photorespiration (Fromm et al. 2016). Indeed, the metabolite profile of this mutant was characterized by lower accumulation of glycine and succinate and lower levels of citrate, malate and fumarate (Fromm et al. 2016). Also of note in this context is the assessment of the metabolite levels of the plastidial GS2 mutant (Pérez-Delgado et al. 2015). This mutant accumulates higher levels of ammonium, glutamine and 2OG concomitant to the accumulation of glycine and serine (Pérez-Delgado et al. 2015). When taken together, these findings provide remarkably conserved changes indicating that the interaction of photorespiration and respiration is largely co-ordinated by the reactions surrounding 2OG with those being involved either in ammonium assimilation (Xu et al. 2012; Zhang et al. 2018) or the GABA shunt (Fait et al. 2008). Intriguingly, photorespiratory metabolites were conversely altered in oxoglutarate dehydrogenase antisense lines (Araujo et al. 2012), confirming the close relationship between 2OG and photorespiration. Intriguingly, when photorespiratory rates were estimated from gas exchange measurements, they were found to increase in citrate synthase and aconitase lines but decline in succinyl-CoA synthetase and succinate dehydrogenase lines (Obata et al. 2016). This observation may be explained by the fact that the former two, but not the latter two enzymes, have extra-mitochondrial isoforms. However, whatever the explanation, this finding hints to the complexity of the relationship between the TCA cycle and photorespiratory pathway.

Furthermore and perhaps unsurprisingly given the intimate connection of the TCA cycle and oxidative phosphorylation mediated by the mitochondrial electron transport chain, other aspects of oxidative phosphorylation are also intertwined with photorespiration. Indeed the mutants of Complex I, *ndufs4* and *ndurfv1*, are characterized by altered levels of both TCA cycle intermediates and photorespiratory intermediates (Kühn et al. 2015). Similarly, inhibition of mitochondrial ATP synthase by oligomycin treatment and the suppression of the δ-subunit of the ATP synthase complex also led to accumulation of serine and glycine (Geisler et al. 2012). Dissipation of the protein gradient across the mitochondrial inner membrane mediated by UCP1 has also been proposed to play a crucial role in maintaining the redox poise of the mitochondrial electron transport chain to facilitate photosynthetic metabolism. As mentioned above, the *ucp1* mutant exhibits decreased levels of glycine and serine and reduced *in vivo* rates of glycine to serine conversion (Sweetlove et al. 2006). Alternative oxidases (AOXs) similarly mediate the dissipation of reducing equivalents by a route that is uncoupled from ATP production (Del-Saz et al. 2018); however, the *aox1a* mutant of Arabidopsis, which lacks the most abundant isoform of AOX, exhibits no change in the levels of photorespiratory, respiratory or GABA shunt metabolites under normal growth conditions (Giraud et al. 2008). Nevertheless, this mutant accumulates more glycine, pyruvate and citrate, and less serine than the wild-type plant when the electron transport via cytochrome oxidase is inhibited by antimycin A treatment (Strodtkötter et al. 2009). Under such conditions, the *aox1*a mutant also enhances the expression of the GCS P-protein (Strodtkötter et al. 2009). The interpretation of this study is however not very clear due to the potential effect of antimycin A particularly on the PGR5/PGRL1-dependent pathway of cyclic electron transport around photosystem I (Yamori and Shikanai 2016). Type II NAD(P)H dehydrogenases are similarly considered to function in the oxidation of NAD(P)H without producing ATP. Two studies characterized the metabolite profiles of RNAi-based suppression mutants of NDA1/NDA2 and NDB1, which are located on the matrix-faced and external surfaces of the mitochondrial inner membrane, respectively (Wallström et al. 2014a; Wallström et al. 2014b). Although the metabolic relationship with photorespiration is less clear for these type II dehydrogenases than it is for UCP and AOX, NDA suppression lines showed increased levels of glycine and serine under high-light conditions, while serine levels were reduced in the NDB line (Wallström et al. 2014a; Wallström et al. 2014b). Altogether, these results indicate a close interaction of mitochondrial electron transport and photorespiration, most likely due to the demand of recycling of NADH to NAD^+ in order to support adequate photorespiratory flux (discussed in Bykova et al. 2014). A further point to note is the likely regulation of the GCS by thioredoxin (Buchanan and

Balmer 2005), since this may represent a further common regulatory mechanism.

The examples above all describe molecular genetic analysis of the interactions of photorespiration and aspects of respiration; however, a wealth of studies have characterized changes in both processes following environmental perturbations. Whilst early studies focussed on the response of individual pathways, two recent studies specifically employed environmental perturbations anticipated to alter the photorespiratory-to-photosynthetic flux ratio in Arabidopsis and surveyed the cellular response at the metabolomic level. In the first of these, Arabidopsis plants growing under normal conditions were transferred to different light and temperature conditions anticipated to affect these parameters (Florian et al. 2014a). Interestingly, their results revealed similar behaviour in response to both treatments – especially with regard to photorespiratory intermediates – and suggest that these metabolic shifts are not mediated at the level of transcription. The second study took the same strategy but rather modified the CO_2 and oxygen partial pressures. Elevated CO_2 provoked the expected decrease in photorespiratory intermediates but also of the organic acids succinate, fumarate and malate and altered amino acid profiles (Florian et al. 2014b). By contrast, increasing oxygen had little effect on the rate of photorespiration but did lead to a rapid increase in photorespiratory metabolite levels but few other clear metabolic changes. However, when analysed alongside one another these two studies clearly reflect the metabolic interactions defined above. Four further papers are of interest in this vein, three taking experiment approaches and the fourth a computation approach to explore metabolic interaction. The experiment papers all involve the characterisation of fluxes. The first of these (Szecowka et al. 2013) does not directly but rather indirectly address photorespiration. It is discussed elsewhere in this article. The other two studies (Abadie et al. 2018; Abadie and Tcherkez 2019) used NMR analysis to characterize carbon fluxes in sunflower leaves. One demonstrated that the in vivo phospho*enol*pyruvate carboxylase catalysed flux from bicarbonate to four-carbon acids is controlled by the mole fractions of CO_2 and oxygen and represents a major flux at high photorespiratory rates (Abadie and Tcherkez 2019). The other looked at the carbon allocation to major metabolites of the sunflower, namely glutamate, alanine, glycine, serine and chlorogenate and looked at their incorporation of ^{13}C following isotopic labelling with $^{13}CO_2$ under different CO_2/oxygen conditions finding that the accumulation of ^{13}C in these compounds is influenced by the gaseous conditions and thereby indicating the interconnection of photorespiration also with the metabolic pathways involved in their biosynthesis (Abadie et al. 2018). In the fourth paper, Huma et al. (2018) constructed a stoichiometric model of photorespiration and subjected it to elementary flux mode analysis – a technique which enumerates all the component minimal pathways of a network. In their study they delineated not only the classical photorespiratory pathway but also modes involving photorespiration coupled to mitochondrial metabolism and ATP production, the glutathione-ascorbate cycle and nitrate reduction. These findings directly mirrored many of the experimental findings described above, and they also provided a demonstration that photorespiration itself only impacts the assimilation quotient (CO_2-to-oxygen ratio) in instances associated with concomitant nitrate reduction.

IV. Past and Future of Photorespiration

Photorespiration has an evolutionary history of several billion years and similar to the Calvin-Benson cycle is a blend of archaeal, proteobacterial and cyanobacterial enzymes (Bauwe et al. 2012; Kern et al. 2015; Gütle et al. 2016; Hagemann et al. 2016).

A. *Early Steps*

The history of glyoxylate and related compounds, such as glyoxal, glycolaldehyde, glycolate and glycine, reaches back to abiotic times (Miller 1953; Menor-Salván 2018). Grounded on prebiotic reactions, first organic life dates back to 4.1–3.9 billion years ago, and it is thought that the last universal common ancestor of archaea, bacteria, and eukaryotes (LUCA) was an autotroph methanogen living in and on hot submarine hydrothermal vents. This organism likely used hydrogen as an energy source in combination with the THF-based chemistry of the acetyl-CoA (or Wood-Ljungdahl) pathway for CO_2 fixation (Cotton et al. 2018). From here, two different domains of life diverged, Archaea and Bacteria, followed by the evolution of eukaryotes from partnership between an Archaeon and a bacterial endosymbiont more than two billion years later (reviewed in Eme et al. 2017).

Complementing autotrophy, heterotrophic pathways evolved that, for example, used ribose moieties derived from RNA degradation. One of these pathways, designated the 'pentose bisphosphate pathway', operates in the archaeon *Thermococcus kodakarensis to* convert ribose into RuBP which is then carboxylated by a type III Rubisco to produce 3PGA (Aono et al. 2015). A related Rubisco-mediated but cyclic and autocatalytic 'reductive hexulose phosphate pathway' operates in in other archaea (Kono et al. 2017). This pathway takes up and releases equimolar amounts of CO_2 and formaldehyde, respectively, driving methane synthesis and/or the reductive acetyl-CoA pathway of CO_2 fixation. The authors speculate that the whole Calvin-Benson cycle may have originated from such an archaeal pathway. Actually, while the Calvin-Benson cycle by far predominates primary production in the extant biosphere, it is is just one, and the youngest in evolutionary terms, of six CO_2 fixation pathways found in nature (Fuchs 2011).

When the dual-photosystem bearing, oxygen-evolving cyanobacteria evolved is actively debated. It could have happened as early as 3.2 billion years ago as indicated by earliest traces of oxygenic photosynthesis; other offered dates span more than a billion years (reviewed in Schirrmeister et al. 2016). Concerning the how, it is broadly accepted that cyanobacteria acquired the ability for oxygenic photosynthesis by lateral gene transfer, PS I from green sulphur bacteria, PS II from anaerobic phototrophs such as purple bacteria, whereas the Rieske/cytochrome *b* complex was inherited vertically (reviewed in Fischer et al. 2016), resulting in the divergence of non-photosynthetic and photosynthetic cyanobacteria from anoxygenic ancestors at least 2.5–2.6 billion years ago (Soo et al. 2017).

These ancestral 'oxyphotobacteria' were the first to be challenged by high to supersaturating daytime oxygen tensions in locally oxic environments, particularly in microbial mats (reviewed in Dick et al. 2018; Hamilton 2019), triggering new metabolic processes and adaptations including massively increased rates of 2PG synthesis and its degradation by an ancient photorespiratory pathway. Reconstruction of this pathway revealed that cyanobacterial 2PG recycling typically comprises two partially redundant pathways, a photorespiratory pathway very similar to that of plants plus the bacterial glycerate pathway (Fig. 12.4; Eisenhut et al. 2008). Both pathways start by 2PG dephosphorylation and the oxidation of glycolate to glyoxylate by glycolate dehydrogenase and/or a GOX-like oxidase (Hackenberg et al. 2011; Hagemann et al. 2016). The glycerate pathway circumvents the glycine-to-serine conversion by directly converting glyoxylate into glycerate using tartronate-semialdehyde synthase and tartronate-semialdehyde reductase. Only few advanced cyanobacteria have a 3PGA-forming GLYK; most use 2PGA-forming glycerate kinases in combination with phosphoglyceromutase for the regener-

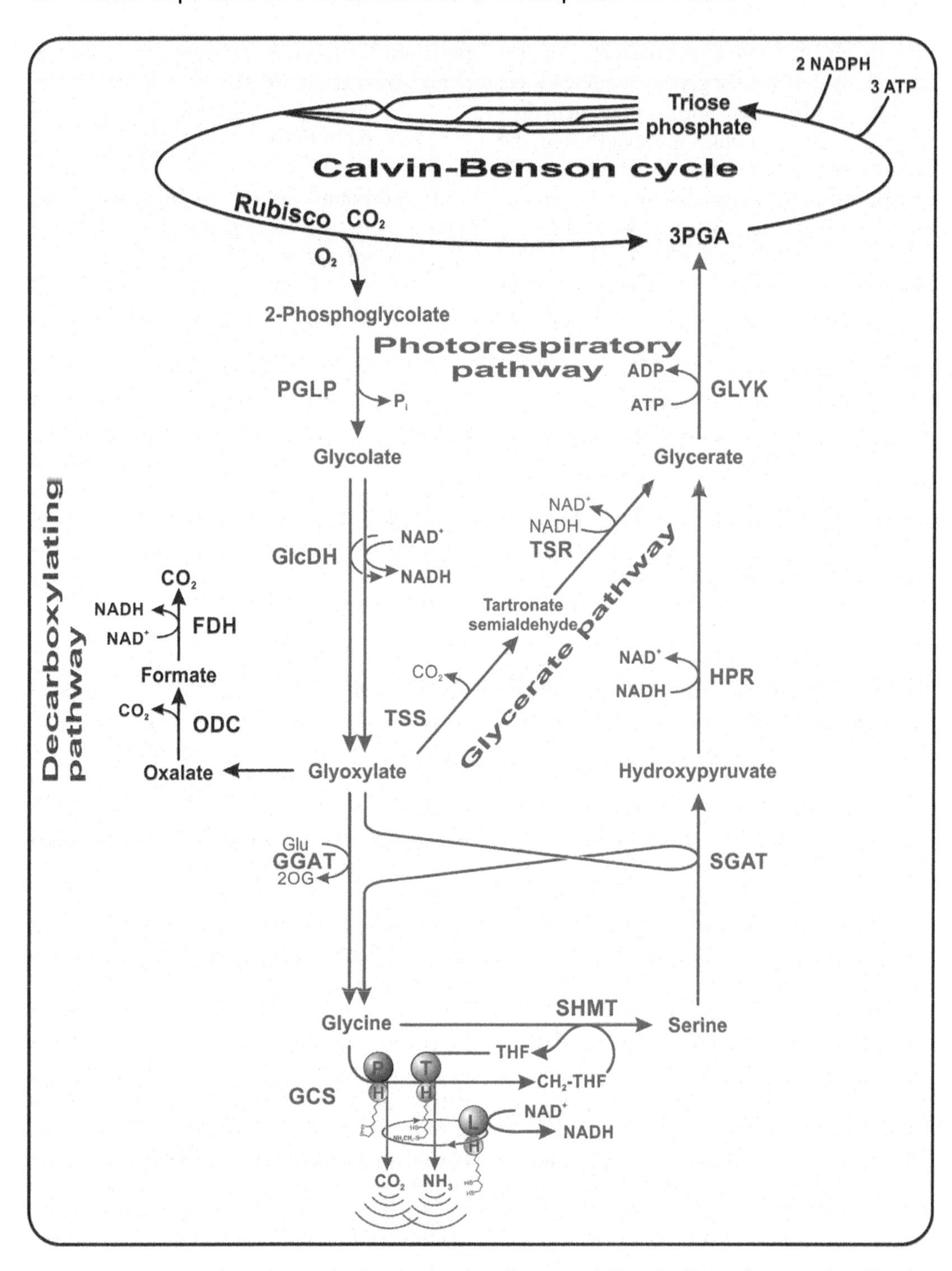

Fig. 12.4. Most cyanobacteria recycle phosphoglycolate via two partially redundant pathways, a plant-like photorespiratory pathway (shown in light blue) and the bacterial glycerate pathway (orange route). Both pathways start with PGLP and glycolate dehydrogenase (GlcDH). The glycerate pathway circumvents the glycine-to-serine conversion by directly converting glyoxylate into glycerate using tartronate-semialdehyde synthase (TSS) and tartronate-semialdehyde reductase (TSR). Some cyanobacteria can also completely decompose glyoxylate to CO_2 (grey route) via oxalate decarboxylase (ODC) and formate dehydrogenase (FDH). Note this scheme is modified from Bauwe (2019)

ation of 3PGA. Some cyanobacteria can also completely decompose glyoxylate to CO_2 via oxalate decarboxylase and formate dehydrogenase. If all three routes are inactivated, the mutant cyanobacteria can no longer grow in normal air and require elevated CO_2 to survive (Eisenhut et al. 2008). Most extant cyanobacteria have highly efficient CO_2 concentrating mechanisms (CCMs) to capture and concentrate CO_2 and hydrogencarbonate even at a very low concentration (Badger and Price 2003). It is remarkable that CCMs do not fully suppress 2PG synthesis – functional photorespiratory metabolism is still essential for photoautotrophic growth. It is also notable that even the relatively small genomes of all marine picocyanobacteria harbor almost all the genes necessary to express a plant-like 2PG metabolism as well as the glycerate pathway (Scanlan et al. 2009).

After these rather local changes, the radiation of cyanobacterial photosynthesis enriched the atmosphere with oxygen and led to the Great Oxidation Event (GOE) about 2.3–2.4 billion years ago, when the oxygen content of the atmosphere rose from essentially anoxic 10^{-5} to about 10^{-2} of its present atmospheric level (PAL). It may have been that increasing oxygen levels destroyed a methane greenhouse and triggered several major glaciations; most importantly, however, they enabled the evolution of aerobic heterotrophic life (see for example, Knoll and Nowak 2017). Overwhelming evidence suggests moderately oxic surface waters in the oceans then coexisted with anoxic deeper strata for a very long time, from 1800 to 800 million years ago (the 'Boring Billion'), during which tight feedback from the oxidation of sulphide by anoxygenic photoautotrophs, nitrification and maybe a biospheric dependence on cyanobacterial N_2-fixation hindered further oxygen accumulation (reviewed in Dick et al. 2018). At the same period of time, low-nutrient pressures may have triggered essential biological innovations, particularly the establishment and radiation of the eukaryotic domain (Mukherjee et al. 2018).

B. *Eukaryote Evolution in a Nutshell*

The multi-step origin of eukaryotes lies within both the Archaea domain and the Proteobacteria (reviewed in Dacks et al. 2016). One of the suggested scenarios involves that a toxic rise in O_2 levels in ancient Archaean microenvironments, due to cyanobacterial photosynthesis, could have driven the establishment of an oxygen-tolerant archaeal lineage and its transformation into the first stem eukaryotes (Gross and Bhattacharya 2010). Molecular clock approaches based on the O_2-intensive process of sterol biosynthesis, which was likely transferred from bacteria to archaea by horizontal gene transfer (HGT), date this event in the time of GOE (Gold et al. 2017).

Whether the acquisition of mitochondria was an earlier or later event in eukaryogenesis is still debated (Roger et al. 2017). Here, it is interesting that mitochondria are not absolutely essential for the viability of a eukaryotic cell (Karnkowska et al. 2016), which favours the view that mitochondrial endosymbiosis at least was not the first event in eukaryogenesis. Presently, a widely advocated hypothesis suggests that the closest relative to mitochondria-containing eukaryotes on the host side was a hydrogen-consuming anaerobic phagotrophic Archaeon of the Asgard supergroup, which around 1.5 billion years ago engulfed a proteobacterium that eventually became the mitochondrion (Eme et al. 2017; Martijn et al. 2018). There is evidence that the pre-mitochondrial endosymbiont diverged from either the Rickettsiales order of the Alphaproteobacteria or from a proteobacterial lineage that branched off before the divergence of Alphaproteobacteria (Martijn et al. 2018). By contrast, the single-membrane-bounded peroxisome did not evolve by endosymbiosis but can be generated de novo by the fusion of endoplasmic reticulum (ER)-derived buds (van der Zand and Tabak 2013) or as hybrids from ER- and mitochondria-derived pre-peroxisomes (Sugiura et al. 2017; Kao et al. 2018). Peroxisomes house diverse functions and with some exceptions, where they have been lost, occur in nearly all eukaryotic cells.

C. *Plant Photorespiration: A Blend of Archaeal and Bacterial Enzymes*

All Archaeplastida trace their origin to the so-called primary endosymbiontic event during the mid-Proterozoic, when a biflagellate phagotrophic eukaryote engulfed a photosynthetic cyanobacterium similar to the extant freshwater cyanobacterium *Gloeomargarita lithophora* to become the primary chloroplast (Ponce-Toledo et al. 2017; review in Nowack and Weber 2018). This was the first major split in eukaryote evolution, during which photosynthesis and the basic frame of the photorespiratory pathway were conveyed to algae and land plants (Eisenhut et al. 2008). The extensive radiation of Rhodophyta and Chlorophyta then occurred between 1600 and 1000 and 1000–542 million years ago, respectively (Sánchez-Baracaldo et al. 2017). Recent data identified the charophytic alga *Chara braunii* at the root of land plants (Nishiyama et al. 2018). While cyanobacteria now are an established prokaryotic model of the function and evolution of photorespiration in embryophytes (Orf et al. 2016), there are some gaps in our knowledge though. To mention just one example, cyanobacterial and plant PGLP belong to different haloacid dehalogenase families, and their evolutionary relationship is still uncertain (Rai et al. 2018). Altogether, it appears that during the evolution of algae and land plants multiple losses and replacements occurred, which resulted in a reticulate provenance of photorespiratory enzymes with different origins in different cellular compartments (Kern et al. 2013).

When land plants appeared on Earth about 500 million years ago (Morris et al. 2018), oxygen levels had increased from likely approximately 0.2–0.3% at the time of the endosymbiotic event leading to chloroplasts to approximately the levels we have today; however, CO_2 was still 15-fold higher than in our present atmosphere (Hetherington and Raven 2005). Massive photosynthetic activity in the Carboniferous period (360–300 million years ago) then resulted in an intermittent drop in CO_2 levels and an intermittent further rise in oxygen levels. During the past 25 million years, CO_2 levels were generally lower than today.

In addition to atmospheric CO_2 levels, temperature was another important factor in some areas because the specificity of Rubisco for CO_2 over oxygen is worse at high temperatures and the solubility of oxygen declines more slowly than that of CO_2. Approximately 30 million years ago, as an adaptation to lower CO_2 levels and higher temperatures, a number of land plants independently from each other evolved CCMs that are known as C_4 photosynthesis (Christin et al. 2008). This pathway is based on a prefixation of CO_2 in the mesophyll by phospho*enol*pyruvate carboxylase to first make a four-carbon (C_4) compound. The enzyme efficiently captures CO_2 even at a very low concentration, creating a large gradient in CO_2 concentration between the inside of the leaf and the outside environment. The C_4 compound then moves to the Rubisco-containing bundle sheath cells, where the CO_2 is released by a C_4 acid decarboxylase. Through this mechanism, Rubisco operates at greatly elevated CO_2 levels and 2PG synthesis is low though not entirely absent. Despite these advantages, C_4 photosynthesis is not competitive in all climates and far most land plants use the C_3 photosynthetic pathway (Sage et al. 2018).

Paradoxically, photorespiration not only provided pressure towards the evolution of C_4 photosynthesis but also triggered this process by inventing the first though not yet very efficient plant CCM (Bauwe 2011). This was worked out by the study of C_3-C_4 intermediate plants. These predecessors of C_4 plants produce glycine in all photosynthetic cells, but, due to a restriction of the GCS to the bundle sheath, the photorespiratory glycine is decarboxylated only in the vein-surrounding bundle sheath cells (Rawsthorne 1992; Schulze et al. 2013; Keerberg et al. 2014). The resulting higher concentration of photorespired CO_2 in the bundle sheath and the need to return the surplus photorespiratory ammonia nitrogen back to the mesophyll prepared the path for C_4 plant evolution (Mallmann et al. 2014).

D. Crop Improvement

Rubisco is responsible for nearly all biological CO_2 fixation on Earth, resulting in a terrestrial gross primary productivity (true photosynthesis minus photorespiration) of about 120 billion tons carbon year^{-1} on land plus a similar productivity of the oceans (Field et al. 1998; Geider et al. 2001). In molecular terms, about every third molecule of RuBP becomes oxygenated instead of carboxylated in a C_3 plant growing under temperate conditions in the present atmosphere (Fig. 12.1; Bauwe et al. 2012). Photorespiration rates accordingly can approach 50% of the net-photosynthesis in most land plants and be even higher in warm and dry environments (Sage and Kubien 2007; Bauwe 2019; Slattery and Ort 2019). Thus, on a global scale, an estimated ~30 billion tons year^{-1} carbon re-liberated from leaf photorespiration and day respiration need to be added to the above numbers (Tcherkez 2013). Additionally, the energy requirements for the recycling of 2PG into RuBP and photorespiratory ammonia into glutamate nitrogen can represent more than one third of the total energetic costs of CO_2 fixation (Walker et al. 2016). Reducing these CO_2 and energetic losses has been a sensible target for molecular breeding for decades but there are no trivial solutions and any such attempt must consider how photorespiration is embedded into photosynthetic and mainline metabolism.

Although real-life photosynthetic carbon metabolism involves additional ancillary metabolism (Huma et al. 2018; Tcherkez and Limami 2019), at first sight, the correlation between photorespiration and (net-) photosynthesis in a C_3 plant is as straightforward as it is simple. The correlation is positive (!) as demonstrated in many studies including a comparison of Arabidopsis ecotypes (Tomeo and Rosenthal 2018) and large-scale field studies with various wheat and soybean genotypes, in which the high-productive genotypes showed high rates of photosynthesis along with high rates of photorespiration (Aliyev 2012). The rash conclusion is that the breeders' primary focus on the physiological level should be net-photosynthesis irrespective of the level of photorespiration, but this view would miss the potential opportunities given by enzyme and metabolic engineering.

Concerning better crops for the future, one focus is on the improvement of Rubisco's catalytic efficiency, including a better ratio of carboxylation versus oxidation of RuBP, by directed evolution and introduction of such an enzyme into the chloroplast of C_3 crops (Carmo-Silva et al. 2015; Sharwood 2017; Flamholz et al. 2019; Zhou and Whitney 2019). A second group of approaches aims at the 2PG production by elevating the CO_2 concentration around Rubisco via the establishment of C_4- and cyanobacteria-like CCMs in C_3 plants, particularly rice (Wang et al. 2016; Long et al. 2018). Third, artificial bypasses to the photorespiratory pathway could reduce or possibly fully avoid the release of photorespiratory ammonia and the resulting high energetic costs of refixation (Betti et al. 2016; Bar-Even 2018). To this end, several complex artificial pathways have been designed that could assimilate CO_2 without photorespiration (Schwander et al. 2016; Bar-Even 2018) or convert glycolate in a CO_2-fixing 3-hydroxypropionate cycle (Shih et al. 2014). The most recent addition involves two newly engineered enzymes to convert glycolate to glycolaldehyde, followed by the condensation of glycolaldehyde and phosphoglyceraldehyde (via aldolase) to form arabinose 5-phosphate and then (via additional enzymes) RuBP (Trudeau et al. 2018). As yet, this latter approach has been tested in vitro while tests in planta are still up in the air. An easier approach is the conversion of glyoxylate to glycerate in the chloroplast, as it happens in bacteria including cyanobacteria (Eisenhut et al. 2006), or in the peroxisome. In either case, glyoxylate would be converted to glycerate without prior conversion to glycine. Both variants were tested by using enzymes of the bacterial glycerate pathway but with inconsistent results. Overexpression of the bacterial glycerate pathway in the chloroplast of several plant species consistently improved photosynthesis and growth under controlled conditions (Kebeish et al. 2007; Bai et al. 2011; Dalal et al. 2015) and in the field (South et al. 2019), whereas plant growth was impaired following peroxisomal overexpression (Carvalho et al. 2011; Chen et al. 2019). Finally, maybe inspired by the photorespiration driven CCM operating in C_3-C_4 plants, several groups tried to achieve full oxidation of glycolate to CO_2 within the chloroplast. Whilst this approach is not fully comprehensible for theoretical reasons (for example Xin et al. 2015), the generated transgenic plants nevertheless displayed improved photosynthesis parameters and better growth (Maier et al. 2012; Shen et al. 2019; South et al. 2019). The chloroplastidal overexpression of only the *E. coli* (Nölke et al. 2014; Chen et al. 2019) or cyanobacterial (Ahmad et al. 2016; Bilal et al. 2019) glycolate dehydrogenase or related enzymes however produced similar growth effects. The functionality of the intended artificial bypasses actually was not directly examined at the molecular level in any of these studies and, altogether, the molecular mechanism of how the higher photosynthetic efficiency was achieved in these lines is not yet known. Potential benefits and drawbacks of the above bypasses were critically tested by using model-based approaches (Xin et al. 2015; Basler et al. 2016). These authors stress the importance of additional factors such as the intra-cellular diffusion of CO_2 and the interaction of photorespiration with other metabolic pathways.

Indeed, exploitation of regulatory feedback from the photorespiratory pathway to the operation of the Calvin-Benson cycle is another way to enhance net photosynthesis. To some extent, this strategy is based on the well-established and already mentioned fact that any impairment of photorespiratory carbon flow results in an impairment of photosynthesis (discussed in Timm et al. 2016). Mechanistically, the underlying regulation could involve photorespiratory metabolites such as 2PG, glycolate, glyoxylate, and glycine. Glycolate itself is not toxic to Chlamydomonas (Taubert et al. 2019) but easily convertible to glyoxylate and glycine, for which inhibitory effects were reported (Chastain and Ogren 1989; Eisenhut et al. 2007). Indeed, overexpression of GCS proteins and other enzymes of the photorespiratory pathway produced plants with improved photosynthetic efficiency and biomass production (Timm et al. 2012a; Timm et al. 2015; Simkin et al. 2017; Lopez-Calcagno et al. 2018). It must be stressed again though that the molecular mechanism of this inter-organellar crosstalk is not known. Better investigated is the regulation of several Calvin-Benson cycle enzymes by the inhibitor 2PG, modulating RuBP regeneration and starch synthesis (Flügel et al. 2017).

During the past 60 years, global CO_2 concentration as continuously measured since 1958 (Keeling and Rakestraw 1960) has increased from ~0.032% to a present level of ~0.041%. This change has decreased the photorespiration/photosynthesis ratio by about 25% worldwide over the twentieth century (Ehlers et al. 2015) and, as the CO_2 concentration continues to increase in the future, photorespiration will continue to fall. The higher efficiency of C_3 plants under these conditions could be levelled to some extent by global warming (discussed in Dusenge et al. 2019), which by and large promotes C_4 more than C_3 plants, but might affect competition between C_3 and C_4 plants and alter the species composition and the diversity in many ecosystems (Ehleringer et al. 1997). Some researchers believe that higher atmospheric CO_2 levels will increase crop yields (for example Walker et al. 2016); however, global climate is difficult to predict and local changes to temperature, rainfall and other determinant factors for agricultural productivity could more than offset any productivity gains at a given location.

V. Outlook

Photorespiration is an important high-flux bearing metabolic process that enables photosynthetic CO_2 fixation based on photochemical water-splitting and, thereby, is a factor that co-determines crop yields and influences the global carbon dynamics. Due to its evolutionary history, plant photorespiration is a blend of archaeal and bacterial enzymes operating in different subcellular compartments. The pathway hence requires transmembrane metabolite passages. As yet, not all involved transporter proteins are known, but several were identified in recent years, such as PLGG1 and BASS6 and the indirectly involved BOU. The detailed function of these and further proteins, such as the ER-ANT1 transporter, remains to be explored.

Plant photorespiration is much more complex than thought still one decade ago, interacting with other day-time metabolism in a variety of ways. For example, interaction with the Calvin-Benson cycle goes much beyond the mere recycling of 2PG and involves regulatory interdependencies operating on multiple levels. Chloroplastidal levels of 2PG but maybe also glycolate and glyoxylate and other metabolites are seemingly central for balancing this web. Important examples of other newly discovered but not yet fully understood interactions concern starch synthesis, mitochondrial respiration, sulphur metabolism, nitrogen assimilation and cyclic electron transport. Future research will likely focus on the post-translational and other biochemical regulation of this network.

Other exciting research of recent years provided an unexpected view of the role played by photorespiration during the evolution of C_4 plants. It appears that this process began with a photorespiration-based CCM, using glycine as a vehicle for freshly assimilated CO_2, followed by the establishment of the C_4 cycle CCM from enzymes that initially balanced the intercellular nitrogen status. Related work explained how the intercellular division of photosynthetic-photorespiratory labour integrates with cellular one-carbon metabolism of C_4 plants.

Due to the ever growing need for high-yield crops, last years have shown an impressive revival of photorespiration research. Particularly, a number of approaches have impressively demonstrated the high potential for better crop yields of optimising the photorespiratory carbon and nitrogen flow. Not all of this work was initially designed to achieve better plant growth but rather served to better understand how photorespiration actually works and is embedded into whole plant metabolism. This is changing. In light of the achieved progress, new and more rational strategies for crop improvement are being designed and tested, which could be one of the most important avenues of photorespiration research for the next decade.

Acknowledgements

Our research on photorespiration was generously supported by the Deutsche Forschungsgemeinschaft (Research Unit FOR 1186 Promics, BA 1177/12 and FE 552/10) and inspired by the collaboration and discussions with many colleagues of the Promics network and beyond.

References

Abadie C, Tcherkez G (2019) In vivo phospho*enol*pyruvate carboxylase activity is controlled by CO_2 and O_2 mole fractions and represents a major flux at high photorespiration rates. New Phytol 221:1843–1852

Abadie C, Boex-Fontvieille ER, Carroll AJ, Tcherkez G (2016) *In vivo* stoichiometry of photorespiratory metabolism. Nat Plants 2:15220

Abadie C, Carroll A, Tcherkez G (2017) Interactions between day respiration, photorespiration, and N and S assimilation in leaves. In: Tcherkez G, Ghashghaie J (eds) Plant Respiration: Metabolic Fluxes and Carbon Balance. Springer International Publishing, Basel, pp 1–18

Abadie C, Bathellier C, Tcherkez G (2018) Carbon allocation to major metabolites in illuminated leaves is not just proportional to photosynthesis when gaseous conditions (CO_2 and O_2) vary. New Phytol 218:94–106

Ahmad R, Bilal M, Jeon J-H, Kim HS, Park Y-I, Shah MM, Kwon S-Y (2016) Improvement of biomass accumulation of potato plants by transformation of cyanobacterial photorespiratory glycolate catabolism pathway genes. Plant Biotechnol Rep 10:269–276

Aliyev JA (2012) Photosynthesis, photorespiration and productivity of wheat and soybean genotypes. Physiol Plant 145:369–383

Allan WL, Clark SM, Hoover GJ, Shelp BJ (2009) Role of plant glyoxylate reductases during stress: a hypothesis. Biochem J 423:15–22

Anderson LE (1971) Chloroplast and cytoplasmic enzymes. 2. Pea leaf triose phosphate isomerases. Biochim Biophys Acta 235:237–244

Andrews M, Condron LM, Kemp PD, Topping JF, Lindsey K, Hodge S, Raven JA (2019) Elevated CO_2 effects on nitrogen assimilation and growth of C_3 vascular plants are similar regardless of N-form assimilated. J Exp Bot 70:683–690

Aono R, Sato T, Imanaka T, Atomi H (2015) A pentose bisphosphate pathway for nucleoside degradation in Archaea. Nat Chem Biol 11:355

Apelt F, Köhl K, Annunziata MG, Stitt M, Lauxmann MA, Feil R, Mengin V, ... Carillo P (2017) Getting back to nature: a reality check for experiments in controlled environments. J Exp Bot 68:4463–4477

Araujo WL, Tohge T, Osorio S, Lohse M, Balbo I, Krahnert I, Sienkiewicz-Porzucek A, ... Fernie AR (2012) Antisense inhibition of the 2-oxoglutarate dehydrogenase complex in tomato demonstrates its importance for plant respiration and during leaf senescence and fruit maturation. Plant Cell 24:2328–2351

Ashida H, Danchin A, Yokota A (2005) Was photosynthetic RuBisCO recruited by acquisitive evolution from RuBisCO-like proteins involved in sulfur metabolism? Res Microbiol 156:611–618

Badger MR, Price GD (2003) CO_2 concentrating mechanisms in cyanobacteria: molecular components, their diversity and evolution. J Exp Bot 54:609–622

Bai XL, Wang D, Wei LJ, Wang Y (2011) Plasmid construction for genetic modification of dicotyledonous plants with a glycolate oxidizing pathway. Genet Mol Res 10:1356–1363

Bar-Even A (2018) Daring metabolic designs for enhanced plant carbon fixation. Plant Sci 273:71–83

Bartsch O, Hagemann M, Bauwe H (2008) Only plant-type (GLYK) glycerate kinases produce D-glycerate 3-phosphate. FEBS Lett 582:3025–3028

Basler G, Küken A, Fernie AR, Nikoloski Z (2016) Photorespiratory bypasses lead to increased growth in *Arabidopsis thaliana*: are predictions consistent with experimental evidence? Front Bioeng Biotechnol 4:31

Bauwe H (2011) Photorespiration – the bridge to C_4 photosynthesis. In: Raghavendra AS, Sage R (eds) C_4 Photosynthesis and Related CO_2 Concentrating Mechanisms. Springer Science+Business Media B.V, New York, pp 81–108

Bauwe H (2018) Photorespiration – damage repair pathway of the Calvin–Benson cycle. In: Logan DC (ed) Plant Mitochondria, 2nd edn. Wiley, Chichester, pp 293–342

Bauwe H (2019) Photorespiration. In: eLS. Wiley, Chichester, pp 1–9. https://doi.org/10.1002/9780470015902.a0001292.pub3

Bauwe H, Hagemann M, Fernie AR (2010) Photorespiration: players, partners and origin. Trends Plant Sci 15:330–336

Bauwe H, Hagemann M, Kern R, Timm S (2012) Photorespiration has a dual origin and manifold links to central metabolism. Curr Opin Plant Biol 15:269–275

Betti M, Bauwe H, Busch FA, Fernie AR, Keech O, Levey M, Ort DR, ... Weber AP (2016) Manipulating photorespiration to increase plant productivity: recent advances and perspectives for crop improvement. J Exp Bot 67:2977–2988

Bilal M, Zeb Abbasi A, Khurshid G, Yiotis C, Hussain J, Shah M, Naqvi T, ... Ahmad R (2019) The expression of cyanobacterial glycolate-decarboxylation pathway genes improves biomass accumulation in *Arabidopsis thaliana* Plant Biotechnol Rep 13:361–373

Bloom AJ, Lancaster KM (2018) Manganese binding to Rubisco could drive a photorespiratory pathway that increases the energy efficiency of photosynthesis. Nat Plants 4:414–422

Bloom AJ, Kasemsap P, Rubio-Asensio JS (2019) Rising atmospheric CO_2 concentration inhibits nitrate assimilation in shoots but enhances it in roots of C_3 plants. Physiol Plant. https://doi.org/10.1111/ppl.13040

Boldt R, Edner C, Kolukisaoglu Ü, Hagemann M, Weckwerth W, Wienkoop S, Morgenthal K, Bauwe H (2005) D-Glycerate 3-kinase, the last unknown enzyme in the photorespiratory cycle in Arabidopsis, belongs to a novel kinase family. Plant Cell 17:2413–2420

Bowes G, Ogren WL, Hageman RH (1971) Phosphoglycolate production catalysed by ribulose diphosphate carboxylase. Biochem Biophys Res Commun 45:716–722

Bracher A, Whitney SM, Hartl FU, Hayer-Hartl M (2017) Biogenesis and metabolic maintenance of Rubisco. Annu Rev Plant Biol 68:29–60

Buchanan BB, Balmer Y (2005) Redox regulation: a broadening horizon. Annu Rev Plant Biol 56:187–220

Budde RJ, Randall DD (1990) Pea leaf mitochondrial pyruvate dehydrogenase complex is inactivated in vivo in a light-dependent manner. Proc Natl Acad Sci U S A 87:673–676

Busch FA, Sage RF, Farquhar GD (2018) Plants increase CO_2 uptake by assimilating nitrogen via the photorespiratory pathway. Nat Plants 4:46–54

Bykova NV, Moller IM, Gardeström P, Igamberdiev AU (2014) The function of glycine decarboxylase complex is optimized to maintain high photorespiratory flux via buffering of its reaction products. Mitochondrion 19:357–364

Carmo-Silva E, Scales JC, Madgwick P, Parry MAJ (2015) Optimising Rubisco and its regulation for greater resource use efficiency. Plant Cell Environ 38:1817–1832

Carvalho J, Madgwick P, Powers S, Keys A, Lea P, Parry M (2011) An engineered pathway for glyoxylate metabolism in tobacco plants aimed to avoid the release of ammonia in photorespiration. BMC Biotechnol 11:111

Chastain CJ, Ogren WL (1989) Glyoxylate inhibition of ribulosebisphosphate carboxylase/oxygenase activation state *in vivo*. Plant Cell Physiol 30:937–944

Chen ZF, Kang XP, Nie HM, Zheng SW, Zhang TL, Zhou D, Xing GM, Sun S (2019) Introduction of exogenous glycolate catabolic pathway can strongly enhances photosynthesis and biomass yield of cucumber grown in a low-CO_2 environment. Front Plant Sci 10:702

Christin P-A, Besnard G, Samaritani E, Duvall MR, Hodkinson TR, Savolainen V, Salamin N (2008) Oligocene CO_2 decline promoted C_4 photosynthesis in grasses. Curr Biol 18:37–43

Collakova E, Goyer A, Naponelli V, Krassovskaya I, Gregory JF, Hanson AD, Shachar-Hill Y (2008) Arabidopsis 10-formyl tetrahydrofolate deformylases are essential for photorespiration. Plant Cell 20:1818–1832

Collard F, Baldin F, Gerin I, Bolsee J, Noel G, Graff J, Veiga-da-Cunha M, … Bommer GT (2016) A conserved phosphatase destroys toxic glycolytic side products in mammals and yeast. Nat Chem Biol 12:601–607

Cotton CAR, Edlich-Muth C, Bar-Even A (2018) Reinforcing carbon fixation: CO_2 reduction replacing and supporting carboxylation. Curr Opin Biotechnol 49:49–56

Cousins AB, Pracharoenwattana I, Zhou W, Smith SM, Badger MR (2008) Peroxisomal malate dehydrogenase is not essential for photorespiration in Arabidopsis but its absence causes an increase in the stoichiometry of photorespiratory CO_2 release. Plant Physiol 148:786–795

Cousins AB, Walker BJ, Pracharoenwattana I, Smith SM, Badger MR (2011) Peroxisomal hydroxypyruvate reductase is not essential for photorespiration in Arabidopsis but its absence causes an increase in the stoichiometry of photorespiratory CO_2 release. Photosynth Res 108:91–100

Dacks JB, Field MC, Buick R, Eme L, Gribaldo S, Roger AJ, Brochier-Armanet C, Devos DP (2016) The changing view of eukaryogenesis – fossils, cells, lineages and how they all come together. J Cell Sci 129:3695–3703

Dalal J, Lopez H, Vasani N, Hu Z, Swift J, Yalamanchili R, Dvora M, … Sederoff H (2015) A photorespiratory bypass increases plant growth and seed yield in biofuel crop *Camelina sativa*. Biotechnol Biofuels 8:1–22

Daloso DM, Müller K, Obata T, Florian A, Tohge T, Bottcher A, Riondet C, … Fernie AR (2015) Thioredoxin, a master regulator of the tricarboxylic acid cycle in plant mitochondria. Proc Natl Acad Sci U S A 112:E1392-E1400

Daloso DM, Medeiros DB, dos Anjos L, Yoshida T, Araújo WL, Fernie AR (2017) Metabolism within the specialized guard cells of plants. New Phytol 216:1018–1033

Del-Saz NF, Ribas-Carbo M, McDonald AE, Lambers H, Fernie AR, Florez-Sarasa I (2018) An in vivo perspective of the role(s) of the alternative oxidase pathway. Trends Plant Sci 23:206–219

Derks A, Schaven K, Bruce D (2015) Diverse mechanisms for photoprotection in photosynthesis. Dynamic regulation of photosystem II excitation in response to rapid environmental change. Biochim Biophys Acta 1847:468–485

Dick GJ, Grim SL, Klatt JM (2018) Controls on O_2 production in cyanobacterial mats and implications for Earth's oxygenation. Annu Rev Earth Planet Sci 46:123–147

Dong H, Bai L, Chang J, Song CP (2018) Chloroplast protein PLGG1 is involved in abscisic acid-regulated lateral root development and stomatal movement in Arabidopsis. Biochem Biophys Res Commun 495:280–285

Dusenge ME, Duarte AG, Way DA (2019) Plant carbon metabolism and climate change: elevated CO_2 and temperature impacts on photosynthesis, photorespiration and respiration. New Phytol 221:32–49

Ehleringer JR, Cerling TE, Helliker BR (1997) C_4 photosynthesis, atmospheric CO_2 and climate. Oecologia 112:285–299

Ehlers I, Augusti A, Betson TR, Nilsson MB, Marshall JD, Schleucher JÃ (2015) Detecting long-term metabolic shifts using isotopomers: CO_2-driven suppression of photorespiration in C_3 plants over the 20th century. Proc Natl Acad Sci U S A 112:15585–15590

Eisenhut M, Kahlon S, Hasse D, Ewald R, Lieman-Hurwitz J, Ogawa T, Ruth W, … Hagemann M (2006) The plant-like C_2 glycolate pathway and the bacterial-like glycerate cycle cooperate in phosphoglycolate metabolism in cyanobacteria. Plant Physiol 142:333–342

Eisenhut M, Bauwe H, Hagemann M (2007) Glycine accumulation is toxic for the cyanobacterium *Synechocystis* sp. strain PCC 6803, but can be compensated by supplementation with magnesium ions. FEMS Microbiol Lett 277:232–237

Eisenhut M, Ruth W, Haimovich M, Bauwe H, Kaplan A, Hagemann M (2008) The photorespiratory glycolate metabolism is essential for cyanobacteria and might have been conveyed endosymbiontically to plants. Proc Natl Acad Sci U S A 105:17199–17204

Eisenhut M, Pick TR, Bordych C, Weber APM (2013a) Towards closing the remaining gaps in photorespiration – the essential but unexplored role of transport proteins. Plant Biol 15:676–685

Eisenhut M, Planchais S, Cabassa C, Guivarc'h A, Justin AM, Taconnat L, Renou JP, … Weber APM (2013b) Arabidopsis A BOUT DE SOUFFLE is a putative mitochondrial transporter involved in photorespiratory metabolism and is required for meristem growth at ambient CO_2 levels. Plant J 73:836–849

Eisenhut M, Hocken N, Weber AP (2015) Plastidial metabolite transporters integrate photorespiration with carbon, nitrogen, and sulfur metabolism. Cell Calcium 58:98–104

Eme L, Spang A, Lombard J, Stairs CW, Ettema TJG (2017) Archaea and the origin of eukaryotes. Nat Rev Microbiol 15:711–723

Facchinelli F, Weber AP (2011) The metabolite transporters of the plastid envelope: an update. Front Plant Sci 2:50

Fahnenstich H, Scarpeci TE, Valle EM, Flugge UI, Maurino VG (2008) Generation of H_2O_2 in chloroplasts of *Arabidopsis thaliana* overexpressing glycolate oxidase as an inducible system to study oxidative stress. Plant Physiol 148:729

Fait A, Fromm H, Walter D, Galili G, Fernie AR (2008) Highway or byway: the metabolic role of the GABA shunt in plants. Trends Plant Sci 13:14–19

Fernie AR, Bauwe H, Weber APM (2017) Photorespiration – Methods and Protocols. Humana Press, New York

Field CB, Behrenfeld MJ, Randerson JT, Falkowski P (1998) Primary production of the biosphere: integrating terrestrial and oceanic components. Science 281:237–240

Fischer WW, Hemp J, Johnson JE (2016) Evolution of oxygenic photosynthesis. Annu Rev Earth Planet Sci 44:647–683

Flamholz AI, Prywes N, Moran U, Davidi D, Bar-On YM, Oltrogge LM, Alves R, Savage D, Milo R (2019) Revisiting trade-offs between Rubisco kinetic parameters. Biochemistry 58:3365–3376

Florian A, Araujo WL, Fernie AR (2013) New insights into photorespiration obtained from metabolomics. Plant Biol 15:656–666

Florian A, Nikoloski Z, Sulpice R, Timm S, Araujo WL, Tohge T, Bauwe H, Fernie AR (2014a) Analysis of short-term metabolic alterations in Arabidopsis following changes in the prevailing environmental conditions. Mol Plant 7:893–911

Florian A, Timm S, Nikoloski Z, Tohge T, Bauwe H, Araujo WL, Fernie AR (2014b) Analysis of metabolic alterations in Arabidopsis following changes in the carbon dioxide and oxygen partial pressures. J Integr Plant Biol 56:941–959

Flügel F, Timm S, Arrivault S, Florian A, Stitt M, Fernie AR, Bauwe H (2017) The photorespiratory metabolite 2-phosphoglycolate regulates photosynthesis and starch accumulation in Arabidopsis. Plant Cell 29:2537–2551

Fromm S, Göing J, Lorenz C, Peterhänsel C, Braun H-P (2016) Depletion of the "gamma-type carbonic anhydrase-like" subunits of complex I affects central mitochondrial metabolism in Arabidopsis thaliana. Biochim Biophys Acta 1857:60–71

Fuchs G (2011) Alternative pathways of carbon dioxide fixation: insights into the early evolution of life? Annu Rev Microbiol 65:631–658

Gardeström P, Igamberdiev AU (2016) The origin of cytosolic ATP in photosynthetic cells. Physiol Plant 157:367–379

Geider RJ, Delucia EH, Falkowski PG, Finzi AC, Grime JP, Grace J, Kana TM, … Woodward FI (2001) Primary productivity of planet earth: biological determinants and physical constraints in terrestrial and aquatic habitats. Glob Chang Biol 7:849–882

Geisler DA, Päpke C, Obata T, Nunes-Nesi A, Matthes A, Schneitz K, Maximova E, … Persson S (2012) Downregulation of the δ-subunit reduces mitochondrial ATP synthase levels, alters respiration, and restricts growth and gametophyte development in Arabidopsis. Plant Cell 24:2792–2811

Giraud E, Ho LHM, Clifton R, Carroll A, Estavillo G, Tan Y-F, Howell KA, … Whelan J (2008) The absence of ALTERNATIVE OXIDASE1a in Arabidopsis results in acute sensitivity to combined light and drought stress. Plant Physiol 147:595–610

Givan CV, Kleczkowski LA (1992) The enzymatic reduction of glyoxylate and hydroxypyruvate in leaves of higher plants. Plant Physiol 100:552–556

Gold DA, Caron A, Fournier GP, Summons RE (2017) Paleoproterozoic sterol biosynthesis and the rise of oxygen. Nature 543:420

Gross J, Bhattacharya D (2010) Uniting sex and eukaryote origins in an emerging oxygenic world. Biol Direct 5:53

Gütle DD, Roret T, Muller SJ, Couturier J, Lemaire SD, Hecker A, Dhalleine T, … Jacquot JP (2016) Chloroplast FBPase and SBPase are thioredoxin-linked enzymes with similar architecture but different evolutionary histories. Proc Natl Acad Sci U S A 113:6779–6784

Hackenberg C, Kern R, Hüge J, Stal LJ, Tsuji Y, Kopka J, Shiraiwa Y, … Hagemann M (2011) Cyanobacterial lactate oxidases serve as essential partners in N_2 fixation and evolved into photorespiratory glycolate oxidases in plants. Plant Cell 23:2978–2990

Hagemann M, Bauwe H (2016) Photorespiration and the potential to improve photosynthesis. Curr Opin Chem Biol 35:109–116

Hagemann M, Kern R, Maurino VG, Hanson DT, Weber AP, Sage RF, Bauwe H (2016) Evolution of photorespiration from cyanobacteria to land plants, considering protein phylogenies and acquisition of carbon concentrating mechanisms. J Exp Bot 67:2963–2976

Hamilton TL (2019) The trouble with oxygen: the ecophysiology of extant phototrophs and implications for the evolution of oxygenic photosynthesis. Free Radic Biol Med 140:233–249

Hanning I, Baumgarten K, Schott K, Heldt HW (1999) Oxaloacetate transport into plant mitochondria. Plant Physiol 119:1025–1031

Hebbelmann I, Selinski J, Wehmeyer C, Goss T, Voss I, Mulo P, Kangasjärvi S, … Scheibe R (2012) Multiple strategies to prevent oxidative stress in Arabidopsis plants lacking the malate valve enzyme NADP-malate dehydrogenase. J Exp Bot 63:1445–1459

Hetherington AM, Raven JA (2005) The biology of carbon dioxide. Curr Biol 15:R406–R410

Heyneke E, Fernie AR (2018) Metabolic regulation of photosynthesis. Biochem Soc Trans 46:321–328

Hoffmann C, Plocharski B, Haferkamp I, Leroch M, Ewald R, Bauwe H, Riemer J, … Neuhaus HE (2013) From endoplasmic reticulum to mitochondria: absence of the Arabidopsis ATP antiporter Endoplasmic Reticulum Adenylate Transporter1 perturbs photorespiration. Plant Cell 25:2647–2660

Huma B, Kundu S, Poolman MG, Kruger NJ, Fell DA (2018) Stoichiometric analysis of the energetics and metabolic impact of photorespiration in C_3 plants. Plant J 96:1228–1241

Jahn TP, Møller ALB, Zeuthen T, Holm LM, Klæke DA, Mohsin B, Kühlbrandt W, Schjoerring JK (2004) Aquaporin homologues in plants and mammals transport ammonia. FEBS Lett 574:31–36

Jiang Y-L, Wang X-P, Sun H, Han S-J, Li W-F, Cui N, Lin GM, … Zhou C-Z (2018) Coordinating carbon and nitrogen metabolic signaling through the cyanobacterial global repressor NdhR. Proc Natl Acad Sci U S A 115:403–408

Kao Y-T, Gonzalez KL, Bartel B (2018) Peroxisome function, biogenesis, and dynamics in plants. Plant Physiol 176:162–177

Karnkowska A, Vacek V, Zubáčová Z, Treitli SC, Petrželková R, Eme L, Novák L, … Hampl V (2016) A eukaryote without a mitochondrial organelle. Curr Biol 26:1274–1284

Kebeish R, Niessen M, Thiruveedhi K, Bari R, Hirsch HJ, Rosenkranz R, Stäbler N, … Peterhänsel C (2007) Chloroplastic photorespiratory bypass increases photosynthesis and biomass production in *Arabidopsis thaliana*. Nat Biotechnol 25:593–599

Keech O, Gardeström P, Kleczkowski LA, Rouhier N (2017) The redox control of photorespiration: from biochemical and physiological aspects to biotechnological considerations. Plant Cell Environ 40:553–569

Keeling CD, Rakestraw NW (1960) The concentration of carbon dioxide in the atmosphere. J Geophys Res 65:2502–2502

Keerberg O, Pärnik T, Ivanova H, Bassüner B, Bauwe H (2014) C_2 photosynthesis generates about 3-fold elevated leaf CO_2 levels in the C_3-C_4 intermediate species *Flaveria pubescens*. J Exp Bot 65:3649–3656

Kern R, Bauwe H, Hagemann M (2011) Evolution of enzymes involved in the photorespiratory 2-phosphoglycolate cycle from cyanobacteria via algae toward plants. Photosynth Res 109:103–114

Kern R, Eisenhut M, Bauwe H, Weber APM, Hagemann M (2013) Does the *Cyanophora paradoxa* genome revise our view on the evolution of photorespiratory enzymes? Plant Biol 15:759–768

Kern R, Bauwe H, Hagemann M (2015) The evolution of glycolate oxidases support the origin of photorespiration among cyanobacteria. Eur J Phycol 50:184–184

Kikuchi G, Motokawa Y, Yoshida T, Hiraga K (2008) Glycine cleavage system: reaction mechanism, physiological significance, and hyperglycinemia. Proc Jpn Acad Ser B Phys Biol Sci 84:246–263

Kinoshita H, Nagasaki J, Yoshikawa N, Yamamoto A, Takito S, Kawasaki M, Sugiyama T, … Taniguchi M (2011) The chloroplastic 2-oxoglutarate/malate transporter has dual function as the malate valve and in carbon/nitrogen metabolism. Plant J 65:15–26

Kleczkowski LA, Randall DD (1983) Purification and partial characterization of spinach leaf glycerate kinase. FEBS Lett 158:313–316

Knoll AH, Nowak MA (2017) The timetable of evolution. Sci Adv 3:e1603076

Kono T, Mehrotra S, Endo C, Kizu N, Matusda M, Kimura H, Mizohata E, … Ashida H (2017) A RuBisCO-mediated carbon metabolic pathway in methanogenic archaea. Nat Commun 8:14007

Kühn K, Obata T, Feher K, Bock R, Fernie AR, Meyer EH (2015) Complete mitochondrial Complex I deficiency induces an up-regulation of respiratory fluxes that is abolished by traces of functional Complex I. Plant Physiol 168:1537–1549

Laing WA, Ogren WL, Hageman RH (1974) Regulation of soybean net photosynthetic CO_2 fixation by the interaction of CO_2, O_2, and ribulose 1,5-diphosphate carboxylase. Plant Physiol 54:678–685

Lee CP, Millar AH (2016) The plant mitochondrial transportome: balancing metabolic demands with energetic constraints. Trends Plant Sci 21:662–676

Leegood RC, Lea PJ, Adcock MD, Hausler RE (1995) The regulation and control of photorespiration. J Exp Bot 46:1397–1414

Li H, Hu B, Chu C (2017) Nitrogen use efficiency in crops: lessons from Arabidopsis and rice. J Exp Bot 68:2477–2488

Li J, Weraduwage SM, Peiser AL, Tietz S, Weise SE, Strand DD, Froehlich JE, … Sharkey TD (2019) A cytosolic bypass and G6P shunt in plants lacking peroxisomal hydroxypyruvate reductase. Plant Physiol 180:783–792

Linka M, Weber APM (2005) Shuffling ammonia between mitochondria and plastids during photorespiration. Trends Plant Sci 10:461–465

Long BM, Hee WY, Sharwood RE, Rae BD, Kaines S, Lim Y-L, Nguyen ND, … Price GD (2018) Carboxysome encapsulation of the CO_2-fixing enzyme Rubisco in tobacco chloroplasts. Nat Commun 9:3570

Lopez-Calcagno PE, Fisk S, Brown KL, Bull SE, South PF, Raines CA (2018) Overexpressing the H-protein of the glycine cleavage system increases biomass yield in glasshouse and field-grown transgenic tobacco plants. Plant Biotechnol J 17:141–151

Maier A, Fahnenstich H, von Caemmerer S, Engqvist MKM, Weber APM, Flügge UI, Maurino VG (2012) Glycolate oxidation in *A. thaliana* chloroplasts improves biomass production. Front Plant Sci 3:38

Mallmann J, Heckmann D, Bräutigam A, Lercher MJ, Weber AP, Westhoff P, Gowik U (2014) The role of photorespiration during the evolution of C_4 photosynthesis in the genus *Flaveria*. elife 3:e02478

Martijn J, Vosseberg J, Guy L, Offre P, Ettema TJG (2018) Deep mitochondrial origin outside the sampled alphaproteobacteria. Nature 557:101–105

Mathur J, Mammone A, Barton KA (2012) Organelle extensions in plant cells. J Integr Plant Biol 54:851–867

Mattsson M, Häusler RE, Leegood RC, Lea PJ, Schjoerring JK (1997) Leaf-atmosphere NH_3 exchange in barley mutants with reduced activities of glutamine synthetase. Plant Physiol 114:1307–1312

Meléndez-Hevia EG, Waddell T, Cascante M (1996) The puzzle of the Krebs citric acid cycle: assembling the pieces of chemically feasible reactions, and opportunism in the design of metabolic pathways during evolution. J Mol Evol 43:293–303

Menor-Salván C (2018) From the dawn of organic chemistry to astrobiology: urea as a foundational component in the origin of nucleobases and nucleotides. In: Menor-Salván C (ed) Prebiotic Chemistry and Chemical Evolution of Nucleic Acids. Springer International Publishing, Cham, pp 85–142

Messant M, Timm S, Fantuzzi A, Weckwerth W, Bauwe H, Rutherford AW, Krieger-Liszkay A (2018) Glycolate induces redox tuning of photosys-

tem II in vivo: study of a photorespiration mutant. Plant Physiol 177:1277–1285
Miller SL (1953) A production of amino acids under possible primitive earth conditions. Science 117:528–529
Modde K, Timm S, Florian A, Michl K, Fernie AR, Bauwe H (2017) High serine:glyoxylate aminotransferase activity lowers leaf daytime serine levels, inducing the phosphoserine pathway in Arabidopsis. J Exp Bot 68:643–656
Monné M, Daddabbo L, Gagneul D, Obata T, Hielscher B, Palmieri L, Miniero DM, … Palmieri F (2018) Uncoupling proteins 1 and 2 (UCP1 and UCP2) from Arabidopsis thaliana are mitochondrial transporters of aspartate, glutamate, and dicarboxylates. J Biol Chem 293: 4213–4227
Morris JL, Puttick MN, Clark JW, Edwards D, Kenrick P, Pressel S, Wellman CH, … Donoghue PCJ (2018) The timescale of early land plant evolution. Proc Natl Acad Sci U S A 115:E2274-E2283
Mukherjee I, Large RR, Corkrey R, Danyushevsky LV (2018) The boring billion, a slingshot for complex life on Earth. Sci Rep 8:4432
Murata N, Nishiyama Y (2018) ATP is a driving force in the repair of photosystem II during photoinhibition. Plant Cell Environ 41:285–299
Murray AJS, Blackwell RD, Lea PJ (1989) Metabolism of hydroxypyruvate in a mutant of barley lacking NADH-dependent hydroxypyruvate reductase, an important photorespiratory enzyme activity. Plant Physiol 91:395–400
Neuburger M, Bourguignon J, Douce R (1986) Isolation of a large complex from the matrix of pea leaf mitochondria involved in the rapid transformation of glycine into serine. FEBS Lett 207:18–22
Nishiyama T, Sakayama H, de Vries J, Buschmann H, Saint-Marcoux D, Ullrich KK et al (2018) The *Chara* genome: secondary complexity and implications for plant terrestrialization. Cell 174:448–464. e424
Nölke G, Houdelet M, Kreuzaler F, Peterhänsel C, Schillberg S (2014) The expression of a recombinant glycolate dehydrogenase polyprotein in potato (*Solanum tuberosum*) plastids strongly enhances photosynthesis and tuber yield. Plant Biotechnol J 12:734–742
Nowack ECM, Weber APM (2018) Genomics-informed insights into endosymbiotic organelle evolution in photosynthetic eukaryotes. Annu Rev Plant Biol 69:51–84
Nunes-Nesi A, Araujo WL, Obata T, Fernie AR (2013) Regulation of the mitochondrial tricarboxylic acid cycle. Curr Opin Plant Biol 16:335–343
Obata T, Florian A, Timm S, Bauwe H, Fernie AR (2016) On the metabolic interactions of (photo)respiration. J Exp Bot 67:3003–3014
Oliver DJ, Neuburger M, Bourguignon J, Douce R (1990) Interaction between the component enzymes of the glycine decarboxylase multienzyme complex. Plant Physiol 94:833–839
Orf I, Timm S, Bauwe H, Fernie AR, Hagemann M, Kopka J, Nikoloski Z (2016) Can cyanobacteria serve as a model of plant photorespiration? – A comparative meta-analysis of metabolite profiles. J Exp Bot 67:2941–2952
Osmond CB, Badger M, Maxwell K, Björkman O, Leegood R (1997) Too many photons: photorespiration, photoinhibition and photooxidation. Trends Plant Sci 2:119–121
Palmieri L, Picault N, Arrigoni R, Besin E, Palmieri F, Hodges M (2008) Molecular identification of three Arabidopsis thaliana mitochondrial dicarboxylate carrier isoforms: organ distribution, bacterial expression, reconstitution into liposomes and functional characterization. Biochem J 410:621–629
Peisker M (1974) A model describing the influence of oxygen on photosynthetic carboxylation. Photosynthetica 8:47–50
Pérez-Delgado CM, García-Calderón M, Márquez AJ, Betti M (2015) Reassimilation of photorespiratory ammonium in *Lotus japonicus* plants deficient in plastidic glutamine synthetase. PLoS One 10:e0130438
Pick TR, Bräutigam A, Schulz MA, Obata T, Fernie AR, Weber APM (2013) PLGG1, a plastidic glycolate glycerate transporter, is required for photorespiration and defines a unique class of metabolite transporters. Proc Natl Acad Sci U S A 110:3185–3190
Ponce-Toledo RI, Deschamps P, López-García P, Zivanovic Y, Benzerara K, Moreira D (2017) An early-branching freshwater cyanobacterium at the origin of plastids. Curr Biol 27:386–391
Porcelli V, Vozza A, Calcagnile V, Gorgoglione R, Arrigoni R, Fontanesi F, Marobbio CMT, … Palmieri L (2018) Molecular identification and functional characterization of a novel glutamate transporter in yeast and plant mitochondria. Biochim Biophys Acta Bioenerg 1859:1249–1258
Pracharoenwattana I, Zhou WX, Smith SM (2010) Fatty acid beta-oxidation in germinating Arabidopsis seeds is supported by peroxisomal hydroxypyruvate reductase when malate dehydrogenase is absent. Plant Mol Biol 72:101–109
Rai S, Lucius S, Kern R, Bauwe H, Kaplan A, Kopka J, Hagemann M (2018) The *Synechocystis* sp. PCC 6803 genome encodes up to four 2-phosphoglycolate phosphatases. Front. Plant Sci 9:1718

Rawsthorne S (1992) C_3-C_4 intermediate photosynthesis – linking physiology to gene expression. Plant J 2:267–274

Renné P, Dreßen U, Hebbeker U, Hille D, Flügge UI, Westhoff P, Weber AP (2003) The *Arabidopsis* mutant *dct* is deficient in the plastidic glutamate/malate translocator DiT2. Plant J 35:316–331

Reumann S (2011) Toward a definition of the complete proteome of plant peroxisomes: where experimental proteomics must be complemented by bioinformatics. Proteomics 11:1764–1779

Roger A, Muñoz-Gómez S, Kamikawa R (2017) The origin and diversification of mitochondria. Curr Biol 27:R1177–R1192

Sage RF, Kubien DS (2007) The temperature response of C_3 and C_4 photosynthesis. Plant Cell Environ 30:1086–1106

Sage RF, Monson RK, Ehleringer JR, Adachi S, Pearcy RW (2018) Some like it hot: the physiological ecology of C_4 plant evolution. Oecologia 187:941–966

Samuilov S, Brilhaus D, Rademacher N, Flachbart S, Arab L, Alfarraj S, Kuhnert F, … Rennenberg H (2018) The photorespiratory *BOU* gene mutation alters sulfur assimilation and its crosstalk with carbon and nitrogen metabolism in *Arabidopsis thaliana*. Front Plant Sci 9:1709

Sánchez-Baracaldo P, Raven JA, Pisani D, Knoll AH (2017) Early photosynthetic eukaryotes inhabited low-salinity habitats. Proc Natl Acad Sci U S A 114:E7737–E7745

Scanlan DJ, Ostrowski M, Mazard S, Dufresne A, Garczarek L, Hess WR, Post AF, … Partensky F (2009) Ecological genomics of marine picocyanobacteria. Microbiol Mol Biol Rev 73:249–299

Schell JC, Rutter J (2013) The long and winding road to the mitochondrial pyruvate carrier. Cancer Metab 1:6–6

Schirch V, Szebenyi DM (2005) Serine hydroxymethyltransferase revisited. Curr Opin Chem Biol 9:482–487

Schirrmeister BE, Sanchez-Baracaldo P, Wacey D (2016) Cyanobacterial evolution during the Precambrian. Int J Astrobiol 15:187–204

Schjoerring JK, Mäck G, Nielsen KH, Husted S, Suzuki A, Driscoll S, Boldt R, Bauwe H (2006) Antisense reduction of serine hydroxymethyltransferase results in diurnal displacement of NH_4^+ assimilation in leaves of *Solanum tuberosum*. Plant J 45:71–82

Schreier TB, Cléry A, Schläfli M, Galbier F, Stadler M, Demarsy E, Albertini D, … Kötting O (2018) Plastidial NAD-dependent malate dehydrogenase: a moonlighting protein involved in early chloroplast development through its interaction with an FtsH12-FtsHi protease complex. Plant Cell 30:1745–1769

Schulze S, Mallmann J, Burscheidt J, Koczor M, Streubel M, Bauwe H, Albertini D, … Westhoff P (2013) Evolution of C_4 photosynthesis in the genus *Flaveria*: establishment of a photorespiratory CO_2 pump. Plant Cell 25:2522–2535

Schwander T, von Borzyskowski LS, Burgener S, Cortina NS, Erb TJ (2016) A synthetic pathway for the fixation of carbon dioxide in vitro. Science 354:900–904

Schwarte S, Bauwe H (2007) Identification of the photorespiratory 2-phosphoglycolate phosphatase, PGLP1, in Arabidopsis. Plant Physiol 144:1580–1586

Selinski J, Scheibe R (2019) Malate valves: old shuttles with new perspectives. Plant Biol (Stuttg) 21:21–30

Sharkey TD, Weise SE (2016) The glucose 6-phosphate shunt around the Calvin–Benson cycle. J Exp Bot 67:4067–4077

Sharwood RE (2017) Engineering chloroplasts to improve Rubisco catalysis: prospects for translating improvements into food and fiber crops. New Phytol 213:494–510

Shen BR, Wang LM, Lin XL, Yao Z, Xu HW, Zhu CH, Teng HY, … Peng XX (2019) Engineering a new chloroplastic photorespiratory bypass to increase photosynthetic efficiency and productivity in rice. Mol Plant 12:199–214

Shih PM, Zarzycki J, Niyogi KK, Kerfeld CA (2014) Introduction of a synthetic CO_2-fixing photorespiratory bypass into a cyanobacterium. J Biol Chem 289:9493–9500

Sienkiewicz-Porzucek A, Nunes-Nesi A, Sulpice R, Lisec J, Centeno DC, Carillo P, Leisse A, … Fernie AR (2008) Mild reductions in mitochondrial citrate synthase activity result in a compromised nitrate assimilation and reduced leaf pigmentation but have no effect on photosynthetic performance or growth. Plant Physiol 147:115–127

Sienkiewicz-Porzucek A, Sulpice R, Osorio A, Krahnert I, Leisse A, Urbanczyk-Wochniak E, Michael H, … Fernie AR (2010) Mild reductions in mitochondrial NAD-dependent isocitrate dehydrogenase activity result in altered nitrate assimilation and pigmentation but do not impact growth. Mol Plant 3:156–173

Simkin AJ, Lopez-Calcagno PE, Davey PA, Headland LR, Lawson T, Timm S, Bauwe H, Raines CA (2017) Simultaneous stimulation of sedoheptulose 1,7-bisphosphatase, fructose 1,6-bisphophate aldolase and the photorespiratory glycine decarboxylase H-protein increases CO_2 assimilation, vegetative biomass and seed yield in Arabidopsis. Plant Biotechnol J 15:805–816

Slattery RA, Ort DR (2019) Carbon assimilation in crops at high temperatures. Plant Cell Environ 42:2750–2758

Slattery RA, Walker BJ, Weber APM, Ort DR (2018) The impacts of fluctuating light on crop performance. Plant Physiol 176:990–1003

Somerville CR, Ogren WL (1979) A phosphoglycolate phosphatase-deficient mutant of *Arabidopsis*. Nature 280:833–836

Soo RM, Hemp J, Parks DH, Fischer WW, Hugenholtz P (2017) On the origins of oxygenic photosynthesis and aerobic respiration in Cyanobacteria. Science 355:1436–1440

South PF, Walker BJ, Cavanagh AP, Rolland V, Badger M, Ort DR (2017) Bile acid sodium symporter BASS6 can transport glycolate and is involved in photorespiratory metabolism in *Arabidopsis thaliana*. Plant Cell 29:808–823

South PF, Cavanagh AP, Liu HW, Ort DR (2019) Synthetic glycolate metabolism pathways stimulate crop growth and productivity in the field. Science 363:eaat9077

Strodtkötter I, Padmasree K, Dinakar C, Speth B, Niazi PS, Wojtera J, Ingo V, … Scheibe R (2009) Induction of the AOX1D Isoform of alternative oxidase in *A. thaliana* T-DNA insertion lines lacking isoform AOX1A is insufficient to optimize photosynthesis when treated with antimycin a. Mol Plant 2:284–297

Sugiura A, Mattie S, Prudent J, McBride HM (2017) Newly born peroxisomes are a hybrid of mitochondrial and ER-derived pre-peroxisomes. Nature 542:251–254

Sweetlove LJ, Lytovchenko A, Morgan M, Nunes-Nesi A, Taylor NL, Baxter CJ, Eickmeier I, Fernie AR (2006) Mitochondrial uncoupling protein is required for efficient photosynthesis. Proc Natl Acad Sci U S A 103:19587–19592

Szecowka M, Heise R, Tohge T, Nunes-Nesi A, Vosloh D, Huege J, Feil R, … Arrivault S (2013) Metabolic fluxes in an illuminated Arabidopsis rosette. Plant Cell 25:694–714

Taira M, Valtersson U, Burkhardt B, Ludwig RA (2004) *Arabidopsis thaliana* GLN2-encoded glutamine synthetase is dual targeted to leaf mitochondria and chloroplasts. Plant Cell 16:2048–2058

Takahashi S, Bauwe H, Badger M (2007) Impairment of the photorespiratory pathway accelerates photoinhibition of Photosystem II by suppression of repair but not acceleration of damage processes in Arabidopsis. Plant Physiol 144:487–494

Taubert A, Jakob T, Wilhelm C (2019) Glycolate from microalgae: an efficient carbon source for biotechnological applications. Plant Biotechnol J 17:1538–1546

Tcherkez G (2013) Is the recovery of (photo) respiratory CO_2 and intermediates minimal? New Phytol 198:334–338

Tcherkez G, Limami AM (2019) Net photosynthetic CO_2 assimilation: more than just CO_2 and O_2 reduction cycles. New Phytol 223:520–529

Tcherkez G, Mahe A, Gauthier P, Mauve C, Gout E, Bligny R, Cornic G, Hodges M (2009) In folio respiratory fluxomics revealed by 13C-isotopic labeling and H/D isotope effects highlight the noncyclic nature of the tricarboxylic acid "cycle" in illuminated leaves. Plant Physiol 151:620–630

Tcherkez G, Boex-Fontvieille E, Mahe A, Hodges M (2012) Respiratory carbon fluxes in leaves. Curr Opin Plant Biol 15:308–314

Tcherkez G, Gauthier P, Buckley TN, Busch FA, Barbour MM, Bruhn D, Heskel MA, … Cornic G (2017) Leaf day respiration: low CO_2 flux but high significance for metabolism and carbon balance. New Phytol 216:986–1001

Timm S, Nunes-Nesi A, Pärnik T, Morgenthal K, Wienkoop S, Keerberg O, Weckwerth W, … Bauwe H (2008) A cytosolic pathway for the conversion of hydroxypyruvate to glycerate during photorespiration in *Arabidopsis*. Plant Cell 20:2848–2859

Timm S, Florian A, Jahnke K, Nunes-Nesi A, Fernie AR, Bauwe H (2011) The hydroxypyruvate-reducing system in *Arabidopsis*: multiple enzymes for the same end. Plant Physiol 155:694–705

Timm S, Florian A, Arrivault S, Stitt M, Fernie AR, Bauwe H (2012a) Glycine decarboxylase controls photosynthesis and plant growth. FEBS Lett 586:3692–3697

Timm S, Mielewczik M, Florian A, Frankenbach S, Dreissen A, Hocken N, Fernie AR, … Bauwe H (2012b) High-to-low CO_2 acclimation reveals plasticity of the photorespiratory pathway and indicates regulatory links to cellular metabolism of Arabidopsis. PLoS One 7:e42809

Timm S, Florian A, Wittmiss M, Jahnke K, Hagemann M, Fernie AR, Bauwe H (2013) Serine acts as a metabolic signal for the transcriptional control of photorespiration-related genes in Arabidopsis. Plant Physiol 162:379–389

Timm S, Wittmiss M, Gamlien S, Ewald R, Florian A, Frank M, Wirtz M, … Bauwe H (2015) Mitochondrial dihydrolipoyl dehydrogenase activity shapes photosynthesis and photorespiration of *Arabidopsis thaliana*. Plant Cell 27:1968–1984

Timm S, Florian A, Fernie AR, Bauwe H (2016) The regulatory interplay between photorespiration and photosynthesis. J Exp Bot 67:2923–2929

Tolbert NE (1997) The C_2 oxidative photosynthetic carbon cycle. Annu Rev Plant Physiol Plant Mol Biol 48:1–25
Tomeo NJ, Rosenthal DM (2018) Photorespiration differs among *Arabidopsis thaliana* ecotypes and is correlated with photosynthesis. J Exp Bot 69:5191–5204
Trudeau DL, Edlich-Muth C, Zarzycki J, Scheffen M, Goldsmith M, Khersonsky O, Avizemer Z, … Bar-Even A (2018) Design and in vitro realization of carbon-conserving photorespiration. Proc Natl Acad Sci U S A 115:E11455-E11464
Ushijima T, Hanada K, Gotoh E, Yamori W, Kodama Y, Tanaka H, Kusano M, … Matsushita T (2017) Light controls protein localization through phytochrome-mediated alternative promoter selection. Cell 171:1316–1325 e1312
van der Zand A, Tabak HF (2013) Peroxisomes: offshoots of the ER. Curr Opin Cell Biol 25:449–454
Voll LM, Jamai A, Renné P, Voll H, McClung CR, Weber APM (2006) The photorespiratory *Arabidopsis shm1* mutant is deficient in SHM1. Plant Physiol 140:59–66
Voon CP, Guan X, Sun Y, Sahu A, Chan MN, Gardestrom P, Wagner S, … Lim BL (2018) ATP compartmentation in plastids and cytosol of *Arabidopsis thaliana* revealed by fluorescent protein sensing. Proc Natl Acad Sci U S A 115:E10778-E10787
Walker BJ, VanLoocke A, Bernacchi CJ, Ort DR (2016) The costs of photorespiration to food production now and in the future. Annu Rev Plant Biol 67:107–129
Wallström SV, Florez-Sarasa I, Araújo WL, Aidemark M, Fernández-Fernández M, Fernie AR, Ribas-Carbó M, Rasmusson AG (2014a) Suppression of the external mitochondrial NADPH Dehydrogenase, NDB1, in *Arabidopsis thaliana* affects central metabolism and vegetative growth. Mol Plant 7:356–368
Wallström SV, Florez-Sarasa I, Araújo WL, Escobar MA, Geisler DA, Aidemark M, Lager I, … Rasmusson AG (2014b) Suppression of NDA-Type alternative mitochondrial NAD(P)H dehydrogenases in *Arabidopsis thaliana* modifies growth and metabolism, but not high light stimulation of mitochondrial electron transport. Plant Cell Physiol 55:881–896
Wang P, Vlad D, Langdale JA (2016) Finding the genes to build C_4 rice. Curr Opin Plant Biol 31:44–50
Wang T, Li S, Chen D, Xi Y, Xu X, Ye N, Zhang J, … Zhu G (2018) Impairment of FtsHi5 function affects cellular redox balance and photorespiratory metabolism in Arabidopsis. Plant Cell Physiol 59:pcy174
Woo KC, Flügge UI, Heldt HW (1987) A two-translocator model for the transport of 2-oxoglutarate and glutamate in chloroplasts during ammonia assimilation in the light. Plant Physiol 84:624–632
Xin CP, Tholen D, Devloo V, Zhu XG (2015) The benefits of photorespiratory bypasses: how can they work? Plant Physiol 167:574–585
Xu G, Fan X, Miller AJ (2012) Plant nitrogen assimilation and use efficiency. Annu Rev Plant Biol 63:153–182
Yamori W, Shikanai T (2016) Physiological functions of cyclic electron transport around photosystem I in sustaining photosynthesis and plant growth. Annu Rev Plant Biol 67:81–106
Ye N, Yang G, Chen Y, Zhang C, Zhang J, Peng X (2014) Two hydroxypyruvate reductases encoded by *OsHPR1* and *OsHPR2* are involved in photorespiratory metabolism in rice. J Integr Plant Biol 56:170–180
Zarei A, Brikis CJ, Bajwa VS, Chiu GZ, Simpson JP, DeEll JR, Bozzo GG, … Shelp BJ (2017) Plant glyoxylate/succinic semialdehyde reductases: comparative biochemical properties, function during chilling stress, and subcellular localization. Front Plant Sci 8:1399
Zelitch I, Walker DA (1964) The role of glycolic acid metabolism in opening of leaf stomata. Plant Physiol 39:856–862
Zhang Q, Lee J, Pandurangan S, Clarke M, Pajak A, Marsolais F (2013) Characterization of Arabidopsis serine:glyoxylate aminotransferase, AGT1, as an asparagine aminotransferase. Phytochemistry 85:30–35
Zhang C-C, Zhou C-Z, Burnap R, Peng L (2018) Carbon/nitrogen metabolic balance: lessons from cyanobacteria. Trends Plant Sci 23:1116–1130
Zhou Y, Whitney S (2019) Directed evolution of an improved Rubisco; *in vitro* analyses to decipher fact from fiction. Int J Mol Sci 20:5019

Index

K. M. Becklin et al. (eds.), *Photosynthesis, Respiration, and Climate Change*, Advances in Photosynthesis and Respiration 48, https://doi.org/10.1007/978-3-030-64926-5

Printed by Printforce, United Kingdom